MIND, BODY, WORLD

OPEL (OPEN PATHS TO ENRICHED LEARNING)

Series Editor: Connor Houlihan

Open Paths to Enriched Learning (OPEL) reflects the continued commitment of Athabasca University to removing barriers — including the cost of course materials — that restrict access to university-level study. The OPEL series offers introductory texts on a broad array of topics, written especially with undergraduate students in mind. Although the books in the series are designed for course use, they also afford lifelong learners an opportunity to enrich their own knowledge. Like all AU Press publications, OPEL course texts are available for free download at www.aupress.ca, as well as for purchase in both print and digital formats.

SERIES TITLES

Open Data Structures: An Introduction
    Pat Morin

Mind, Body, World: Foundations of Cognitive Science
    Michael R.W. Dawson

OPEL   Athabasca University

MIND,

FOUNDATIONS OF COGNITIVE SCIENCE

BODY,

MICHAEL R. W. DAWSON

WORLD

AU PRESS

Published by AU Press, Athabasca University
1200, 10011 – 109 Street, Edmonton, AB T5J 3S8

A volume in OPEL (Open Paths to Enriched Learning)
ISSN 2291-2606 (print)   2291-2614 (digital)

Cover design by Marvin Harder, marvinharder.com.
Interior design by Sergiy Kozakov.
Printed and bound in Canada by Marquis Book Printers.

LIBRARY AND ARCHIVES CANADA CATALOGUING IN PUBLICATION

Dawson, Michael Robert William, 1959—, author
        Mind, body, world: foundations of cognitive science / Michael R. W. Dawson.

(OPEL (Open paths to enriched learning), 2291-2606 ; 2)
Includes bibliographical references and index.
Issued in print and electronic formats.
ISBN 978-1-927356-17-3 (pbk.) — ISBN 978-1-927356-18-0 (pdf) — ISBN 978-1-927356-19-7 (epub)

1. Cognitive science. I. Title. II. Series: Open paths to enriched learning ; 2

BF311.D272 2013   153   C2013-902162-0
                                                            C2013-902163-9

We acknowledge the financial support of the Government of Canada through the Canada Book Fund
(CBF) for our publishing activities.

Canadian   Patrimoine
Heritage   canadien

Assistance provided by the Government of Alberta, Alberta Multimedia Development Fund.

Government

Contents

List of Figures and Tables

Preface

Understanding Cognitive Science (Dawson, 1998) was an attempt to present a particular thread, Marr's (1982) tri-level hypothesis, as a unifying theme for cognitive science. At that time, the 1990s, the primary texts available for survey courses in cognitive science (Gleitman & Liberman, 1995; Green, 1996; Kosslyn & Osherson, 1995; Osherson, 1995; Posner, 1991; Smith & Osherson, 1995; Stillings, 1995) were surveys of research in the many different content areas of cognitive science. A typical text would consist of chapters reflecting different research areas (e.g., concepts and categorization, mental imagery, deductive reasoning), each chapter written by a different specialist. Such texts provided a solid technical introduction to cognitive science and clearly indicated its interdisciplinary nature; over the years, I have used several of these texts in my own courses. However, these works did not successfully provide a "big picture" view of the discipline. Why was it so interdisciplinary? How was it possible for researchers from different disciplines to communicate with one another?

In my opinion, more recent introductions to cognitive science have done little to remedy this situation. Some continue to present a variety of chapters, each written by specialists in different fields (Lepore & Pylyshyn, 1999). A variation of this approach is to produce encyclopedic overviews of the discipline, with many short articles on specific ideas, each written by a different expert (Bechtel, Graham, & Balota, 1998; Wilson & Keil, 1999). Others organize the presentation in terms of diverse proposals about the nature of cognitive information processing (Bermúdez, 2010; Thagard, 2005). This latter approach implies that the breadth of cognitive science leads to its inevitable fragmentation, in a fashion analogous to what has happened in psychology. "One accomplishment that has eluded cognitive science is a unified theory that explains the full range of psychological phenomena, in the way that evolutionary and genetic theory unify biological phenomena, and relativity and quantum theory unify physical phenomena" (Thagard, 2005, p. 133).

The purpose of the current book is to continue the search for unification in cognitive science that was begun with *Understanding Cognitive Science* (Dawson, 1998). This search for unification is made more difficult by the advent of embodied cognitive science; a school of thought that may also be composed of fragmentary trends (Shapiro, 2011). Because of this challenge, unification is pursued in the current work in a more informed and constrained manner than in *Understanding Cognitive Science*. Emphasis is placed on introducing the key ideas that serve as the foundations for each school of thought in cognitive science. An attempt is made to consider whether differences amongst these key ideas can be used to inform conceptions of the cognitive architecture. The hypothesis that I consider in the current book is that the notion of architecture in cognitive science is currently pre-paradigmatic (Kuhn, 1970). One possibility to consider is that this notion can be made paradigmatic by considering a theory of architecture that pays heed to the core ideas of each of the cognitive sciences.

I do not presume to describe or to propose a unified cognitive science. However, I believe that the search for such a science is fundamental, and this search is the thread that runs throughout the current book.

Who Is This Book Written For?

This book is written with a particular audience in mind: the students that I see on a day-to-day basis in my classes. Such students are often senior undergraduates who have already been exposed to one of the core disciplines related to cognitive science. Others are graduate students with a deeper exposure to one of these disciplines. One goal of writing this book is to provide a set of ideas to such students that will help elaborate their understanding of their core discipline and show its relationship to cognitive science. Another is to provide a solid introduction to the foundational ideas of the cognitive sciences.

I will admit from the outset that this book is much more about the ideas in cognitive science than it is about the experimental methodologies, the extant data, or the key facts in the field. This is not to say that these topics are unimportant. My perspective is simply that sometimes an emphasis on the empirical results from different content areas of cognitive science at times obscures the "bigger picture." In my opinion, such results might indicate quite clearly what cognitive science is about, but do not reveal much about what cognitive science *is*. Fortunately, the student of cognitive science has the option of examining a growing array of introductory texts to compensate for the kinds of omissions that the approach taken in the current book necessitates.

Acknowledgements

The writing of this book was the major objective of a sabbatical kindly granted to me by the Faculty of Arts at the University of Alberta for the 2010–2011 academic year. My research is supported by research grants awarded by the Natural Sciences and Engineering Research Council of Canada and by the Social Sciences and Humanities Research Council of Canada. I would like to especially thank my wife Nancy Digdon for her comments and support during writing. This book is dedicated to my two graduate school mentors, Albert Katz and Zenon Pylyshyn. This book is also dedicated to their academic grandchildren: all of the students that I have had the pleasure of supervising in the Biological Computation Project at the University of Alberta.

The Cognitive Sciences: One or Many?

1.0 Chapter Overview

When experimental psychology arose in the nineteenth century, it was a unified discipline. However, as the experimental method began to be applied to a larger and larger range of psychological phenomena, this new discipline fragmented, causing what became known in the 1920s as the "crisis in psychology," a crisis that has persisted to the present day.

Cognitive science arose in the 1950s when it became apparent that a number of different disciplines, including psychology, computer science, linguistics and philosophy, were fragmenting. Some researchers responded to this situation by viewing cognition as a form of information processing. In the 1950s, the only plausible notion of information processing was the kind that was performed by a recent invention, the digital computer. This singular notion of information processing permitted cognitive science to emerge as a highly unified discipline.

A half century of research in cognitive science, though, has been informed by alternative conceptions of both information processing and cognition. As a result, the possibility has emerged that cognitive science itself is fragmenting. The purpose of this first chapter is to note the existence of three main approaches within the discipline: classical cognitive science, connectionist cognitive science, and embodied cognitive science. The existence of these different approaches leads to obvious questions: What are the core assumptions of these three different schools

1

of thought? What are the relationships between these different sets of core assumptions? Is there only one cognitive science, or are there many different cognitive sciences? Chapter 1 sets the stage for asking such questions; the remainder of the book explores possible answers to them.

1.1 A Fragmented Psychology

Modern experimental psychology is rooted in two seminal publications from the second half of the nineteenth century (Schultz & Schultz, 2008), Fechner's (1966) *Elements of Psychophysics*, originally published in 1860, and Wundt's *Principles of Physiological Psychology*, originally published in 1873 (Wundt & Titchener, 1904). Of these two authors, it is Wundt who is viewed as the founder of psychology, because he established the first experimental psychology laboratory—his Institute of Experimental Psychology—in Leipzig in 1879, as well as the first journal devoted to experimental psychology, *Philosophical Studies*, in 1881 (Leahey, 1987).

Fechner's and Wundt's use of experimental methods to study psychological phenomena produced a broad, unified science.

> This general significance of the experimental method is being more and more widely recognized in current psychological investigation; and the definition of experimental psychology has been correspondingly extended beyond its original limits. We now understand by 'experimental psychology' not simply those portions of psychology which are directly accessible to experimentation, but the whole of individual psychology. (Wundt & Titchner, 1904, p. 8)

However, not long after its birth, modern psychology began to fragment into competing schools of thought. The Würzberg school of psychology, founded in 1896 by Oswald Külpe, a former student of Wundt's, challenged Wundt's views on the scope of psychology (Schultz & Schultz, 2008). The writings of the functionalist school being established in North America were critical of Wundt's structuralism (James, 1890a, 1890b). Soon, behaviourism arose as a reaction against both structuralism and functionalism (Watson, 1913).

Psychology's fragmentation soon began to be discussed in the literature, starting with Bühler's 1927 "crisis in psychology" (Stam, 2004), and continuing to the present day (Bower, 1993; Driver-Linn, 2003; Gilbert, 2002; Koch, 1959, 1969, 1976, 1981, 1993; Lee, 1994; Stam, 2004; Valsiner, 2006; Walsh-Bowers, 2009). For one prominent critic of psychology's claim to scientific status,

> psychology is misconceived when seen as a coherent science or as any kind of coherent discipline devoted to the empirical study of human beings. Psychology, in my view, is not a single discipline but a collection of studies of varied cast, some few of which may qualify as science, whereas most do not. (Koch, 1993, p. 902)

The fragmentation of psychology is only made more apparent by repeated attempts to find new approaches to unify the field, or by rebuttals against claims of disunity (Drob, 2003; Goertzen, 2008; Henriques, 2004; Katzko, 2002; Richardson, 2000; Smythe & McKenzie, 2010; Teo, 2010; Valsiner, 2006; Walsh-Bowers, 2009; Watanabe, 2010; Zittoun, Gillespie, & Cornish, 2009).

The breadth of topics being studied by any single psychology department is staggering; psychology correspondingly uses an incredible diversity of methodologies. It is not surprising that Leahey (1987, p. 3) called psychology a "large, sprawling, confusing human undertaking." Because of its diversity, it is likely that psychology is fated to be enormously fragmented, at best existing as a pluralistic discipline (Teo, 2010; Watanabe, 2010).

If this is true of psychology, then what can be expected of a more recent discipline, cognitive science? Cognitive science would seem likely to be even more fragmented than psychology, because it involves not only psychology but also many other disciplines. For instance, the website of the Cognitive Science Society states that the Society,

> brings together researchers from many fields that hold a common goal: understanding the nature of the human mind. The Society promotes scientific interchange among researchers in disciplines comprising the field of Cognitive Science, including Artificial Intelligence, Linguistics, Anthropology, Psychology, Neuroscience, Philosophy, and Education. (Cognitive Science Society, 2013)

The names of all of these disciplines are proudly placed around the perimeter of the Society's logo.

When cognitive science appeared in the late 1950s, it seemed to be far more unified than psychology. Given that cognitive science draws from so many different disciplines, how is this possible?

1.2 A Unified Cognitive Science

When psychology originated, the promise of a new, unified science was fuelled by the view that a coherent object of enquiry (conscious experience) could be studied using a cohesive paradigm (the experimental method). Wundt defined psychological inquiry as "the investigation of conscious processes in the modes of connexion peculiar to them" (Wundt & Titchner, 1904, p. 2). His belief was that using the experimental method would "accomplish a reform in psychological investigation comparable with the revolution brought about in the natural sciences." As experimental psychology evolved the content areas that it studied became markedly differentiated, leading to a proliferation of methodologies. The fragmentation of psychology was a natural consequence.

Cognitive science arose as a discipline in the mid-twentieth century (Boden, 2006; Gardner, 1984; Miller, 2003), and at the outset seemed more unified than psychology. In spite of the diversity of talks presented at the "Special Interest Group in Information Theory" at MIT in 1956, cognitive psychologist George Miller,

> left the symposium with a conviction, more intuitive than rational, that experimental
> psychology, theoretical linguistics, and the computer simulation of cognitive pro-
> cesses were all pieces from a larger whole and that the future would see a progressive
> elaboration and coordination of their shared concerns. (Miller, 2003, p. 143)

The cohesiveness of cognitive science was, perhaps, a natural consequence of its intellectual antecedents. A key inspiration to cognitive science was the digital computer; we see in Chapter 2 that the invention of the computer was the result of the unification of ideas from the diverse fields of philosophy, mathematics, and electrical engineering.

Similarly, the immediate parent of cognitive science was the field known as cybernetics (Ashby, 1956; de Latil, 1956; Wiener, 1948). Cybernetics aimed to study adaptive behaviour of intelligent agents by employing the notions of feedback and information theory. Its pioneers were polymaths. Not only did cyberneticist William Grey Walter pioneer the use of EEG in neurology (Cooper, 1977), he also invented the world's first autonomous robots (Bladin, 2006; Hayward, 2001; Holland, 2003a; Sharkey & Sharkey, 2009). Cybernetics creator Norbert Wiener organized the Macy Conferences (Conway & Siegelman, 2005), which were gatherings of mathematicians, computer scientists, psychologists, psychiatrists, anthropologists, and neuroscientists, who together aimed to determine the general workings of the human mind. The Macy Conferences were the forerunners of the interdisciplinary symposia that inspired cognitive scientists such as George Miller.

What possible glue could unite the diversity of individuals involved first in cybernetics, and later in cognitive science? One answer is that cognitive scientists are united in sharing a key foundational assumption that cognition is information processing (Dawson, 1998). As a result, a critical feature of cognition involves representation or symbolism (Craik, 1943). The early cognitive scientists,

> realized that the integration of parts of several disciplines was possible and desir-
> able, because each of these disciplines had research problems that could be
> addressed by designing 'symbolisms.' Cognitive science is the result of striving
> towards this integration. (Dawson, 1998, p. 5)

Assuming that cognition is information processing provides a unifying principle, but also demands methodological pluralism. Cognitive science accounts for human cognition by invoking an information processing explanation. However, information processors themselves require explanatory accounts framed at very different levels of analysis (Marr, 1982; Pylyshyn, 1984). Each level of analysis involves asking qualitatively different kinds of questions, and also involves using dramatically different methodologies to answer them.

Marr (1982) proposed that information processors require explanations at the computational, algorithmic, and implementational levels. At the computational level, formal proofs are used to determine what information processing problem is being solved. At the algorithmic level, experimental observations and computer simulations are used to determine the particular information processing steps that are being used to solve the information processing problem. At the implementational level, biological or physical methods are used to determine the mechanistic principles that actually instantiate the information processing steps. In addition, a complete explanation of an information processor requires establishing links between these different levels of analysis.

An approach like Marr's is a mandatory consequence of assuming that cognition is information processing (Dawson, 1998). It also makes cognitive science particularly alluring. This is because cognitive scientists are aware not only that a variety of methodologies are required to explain information processing, but also that researchers from a diversity of areas can be united by the goal of seeking such an explanation.

As a result, definitions of cognitive science usually emphasize co-operation across disciplines (Simon, 1980). Cognitive science is "a recognition of a fundamental set of common concerns shared by the disciplines of psychology, computer science, linguistics, economics, epistemology, and the social sciences generally" (Simon, 1980, p. 33). Interviews with eminent cognitive scientists reinforce this theme of interdisciplinary harmony and unity (Baumgartner & Payr, 1995). Indeed, it would appear that cognitive scientists deem it essential to acquire methodologies from more than one discipline.

For instance, philosopher Patricia Churchland learned about neuroscience at the University of Manitoba Medical School by "doing experiments and dissections and observing human patients with brain damage in neurology rounds" (Baumgartner & Payr, 1995, p. 22). Philosopher Daniel Dennett improved his computer literacy by participating in a year-long working group that included two philosophers and four AI researchers. AI researcher Terry Winograd studied linguistics in London before he went to MIT to study computer science. Psychologist David Rumelhart observed that cognitive science has "a collection of methods that have been developed, some uniquely in cognitive science, but some in related disciplines. . . . It is clear that we have to learn to appreciate one another's approaches and understand where our own are weak" (Baumgartner & Payr, 1995, p. 196).

At the same time, as it has matured since its birth in the late 1950s, concerns about cognitive science's unity have also arisen. Philosopher John Searle stated, "I am not sure whether there is such a thing as cognitive science" (Baumgartner & Payr, 1995,

p. 203). Philosopher John Haugeland claimed that "philosophy belongs in cognitive science only because the 'cognitive sciences' have not got their act together yet" (p. 103). AI pioneer Herbert Simon described *cognitive science* as a label "for the fact that there is a lot of conversation across disciplines" (p. 234). For Simon, "cognitive science is the place where they meet. It does not matter whether it is a discipline. It is not really a discipline—yet."

In modern cognitive science there exist intense disagreements about what the assumption "cognition is information processing" really means. From one perspective, modern cognitive science is fragmenting into different schools of thought—classical, connectionist, embodied—that have dramatically different views about what the term *information processing* means. Classical cognitive science interprets this term as meaning rule-governed symbol manipulations of the same type performed by a digital computer. The putative fragmentation of cognitive science begins when this assumption is challenged. John Searle declared, "I think that cognitive science suffers from its obsession with the computer metaphor" (Baumgartner & Payr, 1995, p. 204). Philosopher Paul Churchland declared, "we need to get away from the idea that we are going to achieve Artificial Intelligence by writing clever programs" (p. 37).

Different interpretations of *information processing* produce variations of cognitive science that give the strong sense of being mutually incompatible. One purpose of this book is to explore the notion of information processing at the foundation of each of these varieties. A second is to examine whether these notions can be unified.

1.3 Cognitive Science or the Cognitive Sciences?

One reason that Wilhelm Wundt is seen as the founder of psychology is because he established its first academic foothold at the University of Leipzig. Wundt created the first experimental psychology laboratory there in 1879. Psychology was officially part of the university calendar by 1885. Today, hundreds of psychology departments exist at universities around the world.

Psychology is clearly healthy as an academic discipline. However, its status as a science is less clear. Sigmund Koch, a noted critic of psychology (Koch, 1959, 1969, 1976, 1981, 1993), argued in favor of replacing the term *psychology* with *the psychological studies* because of his view that it was impossible for psychology to exist as a coherent discipline.

Although it is much younger than psychology, cognitive science has certainly matured into a viable *academic* discipline. In the fall of 2010, the website for the Cognitive Science Society listed 77 universities around the world that offered cognitive science as a program of study. Recent developments in cognitive science, though, have raised questions about its *scientific* coherence. To parallel Koch, should we examine "cognitive science," or is it more appropriate to inquire about "the cognitive sciences"? Investigating this issue is one theme of the current book.

According to psychologist George Miller (2003), cognitive science was born on September 11, 1956. At this early stage, the unity of cognitive science was not really an issue. Digital computers were a relatively recent invention (Goldstine, 1993; Lavington, 1980; Williams, 1997; Zuse, 1993). At the time, they presented a unified notion of information processing to be adopted by cognitive science. Digital computers were automatic symbol manipulators (Haugeland, 1985): they were machines that manipulated symbolic representations by applying well-defined rules; they brought symbolic logic to mechanized life. Even though some researchers had already noted that the brain may not work exactly like a computer, the brain was still assumed to be digital, because the all-or-none generation of an action potential was interpreted as being equivalent to assigning a truth value in a Boolean logic (McCulloch & Pitts, 1943; von Neumann, 1958).

Classical cognitive science, which is the topic of Chapter 3, was the first school of thought in cognitive science and continues to dominate the field to this day. It exploited the technology of the day by interpreting "information processing" as meaning "rule-governed manipulation of symbol" (Feigenbaum & Feldman, 1995). This version of the information processing hypothesis bore early fruit, producing major advances in the understanding of language (Chomsky, 1957, 1959b, 1965) and of human problem solving (Newell, Shaw, & Simon, 1958; Newell & Simon, 1961, 1972). Later successes with this approach led to the proliferation of "thinking artifacts": computer programs called expert systems (Feigenbaum & McCorduck, 1983; Kurzweil, 1990). Some researchers have claimed that the classical approach is capable of providing a unified theory of thought (Anderson, 1983; Anderson et al., 2004; Newell, 1990).

The successes of the classical approach were in the realm of well-posed problems, such problems being those with unambiguously defined states of knowledge and goal states, not to mention explicitly defined operations for converting one state of knowledge into another. If a problem is well posed, then its solution can be described as a search through a problem space, and a computer can be programmed to perform this search (Newell & Simon, 1972). However, this emphasis led to growing criticisms of the classical approach. One general issue was whether human cognition went far beyond what could be captured just in terms of solving well-posed problems (Dreyfus, 1992; Searle, 1980; Weizenbaum, 1976).

Indeed, the classical approach was adept at producing computer simulations of game playing and problem solving, but was not achieving tremendous success in such fields as speech recognition, language translation, or computer vision. "An overall pattern had begun to take shape. . . . an early, dramatic success based on the easy performance of simple tasks, or low-quality work on complex tasks, and then diminishing returns, disenchantment, and, in some cases, pessimism" (Dreyfus, 1992, p. 99).

Many abilities that humans are expert at without training, such as speaking, seeing, and walking, seemed to be beyond the grasp of classical cognitive science. These abilities involve dealing with ill-posed problems. An ill-posed problem is deeply ambiguous, has poorly defined knowledge states and goal states, and involves poorly defined operations for manipulating knowledge. As a result, it is not well suited to classical analysis, because a problem space cannot be defined for an ill-posed problem. This suggests that the digital computer provides a poor definition of the kind of information processing performed by humans. "In our view people are smarter than today's computers because the brain employs a basic computational architecture that is more suited to deal with a central aspect of the natural information processing tasks that people are so good at" (Rumelhart & McClelland, 1986c, p. 3).

Connectionist cognitive science reacted against classical cognitive science by proposing a cognitive architecture that is qualitatively different from that inspired by the digital computer metaphor (Bechtel & Abrahamsen, 2002; Churchland, Koch, & Sejnowski, 1990; Churchland & Sejnowski, 1992; Clark, 1989, 1993; Horgan & Tienson, 1996; Quinlan, 1991). Connectionists argued that the problem with the classical notion of information processing was that it ignored the fundamental properties of the brain. Connectionism cast itself as a neuronally inspired, biologically plausible alternative to classical cognitive science (Bechtel & Abrahamsen, 2002; McClelland & Rumelhart, 1986; Rumelhart & McClelland, 1986c). "No serious study of mind (including philosophical ones) can, I believe, be conducted in the kind of biological vacuum to which [classical] cognitive scientists have become accustomed" (Clark, 1989, p. 61).

The architecture proposed by connectionism was the artificial neural network (Caudill & Butler, 1992a, 1992b; Dawson, 2004, 2005; De Wilde, 1997; Muller & Reinhardt, 1990; Rojas, 1996). An artificial neural network is a system of simple processors, analogous to neurons, which operate in parallel and send signals to one another via weighted connections that are analogous to synapses. Signals detected by input processors are converted into a response that is represented as activity in a set of output processors. Connection weights determine the input-output relationship mediated by a network, but they are not programmed. Instead, a learning rule is used to modify the weights. Artificial neural networks learn from example.

Artificial neural networks negate many of the fundamental properties of the digital computer (von Neumann, 1958). Gone was the notion that the brain was a digital symbol manipulator governed by a serial central controller. In its place, the processes of the brain were described as subsymbolic and parallel (Smolensky, 1988); control of these processes was decentralized. Gone was the classical distinction between structure and process, in which a distinct set of explicit rules manipulated discrete symbols stored in a separate memory. In its place, the brain was viewed as a distributed system in which problem solutions emerged from the parallel activity of a large number of simple processors: a network was both structure and process, and networks both stored and modified information at the same time (Hillis, 1985). Gone was the assumption that information processing was akin to doing logic (Oaksford & Chater, 1991). In its place, connectionists viewed the brain as a dynamic, statistical pattern recognizer (Churchland & Sejnowski, 1989; Grossberg, 1980; Smolensky, 1988).

With all such changes, though, connectionism still concerned itself with cognition as information processing—but of a different kind: "These dissimilarities do not imply that brains are not computers, but only that *brains are not serial digital computers*" (Churchland, Koch, & Sejnowski, 1990, p. 48, italics original).

Connectionist models of cognition have had as long a history as have classical simulations (Dawson, 2004; Medler, 1998). McCulloch and Pitts described powerful neural network models in the 1940s (McCulloch, 1988a), and Rosenblatt's (1958, 1962) perceptrons were simple artificial neural networks that were not programmed, but instead learned from example. Such research waned in the late 1960s as the result of proofs about the limitations of simple artificial neural networks (Minsky & Papert, 1988; Papert, 1988).

However, the limitations of early networks were overcome in the mid-1980s, by which time new techniques had been discovered that permitted much more powerful networks to learn from examples (Ackley, Hinton, & Sejnowski, 1985; Rumelhart, Hinton, & Williams, 1986b). Because of these new techniques, modern connectionism has achieved nearly equal status to classical cognitive science. Artificial neural networks have been used to model a wide range of ill-posed problems, have generated many expert systems, and have successfully simulated domains once thought to be exclusive to the classical approach (Bechtel & Abrahamsen, 2002; Carpenter & Grossberg, 1992; Enquist & Ghirlanda, 2005; Gallant, 1993; Gluck & Myers, 2001; Grossberg, 1988; Kasabov, 1996; Pao, 1989; Ripley, 1996; Schmajuk, 1997; Wechsler, 1992).

In a review of a book on neural networks, Hanson and Olson (1991, p. 332) claimed that "the neural network revolution has happened. We are living in the aftermath." This revolution, as is the case with most, has been messy and acrimonious, markedly departing from the sense of unity that cognitive science conveyed at the time of its birth. A serious and angry debate about the merits of classical versus connectionist cognitive science rages in the literature.

On the one hand, classical cognitive scientists view the rise of connectionism as being a rebirth of the associationist and behaviourist psychologies that cognitivism had successfully replaced. Because connectionism eschewed rules and symbols, classicists argued that it was not powerful enough to account for the regularities of thought and language (Fodor & McLaughlin, 1990; Fodor & Pylyshyn, 1988; Pinker, 2002; Pinker & Prince, 1988). "The problem with connectionist models is that all the reasons for thinking that they might be true are reasons for thinking that they couldn't be *psychology*" (Fodor & Pylyshyn, 1988, p. 66). A *Scientific American* news story on a connectionist expert system included Pylyshyn's comparison of connectionism to voodoo: "'People are fascinated by the prospect of getting intelligence by mysterious Frankenstein-like means—by voodoo! And there have been few attempts to do this as successful as neural nets" (Stix, 1994, p. 44). The difficulty with interpreting the internal structure of connectionist networks has been used to argue against their ability to provide models, theories, or even demonstrations to cognitive science (McCloskey, 1991).

On the other hand, and not surprisingly, connectionist researchers have replied in kind. Some of these responses have been arguments about problems that are intrinsic to the classical architecture (e.g., slow, brittle models) combined with claims that the connectionist architecture offers solutions to these problems (Feldman & Ballard, 1982; Rumelhart & McClelland, 1986c). Others have argued that classical models have failed to provide an adequate account of experimental studies of human cognition (Oaksford, Chater, & Stenning, 1990). Connectionist practitioners have gone as far as to claim that they have provided a paradigm shift for cognitive science (Schneider, 1987).

Accompanying claims for a paradigm shift is the view that connectionist cognitive science is in a position to replace an old, tired, and failed classical approach. Searle (1992, p. 247), in a defense of connectionism, has described traditional cognitivist models as being "obviously false or incoherent." Some would claim that classical cognitive science doesn't study the right phenomena. "The idea that human activity is determined by rules is not very plausible when one considers that most of what we do is not naturally thought of as problem solving" (Horgan & Tienson, 1996, p. 31). Paul Churchland noted that "good old-fashioned artificial intelligence was a failure. The contribution of standard architectures and standard programming artificial intelligence was a disappointment" (Baumgartner & Payr, 1995, p. 36). Churchland went on to argue that this disappointment will be reversed with the adoption of more brain-like architectures.

Clearly, the rise of connectionism represents a fragmentation of cognitive science. This fragmentation is heightened by the fact that connectionists themselves freely admit that there are different notions about information processing that fall under the connectionist umbrella (Horgan & Tienson, 1996; Rumelhart & McClelland, 1986c). "It is not clear that anything has appeared that could be called a, let alone the, connectionist conception of cognition" (Horgan & Tienson, 1996, p. 3).

If the only division within cognitive science was between classical and connectionist schools of thought, then the possibility of a unified cognitive science still exists.

Some researchers have attempted to show that these two approaches can be related (Dawson, 1998; Smolensky & Legendre, 2006), in spite of the differences that have been alluded to in the preceding paragraphs. However, the hope for a unified cognitive science is further challenged by the realization that a third school of thought has emerged that represents a reaction to both classical and connectionist cognitive science.

This third school of thought is embodied cognitive science (Chemero, 2009; Clancey, 1997; Clark, 1997; Dawson, Dupuis, & Wilson, 2010; Robbins & Aydede, 2009; Shapiro, 2011). Connectionist cognitive science arose because it felt that classical cognitive science did not pay sufficient attention to a particular part of the body, the brain. Embodied cognitive science critiques both classical and connectionist approaches because both ignore the whole body and its interaction with the world. Radical versions of embodied cognitive science aim to dispense with mental representations completely, and argue that the mind extends outside the brain, into the body and the world (Agre, 1997; Chemero, 2009; Clancey, 1997; Clark, 2008; Clark & Chalmers, 1998; Noë, 2009; Varela, Thompson, & Rosch, 1991; Wilson, 2004).

A key characteristic of embodied cognitive science is that it abandons methodological solipsism (Wilson, 2004). According to methodological solipsism (Fodor, 1980), representational states are individuated only in terms of their relations to other representational states. Relations of the states to the external world—the agent's environment—are not considered. "Methodological solipsism in psychology is the view that psychological states should be construed without reference to anything beyond the boundary of the individual who has those states" (Wilson, 2004, p. 77).

Methodological solipsism is reflected in the sense-think-act cycle that characterizes both classical and connectionist cognitive science (Pfeifer & Scheier, 1999). The sense-think-act cycle defines what is also known as the classical sandwich (Hurley, 2001), in which there is no direct contact between sensing and acting. Instead, thinking—or representations—is the "filling" of the sandwich, with the primary task of planning action on the basis of sensed data. Both classical and connectionist cognitive science adopt the sense-think-act cycle because both have representations standing between perceptual inputs and behavioural outputs. "Representation is an activity that individuals perform in extracting and deploying information that is used in their further actions" (Wilson, 2004, p. 183).

Embodied cognitive science replaces the sense-think-act cycle with sense-act processing (Brooks, 1991, 1999; Clark, 1997, 1999, 2003; Hutchins, 1995; Pfeifer & Scheier, 1999). According to this alternative view, there are direct links between sensing and acting. The purpose of the mind is not to plan action, but is instead to coordinate sense-act relations. "Models of the world simply get in the way. It turns out to be better to use the world as its own model" (Brooks, 1991, p. 139). Embodied cognitive science views the brain as a controller, not as a planner. "The realization was that the so-called central systems of intelligence—or core AI as it has been referred to more recently—was perhaps an unnecessary illusion, and that all the power of intelligence arose from the coupling of perception and actuation systems" (Brooks, 1999, p. viii).

In replacing the sense-think-act cycle with the sense-act cycle, embodied cognitive science distances itself from classical and connectionist cognitive science. This is because sense-act processing abandons planning in particular and the use of representations in general. Brooks (1999, p. 170) wrote: "In particular I have advocated situatedness, embodiment, and highly reactive architectures with no reasoning systems, no manipulable representations, no symbols, and totally decentralized computation." Other theorists make stronger versions of this claim: "I hereby define radical embodied cognitive science as the scientific study of perception, cognition, and action as necessarily embodied phenomena, using explanatory tools that do not posit mental representations" (Chemero, 2009, p. 29).

The focus on sense-act processing leads directly to the importance of embodiment. Embodied cognitive science borrows a key idea from cybernetics: that agents are adaptively linked to their environment (Ashby, 1956; Wiener, 1948). This adaptive link is a source of feedback: an animal's actions on the world can change the world, which in turn will affect later actions. Embodied cognitive science also leans heavily on Gibson's (1966, 1979) theory of direct perception. In particular, the adaptive link between an animal and its world is affected by the physical form of the animal—its embodiment. "It is often neglected that the words *animal* and *environment* make an inseparable pair" (Gibson, 1979, p. 8). Gibson proposed that sensing agents "picked up" properties that indicated potential actions that could be taken on the world. Again, the definition of such affordances requires taking the agent's form into account.

Embodied cognitive science also distances itself from both classical and connectionist cognitive science by proposing the extended mind hypothesis (Clark, 1997, 1999, 2003, 2008; Wilson, 2004, 2005). According to the extended mind hypothesis, the mind is not separated from the world by the skull. Instead, the boundary between the mind and the world is blurred, or has disappeared. A consequence of the extended mind is cognitive scaffolding, where the abilities of "classical" cognition are enhanced by using the external world as support. A simple example of this is extending memory by using external aids, such as notepads. However, full-blown information processing can be placed into the world if appropriate artifacts are used. Hutchins (1995) provided many examples of navigational tools that externalize computation. "It seems that much of the computation was done by the tool, or by its designer. The person somehow could succeed by doing less because the tool did more" (p. 151).

Embodied cognitive science provides another fault line in a fragmenting cognitive science. With notions like the extended mind, the emphasis on action, and the abandonment of representation, it is not clear at first glance whether embodied cognitive science is redefining the notion of information processing or abandoning it altogether. "By failing to understand the source of the computational power in our interactions with simple 'unintelligent' physical devices, we position ourselves well to squander opportunities with so-called intelligent computers" (Hutchins, 1995, p. 171).

Further fragmentation is found within the embodied cognition camp (Robbins & Aydede, 2009; Shapiro, 2011). Embodied cognitive scientists have strong disagreements amongst themselves about the degree to which each of their radical views is to be accepted. For instance, Clark (1997) believed there is room for representation in embodied cognitive science, while Chemero (2009) did not.

In summary, early developments in computer science led to a unitary notion of information processing. When information processing was adopted as a hypothesis about cognition in the 1950s, the result was a unified cognitive science. However, a half century of developments in cognitive science has led to a growing fragmentation of the field. Disagreements about the nature of representations, and even about their necessity, have spawned three strong camps within cognitive science: classical, connectionist, and embodied. Fragmentation within each of these camps can easily be found. Given this situation, it might seem foolish to ask whether there exist any central ideas that can be used to unify cognitive science. However, the asking of that question is an important thread that runs through the current book.

1.4 Cognitive Science: Pre-paradigmatic?

In the short story *The Library of Babel*, Jorge Luis Borges (1962) envisioned the universe as the Library, an infinite set of hexagonal rooms linked together by a spiral staircase. Each room held exactly the same number of books, each book being exactly 410 pages long, all printed in an identical format. The librarians hypothesize that the Library holds all possible books, that is, all possible arrangements of a finite set of orthographic symbols. They believe that "the Library is total and that its shelves register . . . all that is given to express, in all languages" (p. 54).

Borges' librarians spend their lives sorting through mostly unintelligible volumes, seeking those books that explain "humanity's basic mysteries" (Borges, 1962, p. 55). Central to this search is the faith that there exists a language in which to express these answers. "It is verisimilar that these grave mysteries could be explained in words: if the language of philosophers is not sufficient, the multiform Library will have produced the unprecedented language required, with its vocabularies and grammars" (p. 55).

The fictional quest of Borges' librarians mirrors an actual search for ancient texts. Scholasticism was dedicated to reviving ancient wisdom. It was spawned in the tenth century when Greek texts preserved and translated by Islamic scholars made their way to Europe and led to the creation of European universities. It reached its peak in the thirteenth century with Albertus Magnus' and Thomas Aquinas' works on Aristotelian philosophy. A second wave of scholasticism in the fifteenth century was fuelled by new discoveries of ancient texts (Debus, 1978). "The search for new classical texts was intense in the fifteenth century, and each new discovery was hailed as a major achievement" (Debus, 1978, p. 4). These discoveries included Ptolemy's *Geography* and the only copy of Lucretius' *De rerum natura*, which later revived interest in atomism.

Borges' (1962) emphasis on language is also mirrored in the scholastic search for the wisdom of the ancients. The continued discovery of ancient texts led to the Greek revival in the fifteenth century (Debus, 1978), which enabled this treasure trove of texts to be translated into Latin. In the development of modern science, Borges' "unprecedented language" was first Greek and then Latin.

The departure from Latin as the language of science was a turbulent development during the scientific revolution. Paracelsus was attacked by the medical establishment for presenting medical lectures in his native Swiss German in 1527 (Debus, 1978). Galileo published his 1612 *Discourse on Bodies in Water* in Italian, an act that enraged his fellow philosophers of the Florentine Academy (Sobel, 1999). For a long period, scholars who wrote in their vernacular tongue had to preface their writings with apologies and explanations of why this did not represent a challenge to the universities of the day (Debus, 1978).

Galileo wrote in Italian because "I must have everyone able to read it" (Sobel, 1999, p. 47). However, from some perspectives, writing in the vernacular actually produced a communication breakdown, because Galileo was not disseminating knowledge in the scholarly *lingua franca*, Latin. Galileo's writings were examined as part of his trial. It was concluded that "he writes in Italian, certainly not to extend the hand to foreigners or other learned men" (Sobel, 1999, p. 256).

A different sort of communication breakdown is a common theme in modern philosophy of science. It has been argued that some scientific theories are incommensurable with others (Feyerabend, 1975; Kuhn, 1970). Incommensurable scientific theories are theories that are impossible to compare because there is no logical or meaningful relation between some or all of the theories' terms. Kuhn argued that this situation would occur if, within a science, different researchers operated under different paradigms. "Within the new paradigm, old terms, concepts, and experiments fall into new relationships one with the other. The inevitable result is what we must call, though the term is not quite right, a misunderstanding between the two schools" (Kuhn, 1970, p. 149). Kuhn saw holders of different paradigms as being members of different language communities—even if they wrote in the same vernacular tongue! Differences in paradigms caused communication breakdowns.

The modern fragmentation of cognitive science might be an example of communication breakdowns produced by the existence of incommensurable theories. For instance, it is not uncommon to see connectionist cognitive science described as a Kuhnian paradigm shift away from classical cognitive science (Horgan & Tienson, 1996; Schneider, 1987). When embodied cognitive science is discussed in Chapter 5, we see that it too might be described as a new paradigm.

To view the fragmentation of cognitive science as resulting from competing, incommensurable paradigms is also to assume that cognitive science is paradigmatic. Given that cognitive science as a discipline is less than sixty years old (Boden, 2006; Gardner, 1984; Miller, 2003), it is not impossible that it is actually pre-paradigmatic. Indeed, one discipline to which cognitive science is frequently compared—experimental psychology—may also be pre-paradigmatic (Buss, 1978; Leahey, 1992).

Pre-paradigmatic sciences exist in a state of disarray and fragmentation because data are collected and interpreted in the absence of a unifying body of belief. "In the early stages of the development of any science different men confronting the same range of phenomena, but not usually all the same particular phenomena, describe and interpret them in different ways" (Kuhn, 1970, p. 17). My suspicion is that cognitive science has achieved some general agreement about the kinds of phenomena that it believes it should be explaining. However, it is pre-paradigmatic with respect to the kinds of technical details that it believes are necessary to provide the desired explanations.

In an earlier book, I argued that the assumption that cognition is information processing provided a framework for a "language" of cognitive science that made interdisciplinary conversations possible (Dawson, 1998). I demonstrated that when this framework was applied, there were more similarities than differences between classical and connectionist cognitive science. The source of these similarities was the fact that both classical and connectionist cognitive science adopted the information processing hypothesis. As a result, both schools of thought can be examined and compared using Marr's (1982) different levels of analysis. It can be shown that classical and connectionist cognitive sciences are highly related at the computational and algorithmic levels of analysis (Dawson, 1998, 2009).

In my view, the differences between classical and cognitive science concern the nature of the architecture, the primitive set of abilities or processes that are available for information processing (Dawson, 2009). The notion of an architecture is detailed in Chapter 2. One of the themes of the current book is that debates between different schools of thought in cognitive science are pre-paradigmatic discussions about the possible nature of the cognitive architecture.

These debates are enlivened by the modern rise of embodied cognitive science. One reason that classical and connectionist cognitive science can be easily compared is that both are representational (Clark, 1997; Dawson, 1998, 2004). However, some schools of thought in embodied cognitive science are explicitly anti-representational (Brooks, 1999; Chemero, 2009; Noë, 2004). As a result, it is not clear that the information processing hypothesis is applicable to embodied cognitive science. One of the goals of the current book is to examine embodied cognitive science from an information processing perspective, in order to use some of its key departures from both classical and connectionist cognitive science to inform the debate about the architecture.

The search for truth in the Library of Babel had dire consequences. Its librarians "disputed in the narrow corridors, proffered dark curses, strangled each other on the divine stairways, flung the deceptive books into the air shafts, met their death cast down in a similar fashion by the inhabitants of remote regions. Others went mad" (Borges, 1962, p. 55). The optimistic view of the current book is that a careful examination of the three different schools of cognitive science can provide a fruitful, unifying position on the nature of the cognitive architecture.

1.5 A Plan of Action

A popular title for surveys of cognitive science is *What is cognitive science?* (Lepore & Pylyshyn, 1999; von Eckardt, 1995). Because this one is taken, a different title is used for the current book. But steering the reader towards an answer to this excellent question is the primary purpose of the current manuscript.

Answering the question *What is cognitive science?* resulted in the current book being organized around two central themes. One is to introduce key ideas at the foundations of three different schools of thought: classical cognitive science, connectionist cognitive science, and embodied cognitive science. A second is to examine these ideas to see whether these three "flavours" of cognitive science can be unified. As a result, this book is presented in two main parts.

The purpose of Part I is to examine the foundations of the three schools of cognitive science. It begins in Chapter 2, with an overview of the need to investigate cognitive agents at multiple levels. These levels are used to provide a framework for considering potential relationships between schools of cognitive science. Each of these schools is also introduced in Part I. I discuss classical cognitive science in Chapter 3, connectionist cognitive science in Chapter 4, and embodied cognitive science in Chapter 5.

With the foundations of the three different versions of cognitive science laid out in Part I, in Part II, I turn to a discussion of a variety of topics within cognitive science. The purpose of these discussions is to seek points of either contention or convergence amongst the different schools of thought.

The theme of Part II is that the key area of disagreement amongst classical, connectionist, and embodied cognitive science is the nature of the cognitive architecture. However, this provides an opportunity to reflect on the technical details of the architecture as the potential for a unified cognitive science. This is because the properties of the architecture—regardless of the school of thought—are at best vaguely defined. For instance, Searle (1992, p. 15) has observed that "'intelligence,' 'intelligent behavior,' 'cognition' and 'information processing,' for example are not precisely defined notions. Even more amazingly, a lot of very technically sounding notions are poorly defined—notions such as 'computer,' 'computation,' 'program,' and 'symbol'" (Searle, 1992, p. 15).

In Part II, I also present a wide range of topics that permit the different schools of cognitive science to make contact. It is hoped that my treatment of these topics will show how the competing visions of the different schools of thought can be coordinated in a research program that attempts to specify an architecture of cognition inspired by all three schools.

Multiple Levels of Investigation

2.0 Chapter Overview

Cognitive science is an intrinsically interdisciplinary field of study. Why is this so? In the current chapter, I argue that the interdisciplinary nature of cognitive science necessarily emerges because it assumes that cognition is information processing. The position I take is that explanations of information processors require working at four different levels of investigation, with each level involving a different vocabulary and being founded upon the methodologies of different disciplines.

The chapter begins with a historical treatment of logicism, the view that thinking is equivalent to performing mental logic, and shows how this view was converted into the logical analysis of relay circuits by Claude Shannon. Shannon's work is then used to show that a variety of different arrangements of switches in a circuit can perform the same function, and that the same logical abilities can be constructed from different sets of core logical properties. Furthermore, any one of these sets of logical primitives can be brought to life in a variety of different physical realizations.

The consequence of this analysis is that information processors must be explained at four different levels of investigation. At the computational level, one asks what kinds of information processing problems can be solved by a system. At the algorithmic level, one asks what procedures are being used by a system to solve a particular problem of interest. At the architectural level, one asks what basic operations are used as the foundation for a specific algorithm. At the implementational level, one asks what physical mechanisms are responsible for bringing a particular architecture to life.

My goal in this chapter is to introduce these different levels of investigation. Later chapters reveal that different approaches within cognitive science have differing perspectives on the relative importance, and on the particular details, of each level.

2.1 Machines and Minds

Animism is the assignment of lifelike properties to inanimate, but moving, objects. Animism characterizes the thinking of young children, who may believe that a car, for instance, is alive because it can move on its own (Piaget, 1929). Animism was also apparent in the occult tradition of the Renaissance; the influential memory systems of Lull and of Bruno imbued moving images with powerful, magical properties (Yates, 1966).

Animism was important to the development of scientific and mathematical methods in the seventeenth century: "The Renaissance conception of an animistic universe, operated by magic, prepared the way for a conception of a mechanical universe, operated by mathematics" (Yates, 1966, p. 224). Note the animism in the introduction to Hobbes' (1967) *Leviathan*:

> For seeing life is but a motion of limbs, the beginning whereof is in some principal part within; why may we not say, that all *Automata* (Engines that move themselves by means of springs and wheels as doth a watch) have an artificial life? For what is the *Heart*, but a *Spring*; and the *Nerves*, but so many *Springs*; and the *Joynts*, but so many *Wheeles*, giving motion to the whole Body, such as was intended by the Artificer? (Hobbes, 1967, p. 3)

Such appeals to animism raised new problems. How were moving humans to be distinguished from machines and animals? Cartesian philosophy grounded humanity in mechanistic principles, but went on to distinguish humans-as-machines from animals because only the former possessed a soul, whose essence was "only to think" (Descartes, 1960, p. 41).

Seventeenth-century philosophy was the source of the mechanical view of man (Grenville, 2001; Wood, 2002). It was also the home of a reverse inquiry: was it possible for human artifacts, such as clockwork mechanisms, to become alive or intelligent?

By the eighteenth century, such philosophical ponderings were fuelled by "living machines" that had made their appearance to great public acclaim. Between 1768 and 1774, Pierre and Henri-Louis Jaquet-Droz constructed elaborate clockwork androids that wrote, sketched, or played the harpsichord (Wood, 2002). The eighteenth-century automata of Jacques de Vaucanson, on display for a full century, included a flute player and a food-digesting duck. Von Kempelen's infamous chess-playing Turk first appeared in 1770; it was in and out of the public eye until its destruction by fire in 1854 (Standage, 2002).

Wood (2002, p. xxvii) notes that all automata are presumptions "that life can be simulated by art or science or magic. And embodied in each invention is a riddle, a fundamental challenge to our perception of what makes us human." In the eighteenth century, this challenge attracted the attention of the Catholic Church. In 1727, Vaucanson's workshop was ordered destroyed because his clockwork servants, who served dinner and cleared tables, were deemed profane (Wood, 2002). The Spanish Inquisition imprisoned both Pierre Jaquet-Droz *and* his writing automaton!

In spite of the Church's efforts, eighteenth-century automata were popular, tapping into a nascent fascination with the possibility of living machines. This fascination has persisted uninterrupted to the present day, as evidenced by the many depictions of robots and cyborgs in popular fiction and films (Asimov, 2004; Caudill, 1992; Grenville, 2001; Ichbiah, 2005; Levin, 2002; Menzel, D'Aluisio, & Mann, 2000).

Not all modern automata were developed as vehicles of entertainment. The late 1940s saw the appearance of the first autonomous robots, which resembled, and were called, Tortoises (Grey Walter, 1963). These devices provided "mimicry of life" (p. 114) and were used to investigate the possibility that living organisms were simple devices that were governed by basic cybernetic principles. Nonetheless, Grey Walter worried that animism might discredit the scientific merit of his work:

> We are daily reminded how readily living and even divine properties are projected into inanimate things by hopeful but bewildered men and women; and the scientist cannot escape the suspicion that his projections may be psychologically the substitutes and manifestations of his own hope and bewilderment. (Grey Walter, 1963, p. 115)

While Grey Walter's Tortoises were important scientific contributions (Bladin, 2006; Hayward, 2001; Holland, 2003b; Sharkey & Sharkey, 2009), the twentieth century saw the creation of another, far more important, automaton: the digital computer. The computer is rooted in seventeenth-century advances in logic and mathematics. Inspired by the Cartesian notion of rational, logical, mathematical thought, the computer brought logicism to life.

Logicism is the idea that thinking is identical to performing logical operations (Boole, 2003). By the end of the seventeenth century, numerous improvements to Boole's logic led to the invention of machines that automated logical operations; most of these devices were mechanical, but electrical logic machines had also been conceived (Buck & Hunka, 1999; Jevons, 1870; Marquand, 1885; Mays, 1953). If thinking was logic, then thinking machines—machines that could do logic—existed in the late nineteenth century.

The logic machines of the nineteenth century were, in fact, quite limited in ability, as we see later in this chapter. However, they were soon replaced by much more powerful devices. In the first half of the twentieth century, the basic theory of a general computing mechanism had been laid out in Alan Turing's account of his universal machine (Hodges, 1983; Turing, 1936). The universal machine was a device that "could simulate the work done by any machine. . . . It would be a machine to do everything, which was enough to give anyone pause for thought" (Hodges, 1983, p. 104). The theory was converted into working universal machines—electronic computers—by the middle of the twentieth century (Goldstine, 1993; Reid, 2001; Williams, 1997).

The invention of the electronic computer made logicism practical. The computer's general ability to manipulate symbols made the attainment of machine intelligence seem plausible to many, and inevitable to some (Turing, 1950). Logicism was validated every time a computer accomplished some new task that had been presumed to be the exclusive domain of human intelligence (Kurzweil, 1990, 1999). The pioneers of cognitive science made some bold claims and some aggressive predictions (McCorduck, 1979): in 1956, Herbert Simon announced to a mathematical modelling class that "Over Christmas Allen Newell and I invented a thinking machine" (McCorduck, 1979, p. 116). It was predicted that by the late 1960s most theories in psychology would be expressed as computer programs (Simon & Newell, 1958).

The means by which computers accomplished complex information processing tasks inspired theories about the nature of human thought. The basic workings of computers became, at the very least, a metaphor for the architecture of human cognition. This metaphor is evident in philosophy in the early 1940s (Craik, 1943).

> My hypothesis then is that thought models, or parallels, reality—that its essential
> feature is not 'the mind,' 'the self,' 'sense data' nor 'propositions,' but is symbol-
> ism, and that this symbolism is largely of the same kind which is familiar to us in
> mechanical devices which aid thought and calculation. (Craik, 1943, p. 57)

Importantly, many modern cognitive scientists do not see the relationship between cognition and computers as being merely metaphorical (Pylyshyn, 1979a, p. 435): "For me, the notion of computation stands in the same relation to cognition as geometry does to mechanics: It is not a metaphor but part of a literal description of cognitive activity."

Computers are special devices in another sense: in order to explain how they work, one must look at them from several different perspectives. Each perspective requires a radically different vocabulary to describe what computers do. When cognitive science assumes that cognition is computation, it also assumes that human cognition must be explained using multiple vocabularies.

In this chapter, I provide an historical view of logicism and computing to introduce these multiple vocabularies, describe their differences, and explain why all are needed. We begin with the logicism of George Boole, which, when transformed into modern binary logic, defined the fundamental operations of modern digital computers.

2.2 From the Laws of Thought to Binary Logic

In 1854, with the publication of *An Investigation of the Laws of Thought*, George Boole (2003) invented modern mathematical logic. Boole's goal was to move the study of thought from the domain of philosophy into the domain of mathematics:

> There is not only a close analogy between the operations of the mind in general
> reasoning and its operations in the particular science of Algebra, but there is to
> a considerable extent an exact agreement in the laws by which the two classes of
> operations are conducted. (Boole, 2003, p. 6)

Today we associate Boole's name with the logic underlying digital computers (Mendelson, 1970). However, Boole's algebra bears little resemblance to our modern interpretation of it. The purpose of this section is to trace the trajectory that takes us from Boole's nineteenth-century calculus to the twentieth-century invention of truth tables that define logical functions over two binary inputs.

Boole did not create a binary logic; instead he developed an algebra of sets. Boole used symbols such as x, y, and z to represent classes of entities. He then defined "signs of operation, as +, -, ´, standing for those operations of the mind by which the conceptions of things are combined or resolved so as to form new conceptions involving the same elements" (Boole, 2003, p. 27). The operations of his algebra were those of election: they selected subsets of entities from various classes of interest (Lewis, 1918).

For example, consider two classes: x (e.g., "black things") and y (e.g., "birds"). Boole's expression $x + y$ performs an "exclusive or" of the two constituent classes, electing the entities that were "black things," or were "birds," but not those that were "black birds."

Elements of Boole's algebra pointed in the direction of our more modern binary logic. For instance, Boole used multiplication to elect entities that shared properties defined by separate classes. So, continuing our example, the set of "black birds" would be elected by the expression xy. Boole also recognized that if one multiplied a class with itself, the result would simply be the original set again. Boole wrote his *fundamental law of thought* as $xx = x$, which can also be expressed as $x^2 = x$. He realized that if one assigned numerical quantities to x, then this law would only be true for the values 0 and 1. "Thus it is a consequence of the fact that the fundamental equation of thought is of the second degree, that we perform the operation of analysis and classification, by division into pairs of opposites, or, as it is technically said, by *dichotomy*" (Boole, 2003, pp. 50–51). Still, this dichotomy was not to be exclusively interpreted in terms of truth or falsehood, though Boole exploited this representation in his treatment of secondary propositions. Boole typically used o to represent the empty set and 1 to represent the universal set; the expression $1 - x$ elected those entities that did not belong to x.

Boole's operations on symbols were purely formal. That is, the actions of his logical rules were independent of any semantic interpretation of the logical terms being manipulated.

> We may in fact lay aside the logical interpretation of the symbols in the given equation; convert them into quantitative symbols, susceptible only of the values o and 1; perform upon them as such all the requisite processes of solution; and finally restore to them their logical interpretation. (Boole, 2003, p. 70)

This formal approach is evident in Boole's analysis of his fundamental law. Beginning with $x^2 = x$, Boole applied basic algebra to convert this expression into $x - x^2 = 0$. He then simplified this expression to $x(1 - x) = 0$. Note that none of these steps are logical in nature; Boole would not be able to provide a logical justification for his derivation. However, he did triumphantly provide a logical interpretation of his result: o is the empty set, 1 the universal set, x is some set of interest, and $1 - x$ is the negation of this set. Boole's algebraic derivation thus shows that the intersection of x with its negation is the empty set. Boole noted that, in terms of logic, the equation $x(1 - x) = 0$ expressed,

> that it is impossible for a being to possess a quality and not to possess that quality at the same time. But this is identically that 'principle of contradiction' which Aristotle has described as the fundamental axiom of all philosophy. (Boole, 2003, p. 49)

It was important for Boole to link his calculus to Aristotle, because Boole not only held Aristotelian logic in high regard, but also hoped that his new mathematical methods would both support Aristotle's key logical achievements as well as extend Aristotle's work in new directions. To further link his formalism to Aristotle's logic, Boole applied his methods to what he called secondary propositions. A secondary proposition was a statement about a proposition that could be either true or false. As a result, Boole's analysis of secondary propositions provides another glimpse of how his work is related to our modern binary interpretation of it.

Boole applied his algebra of sets to secondary propositions by adopting a temporal interpretation of election. That is, Boole considered that a secondary proposition could be true or false for some duration of interest. The expression xy would now be interpreted as electing a temporal period during which both propositions x and y are true. The symbols o and 1 were also given temporal interpretations, meaning "no time" and "the whole of time" respectively. While this usage differs substantially from our modern approach, it has been viewed as the inspiration for modern binary logic (Post, 1921).

Boole's work inspired subsequent work on logic in two different ways. First, Boole demonstrated that an algebra of symbols was possible, productive, and worthy of exploration: "Boole showed incontestably that it was possible, by the aid of a system of mathematical signs, to deduce the conclusions of all these ancient modes of reasoning, and an indefinite number of other conclusions" (Jevons, 1870, p. 499). Second, logicians noted certain idiosyncrasies of and deficiencies with Boole's calculus, and worked on dealing with these problems. Jevons also wrote that Boole's examples "can be followed only by highly accomplished mathematical minds; and even a mathematician would fail to find any demonstrative force in a calculus which fearlessly employs unmeaning and incomprehensible symbols" (p. 499). Attempts to simplify and correct Boole produced new logical systems that serve as the bridge between Boole's nineteenth-century logic and the binary logic that arose in the twentieth century.

Boole's logic is problematic because certain mathematical operations do not make sense within it (Jevons, 1870). For instance, because addition defined the "exclusive or" of two sets, the expression $x + x$ had no interpretation in Boole's system. Jevons believed that Boole's interpretation of addition was deeply mistaken and corrected this by defining addition as the "inclusive or" of two sets. This produced an interpretable additive law, $x + x = x$, that paralleled Boole's multiplicative fundamental law of thought.

Jevons' (1870) revision of Boole's algebra led to a system that was simple enough to permit logical inference to be mechanized. Jevons illustrated this with a three-class system, in which upper-case letters (e.g., A) picked out those entities that belonged to a set and lower-case letters (e.g., a) picked out those entities that did not belong. He then produced what he called the logical abecedarium, which was the set of possible combinations of the three classes. In his three-class example, the abecedarium consisted of eight combinations: ABC, ABc, AbC, Abc, aBC, aBc, abC, and abc. Note that each of these combinations is a multiplication of three terms in Boole's sense, and thus elects an intersection of three different classes. As well, with the improved definition of logical addition, different terms of the abecedarium could be added together to define some set of interest. For example Jevons (but not Boole!) could elect the class B with the following expression: $B = ABC + ABc + aBC + aBc$.

Jevons (1870) demonstrated how the abecedarium could be used as an inference engine. First, he used his set notation to define concepts of interest, such as in the example A = iron, B = metal, and C = element. Second, he translated propositions into intersections of sets. For instance, the premise "Iron is metal" can be rewritten as "*A* is *B*," which in Boole's algebra becomes *AB*, and "metal is element" becomes *BC*. Third, given a set of premises, Jevons removed the terms that were inconsistent with the premises from the abecedarium: the only terms consistent with the premises *AB* and *BC* are *ABC*, *aBC*, *abC*, and *abc*. Fourth, Jevons inspected and interpreted the remaining abecedarium terms to perform valid logical inferences. For instance, from the four remaining terms in Jevons' example, we can conclude that "all iron is element," because *A* is only paired with *C* in the terms that remain, and "there are some elements that are neither metal nor iron," or *abC*. Of course, the complete set of entities that is elected by the premises is the logical sum of the terms that were not eliminated.

Jevons (1870) created a mechanical device to automate the procedure described above. The machine, known as the "logical piano" because of its appearance, displayed the 16 different combinations of the abecedarium for working with four different classes. Premises were entered by pressing keys; the depression of a pattern of keys removed inconsistent abecedarium terms from view. After all premises had been entered in sequence, the terms that remained on display were interpreted. A simpler variation of Jevons' device, originally developed for four-class problems but more easily extendable to larger situations, was invented by Allan Marquand (Marquand, 1885). Marquand later produced plans for an electric version of his device that used electromagnets to control the display (Mays, 1953). Had this device been constructed, and had Marquand's work come to the attention of a wider audience, the digital computer might have been a nineteenth-century invention (Buck & Hunka, 1999).

With respect to our interest in the transition from Boole's work to our modern interpretation of it, note that the logical systems developed by Jevons, Marquand, and others were binary in two different senses. First, a set and its complement (e.g., *A* and *a*) never co-occurred in the same abecedarium term. Second, when premises were applied, an abecedarium term was either eliminated or not. These binary characteristics of such systems permitted them to be simple enough to be mechanized.

The next step towards modern binary logic was to adopt the practice of assuming that propositions could either be true or false, and to algebraically indicate these states with the values 1 and 0. We have seen that Boole started this approach, but that he did so by applying awkward temporal set-theoretic interpretations to these two symbols.

The modern use of 1 and 0 to represent true and false arises later in the nineteenth century. British logician Hugh McColl's (1880) symbolic logic used this notation, which he borrowed from the mathematics of probability. American logician Charles Sanders Peirce (1885) also explicitly used a binary notation for truth in his famous paper "On the algebra of logic: A contribution to the philosophy of notation." This paper is often cited as the one that introduced the modern usage

(Ewald, 1996). Peirce extended Boole's work on secondary propositions by stipulating an additional algebraic law of propositions: for every element x, either $x = 0$ or $x = 1$, producing a system known as "the two-valued algebra" (Lewis, 1918).

The two-valued algebra led to the invention of truth tables, which are established in the literature in the early 1920s (Post, 1921; Wittgenstein, 1922), but were likely in use much earlier. There is evidence that Bertrand Russell and his then student Ludwig Wittgenstein were using truth tables as early as 1910 (Shosky, 1997). It has also been argued that Charles Peirce and his students probably were using truth tables as early as 1902 (Anellis, 2004).

Truth tables make explicit an approach in which primitive propositions (p, q, r, etc.) that could only adopt values of 0 or 1 are used to produce more complex expressions. These expressions are produced by using logical functions to combine simpler terms. This approach is known as "using truth-value systems" (Lewis & Langford, 1959). Truth-value systems essentially use truth tables to determine the truth of functions of propositions (i.e., of logical combinations of propositions). "It is a distinctive feature of this two-valued system that when the property, 0 or 1, of the elements p, q, etc., is given, any function of the elements which is in the system is thereby determined to have the property 0 or the property 1" (p. 199).

Consider Table 2-1, which provides the values of three different functions (the last three columns of the table) depending upon the truth value of two simple propositions (the first two columns of the table):

p	q	$p \cdot q$	$p + q$	$p \cdot (p + q)$
1	1	1	1	1
1	0	0	1	1
0	1	0	1	0
0	0	0	0	0

Table 2-1. Examples of the truth value system for two elementary propositions and some of their combinations. The possible values of p and q are given in the first two columns. The resulting values of different functions of these propositions are provided in the remaining columns.

Truth-value systems result in a surprising, simplified approach to defining basic or primitive logical functions. When the propositions p and q are interpreted as being only true or false, then there are only four possible combinations of these two propositions that can exist, i.e., the first two columns of Table 2-1. A primitive function can be defined as a function that is defined over p and q, and that takes on a truth value for each combination of these variables.

Given that in a truth-value system a function can only take on the value of 0 or 1, then there are only 16 different primitive functions that can be defined for combinations of the binary inputs p and q (Ladd, 1883). These primitive functions are provided in Table 2-2; each row of the table shows the truth values of each function for each combination of the inputs. An example logical notation for each function is provided in the last column of the table. This notation was used by Warren McCulloch (1988b), who attributed it to earlier work by Wittgenstein.

Not surprisingly, an historical trajectory can also be traced for the binary logic defined in Table 2-2. Peirce's student Christine Ladd actually produced the first five columns of that table in her 1883 paper, including the conversion of the first four numbers in a row from a binary to a base 10 number. However, Ladd did not interpret each row as defining a logical function. Instead, she viewed the columns in terms of set notation and each row as defining a different "universe." The interpretation of the first four columns as the truth values of various logical functions arose later with the popularization of truth tables (Post, 1921; Wittgenstein, 1922).

$p=0$ $q=0$	$p=1$ $q=0$	$p=0$ $q=1$	$p=1$ $q=1$	Number	Notation
0	0	0	0	0	Contradiction
0	0	0	1	1	$p{\cdot}q$
0	0	1	0	2	${\sim}p{\cdot}q$
0	0	1	1	3	q
0	1	0	0	4	$p{\cdot}{\sim}q$
0	1	0	1	5	p
0	1	1	0	6	$p \wedge q$
0	1	1	1	7	$p \vee q$
1	0	0	0	8	${\sim}p{\cdot}{\sim}q$
1	0	0	1	9	$p \equiv q$
1	0	1	0	10	${\sim}p$
1	0	1	1	11	$p \supset q$
1	1	0	0	12	${\sim}q$
1	1	0	1	13	$q \supset p$
1	1	1	0	14	$p \mid q$
1	1	1	1	15	Tautology

Table 2-2. Truth tables for all possible functions of pairs of propositions. Each function has a truth value for each possible combination of the truth values of p and q, given in the first four columns of the table. The *Number* column converts the first four values in a row into a binary number (Ladd, 1883). The logical notation for each function is taken Warren McCulloch (1988b).

Truth tables, and the truth-value system that they support, are very powerful. They can be used to determine whether any complex expression, based on combinations of primitive propositions and primitive logical operations, is true or false (Lewis, 1932). In the next section we see the power of the simple binary truth-value system, because it is the basis of the modern digital computer. We also see that bringing this system to life in a digital computer leads to the conclusion that one must use more than one vocabulary to explain logical devices.

2.3 From the Formal to the Physical

The short story *The Dreams in the Witch-house* by Lovecraft (1933) explored the link between mathematics and magic. The story explained how a student discovers that the act of writing out mathematical equations can alter reality. This alteration provided an explanation of how the accused Salem witch Keziah Mason escaped her seventeenth-century captors:

> She had told Judge Hathorne of lines and curves that could be made to point out directions leading through the walls of space to other spaces and beyond. . . . Then she had drawn those devices on the walls of her cell and vanished. (Lovecraft, 1933, p. 140)

This strange link between the formal and the physical was also central to another paper written in the same era as Lovecraft's story. The author was Claude Shannon, and the paper's title was "A symbolic analysis of relay and switching circuits" (Shannon, 1938). However, his was not a work of fiction. Instead, it was a brief version what is now known as one of the most important master's theses ever written (Goldstine, 1993). It detailed the link between Boolean algebra and electrical circuits, and showed how mathematical logic could be used to design, test, and simplify circuits. "The paper was a landmark in that it helped to change digital circuit design from an art to a science" (p. 120).

Shannon had a lifelong interest in both mathematics and mechanics. While his most influential papers were mathematical in focus (Shannon, 1938, 1948), he was equally famous for his tinkering (Pierce, 1993). His mechanical adeptness led to the invention of a number of famous devices, including Theseus, a mechanical maze-solving mouse. Later in his career Shannon seemed to take more pride in the gadgets that he had created and collected than in his numerous impressive scientific awards (Horgan, 1992).

Shannon's combined love of the mathematical and the mechanical was evident in his education: he completed a double major in mathematics and electrical engineering at the University of Michigan (Calderbank & Sloane, 2001). In 1936, he was hired as a research assistant at MIT, working with the differential analyzer of

Vannevar Bush. This machine was a pioneering analog computer, a complex array of electrical motors, gears, and shafts that filled an entire room. Its invention established Bush as a leader in electrical engineering as well as a pioneer of computing (Zachary, 1997). Bush, like Shannon, was enamored of the link between the formal and the physical. The sight of the differential analyzer at work fascinated Bush "who loved nothing more than to see things work. It was only then that mathematics—his sheer abstractions—came to life" (Zachary, 1997, p. 51).

Because of his work with Bush's analog computer, Shannon was prepared to bring another mathematical abstraction to life when the opportunity arose. The differential analyzer had to be physically reconfigured for each problem that was presented to it, which in part required configuring circuits that involved more than one hundred electromechanical relays, which were used as switches. In the summer of 1937, Shannon worked in Bell Labs and saw that engineers there were confronted with designing more complex systems that involved thousands of relays. At the time, this was labourious work that was done by hand. Shannon wondered if there was a more efficient approach. He discovered one when he realized that there was a direct mapping between switches and Boolean algebra, which Shannon had been exposed to in his undergraduate studies.

An Internet search will lead to many websites suggesting that Shannon recognized that the opening or closing of a switch could map onto the notions of "false" or "true." Actually, Shannon's insight involved the logical properties of *combinations* of switches. In an interview that originally appeared in *Omni* magazine in 1987, he noted "It's not so much that a thing is 'open' or 'closed,' the 'yes' or 'no' that you mentioned. The real point is that two things in series are described by the word 'and' in logic, so you would say this 'and' this, while two things in parallel are described by the word 'or'" (Liversidge, 1993).

In particular, Shannon (1938) viewed a switch (Figure 2-1A) as a source of impedance; when the switch was closed, current could flow and the impedance was 0, but when the switch was open (as illustrated in the figure) the impedance was infinite; Shannon used the symbol 1 to represent this state. As a result, if two switches were connected in series (Figure 2-1B) current would only flow if both switches were closed. Shannon represented this as the sum $x + y$. In contrast, if switch x and switch y were connected in parallel (Figure 2-1C), then current would flow through the circuit if either (i.e., both) of the switches were closed. Shannon represented this circuit as the product xy.

Shannon's (1938) logical representation is a variation of the two-valued logic that was discussed earlier. The Boolean version of this logic represented *false* with 0, *true* with 1, *or* with addition, and *and* with multiplication. Shannon's version represented *false* with 1, *true* with 0, *or* with multiplication, and *and* with addition. But because Shannon's reversal of the traditional logic is complete, the two are equivalent. Shannon noted that the basic properties of the two-valued logic were true of his logical interpretation of switches: "Due to this analogy any theorem of the calculus of propositions is also a true theorem if interpreted in terms of relay circuits" (p. 714).

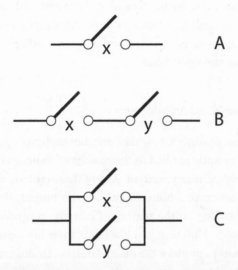

Figure 2-1. (A) An electrical switch, labelled *x*. (B) Switches *x* and *y* in series. (C) Switches *x* and *y* in parallel.

The practical implication of Shannon's (1938) paper was that circuit design and testing was no longer restricted to hands-on work in the physical domain. Instead, one could use pencil and paper to manipulate symbols using Boolean logic, designing a circuit that could be proven to generate the desired input-output behaviour. Logical operations could also be used to ensure that the circuit was as simple as possible by eliminating unnecessary logical terms: "The circuit may then be immediately drawn from the equations" (p. 713). Shannon illustrated this technique with examples that included a "selective circuit" that would permit current when 1, 3, or 4—but not 0 or 2—of its relays were closed, as well as an electric combination lock that would only open when its 5 switches were depressed in a specific order.

Amazingly, Shannon was not the first to see that electrical circuits were logical in nature (Burks, 1975)! In 1886, Charles Peirce wrote a letter to his student Alan Marquand suggesting how the latter's logic machine (Marquand, 1885) could be improved by replacing its mechanical components with electrical ones. Peirce provided diagrams of a serial 3-switch circuit that represented logical conjunction (*and*) and a parallel 3-switch circuit that represented logical disjunction (*or*). Peirce's nineteenth-century diagrams would not have been out of place in Shannon's twentieth-century paper.

In Lovecraft's (1933) story, the witch Keziah "might have had excellent reasons for living in a room with peculiar angles; for was it not through certain angles that she claimed to have gone outside the boundaries of the world of space we know?" Shannon's (1938) scholarly paper led to astonishing conclusions for similar reasons: it detailed equivalence between the formal and the physical. It proved that electric circuits could be described in two very different vocabularies: one the physical vocabulary of current, contacts, switches and wires; the other the abstract vocabulary of logical symbols and operations.

2.4 Multiple Procedures and Architectures

According to a Chinese proverb, we all like lamb, but each has a different way to cook it. This proverb can be aptly applied to the circuits of switches for which Shannon (1938) developed a logical interpretation. Any of these circuits can be described as defining a logical function that maps inputs onto an output: the circuit outputs a current (or not) depending on the pattern of currents controlled by one or more switches that flow into it. However, just like lamb, there are many different ways to "cook" the input signals to produce the desired output. In short, many different circuits can be constructed to compute the same input-output function.

To illustrate this point, let us begin by considering Shannon's (1938) selective circuit, which would be off when 0 or 2 of its 4 relays were closed, but which would be on when any other number of its relays was closed. In Shannon's original formulation, 20 components—an arrangement of 20 different switches—defined a circuit that would behave in the desired fashion. After applying logical operations to simplify the design, Shannon reduced the number of required components from 20 to 14. That is, a smaller circuit that involved an arrangement of only 14 different switches delivered the same input-output behaviour as did the 20-switch circuit.

Reflecting on these two different versions of the selective circuit, it's clear that if one is interested in comparing them, the result of the comparison depends on the perspective taken. On the one hand, they are quite different: they involve different numbers of components, related to one another by completely different patterns of wiring. On the other hand, in spite of these obvious differences in details, at a more abstract level the two designs are identical, in the sense that both designs produce the same input-output mapping. That is, if one built a truth table for either circuit that listed the circuit's conductivity (output) as a function of all possible combinations of its 4 relays (inputs), the two truth tables would be identical. One might say that the two circuits use markedly different procedures (i.e., arrangements of internal components) to compute the same input-output function. They generate the same behaviour, but for different reasons.

Comparisons between different devices are further complicated by introducing the notion of an architecture (Brooks, 1962). In computer science, the term *architecture* was originally used by Frederick P. Brooks Jr., a pioneering force in the creation of IBM's early computers. As digital computers evolved, computer designers faced changing constraints imposed by new hardware technologies. This is because new technologies defined anew the basic information processing properties of a computer, which in turn determined what computers could and could not do. A computer's architecture is its set of basic information processing properties (Blaauw & Brooks, 1997, p. 3): "The architecture of a computer system we define as the minimal set of properties that determine what programs will run and what results they will produce."

The two different versions of Shannon's (1938) selective circuit were both based on the same architecture: the architecture's primitives (its basic components) were parallel and serial combinations of pairs of switches. However, other sets of primitives could be used.

An alternative architecture could use a larger number of what Shannon (1938) called special types of relays or switches. For instance, we could take each of the 16 logical functions listed in Table 2-2 and build a special device for each. Each device would take two currents as input, and would convert them into an appropriate output current. For example, the XOR device would only deliver a current if only one of its input lines was active; it would *not* deliver a current if both its input lines were either active or inactive—behaving exactly as it is defined in Table 2-2. It is easy to imagine building some switching circuit that used all of these logic gates as primitive devices; we could call this imaginary device "circuit x."

The reason that the notion of architecture complicates (or enriches!) the comparison of devices is that the same circuit can be created from different primitive components. Let us define one additional logic gate, the NOT gate, which does not appear in Table 2-2 because it has only one input signal. The NOT gate reverses or inverts the signal that is sent into it. If a current is sent into a NOT gate, then the NOT gate does not output a current. If a current is not sent into a NOT gate, then the gate outputs a current. The first NOT gate—the first electromechanical relay— was invented by American physicist Joseph Henry in 1835. In a class demonstration, Henry used an input signal to turn off an electromagnet from a distance, startling his class when the large load lifted by the magnet crashed to the floor (Moyer, 1997).

The NOT gate is important, because it can be used to create any of the Table 2-2 operations when combined with two other operators that are part of that table: AND, which McCulloch represented as $p \cdot q$, and OR, which McCulloch represented as $p \vee q$. To review, if the only special relays available are NOT, A,ND and OR, then one can use these three primitive logic blocks to create any of the other logical operations that are given in Table 2-2 (Hillis, 1998). "This idea of a universal set of blocks is important: it means that the set is general enough to build anything" (p. 22).

To consider the implications of the universal set of logic gates to comparing circuits, let us return to our imaginary circuit x. We could have two different versions of this circuit, based on different architectures. In one, the behaviour of the circuit would depend upon wiring up some arrangement of all the various logical operations given in Table 2-2, where each operation is a primitive—that is, carried out by its own special relay. In the other, the arrangement of the logical operations would be identical, but the logical operations in Table 2-2 would not be primitive. Instead, we would replace each special relay from the first circuit with a circuit involving NOT, AND, and OR that would produce the desired behaviour.

Let us compare these two different versions of circuit x. At the most abstract level, they are identical, because they are generating the same input-output behaviour. At a more detailed level—one that describes how this behaviour is generated in terms of how the logical operations of Table 2-2 are combined together—the two are also identical. That is, the two circuits are based on the same combinations of the Table 2-2 operations. However, at a more detailed level, the level of the architecture, the two circuits are different. For the first circuit, each logical operation from Table 2-2 would map onto a physical device, a special relay. This would not be true for the second circuit. For it, each logical operation from Table 2-2 could be decomposed into a combination of simpler logical operations—NOT, AND, OR—which in turn could be implemented by simple switches. The two circuits are different in the sense that they use different architectures, but these different architectures are used to create the same logical structure to compute the same input-output behaviour.

We now can see that Shannon's (1938) discoveries have led us to a position where we can compare two different electrical circuits by asking three different questions. First, do the two circuits compute the same input-output function? Second, do the two circuits use the same arrangement of logical operations used to compute this function? Third, do the two circuits use the same architecture to bring these logical operations to life? Importantly, the comparison between two circuits can lead to affirmative answers to some of these questions, and negative answers to others. For instance, Shannon's two selective circuits use different arrangements of logical operations, but are based on the same architecture, and compute the same input-output function. The two versions of our imaginary circuit x compute the same input-output function, and use the same arrangement of logical operations, but are based on different architectures.

Ultimately, all of the circuits we have considered to this point are governed by the same physical laws: the laws of electricity. However, we will shortly see that it is possible to have two systems that have affirmative answers to the three questions listed in the previous paragraph, but are governed by completely different physical laws.

2.5 Relays and Multiple Realizations

Many of the ideas that we have been considering in this chapter have stemmed from Shannon's (1938) logical interpretation of relay circuits. But what is a relay?

A relay is essentially a remote-controlled switch that involves two separate circuits (Gurevich, 2006). One of these circuits involves a source of current, which can be output through the relay's drain. The second circuit controls the relay's gate. In an electromechanical relay, the gate is an electromagnet (Figure 2-2). When a signal flows through the gate, the magnet becomes active and pulls a switch closed so that the source flows through the drain. When the gate's signal is turned off, a spring pulls the switch open, breaking the first circuit, and preventing the source from flowing through the drain.

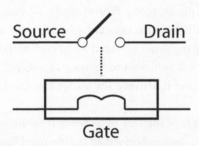

Figure 2-2. A relay, in which a signal through an electromagnetic gate controls a switch that determines whether the current from the source will flow through the drain.

The relay shown in Figure 2-2 can be easily reconfigured to convert it into a NOT gate. This is accomplished by having the switch between the source and the drain pulled open by the gate, and having it closed by a spring when the gate is not active. This was how, in 1835, Joseph Henry turned the power off to a large electromagnet, causing it to drop its load and startle his class (Moyer, 1997).

The type of relay shown in Figure 2-2 was critically important to the development of the telegraph in the mid-nineteenth century. Telegraphs worked by sending electrical pulses—dots and dashes—long distances over copper wire. As the signals travelled, they weakened in intensity. In order to permit a message to be communicated over a long distance, the signal would have to be re-amplified at various points along its journey. Relays were the devices that accomplished this. The weak incoming signals were still strong enough to activate a relay's magnet. When this happened, a stronger current—provided by the source—was sent along the telegraph wire, which was connected to the relay's drain. The relay mechanism ensured that the pattern of pulses being sent along the drain matched the pattern of pulses that turned the gate on and off. That is, the periods of time during which the relay's switch was closed matched the durations of the dots and dashes that operated the relay's magnet. The ability of a telegraph company to communicate messages over very long distances depended completely on the relays that were interspersed along the company's network.

This dependence upon relays played a critical role in the corporate warfare between competing telegraph companies. In 1874, the only relay in use in the telegraph industry was an electromagnetic one invented by Charles Grafton Page; the patent for this device was owned by Western Union. An imminent court decision was going to prevent the Automatic Telegraph Company from using this device in its own telegraph system because of infringement on the patent.

The Automatic Telegraph Company solved this problem by commissioning Thomas Edison to invent a completely new relay, one that avoided the Page patent by not using magnets (Josephson, 1961). Edison used a rotating chalk drum to replace the electromagnet. This is because Edison had earlier discovered that the friction of a wire dragging along the drum changed when current flowed through the wire. This change in friction was sufficient to be used as a signal that could manipulate the gate controlling the circuit between the source and the drain. Edison's relay was called a motograph.

Edison's motograph is of interest to us when it is compared to the Page relay. On the one hand, the two devices performed the identical function; indeed, Edison's relay fit exactly into the place of the page relay:

> First he detached the Page sounder from the instrument, an intensely interested crowd watching his every movement. From one of his pockets he took a pair of pliers and fitted [his own motograph relay] precisely where the Page sounder had been previously connected, and tapped the key. The clicking—and it was a joyful sound—could be heard all over the room. There was a general chorus of surprise. 'He's got it! He's got it!' (Josephson, 1961, p. 118)

On the other hand, the physical principles governing the two relays were completely different. The key component of one was an electromagnet, while the critical part of the other was a rotating drum of chalk. In other words, the two relays were functionally identical, but physically different. As a result, if one were to describe the purpose, role, or function of each relay, then the Page relay and the Edison motograph would be given the same account. However, if one were to describe the physical principles that accomplished this function, the account of the Page relay would be radically different from the account of the Edison motograph—so different, in fact, that the same patent did not apply to both. *Multiple realization* is the term used to recognize that different physical mechanisms can bring identical functions to life.

The history of advances in communications and computer technology can be described in terms of evolving multiple realizations of relays and switches. Electromagnetic relays were replaced by vacuum tubes, which could be used to rapidly switch currents on and off and to amplify weak signals (Reid, 2001). Vacuum tubes were replaced by transistors built from semiconducting substances such as silicon. Ultimately, transistors were miniaturized to the point that millions could be etched into a single silicon chip.

One might suggest that the examples listed above are not as physically different as intended, because all are electrical in nature. But relays can be implemented in many nonelectrical ways as well. For example, nanotechnology researchers are exploring various molecular ways in which to create logic gates (Collier et al., 1999; Okamoto, Tanaka, & Saito, 2004). Similarly, Hillis (1998) described in detail a hydraulic relay, in which the source and drain involve a high-pressure water line and a weaker input flow controls a valve. He pointed out that his hydraulic relay is functionally identical to a transistor, and that it could therefore be used as the basic building block for a completely hydraulic computer. "For most purposes, we can forget about technology [physical realization]. This is wonderful, because it means that almost everything that we say about computers will be true even when transistors and silicon chips become obsolete" (p. 19).

Multiple realization is a key concept in cognitive science, particularly in classical cognitive science, which is the topic of Chapter 3. Multiple realization is in essence an argument that while an architectural account of a system is critical, it really doesn't matter what physical substrate is responsible for bringing the architecture into being. Methodologically this is important, because it means that computer simulation is a viable tool in cognitive science. If the physical substrate doesn't matter, then it is reasonable to emulate the brain-based architecture of human cognition using completely different hardware—the silicon chips of the digital computer.

Theoretically, multiple realization is also important because it raises the possibility that non-biological systems could be intelligent and conscious. In a famous thought experiment (Pylyshyn, 1980), each neuron in a brain is replaced with a silicon chip that is functionally equivalent to the replaced neuron. Does the person experience any changes in consciousness because of this change in hardware? The logical implication of multiple realization is that no change should be experienced. Indeed, the assumption that intelligence results from purely biological or neurological processes in the human brain may simply be a dogmatic attempt to make humans special when compared to lower animals or machines (Wiener, 1964, p. 31): "Operative images, which perform the functions of their original, may or may not bear a pictorial likeness to it. Whether they do or not, they may replace the original in its action, and this is a much deeper similarity."

2.6 Multiple Levels of Investigation and Explanation

Imagine bringing several different calculating devices into a class, with the goal of explaining how they work. How would you explain those devices? The topics that have been covered in the preceding pages indicate that several different approaches could—and likely should—be taken.

One approach would be to explain what was going on at a physical or implementational level. For instance, if one of the devices was an old electronic calculator, then you would feel comfortable in taking it apart to expose its internal workings. You would likely see an internal integrated circuit. You might explain how such circuits work by talking about the properties of semiconductors and how different layers of a silicon semiconductor can be doped with elements like arsenic or boron to manipulate conductivity (Reid, 2001) in order to create components like transistors and resistors.

Interestingly, the physical account of one calculator will not necessarily apply to another. Charles Babbage's difference engine was an automatic calculator, but was built from a set of geared columns (Swade, 1993). Slide rules were the dominant method of calculation prior to the 1970s (Stoll, 2006) and involved aligning rulers that represented different number scales. The abacus is a set of moveable beads mounted on vertical bars and can be used by experts to perform arithmetic calculations extremely quickly (Kojima, 1954). The physical accounts of each of these three calculating devices would be quite different from the physical account of any electronic calculator.

A second approach to explaining a calculating device would be to describe its basic architecture, which might be similar for two different calculators that have obvious physical differences. For example, consider two different machines manufactured by Victor. One, the modern 908 pocket calculator, is a solar-powered device that is approximately 3" × 4" × ½" in size and uses a liquid crystal display. The other is the 1800 desk machine, which was introduced in 1971 with the much larger dimensions of 9" × 11" × 4½". One reason for the 1800's larger size is the nature of its power supply and display: it plugged into a wall socket, and it had to be large enough to enclose two very large (inches-high!) capacitors and a transformer. It also used a gas discharge display panel instead of liquid crystals. In spite of such striking physical differences between the 1800 and the 908, the "brains" of each calculator are integrated circuits that apply arithmetic operations to numbers represented in binary format. As a result, it would not be surprising to find many similarities between the architectures of these two devices.

Of course, there can be radical differences between the architectures of different calculators. The difference engine did not use binary numbers, instead representing values in base 10 (Swade, 1993). Claude Shannon's THROBACK computer's input, output, and manipulation processes were all designed for quantities represented as Roman numerals (Pierce, 1993). Given that they were designed to work with different number systems, it would be surprising to find many architectural similarities between the architectures of THROBACK, the difference engine, and the Victor electronic machines.

A third approach to explaining various calculators would be to describe the procedures or algorithms that these devices use to accomplish their computations. For instance, what internal procedures are used by the various machines to manipulate numerical quantities? Algorithmic accounts could also describe more external elements, such as the activities that a user must engage in to instruct a machine to perform an operation of interest. Different electronic calculators may require different sequences of key presses to compute the same equation.

For example, my own experience with pocket calculators involves typing in an arithmetic expression by entering symbols in the same order in which they would be written down in a mathematical expression. For instance, to subtract 2 from 4, I would enter "4 – 2 =" and expect to see 2 on display as the result. However, when I tested to see if the Victor 1800 that I found in my lab still worked, I couldn't type that equation in and get a proper response. This is because this 1971 machine was designed to be easily used by people who were more familiar with mechanical adding machines. To subtract 2 from 4, the following expression had to be entered: "4 + 2 –". Apparently the "=" button is only used for multiplication and division on this machine!

More dramatic procedural differences become evident when comparing devices based on radically different architectures. A machine such as the Victor 1800 adds two numbers together by using its logic gates to combine two memory registers that represent digits in binary format. In contrast, Babbage's difference engine represents numbers in decimal format, where each digit in a number is represented by a geared column. Calculations are carried out by setting up columns to represent the desired numbers, and then by turning a crank that rotates gears. The turning of the crank activates a set of levers and racks that raise and lower and rotate the numerical columns. Even the algorithm for processing columns proceeds in a counterintuitive fashion. During addition, the difference engine first adds the odd-numbered columns to the even-numbered columns, and then adds the even-numbered columns to the odd-numbered ones (Swade, 1993).

A fourth approach to explaining the different calculators would be to describe them in terms of the relation between their inputs and outputs. Consider two of our example calculating devices, the Victor 1800 and Babbage's difference engine. We have already noted that they differ physically, architecturally, and procedurally. Given these differences, what would classify both of these machines as calculating devices? The answer is that they are both calculators in the sense that they generate the same input-output pairings. Indeed, all of the different devices that have been mentioned in the current section are considered to be calculators for this reason. In spite of the many-levelled differences between the abacus, electronic calculator, difference engine, THROBACK, and slide rule, at a very abstract level—the level concerned with input-output mappings—these devices are equivalent.

To summarize the discussion to this point, how might one explain calculating devices? There are at least four different approaches that could be taken, and each approach involves answering a different question about a device. What is its physical nature? What is its architecture? What procedures does it use to calculate? What input-output mapping does it compute?

Importantly, answering each question involves using very different vocabularies and methods. The next few pages explore the diversity of these vocabularies. This diversity, in turn, accounts for the interdisciplinary nature of cognitive science.

2.7 Formal Accounts of Input-Output Mappings

For a cyberneticist, a machine was simply a device for converting some input into some output—and nothing more (Ashby, 1956, 1960; Wiener, 1948, 1964). A cyberneticist would be concerned primarily with describing a machine such as a calculating device in terms of its input-output mapping. However, underlying this simple definition was a great deal of complexity.

First, cybernetics was not interested in the relation between a particular input and output, but instead was interested in a general account of a machine's *possible* behaviour "by asking not 'what individual act will it produce here and now?' but 'what are all the possible behaviours that it can produce?'" (Ashby, 1956, p. 3).

Second, cybernetics wanted not only to specify what possible input-outputs could be generated by a device, but also to specify what behaviours could *not* be generated, and *why*: "Cybernetics envisages a set of possibilities much wider than the actual, and then asks why the particular case should conform to its usual particular restriction" (Ashby, 1956, p. 3).

Third, cybernetics was particularly concerned about machines that were nonlinear, dynamic, and adaptive, which would result in very complex relations between input and output. The nonlinear relationships between four simple machines that interact with each other in a network are so complex that they are mathematically intractable (Ashby, 1960).

Fourth, cybernetics viewed machines in a general way that not only ignored their physical nature, but was not even concerned with whether a particular machine had been (or could be) constructed or not. "What cybernetics offers is the framework on which all individual machines may be ordered, related and understood" (Ashby, 1956, p. 2).

How could cybernetics study machines in such a way that these four different perspectives could be taken? To accomplish this, the framework of cybernetics was exclusively mathematical. Cyberneticists investigated the input-output mappings of machines by making general statements or deriving proofs that were expressed in some logical or mathematical formalism.

By the late 1950s, research in cybernetics proper had begun to wane (Conway & Siegelman, 2005); at this time cybernetics began to evolve into the modern field of cognitive science (Boden, 2006; Gardner, 1984; Miller, 2003). Inspired by advances in digital computers, cognitive science was not interested in generic "machines" as such, but instead focused upon particular devices that could be described as information processors or symbol manipulators.

Given this interest in symbol manipulation, one goal of cognitive science is to describe a device of interest in terms of the specific information processing problem that it is solving. Such a description is the result of performing an analysis at the computational level (Dawson, 1998; Marr, 1982; Pylyshyn, 1984).

A computational analysis is strongly related to the formal investigations carried out by a cyberneticist. At the computational level of analysis, cognitive scientists use formal methods to prove what information processing problems a system can—and cannot—solve. The formal nature of computational analyses lend them particular authority: "The power of this type of analysis resides in the fact that the discovery of valid, sufficiently universal constraints leads to conclusions . . . that have the same permanence as conclusions in other branches of science" (Marr, 1982, p. 331).

However, computational accounts do not capture all aspects of information processing. A proof that a device is solving a particular information processing problem is only a proof concerning the device's input-output mapping. It does not say what algorithm is being used to compute the mapping or what physical aspects of the device are responsible for bringing the algorithm to life. These missing details must be supplied by using very different methods and vocabularies.

2.8 Behaviour by Design and by Artifact

What vocabulary is best suited to answer questions about the how a particular input-output mapping is calculated? To explore this question, let us consider an example calculating device, a Turing machine (Turing, 1936). This calculator processes symbols that are written on a ticker-tape memory divided into cells, where each cell can hold a single symbol. To use a Turing machine to add (Weizenbaum, 1976), a user would write a question on the tape, that is, the two numbers to be added together. They would be written in the format that could be understood by the machine. The Turing machine would answer the input question by reading and rewriting the tape. Eventually, it would write the sum of the two numbers on the tape—its answer—and then halt.

How does a Turing machine generate answers to the written questions? A Turing machine consists of a machine head whose actions are governed by a set of instructions called the machine table. The machine head will also be in one of a set of possible physical configurations called machine states. The machine head reads a symbol on the tape. This symbol, in combination with the current machine state, determines which machine table instruction to execute next. An instruction might tell the machine head to write a symbol, or to move one cell to the left or the right along the tickertape. The instruction will also change the machine head's machine state.

A Turing machine does not answer questions instantly. Instead, it takes its time, moving back and forth along the tape, reading and writing symbols as it works. A long sequence of actions might be observed and recorded, such as "First the machine head moves four cells to the right. Then it stops, and replaces the 1 on the tape with a 0. Then it moves three cells to the left."

The record of the observed Turing machine behaviours would tell us a great deal about its design. Descriptions such as "When given Question A, the machine generated Answer X" would provide information about the input-output mapping that the Turing machine was designed to achieve. If we were also able to watch changes in machine states, more detailed observations would be possible, such as "If the machine head is in State 1 and reads a '1' on the tape, then it moves one cell left and adopts State 6." Such observations would provide information about the machine table that was designed for this particular device's machine head.

Not all Turing machine behaviours occur by design; some behaviours are artifacts. Artifacts occur because of the device's design but are not explicitly part of the design (Pylyshyn, 1980, 1984). They are unintentional consequences of the designed procedure.

For instance, the Turing machine takes time to add two numbers together; the time taken will vary from question to question. The amount of time taken to answer a question is a consequence of the machine table, but is not intentionally designed into it. The time taken is an artifact because Turing machines are designed to answer questions (e.g., "What is the sum of these two integers?"); they are not explicitly designed to answer questions in a particular amount of time.

Similarly, as the Turing machine works, the ticker tape adopts various intermediate states. That is, during processing the ticker tape will contain symbols that are neither the original question nor its eventual answer. Answering a particular question will produce a sequence of intermediate tape states; the sequence produced will also vary from question to question. Again, the sequence of symbol states is an artifact. The Turing machine is not designed to produce a particular sequence of intermediate states; it is simply designed to answer a particular question.

One might think that artifacts are not important because they are not explicit consequences of a design. However, in many cases artifacts are crucial sources of information that help us reverse engineer an information processor that is a "black box" because its internal mechanisms are hidden from view.

2.9 Algorithms from Artifacts

Neuroscientist Valentino Braitenberg imagined a world comprising domains of both water and land (Braitenberg, 1984). In either of these domains one would find a variety of agents who sense properties of their world, and who use this information to guide their movements through it. Braitenberg called these agents "vehicles." In Braitenberg's world of vehicles, scientists encounter these agents and attempt to explain the internal mechanisms that are responsible for their diverse movements. Many of these scientists adopt what Braitenberg called an analytic perspective: they infer internal mechanisms by observing how external behaviours are altered as a function of specific changes in a vehicle's environment. What Braitenberg called analysis is also called reverse engineering.

We saw earlier that a Turing machine generates observable behaviour as it calculates the answer to a question. A description of a Turing machine's behaviours— be they by design or by artifact—would provide the sequence of operations that were performed to convert an input question into an output answer. Any sequence of steps which, when carried out, accomplishes a desired result is called an algorithm (Berlinski, 2000). The goal, then, of reverse engineering a Turing machine or any other calculating device would be to determine the algorithm it was using to transform its input into a desired output.

Calculating devices exhibit two properties that make their reverse engineering difficult. First, they are often what are called black boxes. This means that we can observe external behaviour, but we are unable to directly observe internal properties. For instance, if a Turing machine was a black box, then we could observe its movements along, and changing of symbols on, the tape, but we could not observe the machine state of the machine head.

Second, and particularly if we are faced with a black box, another property that makes reverse engineering challenging is that there is a many-to-one relationship between algorithm and mapping. This means that, in practice, a single input-output mapping can be established by one of several different algorithms. For example, there are so many different methods for sorting a set of items that hundreds of pages are required to describe the available algorithms (Knuth, 1997). In principle, an infinite number of different algorithms exist for computing a single input-output mapping of interest (Johnson-Laird, 1983).

The problem with reverse engineering a black box is this: if there are potentially many different algorithms that can produce the same input-output mapping, then mere observations of input-output behaviour will not by themselves indicate which particular algorithm is used in the device's design. However, reverse engineering a black box is not impossible. In addition to the behaviours that it was designed to produce, the black box will also produce artifacts. Artifacts can provide great deal of information about internal and unobservable algorithms.

Imagine that we are faced with reverse engineering an arithmetic calculator that is also a black box. Some of the artifacts of this calculator provide relative complexity evidence (Pylyshyn, 1984). To collect such evidence, one could conduct an experiment in which the problems presented to the calculator were systematically varied (e.g., by using different numbers) and measurements were made of the amount of time taken for the correct answer to be produced. To analyze this relative complexity evidence, one would explore the relationship between characteristics of problems and the time required to solve them.

For instance, suppose that one observed a linear increase in the time taken to solve the problems $9 \times 1, 9 \times 2, 9 \times 3$, et cetera. This could indicate that the device was performing multiplication by doing repeated addition ($9, 9 + 9, 9 + 9 + 9$, and so on) and that every "+ 9" operation required an additional constant amount of time to be carried out. Psychologists have used relative complexity evidence to investigate cognitive algorithms since Franciscus Donders invented his subtractive method in 1869 (Posner, 1978).

Artifacts can also provide intermediate state evidence (Pylyshyn, 1984). Intermediate state evidence is based upon the assumption that an input-output mapping is not computed directly, but instead requires a number of different stages of processing, with each stage representing an intermediate result in a different way. To collect intermediate state evidence, one attempts to determine the number and nature of these intermediate results.

For some calculating devices, intermediate state evidence can easily be collected. For instance, the intermediate states of the Turing machine›s tape, the abacus' beads or the difference engine's gears are in full view. For other devices, though, the intermediate states are hidden from direct observation. In this case, clever techniques must be developed to measure internal states as the device is presented with different inputs. One might measure changes in electrical activity in different components of an electronic calculator as it worked, in an attempt to acquire intermediate state evidence.

Artifacts also provide error evidence (Pylyshyn, 1984), which may also help to explore intermediate states. When extra demands are placed on a system's resources, it may not function as designed, and its internal workings are likely to become more evident (Simon, 1969). This is not just because the overtaxed system makes errors in general, but because these errors are often systematic, and their systematicity reflects the underlying algorithm.

Because we rely upon their accuracy, we would hope that error evidence would be difficult to collect for most calculating devices. However, error evidence should be easily available for calculators that might be of particular interest to us: humans doing mental arithmetic. We might find, for instance, that overtaxed human cal-culators make mistakes by forgetting to carry values from one column of numbers to the next. This would provide evidence that mental arithmetic involved repre-senting numbers in columnar form, and performing operations column by column (Newell & Simon, 1972). Very different kinds of errors would be expected if a dif-ferent approach was taken to perform mental arithmetic, such as imagining and manipulating a mental abacus (Hatano, Miyake, & Binks, 1977).

In summary, discovering and describing what algorithm is being used to cal-culate an input-output mapping involves the systematic examination of behaviour. That is, one makes and interprets measurements that provide relative complexity evidence, intermediate state evidence, and error evidence. Furthermore, the algo-rithm that will be inferred from such measurements is in essence a sequence of actions or behaviours that will produce a desired result.

The discovery and description of an algorithm thus involves empirical methods and vocabularies, rather than the formal ones used to account for input-output reg-ularities. Just as it would seem likely that input-output mappings would be the topic of interest for formal researchers such as cyberneticists, logicians, or mathemati-cians, algorithmic accounts would be the topic of interest for empirical researchers such as experimental psychologists.

The fact that computational accounts and algorithmic accounts are presented in different vocabularies suggests that they describe very different properties of a device. From our discussion of black boxes, it should be clear that a computational account does not provide algorithmic details: knowing *what* input-output mapping is being computed is quite different from knowing *how* it is being computed. In a similar vein, algorithmic accounts are silent with respect to the computation being carried out.

For instance, in *Understanding Cognitive Science*, Dawson (1998) provides an example machine table for a Turing machine that adds pairs of integers. Dawson also provides examples of questions to this device (e.g., strings of blanks, os, and 1s) as well as the answers that it generates. Readers of *Understanding Cognitive Science* can pretend to be the machine head by following the instructions of the machine table, using pencil and paper to manipulate a simulated ticker tape. In this fashion they can easily convert the initial question into the final answer—they fully understand the algorithm. However, they are unable to say what the algorithm accomplishes until they read further in the book.

2.10 Architectures against Homunculi

We have described an algorithm for calculating an input-output mapping as a sequence of operations or behaviours. This description is misleading, though, because the notion of sequence gives the impression of a linear ordering of steps. However, we would not expect most algorithms to be linearly organized. For instance, connectionist cognitive scientists would argue that more than one step in an algorithm can be carried out at the same time (Feldman & Ballard, 1982). As well, most algorithms of interest to classical cognitive scientists would likely exhibit a markedly hierarchical organization (Miller, Galanter, & Pribram, 1960; Simon, 1969). In this section, I use the notion of hierarchical organization to motivate the need for an algorithm to be supported by an architecture.

What does it mean for an algorithm to be hierarchical in nature? To answer this question, let us again consider the situation in which behavioural measurements are being used to reverse engineer a calculating black box. Initial experiments could suggest that an input-output mapping is accomplished by an algorithm that involves three steps (Step 1 → Step 2 → Step 3). However, later studies could also indicate that each of these steps might themselves be accomplished by sub-algorithms.

For instance, it might be found that Step 1 is accomplished by its own four-step sub-algorithm (Step a → Step b → Step c → Step d). Even later it could be discovered that one of these sub-algorithms is itself the product of another sub-sub-algorithm. Such hierarchical organization is common practice in the development of algorithms for digital computers, where most programs are organized systems of functions, subfunctions, and sub-subfunctions. It is also a common characteristic of cognitive theories (Cummins, 1983).

The hierarchical organization of algorithms can pose a problem, though, if an algorithmic account is designed to *explain* a calculating device. Consider our example where Step 1 of the black box's algorithm is explained by being hierarchically decomposed into the sub-algorithm "Step a → Step b → Step c → Step d." On closer examination, it seems that nothing has really been explained at all. Instead, we have replaced Step 1 with a sequence of four new steps, each of which requires further explanation. If each of these further explanations is of the same type as the one to account for Step 1, then this will in turn produce even more steps requiring explanation. There seems to be no end to this infinite proliferation of algorithmic steps that are appearing in our account of the calculating device.

This situation is known as Ryle's regress. The philosopher Gilbert Ryle raised it as a problem with the use of mentalistic terms in explanations of intelligence:

> Must we then say that for the hero's reflections how to act to be intelligent he must first reflect how best to reflect to act? The endlessness of this implied regress shows

that the application of the criterion of appropriateness does not entail the occurrence of a process of considering this criterion. (Ryle, 1949, p. 31)

Ryle's regress occurs when we explain outer intelligence by appealing to inner intelligence.

Ryle's regress is also known as the homunculus problem, where a homunculus is an intelligent inner agent. The homunculus problem arises when one explains outer intelligence by appealing to what is in essence an inner homunculus. For instance, Fodor noted the obvious problems with a homuncular explanation of how one ties their shoes:

> And indeed there would be something wrong with an explanation that said, 'This is the way we tie our shoes: we notify a little man in our head who does it for us.' This account invites the question: 'How does the little man do it?' but, *ex hypothesi*, provides no conceptual mechanisms for answering such questions. (Fodor, 1968a, p. 628)

Indeed, if one proceeds to answer the invited question by appealing to another homunculus within the "little man," then the result is an infinite proliferation of homunculi.

To solve Ryle's regress an algorithm must be analyzed into steps that do not require further decomposition in order to be explained. This means when some function is decomposed into a set of subfunctions, it is critical that each of the subfunctions be simpler than the overall function that they work together to produce (Cummins, 1983; Dennett, 1978; Fodor, 1968a). Dennett (1978, p. 123) noted that "homunculi are *bogeymen* only if they duplicate *entire* the talents they are rung in to explain." Similarly, Fodor (1968a, p. 629) pointed out that "we refine a psychological theory by replacing global little men by less global little men, each of whom has fewer unanalyzed behaviors to perform than did his predecessors."

If the functions produced in a first pass of analysis require further decomposition in order to be themselves explained, then the subfunctions that are produced must again be even simpler. At some point, the functions become so simple—the homunculi become so stupid—that they can be replaced by machines. This is because at this level all they do is answer "yes" or "no" to some straightforward question. "One *discharges* fancy homunculi from one's scheme by organizing armies of such idiots to do the work" (Dennett, 1978, p. 124).

The set of subfunctions that exist at this final level of decomposition belongs to what computer scientists call the device's architecture (Blaauw & Brooks, 1997; Brooks, 1962; Dasgupta, 1989). The architecture defines what basic abilities are built into the device. For a calculating device, the architecture would specify three different types of components: the basic operations of the device, the objects to which these operations are applied, and the control scheme that decides which operation to carry out at any given time (Newell, 1980; Simon, 1969). To detail the architecture is to specify "what operations are primitive, how memory is organized and accessed, what sequences are allowed, what limitations exist on the passing of arguments and on the capacities of various buffers, and so on" (Pylyshyn, 1984, p. 92).

What is the relationship between an algorithm and its architecture? In general, the architecture provides the programming language in which an algorithm is written. "Specifying the functional architecture of a system is like providing a manual that defines some programming language. Indeed, defining a programming language is equivalent to specifying the functional architecture of a virtual machine" (Pylyshyn, 1984, p. 92).

This means that algorithms and architectures share many properties. Foremost of these is that they are both described as operations, behaviours, or functions, and not in terms of physical makeup. An algorithm is a set of functions that work together to accomplish a task; an architectural component is one of the simplest functions—a primitive operation—from which algorithms are composed. In order to escape Ryle's regress, one does not have to replace an architectural function with its physical account. Instead, one simply has to be sure that such a replacement is available if one wanted to explain how the architectural component works. It is no accident that Pylyshyn (1984) uses the phrase *functional architecture* in the quote given above.

Why do we insist that the architecture is functional? Why don't we appeal directly to the physical mechanisms that bring an architecture into being? Both of these questions are answered by recognizing that multiple physical realizations are possible for any functional architecture. For instance, simple logic gates are clearly the functional architecture of modern computers. But we saw earlier that functionally equivalent versions of these gates could be built out of wires and switches, vacuum tubes, semiconductors, or hydraulic valves.

To exit Ryle's regress, we have to discharge an algorithm's homunculi. We can do this by identifying the algorithm's programming language—by saying what its architecture *is*. Importantly, this does not require us to say how, or from what physical stuff, the architecture *is made*! "Whether you build a computer out of transistors, hydraulic valves, or a chemistry set, the principles on which it operates are much the same" (Hillis, 1998, p. 10).

2.11 Implementing Architectures

At the computational level, one uses a formal vocabulary to provide a rigorous description of input-output mappings. At the algorithmic level, a procedural or behavioural vocabulary is employed to describe the algorithm being used to calculate a particular input-output mapping. The functional architecture plays a special role at the algorithmic level, for it provides the primitive operations from which algorithms are created. Thus we would expect that the behavioural vocabulary used for algorithms to also be applied to the architecture.

The special nature of the architecture means that additional behavioural descriptions are required. A researcher must also collect behavioural evidence to support his or her claim that some algorithmic component is in fact an architectural primitive. One example of this, which appears when the ideas that we have been developing in this chapter are applied to the science of human cognition, is to conduct behavioural experiments to determine whether a function is cognitively impenetrable (Pylyshyn, 1984; Wright & Dawson, 1994). We return to this kind of evidence in Chapter 3.

Of course, the fundamental difference between algorithm and architecture is that only the latter can be described in terms of physical properties. Algorithms are explained in terms of the architectural components in which they are written. Architectural components are explained by describing how they are implemented by some physical device. At the implementational level a researcher uses a physical vocabulary to explain how architectural primitives are brought to life.

An implementational account of the logic gates illustrated in Figure 2-1 would explain their function by appealing to the ability of metal wires to conduct electricity, to the nature of electric circuits, and to the impedance of the flow of electricity through these circuits when switches are open (Shannon, 1938). An implementational account of how a vacuum tube creates a relay of the sort illustrated in Figure 2-2 would appeal to what is known as the Edison effect, in which electricity can mysteriously flow through a vacuum and the direction of this flow can be easily and quickly manipulated to manipulate the gate between the source and the drain (Josephson, 1961; Reid, 2001).

That the architecture has dual lives, both physical and algorithmic (Haugeland, 1985), leads to important philosophical issues. In the philosophy of science there is a great deal of interest in determining whether a theory phrased in one vocabulary (e.g., chemistry) can be reduced to another theory laid out in a different vocabulary (e.g., physics). One approach to reduction is called the "new wave" (Churchland, 1985; Hooker, 1981). In a new wave reduction, the translation of one theory into another is accomplished by creating a third, intermediate theory that serves as a bridge between the two. The functional architecture is a bridge between the algorithmic and the implementational. If one firmly believed that a computational or algorithmic account could be reduced to an implementational one (Churchland, 1988), then a plausible approach to doing so would be to use the bridging properties of the architecture.

The dual nature of the architecture plays a role in another philosophical discussion, the famous "Chinese room argument" (Searle, 1980). In this thought experiment, people write questions in Chinese symbols and pass them through a slot into a room. Later, answers to these questions, again written in Chinese symbols, are passed back to the questioner. The philosophical import of the Chinese room arises when one looks into the room to see how it works.

Inside the Chinese room is a native English speaker—Searle himself—who knows no Chinese, and for whom Chinese writing is a set of meaningless squiggles. The room contains boxes of Chinese symbols, as well as a manual for how to put these together in strings. The English speaker is capable of following these instructions, which are the room's algorithm. When a set of symbols is passed into the room, the person inside can use the instructions and put together a new set of symbols to pass back outside. This is the case even though the person inside the room does not understand what the symbols mean, and does not even know that the inputs are questions and the outputs are answers. Searle (1980) uses this example to challengingly ask where in this room is the knowledge of Chinese? He argues that it is not to be found, and then uses this point to argue against strong claims about the possibility of machine intelligence.

But should we expect to see such knowledge if we were to open the door to the Chinese room and peer inside? Given our current discussion of the architecture, it would perhaps be unlikely to answer this question affirmatively. This is because if we could look inside the "room" of a calculating device to see how it works—to see how its physical properties bring its calculating abilities to life—we would not see the input-output mapping, nor would we see a particular algorithm in its entirety. At best, we would see the architecture and how it is physically realized in the calculator. The architecture of a calculator (e.g., the machine table of a Turing machine) would look as much like the knowledge of arithmetic calculations as Searle and the instruction manual would look like knowledge of Chinese. However, we would have no problem recognizing the possibility that the architecture is responsible for producing calculating behaviour!

Because the architecture is simply the primitives from which algorithms are constructed, it is responsible for algorithmic behaviour—but doesn't easily reveal this responsibility on inspection. That the holistic behaviour of a device would not be easily seen in the actions of its parts was recognized in Leibniz' mill, an early eighteenth-century ancestor to the Chinese room.

In his *Monadology*, Gottfried Leibniz wrote:

> Supposing there were a machine whose structure produced thought, sensation, and perception, we could conceive of it as increased in size with the same proportions until one was able to enter into its interior, as he would into a mill. Now, on going into it he would find only pieces working upon one another, but never would he find anything to explain Perception. It is accordingly in the simple substance,

and not in the composite nor in a machine that the Perception is to be sought.
(Leibniz, 1902, p. 254)

Leibniz called these simple substances monads and argued that all complex experiences were combinations of monads. Leibniz' monads are clearly an antecedent of the architectural primitives that we have been discussing over the last few pages. Just as thoughts are composites in the sense that they can be built from their component monads, an algorithm is a combination or sequence of primitive processing steps. Just as monads cannot be further decomposed, the components of an architecture are not explained by being further decomposition, but are instead explained by directly appealing to physical causes. Just as the Leibniz mill's monads would look like working pieces, and not like the product they created, the architecture produces, but does not resemble, complete algorithms.

The Chinese room would be a more compelling argument against the possibility of machine intelligence if one were to look inside it and actually see its knowledge. This would mean that its homunculi were not discharged, and that intelligence was not the product of basic computational processes that could be implemented as physical devices.

2.12 Levelling the Field

The logic machines that arose late in the nineteenth century, and the twentieth-century general-purpose computers that they evolved into, are examples of information processing devices. It has been argued in this chapter that in order to explain such devices, four different vocabularies must be employed, each of which is used to answer a different kind of question. At the computational level, we ask what information processing problem is being solved by the device. At the algorithmic level, we ask what procedure or program is being used to solve this problem. At the architectural level, we ask from what primitive information capabilities is the algorithm composed. At the implementational level, we ask what physical properties are responsible for instantiating the components of the architecture.

As we progress from the computational question through questions about algorithm, architecture, and implementation we are moving in a direction that takes us from the very abstract to the more concrete. From this perspective each of these questions defines a different level of analysis, where the notion of level is to be taken as "level of abstractness." The main theme of this chapter, then, is that to fully explain an information processing device one must explain it at four different levels of analysis.

The theme that I've developed in this chapter is an elaboration of an approach with a long history in cognitive science that has been championed in particular by Pylyshyn (1984) and Marr (1982). This historical approach, called the tri-level hypothesis (Dawson, 1998), is used to explain information devices by performing analyses at three different levels: computational, algorithmic, and implementational. The approach that has been developed in this chapter agrees with this view, but adds to it an additional level of analysis: the architectural. We will see throughout this book that an information processing architecture has properties that separate it from both algorithm and implementation, and that treating it as an independent level is advantageous.

The view that information processing devices must be explained by multiple levels of analysis has important consequences for cognitive science, because the general view in cognitive science is that cognition is also the result of information processing. This implies that a full explanation of human or animal cognition also requires multiple levels of analysis.

Not surprisingly, it is easy to find evidence of all levels of investigation being explored as cognitive scientists probe a variety of phenomena. For example, consider how classical cognitive scientists explore the general phenomenon of human memory.

At the computational level, researchers interested in the formal characterization of cognitive processes (such as those who study cognitive informatics [Wang, 2003, 2007]), provide abstract descriptions of what it means to memorize, including attempts to mathematically characterize the capacity of human memory (Lopez, Nunez, & Pelayo, 2007; Wang, 2009; Wang, Liu, & Wang, 2003).

At the algorithmic level of investigation, the performance of human subjects in a wide variety of memory experiments has been used to reverse engineer "memory" into an organized system of more specialized functions (Baddeley, 1990) including working memory (Baddeley, 1986, 2003), declarative and nondeclarative memory (Squire, 1992), semantic and episodic memory (Tulving, 1983), and verbal and imagery stores (Paivio, 1971, 1986). For instance, the behaviour of the serial position curve obtained in free recall experiments under different experimental conditions was used to pioneer cognitive psychology's proposal of the modal memory model, in which memory was divided into a limited-capacity, short-term store and a much larger-capacity, long-term store (Waugh & Norman, 1965). The algorithmic level is also the focus of the art of memory (Yates, 1966), in which individuals are taught mnemonic techniques to improve their ability to remember (Lorayne, 1998, 2007; Lorayne & Lucas, 1974).

That memory can be reverse engineered into an organized system of subfunctions leads cognitive scientists to determine the architecture of memory. For instance, what kinds of encodings are used in each memory system, and what primitive processes are used to manipulate stored information? Richard Conrad's (1964a, 1964b) famous studies of confusion in short-term memory indicated that it represented information using an acoustic code. One of the most controversial topics in classical cognitive science, the "imagery debate," concerns whether the primitive form of spatial information is imagery, or whether images are constructed from more primitive propositional codes (Anderson, 1978; Block, 1981; Kosslyn, Thompson, & Ganis, 2006; Pylyshyn, 1973, 1981a, 2003b).

Even though classical cognitive science is functionalist in nature and (in the eyes of its critics) shies away from biology, it also appeals to implementational evidence in its study of memory. The memory deficits revealed in patient Henry Molaison after his hippocampus was surgically removed to treat his epilepsy (Scoville & Milner, 1957) provided pioneering biological support for the functional separations of short-term from long-term memory and of declarative memory from nondeclarative memory. Modern advances in cognitive neuroscience have provided firm biological foundations for elaborate functional decompositions of memory (Cabeza & Nyberg, 2000; Poldrack et al., 2001; Squire, 1987, 2004). Similar evidence has been brought to bear on the imagery debate as well (Kosslyn, 1994; Kosslyn et al., 1995; Kosslyn et al., 1999; Kosslyn, Thompson, & Alpert, 1997).

In the paragraphs above I have taken one tradition in cognitive science (the classical) and shown that its study of one phenomenon (human memory) reflects the use of all of the levels of investigation that have been the topic of the current chapter. However, the position that cognitive explanations require multiple levels of analysis (e.g., Marr, 1982) has not gone unchallenged. Some researchers have suggested that this process is not completely appropriate for explaining cognition or intelligence in biological agents (Churchland, Koch, & Sejnowski 1990; Churchland & Sejnowski, 1992).

For instance, Churchland, Koch, & Sejnowski (1990, p. 52) observed that "when we measure Marr's three levels of analysis against levels of organization in the nervous system, the fit is poor and confusing." This observation is based on the fact that there appear to be a great many different spatial levels of organization in the brain, which suggests to Churchland, Koch, & Sejnowski that there must be many different implementational levels, which implies in turn that there must be many different algorithmic levels.

The problem with this argument is that it confuses ontology with epistemology. That is, Churchland, Koch, & Sejnowski (1990) seemed to be arguing that Marr's levels are accounts of the way nature is—that information processing devices are literally organized into the three different levels. Thus when a system appears to exhibit, say, multiple levels of physical organization, this brings Marr-as-ontology into question. However, Marr's levels do not attempt to explain the nature of devices, but instead provide an epistemology—a way to inquire about the nature of the world. From this perspective, a system that has multiple levels of physical organization would not challenge Marr, because Marr and his followers would be comfortable applying their approach to the system at each of its levels of physical organization.

Other developments in cognitive science provide deeper challenges to the multiple-levels approach. As has been outlined in this chapter, the notion of multiple levels of explanation in cognitive science is directly linked to two key ideas: 1) that information processing devices invite and require this type of explanation, and 2) that cognition is a prototypical example of information processing. Recent developments in cognitive science represent challenges to these key ideas. For instance, embodied cognitive science takes the position that cognition is not information processing of the sort that involves the rule-governed manipulation of mentally represented worlds; it is instead the control of action on the world (Chemero, 2009; Clark, 1997, 1999; Noë, 2004, 2009; Robbins & Aydede, 2009). Does the multiple-levels approach apply if the role of cognition is radically reconstrued?

Churchland, Koch, & Sejnowski. (1990, p. 52) suggested that "[']which really are the levels relevant to explanation in the nervous system['] is an empirical, not an a priori, question." One of the themes of the current book is to take this suggestion to heart by seeing how well the same multiple levels of investigation can be applied to the three major perspectives in modern cognitive science: classical, connectionist, and embodied. In the next three chapters, I begin this pursuit by using the multiple levels introduced in Chapter 2 to investigate the nature of classical cognitive science (Chapter 3), connectionist cognitive science (Chapter 4), and embodied cognitive science (Chapter 5). Can the multiple levels of investigation be used to reveal principles that unify these three different and frequently mutually antagonistic approaches? Or is modern cognitive science beginning to fracture in a fashion similar to what has been observed in experimental psychology?

Elements of Classical Cognitive Science

3.0 Chapter Overview

When cognitive science arose in the late 1950s, it did so in the form of what is now known as the classical approach. Inspired by the nature of the digital electronic computer, classical cognitive science adopted the core assumption that cognition was computation. The purpose of the current chapter is to explore the key ideas of classical cognitive science that provide the core elements of this assumption.

The chapter begins by showing that the philosophical roots of classical cognitive science are found in the rationalist perspective of Descartes. While classical cognitive scientists agree with the Cartesian view of the infinite variety of language, they do not use this property to endorse dualism. Instead, taking advantage of modern formal accounts of information processing, they adopt models that use recursive rules to manipulate the components of symbolic expressions. As a result, finite devices—physical symbol systems—permit an infinite behavioural potential. Some of the key properties of physical symbol systems are reviewed.

One consequence of viewing the brain as a physical substrate that brings a universal machine into being is that this means that cognition can be simulated by other universal machines, such as digital computers. As a result, the computer simulation of human cognition becomes a critical methodology of the classical approach. One issue that arises is validating such simulations. The notions of weak and strong equivalence are reviewed, with the latter serving as the primary goal of classical cognitive science.

To say that two systems—such as a simulation and a human subject—are strongly equivalent is to say that both are solving the same information processing problem, using the same algorithm, based on the same architecture. Establishing strong equivalence requires collecting behavioural evidence of the types introduced in Chapter 2 (relative complexity, intermediate state, and error evidence) to reverse engineer a subject's algorithm. It also requires discovering the components of a subject's architecture, which involves behavioural evidence concerning cognitive impenetrability as well as biological evidence about information processing in the brain (e.g., evidence about which areas of the brain might be viewed as being information processing modules). In general, the search for strong equivalence by classical cognitive scientists involves conducting a challenging research program that can be described as functional analysis or reverse engineering.

The reverse engineering in which classical cognitive scientists are engaged involves using a variety of research methods adopted from many different disciplines. This is because this research strategy explores cognition at all four levels of investigation (computational, algorithmic, architectural, and implementational) that were introduced in Chapter 2. The current chapter is organized in a fashion that explores computational issues first, and then proceeds through the remaining levels to end with some considerations about implementational issues of importance to classical cognitive science.

3.1 Mind, Disembodied

In the seventh century, nearly the entire Hellenistic world had been conquered by Islam. The Greek texts of philosophers such as Plato and Aristotle had already been translated into Syriac; the new conquerors translated these texts into Arabic (Kuhn, 1957). Within two centuries, these texts were widely available in educational institutions that ranged from Baghdad to Cordoba and Toledo. By the tenth century, Latin translations of these Arabic texts had made their way to Europe. Islamic civilization "preserved and proliferated records of ancient Greek science for later European scholars" (Kuhn, 1957, p. 102).

The availability of the ancient Greek texts gave rise to scholasticism in Europe during the middle ages. Scholasticism was central to the European universities that arose in the twelfth century, and worked to integrate key ideas of Greek philosophy into the theology of the Church. During the thirteenth century, scholasticism achieved its zenith with the analysis of Aristotle's philosophy by Albertus Magnus and Thomas Aquinas.

Scholasticism, as a system of education, taught its students the wisdom of the ancients. The scientific revolution that took flight in the sixteenth and seventeenth centuries arose in reaction to this pedagogical tradition. The discoveries of such luminaries as Newton and Leibniz were only possible when the ancient wisdom was directly questioned and challenged.

The seventeenth-century philosophy of René Descartes (1996, 2006) provided another example of fundamental insights that arose from a reaction against scholasticism. Descartes' goal was to establish a set of incontestable truths from which a rigorous philosophy could be constructed, much as mathematicians used methods of deduction to derive complete geometries from a set of foundational axioms. "The only order which I could follow was that normally employed by geometers, namely to set out all the premises on which a desired proposition depends, before drawing any conclusions about it" (Descartes, 1996, p. 9).

Descartes began his search for truth by applying his own, new method of inquiry. This method employed extreme skepticism: any idea that could possibly be doubted was excluded, including the teachings of the ancients as endorsed by scholasticism. Descartes, more radically, also questioned ideas supplied by the senses because "from time to time I have found that the senses deceive, and it is prudent never to trust completely those who have deceived us even once" (Descartes, 1996, p. 12). Clearly this approach brought a vast number of concepts into question, and removed them as possible foundations of knowledge.

What ideas were removed? All notions of the external world could be false, because knowledge of them is provided by unreliable senses. Also brought into question is the existence of one's physical body, for the same reason. "I shall consider myself as not having hands or eyes, or flesh, or blood or senses, but as falsely believing that I have all these things" (Descartes, 1996, p. 15).

Descartes initially thought that basic, self-evident truths from mathematics could be spared, facts such as $2 + 3 = 5$. But he then realized that these facts too could be reasonably doubted.

> How do I know that God has not brought it about that I too go wrong every time I add two and three or count the sides of a square, or in some even simpler matter, if that is imaginable? (Descartes, 1996, p. 14)

With the exclusion of the external world, the body, and formal claims from mathematics, what was left for Descartes to believe in? He realized that in order to doubt, or even to be deceived by a malicious god, he must exist as a thinking thing. "I must finally conclude that this proposition, *I am, I exist*, is necessarily true whenever it is put forward by me or conceived in my mind" (Descartes, 1996, p. 17). And what is a thinking thing? "A thing that doubts, understands, affirms, denies, is willing, is unwilling, and also imagines and has sensory perceptions" (p. 19).

After establishing his own existence as incontestably true, Descartes used this fact to prove the existence of a perfect God who would not deceive. He then established the existence of an external world that was imperfectly sensed.

However, a fundamental consequence of Descartes' analysis was a profound division between mind and body. First, Descartes reasoned that mind and body must be composed of different "stuff." This had to be the case, because one could imagine that the body was divisible (e.g., through losing a limb) but that the mind was impossible to divide.

> Indeed the idea I have of the human mind, in so far as it is a thinking thing, which is not extended in length, breadth or height and has no other bodily characteristics, is much more distinct than the idea of any corporeal thing. (Descartes, 1996, p. 37)

Further to this, the mind was literally disembodied—the existence of the mind did not depend upon the existence of the body.

> Accordingly this 'I,' that is to say, the Soul by which I am what I am, is entirely distinct from the body and is even easier to know than the body; and would not stop being everything it is, even if the body were not to exist. (Descartes, 2006, p. 29)

Though Descartes' notion of mind was disembodied, he acknowledged that mind and body had to be linked in some way. The interaction between mind and brain was famously housed in the pineal gland: "The mind is not immediately affected by all parts of the body, but only by the brain, or perhaps just by one small part of the brain, namely the part which is said to contain the 'common' sense" (Descartes, 1996, p. 59). What was the purpose of this type of interaction? Descartes noted that the powers of the mind could be used to make decisions beneficial to the body, to which the mind is linked: "For the proper purpose of the sensory perceptions given me by nature is simply to inform the mind of what is beneficial or harmful for the composite of which the mind is a part" (p. 57).

For Descartes the mind, as a thinking thing, could apply various rational operations to the information provided by the imperfect senses: sensory information could be doubted, understood, affirmed, or denied; it could also be elaborated via imagination. In short, these operations could not only inform the mind of what would benefit or harm the mind-body composite, but could also be used to plan a course of action to obtain the benefits or avoid the harm. Furthermore, the mind—via its capacity for willing—could cause the body to perform the desired actions to bring this plan into fruition. In Cartesian philosophy, the disembodied mind was responsible for the "thinking" in a sense-think-act cycle that involved the external world and the body to which the mind was linked.

Descartes' disembodiment of the mind—his claim that the mind is composed of different "stuff" than is the body or the physical world—is a philosophical position called dualism. Dualism has largely been abandoned by modern science, including cognitive science. The vast majority of cognitive scientists adopt a very different philosophical position called materialism. According to materialism, the mind is caused by the brain. In spite of the fact that it has abandoned Cartesian dualism, most of the core ideas of classical cognitive science are rooted in the ideas that Descartes wrote about in the seventeenth century. Indeed, classical cognitive science can be thought of as a synthesis between Cartesian philosophy and materialism. In classical cognitive science, this synthesis is best expressed as follows: cognition is the product of a physical symbol system (Newell, 1980). The physical symbol system hypothesis is made plausible by the existence of working examples of such devices: modern digital computers.

3.2 Mechanizing the Infinite

We have seen that the disembodied Cartesian mind is the thinking thing that mediates the sensing of, and acting upon, the world. It does so by engaging in such activities as doubting, understanding, affirming, denying, perceiving, imagining, and willing. These activities were viewed by Descartes as being analogous to a geometer's use of rules to manipulate mathematical expressions. This leads us to ask, in what medium is thought carried out? What formal rules does it employ? What symbolic expressions does it manipulate?

Many other philosophers were sympathetic to the claim that mental activity was some sort of symbol manipulation. Thomas Hobbes is claimed as one of the philosophical fathers of classical cognitive science because of his writings on the nature of the mind:

> When a man *Reasoneth*, hee does nothing else but conceive a summe totall, from *Addition* of parcels; or conceive a Remainder, from *Substraction* of one summe from another." Such operations were not confined to numbers: "These operations are not incident to Numbers only, but to all manner of things that can be added together, and taken one out of another. (Hobbes, 1967; p. 32)

Hobbes noted that geometricians applied such operations to lines and figures, and that logicians applied these operations to words. Thus it is not surprising that Hobbes described thought as mental discourse—thinking, for him, was language-like.

Why were scholars taken by the idea that language was the medium in which thought was conducted? First, they agreed that thought was exceptionally powerful, in the sense that there were no limits to the creation of ideas. In other words, man in principle was capable of an infinite variety of different thoughts. "Reason is a universal instrument which can operate in all sorts of situations" (Descartes, 2006, p. 47). Second, language was a medium in which thought could be expressed, because it too was capable of infinite variety. Descartes expressed this as follows:

> For it is a very remarkable fact that there are no men so dull-witted and stupid, not
> even madmen, that they are incapable of stringing together different words, and
> composing them into utterances, through which they let their thoughts be known.
> (Descartes, 2006, p. 47)

Modern linguists describe this as the creative aspect of language (Chomsky, 1965, 1966). "An essential property of language is that it provides the means for expressing indefinitely many thoughts and for reacting appropriately in an indefinite range of new situations" (Chomsky, 1965, p. 6).

While Descartes did not write a great deal about language specifically (Chomsky, 1966), it is clear that he was sympathetic to the notion that language was the medium for thought. This is because he used the creative aspect of language to argue in favor of dualism. Inspired by the automata that were appearing in Europe in his era, Descartes imagined the possibility of having to prove that sophisticated future devices were not human. He anticipated the Turing test (Turing, 1950) by more than three centuries by using language to separate man from machine.

> For we can well conceive of a machine made in such a way that it emits words,
> and even utters them about bodily actions which bring about some correspond-
> ing change in its organs . . . but it is not conceivable that it should put these words
> in different orders to correspond to the meaning of things said in its presence.
> (Descartes, 2006, p. 46)

Centuries later, similar arguments still appear in philosophy. For instance, why is a phonograph recording of someone's entire life of speech an inadequate simulation of that speech (Fodor, 1968b)? "At the very best, phonographs do what speakers *do*, not what speakers *can do*" (p. 129).

Why might it be impossible for a device to do what speakers can do? For Descartes, language-producing machines were inconceivable because machines were physical and therefore finite. Their finite nature made it impossible for them to be infinitely variable.

> Although such machines might do many things as well or even better than any of
> us, they would inevitably fail to do some others, by which we would discover that
> they did not act consciously, but only because their organs were disposed in a cer-
> tain way. (Descartes, 2006, pp. 46-47)

In other words, the creativity of thought or language was only possible in the infinite, nonphysical, disembodied mind.

It is this conclusion of Descartes' that leads to a marked distinction between Cartesian philosophy and classical cognitive science. Classical cognitive science embraces the creative aspect of language. However, it views such creativity from a materialist, not a dualist, perspective. Developments in logic and in computing that have occurred since the seventeenth century have produced a device that

Descartes did not have at his disposal: the physical symbol system. And—seemingly magically—a physical symbol system is a finite artifact that is capable of an infinite variety of behaviour.

By the nineteenth century, the notion of language as a finite system that could be infinitely expressive was well established (Humboldt, 1999, p. 91): "For language is quite peculiarly confronted by an unending and truly boundless domain, the essence of all that can be thought. It must therefore make infinite employment of finite means." While Humboldt's theory of language has been argued to presage many of the key properties of modern generative grammars (Chomsky, 1966), it failed to provide a specific answer to the foundational question that it raised: how can a finite system produce the infinite? The answer to that question required advances in logic and mathematics that came after Humboldt, and which in turn were later brought to life by digital computers.

While it had been suspected for centuries that all traditional pure mathematics can be derived from the basic properties of natural numbers, confirmation of this suspicion was only obtained with advances that occurred in the nineteenth and twentieth centuries (Russell, 1993). The "arithmetisation" of mathematics was established in the nineteenth century, in what are called the Dedekind-Peano axioms (Dedekind, 1901; Peano, 1973). This mathematical theory defines three primitive notions: 0, number, and successor. It also defines five basic propositions: 0 is a number; the successor of any number is a number; no two numbers have the same successor; 0 is not the successor of any number; and the principle of mathematical induction. These basic ideas were sufficient to generate the entire theory of natural numbers (Russell, 1993).

Of particular interest to us is the procedure that is used in this system to generate the set of natural numbers. The set begins with 0. The next number is 1, which can be defined as the successor of 0, as $s(0)$. The next number is 2, which is the successor of 1, $s(1)$, and is also the successor of the successor of 0, $s(s(0))$. In other words, the successor function can be used to create the entire set of natural numbers: 0, $s(0)$, $s(s(0))$, $s(s(s(0)))$, and so on.

The definition of natural numbers using the successor function is an example of simple recursion; a function is recursive when it operates by referring to itself. The expression $s(s(0))$ is recursive because the first successor function takes as input another version of itself. Recursion is one method by which a finite system (such as the Dedekind-Peano axioms) can produce infinite variety, as in the set of natural numbers.

Recursion is not limited to the abstract world of mathematics, nor is its only role to generate infinite variety. It can work in the opposite direction, transforming the large and complex into the small and simple. For instance, recursion can be used to solve a complex problem by reducing it to a simple version of itself. This problem-solving approach is often called divide and conquer (Knuth, 1997).

One example of this is the famous Tower of Hanoi problem (see Figure 3-1), first presented to the world as a wooden puzzle by French mathematician Edouard Lucas in 1883. In this puzzle, there are three locations, A, B, and C. At the start of this problem there is a set of differently sized wooden discs stacked upon one another at location A. Let us number these discs 0, 1, 2, and so on, where the number assigned to a disc indicates its size. The goal for the problem is to move this entire stack to location C, under two restrictions: first, only one disc can be moved at a time; second, a larger disc can never be placed upon a smaller disc.

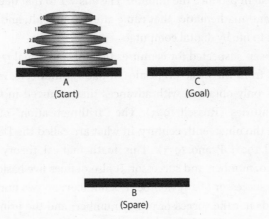

Figure 3-1. The starting configuration for a five-disc version of the Tower of Hanoi problem.

The simplest version of the Tower of Hanoi problem starts with only disc 0 at location A. Its solution is completely straightforward: disc 0 is moved directly to location C, and the problem is solved. The problem is only slightly more complicated if it starts with two discs stacked on location A. First, disc 0 is moved to location B. Second, disc 1 is moved to location C. Third, disc 0 is moved from A to C, stacked on top of disc 1, and the problem has been solved.

What about a Tower of Hanoi problem that begins with three discs? To solve this more complicated problem, we can first define a simpler subproblem: stacking discs 0 and 1 on location B. This is accomplished by doing the actions defined in the preceding paragraph, with the exception that the goal location is B for the subproblem. Once this subtask is accomplished, disc 2 can be moved directly to the final goal, location C. Now, we solve the problem by moving discs 0 and 1, which are stacked on B, to location C, by again using a procedure like the one described in the preceding paragraph.

This account of solving a more complex version of the Tower of Hanoi problem points to the recursive nature of divide and conquer: we solve the bigger problem by first solving a smaller version of the same kind of problem. To move a stack of n discs to location C, we first move the smaller stack of $n - 1$ discs to location B. "Moving the stack" is the same kind of procedure for the n discs and for the $n - 1$ discs. The whole approach is recursive in the sense that to move the big stack, the same procedure must first be used to move the smaller stack on top of the largest disc.

The recursive nature of the solution to the Tower of Hanoi is made obvious if we write a pseudocode algorithm for moving the disks. Let us call our procedure MoveStack (). It will take four arguments: the number of discs in the stack to be moved, the starting location, the "spare" location, and the goal location. So, if we had a stack of three discs at location A, and wanted to move the stack to location C using location B as the spare, we would execute MoveStack (3, A, B, C).

The complete definition of the procedure is as follows:

MoveStack (N, Start, Spare, Goal)

If N = 0

Exit

Else

MoveStack (N – 1, Start, Goal, Spare)

MoveStack (1, Start, Spare, Goal)

MoveStack (N – 1, Spare, Start, Goal)

EndIf

Note the explicit recursion in this procedure, because MoveStack () calls itself to move a smaller stack of disks stacked on top of the disk that it is going to move. Note too that the recursive nature of this program means that it is flexible enough to work with any value of N. Figure 3-2 illustrates an intermediate state that occurs when this procedure is applied to a five-disc version of the problem.

Figure 3-2. An intermediate state that occurs when MoveStack () is applied to a five-disc version of the Tower of Hanoi.

In the code given above, recursion was evident because MoveStack () called itself. There are other ways in which recursion can make itself evident. For instance, recursion can produce hierarchical, self-similar structures such as fractals (Mandelbrot, 1983), whose recursive nature is immediately evident through visual inspection. Consider the Sierpinski triangle (Mandelbrot, 1983), which begins as an equilateral triangle (Figure 3-3).

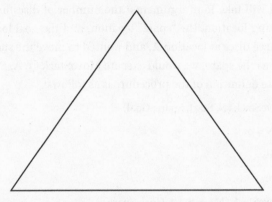

Figure 3-3. The root of the Sierpinski triangle is an equilateral triangle.

The next step in creating the Sierpinski triangle is to take Figure 3-3 and reduce it to exactly half of its original size. Three of these smaller triangles can be inscribed inside of the original triangle, as is illustrated in Figure 3-4.

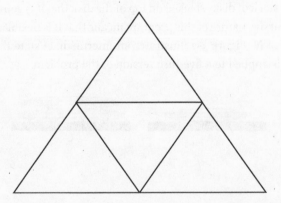

Figure 3-4. The second step of constructing a Sierpinski triangle.

The rule used to create Figure 3-4 can be applied recursively and (in principle) infinitely. One takes the smaller triangle that was used to create Figure 3-4, makes it exactly half of its original size, and inscribes three copies of this still smaller triangle into each of the three triangles that were used to create Figure 3-4. This rule can be applied recursively to inscribe smaller triangles into any of the triangles that were added to the figure in a previous stage of drawing. Figure 3-5 shows the result when this rule is applied four times to Figure 3-4.

Figure 3-5. The Sierpinski triangle that results when the recursive rule is applied four times to Figure 3-4.

The Sierpinski triangle, and all other fractals that are created by recursion, are intrinsically self-similar. That is, if one were to take one of the smaller triangles from which Figure 3-4 is constructed and magnify it, one would see still see the hierarchical structure that is illustrated above. The structure of the whole is identical to the (smaller) structure of the parts. In the next section, we see that the recursive nature of human language reveals itself in the same way.

3.3 Phrase Markers and Fractals

Consider a finite set of elements (e.g., words, phonemes, morphemes) that can, by applying certain rules, be combined to create a sentence or expression that is finite in length. A language can be defined as the set of all of the possible expressions that can generated in this way from the same set of building blocks and the same set of rules (Chomsky, 1957). From this perspective, one can define a grammar as a device that can distinguish the set of grammatical expressions from all other expressions, including those that are generated from the same elements but which violate the rules that define the language. In modern linguistics, a basic issue to investigate is the nature of the grammar that defines a natural human language.

Chomsky (1957) noted that one characteristic of a natural language such as English is that a sentence can be lengthened by inserting a clause into its midst. As we see in the following section, this means that the grammar of natural languages is complicated enough that simple machines, such as finite state automata, are not powerful enough to serve as grammars for them.

The complex, clausal structure of a natural language is instead captured by a more powerful device—a Turing machine—that can accommodate the regularities of a context-free grammar (e.g., Chomsky, 1957, 1965). A context-free grammar can be described as a set of rewrite rules that convert one symbol into one or more other symbols. The application of these rewrite rules produces a hierarchically organized symbolic structure called a phrase marker (Radford, 1981). A phrase marker is a set of points or labelled nodes that are connected by branches. Nonterminal nodes represent lexical categories; at the bottom of a phrase marker are the terminal nodes that represent lexical categories (e.g., words). A phrase marker for the simple sentence *Dogs bark* is illustrated in Figure 3-6.

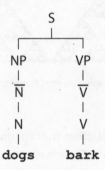

Figure 3-6. A phrase marker for the sentence *Dogs bark*.

The phrase marker for a sentence can be illustrated as an upside-down tree whose structure is grown from the root node S (for sentence). The application of the rewrite rule S → NP VP produces the first layer of the Figure 3-6 phrase marker, showing how the nodes NP (noun phrase) and VP (verb phrase) are grown from S. Other rewrite rules that are invoked to create that particular phrase marker are NP → $\overline{N}$, $\overline{N}$ → N, N → dogs, VP → $\overline{V}$, $\overline{V}$ → V, and V → bark. When any of these rewrite rules are applied, the symbol to the left of the → is rewritten as the symbol or symbols to the right. In the phrase marker, this means the symbols on the right of the → are written as nodes below the original symbol, and are connected to the originating node above, as is shown in Figure 3-6.

In a modern grammar called x-bar syntax (Jackendoff, 1977), nodes like NP and VP in Figure 3-6 are symbols that represent phrasal categories, nodes like $\overline{N}$ and $\overline{V}$ are symbols that represent lexical categories, and nodes like "and" are symbols that represent categories that are intermediates between lexical categories and phrasal categories. Such intermediate categories are required to capture some regularities in the syntax of natural human languages.

In some instances, the same symbol can be found on both sides of the → in a rewrite rule. For instance, one valid rewrite rule for the intermediate node of a noun phrase is $\overline{N} \rightarrow AP$, where AP represents an adjective phrase. Because the same symbol occurs on each side of the equation, the context-free grammar is recursive. One can apply this rule repeatedly to insert clauses of the same type into a phrase. This is shown in Figure 3-7, which illustrates phrase markers for noun phrases that might apply to my dog Rufus. The basic noun phrase is *the dog*. If this recursive rule is applied once, it permits a more elaborate noun phrase to be created, as in *the cute dog*. Recursive application of this rule permits the noun phrase to be elaborated indefinitely, (e.g., *the cute brown scruffy dog*).

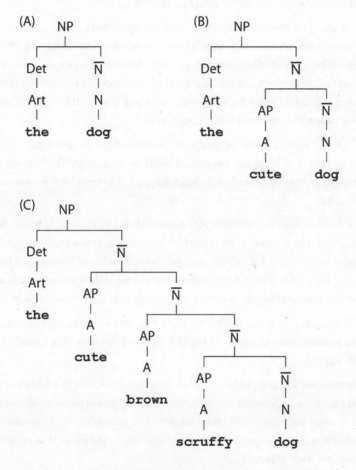

Figure 3-7. Phrase markers for three noun phrases: (A) *the dog*, (B) *the cute dog*, and (C) *the cute brown scruffy dog*. Note the recursive nature of (C).

The recursive nature of a context-free grammar is revealed in a visual inspection of a phrase marker like the one illustrated in Figure 3-7C. As one inspects the figure, one sees the same pattern recurring again and again, as was the case with the Sierpinski triangle. The recursive nature of a context-free grammar produces self-similarity within a phrase marker. The recursion of such a grammar is also responsible for its ability to use finite resources (a finite number of building blocks and a finite number of rewrite rules) to produce a potentially infinite variety of expressions, as in the sentences of a language, each of which is represented by its own phrase marker.

3.4 Behaviourism, Language, and Recursion

Behaviourism viewed language as merely being observable behaviour whose development and elicitation was controlled by external stimuli:

> A speaker possesses a verbal repertoire in the sense that responses of various forms appear in his behavior from time to time in relation to identifiable conditions. A repertoire, as a collection of verbal operants, describes the potential behavior of a speaker. To ask where a verbal operant is when a response is not in the course of being emitted is like asking where one's knee-jerk is when the physician is not tapping the patellar tendon. (Skinner, 1957, p. 21)

Skinner's (1957) treatment of language as verbal behaviour explicitly rejected the Cartesian notion that language expressed ideas or meanings. To Skinner, explanations of language that appealed to such unobservable internal states were necessarily unscientific:

> It is the function of an explanatory fiction to allay curiosity and to bring inquiry to an end. The doctrine of ideas has had this effect by appearing to assign important problems of verbal behavior to a psychology of ideas. The problems have then seemed to pass beyond the range of the techniques of the student of language, or to have become too obscure to make further study profitable. (Skinner, 1957, p. 7)

Modern linguistics has explicitly rejected the behaviourist approach, arguing that behaviourism cannot account for the rich regularities that govern language (Chomsky, 1959b).

> The composition and production of an utterance is not strictly a matter of stringing together a sequence of responses under the control of outside stimulation and intraverbal association, and that the syntactic organization of an utterance is not something directly represented in any simple way in the physical structure of the utterance itself. (Chomsky, 1959b, p. 55)

Modern linguistics has advanced beyond behaviourist theories of verbal behaviour by adopting a particularly technical form of logicism. Linguists assume that verbal behaviour is the result of sophisticated symbol manipulation: an internal generative grammar.

> By a generative grammar I mean simply a system of rules that in some explicit and well-defined way assigns structural descriptions to sentences. Obviously, every

speaker of a language has mastered and internalized a generative grammar that expresses his knowledge of his language. (Chomsky, 1965, p. 8)

A sentence's structural description is represented by using a phrase marker, which is a hierarchically organized symbol structure that can be created by a recursive set of rules called a context-free grammar. In a generative grammar another kind of rule, called a transformation, is used to convert one phrase marker into another.

The recursive grammars that have been developed in linguistics serve two purposes. First, they formalize key structural aspects of human languages, such as the embedding of clauses within sentences. Second, they explain how finite resources are capable of producing an infinite variety of potential expressions. This latter accomplishment represents a modern rebuttal to dualism; we have seen that Descartes (1996) used the creative aspect of language to argue for the separate, non-physical existence of the mind. For Descartes, machines were not capable of generating language because of their finite nature.

Interestingly, a present-day version of Descartes' (1996) analysis of the limitations of machines is available. It recognizes that a number of different information processing devices exists that vary in complexity, and it asks which of these devices are capable of accommodating modern, recursive grammars. The answer to this question provides additional evidence against behaviourist or associationist theories of language (Bever, Fodor, & Garrett, 1968).

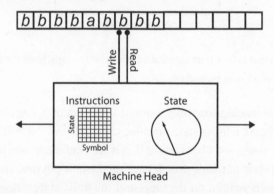

Figure 3-8. How a Turing machine processes its tape.

In Chapter 2, we were introduced to one simple—but very powerful—device, the Turing machine (Figure 3-8). It consists of a machine head that manipulates the symbols on a ticker tape, where the ticker tape is divided into cells, and each cell is capable of holding only one symbol at a time. The machine head can move back and forth along the tape, one cell at a time. As it moves it can read the symbol on the current cell, which can cause the machine head to change its physical state. It is also capable of writing a new symbol on the tape. The behaviour of the machine head—its new physical state, the direction it moves, the symbol that it writes—is controlled by a machine table that depends only upon the current symbol being read and the current state of the device. One uses a Turing machine by writing a question on its tape, and setting the machine head into action. When the machine head halts, the Turing machine's answer to the question has been written on the tape.

What is meant by the claim that different information processing devices are available? It means that systems that are different from Turing machines must also exist. One such alternative to a Turing machine is called a finite state automaton (Minsky, 1972; Parkes, 2002), which is illustrated in Figure 3-9. Like a Turing machine, a finite state automaton can be described as a machine head that interacts with a ticker tape. There are two key differences between a finite state machine and a Turing machine.

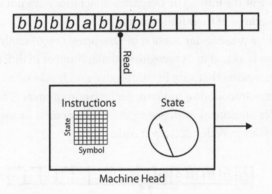

Figure 3-9. How a finite state automaton processes the tape. Note the differences between Figures 3-9 and 3-8.

First, a finite state machine can only move in one direction along the tape, again one cell at a time. Second, a finite state machine can only read the symbols on the tape; it does not write new ones. The symbols that it encounters, in combination with the current physical state of the device, determine the new physical state of the device. Again, a question is written on the tape, and the finite state automaton is started. When it reaches the end of the question, the final physical state of the finite state automaton represents its answer to the original question on the tape.

It is obvious that a finite state automaton is a simpler device than a Turing machine, because it cannot change the ticker tape, and because it can only move in one direction along the tape. However, finite state machines are important information processors. Many of the behaviours in behaviour-based robotics are produced using finite state machines (Brooks, 1989, 1999, 2002). It has also been argued that such devices are all that is required to formalize behaviourist or associationist accounts of behaviour (Bever, Fodor, & Garrett., 1968).

What is meant by the claim that an information processing device can "accommodate" a grammar? In the formal analysis of the capabilities of information processors (Gold, 1967), there are two answers to this question. Assume that knowledge of some grammar has been built into a device's machine head. One could then ask whether the device is capable of accepting a grammar. In this case, the "question" on the tape would be an expression, and the task of the information processor would be to accept the string, if it is grammatical according to the device's grammar, or to reject the expression, if it does not belong to the grammar. Another question to ask would be whether the information processor is capable of generating the grammar. That is, given a grammatical expression, can the device use its existing grammar to replicate the expression (Wexler & Culicover, 1980)?

In Chapter 2, it was argued that one level of investigation to be conducted by cognitive science was computational. At the computational level of analysis, one uses formal methods to investigate the kinds of information processing problems a device is solving. When one uses formal methods to determine whether some device is capable of accepting or generating some grammar of interest, one is conducting an investigation at the computational level.

One famous example of such a computational analysis was provided by Bever, Fodor, and Garrett (1968). They asked whether a finite state automaton was capable of accepting expressions that were constructed from a particular artificial grammar. Expressions constructed from this grammar were built from only two symbols, a and b. Grammatical strings in the sentence were "mirror images," because the pattern used to generate expressions was $b^N a b^N$ where N is the number of bs in the string. Valid expressions generated from this grammar include a, $bbbbabbbb$, and $bbabb$. Expressions that cannot be generated from the grammar include ab, $babb$, bb, and $bbbabb$.

While this artificial grammar is very simple, it has one important property: it is recursive. That is, a simple context-free grammar can be defined to generate its potential expressions. This context-free grammar consists of two rules, where Rule 1 is S → a, and Rule 2 is a → bab. A string is begun by using Rule 1 to generate an a. Rule 2 can then be applied to generate the string bab. If Rule 2 is applied recursively to the central bab then longer expressions will be produced that will always be consistent with the pattern $b^N a b^N$.

Bever, Fodor, and Garrett (1968) proved that a finite state automaton was not capable of accepting strings generated from this recursive grammar. This is because a finite state machine can only move in one direction along the tape, and cannot write to the tape. If it starts at the first symbol of a string, then it is not capable of keeping track of the number of bs read before the a, and comparing this to the number of bs read after the a. Because it can't go backwards along the tape, it can't deal with recursive languages that have embedded clausal structure.

Bever, Fodor, and Garrett (1968) used this result to conclude that association-ism (and radical behaviourism) was not powerful enough to deal with the embed-ded clauses of natural human language. As a result, they argued that associationism should be abandoned as a theory of mind. The impact of this proof is measured by the lengthy responses to this argument by associationist memory researchers (Anderson & Bower, 1973; Paivio, 1986). We return to the implications of this argu-ment when we discuss connectionist cognitive science in Chapter 4.

While finite state automata cannot accept the recursive grammar used by Bever, Fodor, and Garrett (1968), Turing machines can (Révész, 1983). Their ability to move in both directions along the tape provides them with a memory that enables them to match the number of leading bs in a string with the number of trailing bs.

Modern linguistics has concluded that the structure of human language must be described by grammars that are recursive. Finite state automata are not powerful-enough devices to accommodate grammars of this nature, but Turing machines are. This suggests that an information processing architecture that is sufficiently rich to explain human cognition must have the same power—must be able to answer the same set of questions—as do Turing machines. This is the essence of the physical symbol system hypothesis (Newell, 1980), which are discussed in more detail below. The Turing machine, as we saw in Chapter 2 and further discuss below, is a univer-sal machine, and classical cognitive science hypothesizes that "this notion of symbol system will prove adequate to all of the symbolic activity this physical universe of ours can exhibit, and in particular all the symbolic activities of the human mind" (Newell, 1980, p. 155).

3.5 Underdetermination and Innateness

The ability of a device to accept or generate a grammar is central to another com-putational level analysis of language (Gold, 1967). Gold performed a formal analy-sis of language learning which revealed a situation that is known as Gold's paradox (Pinker, 1979). One solution to this paradox is to adopt a position that is characteristic of classical cognitive science, and which we have seen is consistent with its Cartesian roots. This position is that a good deal of the architecture of cognition is innate.

Gold (1967) was interested in the problem of how a system could learn the grammar of a language on the basis of a finite set of example expressions. He con-sidered two different situations in which the learning system could be presented with expressions. In informant learning, the learner is presented with either valid or invalid expressions, and is also told about their validity, i.e., told whether they belong to the grammar or not. In text learning, the only expressions that are pre-sented to the learner are grammatical.

Whether a learner is undergoing informant learning or text learning, Gold (1967) assumed that learning would proceed as a succession of presentations of expressions. After each expression was presented, the language learner would generate a hypothesized grammar. Gold proposed that each hypothesis could be described as being a Turing machine that would either accept the (hypothesized) grammar or generate it. In this formalization, the notion of "learning a language" has become "selecting a Turing machine that represents a grammar" (Osherson, Stob, & Weinstein, 1986).

According to Gold's (1967) algorithm, a language learner would have a current hypothesized grammar. When a new expression was presented to the learner, a test would be conducted to see if the current grammar could deal with the new expression. If current grammar succeeded, then it remained. If the current grammar failed, then a new grammar—a new Turing machine—would have to be selected.

Under this formalism, when can we say that a grammar has been learned? Gold defined language learning as the identification of the grammar in the limit. When a language is identified in the limit, this means that the current grammar being hypothesized by the learner does not change even as new expressions are encountered. Furthermore, it is expected that this state will occur after a finite number of expressions have been encountered during learning.

In the previous section, we considered a computational analysis in which different kinds of computing devices were presented with the same grammar. Gold (1967) adopted an alternative approach: he kept the information processing constant—that is, he always studied the algorithm sketched above—but he varied the complexity of the grammar that was being learned, and he varied the conditions under which the grammar was presented, i.e., informant learning versus text learning.

In computer science, a formal description of any class of languages (human or otherwise) relates its complexity to the complexity of a computing device that could generate or accept it (Hopcroft & Ullman, 1979; Révész, 1983). This has resulted in a classification of grammars known as the Chomsky hierarchy (Chomsky, 1959a). In the Chomsky hierarchy, the simplest grammars are regular, and they can be accommodated by finite state automata. The next most complicated are context-free grammars, which can be processed by pushdown automata (a device that is a finite state automaton with a finite internal memory). Next are the context-sensitive grammars, which are the domain of linear bounded automata (i.e., a device like a Turing machine, but with a ticker tape of bounded length). The most complex grammars are the generative grammars, which can only be dealt with by Turing machines.

Gold (1967) used formal methods to determine the conditions under which each class of grammars could be identified in the limit. He was able to show that text learning could only be used to acquire the simplest grammar. In contrast, Gold found that informant learning permitted context-sensitive and context-free grammars to be identified in the limit.

Gold's (1967) research was conducted in a relatively obscure field of theoretical computer science. However, Steven Pinker brought it to the attention of cognitive science more than a decade later (Pinker, 1979), where it sparked a great deal of interest and research. This is because Gold's computational analysis revealed a paradox of particular interest to researchers who studied how human children acquire language.

Gold's (1967) proofs indicated that informant learning was powerful enough that a complex grammar can be identified in the limit. Such learning was not possible with text learning. Gold's paradox emerged because research strongly suggests that children are text learners, *not* informant learners (Pinker, 1979, 1994, 1999). It is estimated that 99.93 percent of the language to which children are exposed is grammatical (Newport, Gleitman, & Gleitman, 1977). Furthermore, whenever feedback about language grammaticality is provided to children, it is not systematic enough to be used to select a grammar (Marcus, 1993).

Gold's paradox is that while he proved that grammars complex enough to model human language could not be text learned, children learn such grammars—and do so via text learning! How is this possible?

Gold's paradox is an example of a problem of underdetermination. In a problem of underdetermination, the information available from the environment is not sufficient to support a unique interpretation or inference (Dawson, 1991). For instance, Gold (1967) proved that a finite number of expressions presented during text learning were not sufficient to uniquely determine the grammar from which these expressions were generated, provided that the grammar was more complicated than a regular grammar.

There are many approaches available for solving problems of underdetermination. One that is most characteristic of classical cognitive science is to simplify the learning situation by assuming that some of the to-be-learned information is already present because it is innate. For instance, classical cognitive scientists assume that much of the grammar of a human language is innately available before language learning begins.

> The child has an innate theory of potential structural descriptions that is sufficiently rich and fully developed so that he is able to determine, from a real situation in which a signal occurs, which structural descriptions may be appropriate to this signal. (Chomsky, 1965, p. 32)

If the existence of an innate, universal base grammar—a grammar used to create phrase markers—is assumed, then a generative grammar of the type proposed by Chomsky can be identified in the limit (Wexler & Culicover, 1980). This is because learning the language is simplified to the task of learning the set of transformations that can be applied to phrase markers. More modern theories of transformational grammars have reduced the number of transformations to one, and have described language learning as the setting of a finite number of parameters that determine grammatical structure (Cook & Newson, 1996). Again, these grammars can be identified in the limit on the basis of very simple input expressions (Lightfoot, 1989). Such proofs are critical to cognitive science and to linguistics, because if a theory of language is to be explanatorily adequate, then it must account for how language is acquired (Chomsky, 1965).

Rationalist philosophers assumed that some human knowledge must be innate. This view was reacted against by empiricist philosophers who viewed experience as the only source of knowledge. For the empiricists, the mind was a *tabula rasa*, waiting to be written upon by the world. Classical cognitive scientists are comfortable with the notion of innate knowledge, and have used problems of underdetermination to argue against the modern tabula rasa assumed by connectionist cognitive scientists (Pinker, 2002, p. 78): "The connectionists, of course, do not believe in a blank slate, but they do believe in the closest mechanistic equivalent, a general-purpose learning device." The role of innateness is an issue that separates classical cognitive science from connectionism, and will be encountered again when connectionism is explored in Chapter 4.

3.6 Physical Symbol Systems

Special-purpose logic machines had been developed by philosophers in the late nineteenth century (Buck & Hunka, 1999; Jevons, 1870; Marquand, 1885). However, abstract descriptions of how devices could perform general-purpose symbol manipulation did not arise until the 1930s (Post, 1936; Turing, 1936). The basic properties laid out in these mathematical theories of computation define what is now known as a physical symbol system (Newell, 1980; Newell & Simon, 1976). The concept *physical symbol system* defines "a broad class of systems that is capable of having and manipulating symbols, yet is also realizable within our physical universe" (Newell, 1980, p. 136).

A physical symbol system operates on a finite set of physical tokens called symbols. These are components of a larger physical entity called a symbol structure or a symbolic expression. It also consists of a set of operators that can create, modify, duplicate, or destroy symbols. Some sort of control is also required to select at any given time some operation to apply. A physical symbol system produces, over time, an evolving or changing collection of expressions. These expressions represent or designate entities in the world (Newell, 1980; Newell & Simon, 1976). As a result, the symbol manipulations performed by such a device permit new meanings to be derived, in the same way as new knowledge is arrived at in the proofs discovered by logicians and mathematicians (Davis & Hersh, 1981).

The abstract theories that describe physical symbol systems were not developed into working artifacts until nearly the midpoint of the twentieth century. "Our deepest insights into information processing were achieved in the thirties, before modern computers came into being. It is a tribute to the genius of Alan Turing" (Newell & Simon, 1976, p. 117). The first digital computer was the Z3, invented in Germany in 1941 by Konrad Zuse (1993). In the United States, the earliest computers were University of Pennsylvania's ENIAC (created 1943–1946) and EDVAC (created 1945–1950), Harvard's MARK I (created 1944), and Princeton's IAS or von Neumann computer (created 1946–1951) (Burks, 2002; Cohen, 1999). The earliest British computer was University of Manchester's "Baby," the small-scale experimental machine (SSEM) that was first activated in June, 1948 (Lavington, 1980).

Although specific details vary from machine to machine, all digital computers share three general characteristics (von Neumann, 1958). First, they have a memory for the storage of symbolic structures. In what is now known as the von Neumann architecture, this is a random access memory (RAM) in which any memory location can be immediately accessed—without having to scroll through other locations, as in a Turing machine—by using the memory's address. Second, they have a mechanism separate from memory that is responsible for the operations that manipulate stored symbolic structures. Third, they have a controller for determining which operation to perform at any given time. In the von Neumann architecture, the control mechanism imposes serial processing, because only one operation will be performed at a time.

Perhaps the earliest example of serial control is the nineteenth-century punched cards used to govern the patterns in silk that were woven by Joseph Marie Jacquard's loom (Essinger, 2004). During weaving, at each pass of the loom's shuttle, holes in a card permitted some thread-controlling rods to be moved. When a rod moved, the thread that it controlled was raised; this caused the thread to be visible in that row of the pattern. A sequence of cards was created by tying cards together end to end. When this "chain" was advanced to the next card, the rods would be altered to create the appropriate appearance for the silk pattern's next row.

The use of punched cards turned the Jacquard loom into a kind of universal machine: one changed the pattern being produced not by changing the loom, but simply by loading it with a different set of punched cards. Thus not only did Jacquard invent a new loom, but he also invented the idea of using a program to control the actions of a machine. Jacquard's program was, of course, a sequence of punched cards. Their potential for being applied to computing devices in general was recognized by computer pioneer Charles Babbage, who was inspired by Jacquard's invention (Essinger, 2004).

By the late 1950s, it became conventional to load the program—then known as the "short code" (von Neumann, 1958)—into memory. This is called memory-stored control; the first modern computer to use this type of control was Manchester's "Baby" (Lavington, 1980). In Chapter 2 we saw an example of this type of control in the universal Turing machine, whose ticker tape memory holds both the data to be manipulated and the description of a special-purpose Turing machine that will do the manipulating. The universal Turing machine uses the description to permit it to pretend to be the specific machine that is defined on its tape (Hodges, 1983).

In a physical symbol system that employs memory-stored control, internal characteristics will vary over time. However, the time scale of these changes will not be uniform (Newell, 1990). The data that is stored in memory will likely be changed rapidly. However, some stored information—in particular, the short code, or what cognitive scientists would call the virtual machine (Pylyshyn, 1984, 1991), that controls processing would be expected to be more persistent. Memory-stored control in turn chooses which architectural operation to invoke at any given time. In a digital computer, the architecture would not be expected to vary over time at all because it is fixed, that is, literally built into the computing device.

The different characteristics of a physical symbol system provide a direct link back to the multiple levels of investigation that were the topic of Chapter 2. When such a device operates, it is either computing some function or solving some information processing problem. Describing this aspect of the system is the role of a computational analysis. The computation being carried out is controlled by an algorithm: the program stored in memory. Accounting for this aspect of the system is the aim of an algorithmic analysis. Ultimately, a stored program results in the device executing a primitive operation on a symbolic expression stored in memory. Identifying the primitive processes and symbols is the domain of an architectural analysis. Because the device is a physical symbol system, primitive processes and symbols must be physically realized. Detailing the physical nature of these components is the goal of an implementational analysis.

The invention of the digital computer was necessary for the advent of classical cognitive science. First, computers are general symbol manipulators. Their existence demonstrated that finite devices could generate an infinite potential of symbolic behaviour, and thus supported a materialist alternative to Cartesian dualism. Second, the characteristics of computers, and of the abstract theories of computation that led to their development, in turn resulted in the general notion of physical symbol system, and the multiple levels of investigation that such systems require.

The final link in the chain connecting computers to classical cognitive science is the logicist assumption that cognition is a rule-governed symbol manipulation of the sort that a physical symbol system is designed to carry out. This produces the physical symbol system hypothesis: "the necessary and sufficient condition for a physical system to exhibit general intelligent action is that it be a physical symbol system" (Newell, 1980, p. 170). By *necessary*, Newell meant that if an artifact exhibits general intelligence, then it must be an instance of a physical symbol system. By *sufficient*, Newell claimed that any device that is a physical symbol system can be configured to exhibit general intelligent action—that is, he claimed the plausibility of machine intelligence, a position that Descartes denied.

What did Newell (1980) mean by *general intelligent action*? He meant,

> the same scope of intelligence seen in human action: that in real situations behavior appropriate to the ends of the system and adaptive to the demands of the environment can occur, within some physical limits. (Newell, 1980, p. 170)

In other words, human cognition must be the product of a physical symbol system. Thus human cognition must be explained by adopting all of the different levels of investigation that were described in Chapter 2.

3.7 Componentiality, Computability, and Cognition

In 1840, computer pioneer Charles Babbage displayed a portrait of loom inventor Joseph Marie Jacquard for the guests at the famous parties in his home (Essinger, 2004). The small portrait was incredibly detailed. Babbage took great pleasure in the fact that most people who first saw the portrait mistook it to be an engraving. It was instead an intricate fabric woven on a loom of the type that Jacquard himself invented.

The amazing detail of the portrait was the result of its being composed of 24,000 rows of weaving. In a Jacquard loom, punched cards determined which threads would be raised (and therefore visible) for each row in the fabric. Each thread in the loom was attached to a rod; a hole in the punched card permitted a rod to move, raising its thread. The complexity of the Jacquard portrait was produced by using 24,000 punched cards to control the loom.

Though Jacquard's portrait was impressively complicated, the process used to create it was mechanical, simple, repetitive—and local. With each pass of the loom's shuttle, weaving a set of threads together into a row, the only function of a punched card was to manipulate rods. In other words, each punched card only controlled small components of the overall pattern. While the entire set of punched cards represented the total pattern to be produced, this total pattern was neither contained in, nor required by, an individual punched card as it manipulated the loom's rods. The portrait of Jacquard was a global pattern that emerged from a long sequence of simple, local operations on the pattern's components.

In the Jacquard loom, punched cards control processes that operate on local components of the "expression" being weaved. The same is true of the physical symbol systems. Physical symbol systems are finite devices that are capable of producing an infinite variety of potential behaviour. This is possible because the operations of a physical symbol system are recursive. However, this explanation is not complete. In addition, the rules of a physical symbol system are local or compownential, in the sense that they act on local components of an expression, not on the expression as a whole.

For instance, one definition of a language is the set of all of its grammatical expressions (Chomsky, 1957). Given this definition, it is logically possible to treat each expression in the set as an unanalyzed whole to which some operation could be applied. This is one way to interpret a behaviourist theory of language (Skinner, 1957): each expression in the set is a holistic verbal behaviour whose likelihood of being produced is a result of reinforcement and stimulus control of the expression as a whole.

However, physical symbol systems do not treat expressions as unanalyzed wholes. Instead, the recursive rules of a physical symbol system are sensitive to the atomic symbols from which expressions are composed. We saw this previously in the example of context-free grammars that were used to construct the phrase markers of Figures 3-6 and 3-7. The rules in such grammars do not process whole phrase markers, but instead operate on the different components (e.g., nodes like S, N, VP) from which a complete phrase marker is constructed.

The advantage of operating on symbolic components, and not on whole expressions, is that one can use a sequence of very basic operations—writing, changing, erasing, or copying a symbol—to create an overall effect of far greater scope than might be expected. As Henry Ford said, nothing is particularly hard if you divide it into small jobs. We saw the importance of this in Chapter 2 when we discussed Leibniz' mill (Leibniz, 1902), the Chinese room (Searle, 1980), and the discharging of homunculi (Dennett, 1978). In a materialist account of cognition, thought is produced by a set of apparently simple, mindless, unintelligent actions—the primitives that make up the architecture.

The small jobs carried out by a physical symbol system reveal that such a system has a dual nature (Haugeland, 1985). On the one hand, symbol manipulations are purely syntactic—they depend upon identifying a symbol's type, and not upon semantically interpreting what the symbol stands for. On the other hand, a physical symbol system's manipulations are semantic—symbol manipulations preserve meanings, and can be used to derive new, sensible interpretations.

> Interpreted formal tokens lead two lives: syntactical lives, in which they are meaningless markers, moved according to the rules of some self-contained game; and

semantic lives, in which they have meanings and symbolic relations to the outside world. (Haugeland, 1985, p. 100)

Let us briefly consider these two lives. First, we have noted that the rules of a physical symbol system operate on symbolic components of a whole expression. For this to occur, all that is required is that a rule identifies a particular physical entity as being a token or symbol of a particular type. If the symbol is of the right type, then the rule can act upon it in some prescribed way.

For example, imagine a computer program that is playing chess. For this program, the "whole expression" is the total arrangement of game pieces on the chess board at any given time. The program analyzes this expression into its components: individual tokens on individual squares of the board. The physical characteristics of each component token can then be used to identify to what symbol class it belongs: queen, knight, bishop, and so on. Once a token has been classified in this way, appropriate operations can be applied to it. If a game piece has been identified as being a "knight," then only knight moves can be applied to it—the operations that would move the piece like a bishop cannot be applied, because the token has not been identified as being of the type "bishop."

Similar syntactic operations are at the heart of a computing device like a Turing machine. When the machine head reads a cell on the ticker tape (another example of componentiality!), it uses the physical markings on the tape to determine that the cell holds a symbol of a particular type. This identification—in conjunction with the current physical state of the machine head—is sufficient to determine which instruction to execute.

To summarize, physical symbol systems are syntactic in the sense that their rules are applied to symbols that have been identified as being of a particular type on the basis of their physical shape or form. Because the shape or form of symbols is all that matters for the operations to be successfully carried out, it is natural to call such systems formal. Formal operations are sensitive to the shape or form of individual symbols, and are not sensitive to the semantic content associated with the symbols.

However, it is still the case that formal systems can produce meaningful expressions. The punched cards of a Jacquard loom only manipulate the positions of thread-controlling rods. Yet these operations can produce an intricate woven pattern such as Jacquard's portrait. The machine head of a Turing machine reads and writes individual symbols on a ticker tape. Yet these operations permit this device to provide answers to any computable question. How is it possible for formal systems to preserve or create semantic content?

In order for the operations of a physical symbol system to be meaningful, two properties must be true. First, the symbolic structures operated on must have semantic content. That is, the expressions being manipulated must have some relationship to states of the external world that permits the expressions to represent these states. This relationship is a basic property of a physical symbol system, and is called designation (Newell, 1980; Newell & Simon, 1976). "An expression designates an object if, given the expression, the system can either affect the object itself or behave in ways dependent on the object" (Newell & Simon, 1976, p. 116).

Explaining designation is a controversial issue in cognitive science and philosophy. There are many different proposals for how designation, which is also called the problem of representation (Cummins, 1989) or the symbol grounding problem (Harnad, 1990), occurs. The physical symbol system hypothesis does not propose a solution, but necessarily assumes that such a solution exists. This assumption is plausible to the extent that computers serve as existence proofs that designation is possible.

The second semantic property of a physical symbol system is that not only are individual expressions meaningful (via designation), but the evolution of expressions—the rule-governed transition from one expression to another—is also meaningful. That is, when some operation modifies an expression, this modification is not only syntactically correct, but it will also make sense semantically. As rules modify symbolic structures, they preserve meanings in the domain that the symbolic structures designate, even though the rules themselves are purely formal. The application of a rule should not produce an expression that is meaningless. This leads to what is known as the formalist's motto: "If you take care of the syntax, then the semantics will take care of itself" (Haugeland, 1985, p. 106).

The assumption that applying a physical symbol system's rules preserves meaning is a natural consequence of classical cognitive science's commitment to logicism. According to logicism, thinking is analogous to using formal methods to derive a proof, as is done in logic or mathematics. In these formal systems, when one applies rules of the system to true expressions (e.g., the axioms of a system of mathematics which by definition are assumed to be true [Davis & Hersh, 1981]), the resulting expressions must also be true. An expression's truth is a critical component of its semantic content.

It is necessary, then, for the operations of a formal system to be defined in such a way that 1) they only detect the form of component symbols, and 2) they are constrained in such a way that manipulations of expressions are meaningful (e.g., truth preserving). This results in classical cognitive science's interest in universal machines.

A universal machine is a device that is maximally flexible in two senses (Newell, 1980). First, its behaviour is responsive to its inputs; a change in inputs will be capable of producing a change in behaviour. Second, a universal machine must be able compute the widest variety of input-output functions that is possible. This "widest variety" is known as the set of computable functions.

A device that can compute every possible input-output function does not exist. The Turing machine was invented and used to prove that there exist some functions that are not computable (Turing, 1936). However, the subset of functions that are computable is large and important:

> It can be proved mathematically that there are infinitely more functions than programs. Therefore, for most functions there is no corresponding program that can compute them. . . . Fortunately, almost all these noncomputable functions are useless, and virtually all the functions we might want to compute are computable. (Hillis, 1998, p. 71)

A major discovery of the twentieth century was that a number of seemingly different symbol manipulators were all identical in the sense that they all could compute the same maximal class of input-output pairings (i.e., the computable functions). Because of this discovery, these different proposals are all grouped together into the class "universal machine," which is sometimes called the "effectively computable procedures." This class is "a large zoo of different formulations" that includes "Turing machines, recursive functions, Post canonical systems, Markov algorithms, all varieties of general purpose digital computers, [and] most programming languages" (Newell, 1980, p. 150).

Newell (1980) proved that a generic physical symbol system was also a universal machine. This proof, coupled with the physical symbol system hypothesis, leads to a general assumption in classical cognitive science: cognition is computation, the brain implements a universal machine, and the products of human cognition belong to the class of computable functions.

The claim that human cognition is produced by a physical symbol system is a scientific hypothesis. Evaluating the validity of this hypothesis requires fleshing out many additional details. What is the organization of the program that defines the physical symbol system for cognition (Newell & Simon, 1972)? In particular, what kinds of symbols and expressions are being manipulated? What primitive operations are responsible for performing symbol manipulation? How are these operations controlled? Classical cognitive science is in the business of fleshing out these details, being guided at all times by the physical symbol system hypothesis.

3.8 The Intentional Stance

According to the formalist's motto (Haugeland, 1985) by taking care of the syntax, one also takes care of the semantics. The reason for this is that, like the rules in a logical system, the syntactic operations of a physical symbol system are constrained to preserve meaning. The symbolic expressions that a physical symbol system evolves will have interpretable designations.

We have seen that the structures a physical symbol system manipulates have two different lives, syntactic and semantic. Because of this, there is a corollary to the formalist's motto, which might be called the semanticist's motto: "If you understand the semantics, then you can take the syntax for granted." That is, if you have a semantic interpretation of a physical symbol system's symbolic expressions, then you can use this semantic interpretation to predict the future behaviour of the system—the future meanings that it will generate—without having to say anything about the underlying physical mechanisms that work to preserve the semantics.

We have seen that one of the fundamental properties of a physical symbol system is designation, which is a relation between the system and the world that provides interpretations to its symbolic expressions (Newell, 1980; Newell & Simon, 1976). More generally, it could be said that symbolic expressions are intentional—they are about some state of affairs in the world. This notion of intentionality is rooted in the philosophy of Franz Brentano (Brentano, 1995). Brentano used intentionality to distinguish the mental from the physical: "We found that the intentional in-existence, the reference to something as an object, is a distinguishing characteristic of all mental phenomena. No physical phenomenon exhibits anything similar" (p. 97).

To assume that human cognition is the product of a physical symbol system is to also assume that mental states are intentional in Brentano's sense. In accord with the semanticist's motto, the intentionality of mental states can be used to generate a theory of other people, a theory that can be used to predict the behaviour of another person. This is accomplished by adopting what is known as the intentional stance (Dennett, 1987).

The intentional stance uses the presumed contents of someone's mental states to predict their behaviour. It begins by assuming that another person possesses intentional mental states such as beliefs, desires, or goals. As a result, the intentional stance involves describing other people with propositional attitudes.

A propositional attitude is a statement that relates a person to a proposition or statement of fact. For example, if I said to someone "Charles Ives' music anticipated minimalism," they could describe me with the propositional attitude "Dawson believes that Charles Ives' music anticipated minimalism." Propositional attitudes are of interest to philosophy because they raise a number of interesting logical problems. For example, the propositional attitude describing me could be true, but at the same time its propositional component could be false (for instance, if Ives' music bore no relationship to minimalism at all!). Propositional attitudes are found everywhere in our language, suggesting that a key element of our understanding of others is the use of the intentional stance.

In addition to describing other people with propositional attitudes, the intentional stance requires that other people are assumed to be rational. To assume that a person is rational is to assume that there are meaningful relationships between the contents of mental states and behaviour. To actually use the contents of mental states to predict behaviour—assuming rationality—is to adopt the intentional stance.

For instance, given the propositional attitudes "Dawson believes that Charles Ives' music anticipated minimalism" and "Dawson desires to only listen to early minimalist music," and assuming that Dawson's behaviour rationally follows from the contents of his intentional states, one might predict that "Dawson often listens to Ives' compositions." The assumption of rationality, "in combination with home truths about our needs, capacities and typical circumstances, generates both an intentional interpretation of us as believers and desirers and actual predictions of behavior in great profusion" (Dennett, 1987, p. 50).

Adopting the intentional stance is also known as employing common-sense psychology or folk psychology. The status of folk psychology, and of its relation to cognitive science, provides a source of continual controversy (Christensen & Turner, 1993; Churchland, 1988; Fletcher, 1995; Greenwood, 1991; Haselager, 1997; Ratcliffe, 2007; Stich, 1983). Is folk psychology truly predictive? If so, should the theories of cognitive science involve lawful operations on propositional attitudes? If not, should folk psychology be expunged from cognitive science? Positions on these issues range from eliminative materialism's argument to erase folk-psychological terms from cognitive science (Churchland, 1988), to experimental philosophy's position that folk concepts are valid and informative, and therefore should be empirically examined to supplant philosophical concepts that have been developed from a purely theoretical or analytic tradition (French & Wettstein, 2007; Knobe & Nichols, 2008).

In form, at least, the intentional stance or folk psychology has the appearance of a scientific theory. The intentional stance involves using a set of general, abstract laws (e.g., the principle of rationality) to predict future events. This brings it into contact with an important view of cognitive development known as the theory-theory (Gopnik & Meltzoff, 1997; Gopnik, Meltzoff, & Kuhl, 1999; Gopnik & Wellman, 1992; Wellman, 1990). According to the theory-theory, children come to understand the world by adopting and modifying theories about its regularities. That is, the child develops intuitive, representational theories in a fashion that is analogous to a scientist using observations to construct a scientific theory. One of the theories that a child develops is a theory of mind that begins to emerge when a child is three years old (Wellman, 1990).

The scientific structure of the intentional stance should be of no surprise, because this is another example of the logicism that serves as one of the foundations of classical cognitive science. If cognition really is the product of a physical symbol system, if intelligence really does emerge from the manipulation of intentional representations according to the rules of some mental logic, then the semanticist's motto should hold. A principle of rationality, operating on propositional attitudes, should offer real predictive power.

However, the logicism underlying the intentional stance leads to a serious problem for classical cognitive science. This is because a wealth of experiments has shown that human reasoners deviate from principles of logic or rationality (Hastie, 2001; Tversky & Kahneman, 1974; Wason, 1966; Wason & Johnson-Laird, 1972). "A purely formal, or syntactic, approach to [reasoning] may suffer from severe limitations" (Wason & Johnson-Laird, 1972, p. 244). This offers a severe challenge to classical cognitive science's adherence to logicism: if thinking is employing mental logic, then how is it possible for thinkers to be illogical?

It is not surprising that many attempts have been made to preserve logicism by providing principled accounts of deviations from rationalism. Some of these attempts have occurred at the computational level and have involved modifying the definition of rationality by adopting a different theory about the nature of mental logic. Such attempts include rational analysis (Chater & Oaksford, 1999) and probabilistic theories (Oaksford & Chater, 1998, 2001). Other, not unrelated approaches involve assuming that ideal mental logics are constrained by algorithmic and architectural-level realities, such as limited memory and real time constraints. The notion of bounded rationality is a prototypical example of this notion (Chase, Hertwig, & Gigerenzer, 1998; Evans, 2003; Hastie, 2001; Rubinstein, 1998; Simon, Egidi, & Marris, 1995).

The attempts to preserve logicism reflect the importance of the intentional stance, and the semanticist's motto, to cognitive science. Classical cognitive science is committed to the importance of a cognitive vocabulary, a vocabulary that invokes the contents of mental states (Pylyshyn, 1984).

3.9 Structure and Process

The physical symbol systems of classical cognitive science make a sharp distinction between symbols and the rules that manipulate them. This is called the structure/process distinction. For instance, in a Turing machine the symbols reside in one medium (the ticker tape) that is separate from another medium (the machine head) that houses the operators for manipulating symbols. Whatever the specific nature of cognition's universal machine, if it is a classical physical symbol system, then it will exhibit the structure/process distinction.

In general, what can be said about the symbols that define the structure that is manipulated by a physical symbol system? It has been argued that cognitive science's notion of symbol is ill defined (Searle, 1992). Perhaps this is because apart from the need that symbols be physically distinctive, so that they can be identified as being tokens of a particular type, symbols do not have definitive properties. Symbols are arbitrary, in the sense that anything can serve as a symbol.

The arbitrary nature of symbols is another example of the property of multiple realization that was discussed in Chapter 2.

> What we had no right to expect is the immense variety of physical ways to realize any fixed symbol system. What the generations of digital technology have demonstrated is that an indefinitely wide array of physical phenomena can be used to develop a digital technology to produce a logical level of essentially identical character. (Newell, 1980, p. 174)

This is why universal machines can be built out of gears (Swade, 1993), LEGO (Agulló et al., 2003), electric train sets (Stewart, 1994), hydraulic valves, or silicon chips (Hillis, 1998).

The arbitrariness of symbols, and the multiple realization of universal machines, is rooted in the relative notion of universal machine. By definition, a machine is universal if it can simulate any other universal machine (Newell, 1980). Indeed, this is the basic idea that justifies the use of computer simulations to investigate cognitive and neural functioning (Dutton & Starbuck, 1971; Gluck & Myers, 2001; Lewandowsky, 1993; Newell & Simon, 1961; O'Reilly & Munakata, 2000).

> For any class of machines, defined by some way of describing its operational structure, a machine of that class is defined to be universal if it can behave like any machine of the class. This puts simulation at the center of the stage. (Newell, 1980, p. 149)

If a universal machine can be simulated by any other, and if cognition is the product of a universal machine, then why should we be concerned about the specific details of the information processing architecture for cognition? The reason for this concern is that the internal aspects of an architecture—the relations between a particular structure-process pairing—are *not* arbitrary. The nature of a particular structure is such that it permits some, but not all, processes to be easily applied. Therefore some input-output functions will be easier to compute than others because of the relationship between structure and process. Newell and Simon (1972, p. 803) called these second-order effects.

Consider, for example, one kind of representation: a table of numbers, such as Table 3-1, which provides the distances in kilometres between pairs of cities in Alberta (Dawson, Boechler, & Valsangkar-Smyth, 2000). One operation that can easily be applied to symbols that are organized in such a fashion is table lookup. For instance, perhaps I was interested in knowing the distance that I would travel if I drove from Edmonton to Fort McMurray. Applying table lookup to Table 3-1, by looking for the number at the intersection between the *Edmonton* row and the *Fort McMurray* column, quickly informs me that the distance is 439 kilometres.

This is because the tabular form of this information makes distances between places explicit, so that they can be "read off of" the representation in a seemingly effortless manner.

Other information cannot be so easily gleaned from a tabular representation. For instance, perhaps I am interested in determining the compass direction that points from Edmonton to Fort McMurray. The table does not make this information explicit—directions between cities cannot be simply read off of Table 3-1.

	BAN	CAL	CAM	DRU	EDM	FMC	GRA	JAS	LET	LLO	MED	RED	SLA
BANFF	0	128	381	263	401	840	682	287	342	626	419	253	652
CALGARY	128	0	274	138	294	733	720	412	216	519	293	145	545
CAMROSE	381	274	0	182	97	521	553	463	453	245	429	129	348
DRUMHELLER	263	138	182	0	279	703	735	547	282	416	247	165	530
EDMONTON	401	294	97	279	0	439	456	366	509	251	526	148	250
FORT MCMURRAY	840	733	521	703	439	0	752	796	948	587	931	587	436
GRANDE PRAIRIE	682	720	553	735	456	752	0	397	935	701	982	586	318
JASPER	287	412	463	547	366	796	397	0	626	613	703	413	464
LETHBRIDGE	342	216	453	282	509	948	935	626	0	605	168	360	760
LLOYDMINSTER	626	519	245	416	251	587	701	613	605	0	480	374	496
MEDICINE HAT	419	293	429	247	526	931	982	703	168	480	0	409	777
RED DEER	253	145	129	165	148	587	586	413	360	374	409	0	399
SLAVE LAKE	652	545	348	530	250	436	318	464	760	496	777	399	0

Table 3-1. Distances in kilometres between cities in Alberta, Canada.

However, this does not mean that the table does not contain information about direction. Distance-like data of the sort provided by Table 3-1 can be used as input to a form of factor analysis called multidimensional scaling (MDS) (Romney, Shepard, & Nerlove, 1972; Shepard, Romney, & Nerlove, 1972). This statistical analysis converts the table of distances into a map-like representation of objects that would produce the set of distances in the table. Dawson et al. (2000) performed such an analysis on the Table 3-1 data and obtained the map that is given in Figure 3-10. This map makes the relative spatial locations of the cities obvious; it could be used to simply "read off" compass directions between pairs of places.

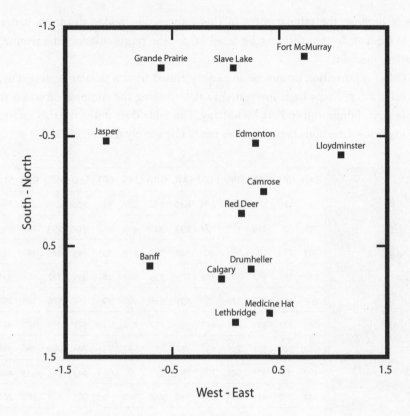

Figure 3-10. Results of applying MDS to Table 3-1.

"Reading off" information from a representation intuitively means accessing this information easily—by using a small number of primitive operations. If this is not possible, then information might be still be accessed by applying a larger number of operations, but this will take more time. The ease of accessing information is a result of the relationship between structure and process.

The structure-process relationship, producing second-order effects, underscores the value of using relative complexity evidence, a notion that was introduced in Chapter 2. Imagine that a physical symbol system uses a tabular representation of distances. Then we would expect it to compute functions involving distance very quickly, but it would be much slower to answer questions about direction. In contrast, if the device uses a map-like representation, then we would expect it to answer questions about direction quickly, but take longer to answer questions about distance (because, for instance, measuring operations would have to be invoked).

In summary, while structures are arbitrary, structure-process relations are not. They produce second-order regularities that can affect such measures as relative complexity evidence. Using such measures to investigate structure-process relations provides key information about a system's algorithms and architecture.

3.10 A Classical Architecture for Cognition

The physical symbol system hypothesis defines classical cognitive science. This school of thought can be thought of as the modern derivative of Cartesian philosophy. It views cognition as computation, where computation is the rule-governed manipulation of symbols. Thus thinking and reasoning are viewed as the result of performing something akin to logical or mathematical inference. A great deal of this computational apparatus must be innate.

However, classical cognitive science crucially departs from Cartesian philosophy by abandoning dualism. Classical cognitive science instead adopts a materialist position that mechanizes the mind. The technical notion of computation is the application of a finite set of recursive rules to a finite set of primitives to evolve a set of finite symbolic structures or expressions. This technical definition of computation is beyond the capabilities of some devices, such as finite state automata, but can be accomplished by universal machines such as Turing machines or electronic computers. The claim that cognition is the product of a device that belongs to the same class of artifacts such as Turing machines or digital computers is the essence of the physical symbol system hypothesis, and the foundation of classical cognitive science.

Since the invention of the digital computer, scholars have seriously considered the possibility that the brain was also a computer of this type. For instance, the all-or-none nature of a neuron's action potential has suggested that the brain is also digital in nature (von Neumann, 1958). However, von Neumann went on to claim that the small size and slow speed of neurons, in comparison to electronic components, suggested that the brain would have a different architecture than an electronic computer. For instance, von Neumann speculated that the brain's architecture would be far more parallel in nature.

Von Neumann's (1958) speculations raise another key issue. While classical cognitive scientists are confident that brains belong to the same class as Turing machines and digital computers (i.e., all are physical symbol systems), they do not expect the brain to have the same architecture. If the brain is a physical symbol system, then what might its architecture be like?

Many classical cognitive scientists believe that the architecture of cognition is some kind of production system. The model of production system architecture was invented by Newell and Simon (Newell, 1973; Newell & Simon, 1961, 1972) and has been used to simulate many psychological phenomena (Anderson, 1983; Anderson et al., 2004; Anderson & Matessa, 1997; Meyer et al. 2001; Meyer & Kieras, 1997a, 1997b; Newell, 1990; Newell & Simon, 1972). Production systems have a number of interesting properties, including an interesting mix of parallel and serial processing.

A production system is a general-purpose symbol manipulator (Anderson, 1983; Newell, 1973; Newell & Simon, 1972). Like other physical symbol systems, production systems exhibit a marked distinction between symbolic expressions and the rules for manipulating them. They include a working memory that is used to store one or more symbolic structures, where a symbolic structure is an expression that is created by combining a set of atomic symbols. In some production systems (e.g., Anderson, 1983) a long-term memory, which also stores expressions, is present as well. The working memory of a production system is analogous to the ticker tape of a Turing machine or to the random access memory of a von Neumann computer.

The process component of a production system is a finite set of symbol-manipulating rules that are called productions. Each production is a single rule that pairs a triggering condition with a resulting action. A production works by scanning the expressions in working memory for a pattern that matches its condition. If such a match is found, then the production takes control of the memory and performs its action. A production's action is some sort of symbol manipulation—adding, deleting, copying, or moving symbols or expressions in the working memory.

A typical production system is a parallel processor in the sense that all of its productions search working memory simultaneously for their triggering patterns. However, it is a serial processor—like a Turing machine or a digital computer—when actions are performed to manipulate the expressions in working memory. This is because in most production systems only one production is allowed to operate on memory at any given time. That is, when one production finds its triggering condition, it takes control for a moment, disabling all of the other productions. The controlling production manipulates the symbols in memory, and then releases its control, which causes the parallel scan of working memory to recommence.

We have briefly described two characteristics, structure and process, that make production systems examples of physical symbol systems. The third characteristic, control, reveals some additional interesting properties of production systems.

On the one hand, stigmergy is used to control a production system, that is, to choose which production acts at any given time. Stigmergic control occurs when different agents (in this case, productions) do not directly communicate with each other, but conduct indirect communication by modifying a shared environment (Theraulaz & Bonabeau, 1999). Stigmergy has been used to explain how a colony of social insects might coordinate their actions to create a nest (Downing & Jeanne, 1988; Karsai, 1999). The changing structure of the nest elicits different nest-building behaviours; the nest itself controls its own construction. When one insect adds a new piece to the nest, this will change the later behaviour of other insects without any direct communication occurring.

Production system control is stigmergic if the working memory is viewed as being analogous to the insect nest. The current state of the memory causes a particular production to act. This changes the contents of the memory, which in turn can result in a different production being selected during the next cycle of the architecture.

On the other hand, production system control is usually not completely stigmergic. This is because the stigmergic relationship between working memory and productions is loose enough to produce situations in which conflicts occur. Examples of this type of situation include instances in which more than one production finds its triggering pattern at the same time, or when one production finds its triggering condition present at more than one location in memory at the same time. Such situations must be dealt with by additional control mechanisms. For instance, priorities might be assigned to productions so that in a case where two or more productions were in conflict, only the production with the highest priority would perform its action.

Production systems have provided an architecture—particularly if that architecture is classical in nature—that has been so successful at simulating higher-order cognition that some researchers believe that production systems provide the foundation for a unified theory of cognition (Anderson, 1983; Anderson et al., 2004; Newell, 1990). Production systems illustrate another feature that is also typical of this approach to cognitive science: the so-called classical sandwich (Hurley, 2001).

Imagine a very simple agent that was truly incapable of representation and reasoning. Its interactions with the world would necessarily be governed by a set of reflexes that would convert sensed information directly into action. These reflexes define a sense-act cycle (Pfeifer & Scheier, 1999).

In contrast, a more sophisticated agent could use internal representations to decide upon an action, by reasoning about the consequences of possible actions and choosing the action that was reasoned to be most beneficial (Popper, 1978, p. 354): "While an uncritical animal may be eliminated altogether with its dogmatically held hypotheses, we may *formulate* our hypotheses, and criticize them. Let our conjectures, our theories die in our stead!" In this second scenario, thinking stands as an intermediary between sensation and action. Such behaviour is not governed by a sense-act cycle, but is instead the product of a sense-think-act cycle (Pfeifer & Scheier, 1999).

Hurley (2001) has argued that the sense-think-act cycle is the stereotypical form of a theory in classical cognitive science; she called this form the classical sandwich. In a typical classical theory, perception can only indirectly inform action, by sending information to be processed by the central representational processes, which in turn decide which action is to be performed.

Production systems exemplify the classical sandwich. The first production systems did not incorporate sensing or acting, in spite of a recognized need to do so. "One problem with psychology's attempt at cognitive theory has been our persistence in thinking about cognition without bringing in perceptual and motor processes" (Newell, 1990, p. 15). This was also true of the next generation of production systems, the adaptive control of thought (ACT) architecture (Anderson, 1983). ACT "historically was focused on higher level cognition and not perception or action" (Anderson et al., 2004, p. 1038).

More modern production systems, such as EPIC (executive-process interactive control) (Meyer & Kieras, 1997a, 1997b), have evolved to include sensing and acting. EPIC simulates the performance of multiple tasks and can produce the psychological refractory period (PRP). When two tasks can be performed at the same time, the stimulus onset asynchrony (SOA) between the tasks is the length of time from the start of the first task to the start of the second task. When SOAs are long, the time taken by a subject to make a response is roughly the same for both tasks. However, for SOAs of half a second or less, it takes a longer time to perform the second task than it does to perform the first. This increase in response time for short SOAs is the PRP.

EPIC is an advanced production system. One of its key properties is that productions in EPIC can act in parallel. That is, at any time cycle in EPIC processing, *all* productions that have matched their conditions in working memory will act to alter working memory. This is important; when multiple tasks are modelled there will be two different sets of productions in action, one for each task. EPIC also includes sensory processors (such as virtual eyes) and motor processors, because actions can constrain task performance. For example, EPIC uses a single motor processor to control two "virtual hands." This results in interference between two tasks that involve making responses with different hands.

While EPIC (Meyer & Kieras, 1997a, 1997b) explicitly incorporates sensing, acting, and thinking, it does so in a fashion that still exemplifies the classical sandwich. In EPIC, sensing transduces properties of the external world into symbols to be added to working memory. Working memory provides symbolic expressions that guide the actions of motor processors. Thus working memory centralizes the "thinking" that maps sensations onto actions. There are no direct connections between sensing and acting that bypass working memory. EPIC is an example of sense-think-act processing.

Radical embodied cognitive science, which is discussed in Chapter 5, argues that intelligence is the result of situated action; it claims that sense-think-act processing can be replaced by sense-act cycles, and that the rule-governed manipulation of expressions is unnecessary (Chemero, 2009). In contrast, classical researchers claim that production systems that include sensing and acting are sufficient to explain human intelligence and action, and that embodied theories are not necessary (Vera & Simon, 1993).

> It follows that there is no need, contrary to what followers of SA [situated action] seem sometimes to claim, for cognitive psychology to adopt a whole new language

and research agenda, breaking completely from traditional (symbolic) cognitive theories. SA is not a new approach to cognition, much less a new school of cognitive psychology. (Vera & Simon, 1993, p. 46)

We see later in this book that production systems provide an interesting medium that can be used to explore the relationship between classical, connectionist, and embodied cognitive science.

3.11 Weak Equivalence and the Turing Test

There are two fundamentals that follow from accepting the physical symbol system hypothesis (Newell, 1980; Newell & Simon, 1976). First, general human intelligence is the product of rule-governed symbol manipulation. Second, because they are universal machines, any particular physical symbol system can be configured to simulate the behaviour of another physical symbol system.

A consequence of these fundamentals is that digital computers, which are one type of physical symbol system, can simulate another putative member of the same class, human cognition (Newell & Simon, 1961, 1972; Simon, 1969). More than fifty years ago it was predicted "that within ten years most theories in psychology will take the form of computer programs, or of qualitative statements about the characteristics of computer programs" (Simon & Newell, 1958, pp. 7–8). One possible measure of cognitive science's success is that a leading critic of artificial intelligence has conceded that this particular prediction has been partially fulfilled (Dreyfus, 1992).

There are a number of advantages to using computer simulations to study cognition (Dawson, 2004; Lewandowsky, 1993). The difficulties in converting a theory into a working simulation can identify assumptions that the theory hides. The formal nature of a computer program provides new tools for studying simulated concepts (e.g., proofs of convergence). Programming a theory forces a researcher to provide rigorous definitions of the theory's components. "Programming is, again like any form of writing, more often than not experimental. One programs, just as one writes, not because one understands, but in order to come to understand." (Weizenbaum, 1976, p. 108).

However, computer simulation research provides great challenges as well. Chief among these is validating the model, particularly because one universal machine can simulate any other. A common criticism of simulation research is that it is possible to model anything, because modelling is unconstrained:

> Just as we may wonder how much the characters in a novel are drawn from real life and how much is artifice, we might ask the same of a model: How much is based on observation and measurement of accessible phenomena, how much is

based on informed judgment, and how much is convenience? (Oreskes, Shrader-Frechette, & Belitz, 1994, p. 644)

Because of similar concerns, mathematical psychologists have argued that computer simulations are impossible to validate in the same way as mathematical models of behaviour (Estes, 1975; Luce, 1989, 1999). Evolutionary biologist John Maynard Smith called simulation research "fact free science" (Mackenzie, 2002).

Computer simulation researchers are generally puzzled by such criticisms, because their simulations of cognitive phenomena must conform to a variety of challenging constraints (Newell, 1980, 1990; Pylyshyn, 1984). For instance, Newell's (1980, 1990) production system models aim to meet a number of constraints that range from behavioural (flexible responses to environment, goal-oriented, operate in real time) to biological (realizable as a neural system, develop via embryological growth processes, arise through evolution).

In validating a computer simulation, classical cognitive science becomes an intrinsically comparative discipline. Model validation requires that theoretical analyses and empirical observations are used to evaluate both the relationship between a simulation and the subject being simulated. In adopting the physical symbol system hypothesis, classical cognitive scientists are further committed to the assumption that this relation is complex, because it can be established (as argued in Chapter 2) at many different levels (Dawson, 1998; Marr, 1982; Pylyshyn, 1984). Pylyshyn has argued that model validation can take advantage of this and proceed by imposing severe empirical constraints. These empirical constraints involve establishing that a model provides an appropriate account of its subject at the computational, algorithmic, and architectural levels of analysis. Let us examine this position in more detail.

First, consider a relationship between model and subject that is not listed above—a relationship at the implementational level of analysis. Classical cognitive science's use of computer simulation methodology is a tacit assumption that the physical structure of its models does not need to match the physical structure of the subject being modelled.

The basis for this assumption is the multiple realization argument that we have already encountered. Cognitive scientists describe basic information processes in terms of their functional nature and ignore their underlying physicality. This is because the same function can be realized in radically different physical media. For instance, AND-gates can be created using hydraulic channels, electronic components, or neural circuits (Hillis, 1998). If hardware or technology were relevant—if the multiple realization argument was false—then computer simulations of cognition would be absurd. Classical cognitive science ignores the physical when models are validated. Let us now turn to the relationships between models and subjects that classical cognitive science cannot and does not ignore.

In the most abstract sense, both a model and a modelled agent can be viewed as opaque devices, black boxes whose inner workings are invisible. From this perspective, both are machines that convert inputs or stimuli into outputs or responses; their behaviour computes an input-output function (Ashby, 1956, 1960). Thus the most basic point of contact between a model and its subject is that the input-output mappings produced by one must be identical to those produced by the other. Establishing this fact is establishing a relationship between model and subject at the computational level.

To say that a model and subject are computing the same input-output function is to say that they are weakly equivalent. It is a weak equivalence because it is established by ignoring the internal workings of both model and subject. There are an infinite number of different algorithms for computing the same input-output function (Johnson-Laird, 1983). This means that weak equivalence can be established between two different systems that use completely different algorithms. Weak equivalence is not concerned with the possibility that two systems can produce the right behaviours but do so for the wrong reasons.

Weak equivalence is also sometimes known as Turing equivalence. This is because weak equivalence is at the heart of a criterion proposed by computer pioneer Alan Turing, to determine whether a computer program had achieved intelligence (Turing, 1950). This criterion is called the Turing test.

Turing (1950) believed that a device's ability to participate in a meaningful conversation was the strongest test of its general intelligence. His test involved a human judge conducting, via teletype, a conversation with an agent. In one instance, the agent was another human. In another, the agent was a computer program. Turing argued that if the judge could not correctly determine which agent was human then the computer program must be deemed to be intelligent. A similiar logic was subscribed to by Descartes (2006). Turing and Descartes both believed in the power of language to reveal intelligence; however, Turing believed that machines could attain linguistic power, while Descartes did not.

A famous example of the application of the Turing test is provided by a model of paranoid schizophrenia, PARRY (Kosslyn, Ball, & Reiser, 1978). This program interacted with a user by carrying on a conversation—it was a natural language communication program much like the earlier ELIZA program (Weizenbaum, 1966). However, in addition to processing the structure of input sentences, PARRY also computed variables related to paranoia: fear, anger, and mistrust. PARRY's responses were thus affected not only by the user's input, by also by its evolving affective states. PARRY's contributions to a conversation became more paranoid as the interaction was extended over time.

A version of the Turing test was used to evaluate PARRY's performance (Colby et al., 1972). Psychiatrists used teletypes to interview PARRY as well as human paranoids. Forty practising psychiatrists read transcripts of these interviews in order to distinguish the human paranoids from the simulated ones. They were only able to do this at chance levels. PARRY had passed the Turing test: "We can conclude that psychiatrists using teletyped data do not distinguish real patients from our simulation of a paranoid patient" (p. 220).

The problem with the Turing test, though, is that in some respects it is too easy to pass. This was one of the points of the pioneering conversation-making program, ELIZA (Weizenbaum, 1966), which was developed to engage in natural language conversations. Its most famous version, DOCTOR, modelled the conversational style of an interview with a humanistic psychotherapist. ELIZA's conversations were extremely compelling. "ELIZA created the most remarkable illusion of having understood the minds of the many people who conversed with it" (Weizenbaum, 1976, p. 189). Weizenbaum was intrigued by the fact that "some subjects have been very hard to convince that ELIZA is not human. This is a striking form of Turing's test" (Weizenbaum, 1966, p. 42).

However, ELIZA's conversations were not the product of natural language understanding. It merely parsed incoming sentences, and then put fragments of these sentences into templates that were output as responses. Templates were ranked on the basis of keywords that ELIZA was programmed to seek during a conversation; this permitted ELIZA to generate responses rated as being highly appropriate. "A large part of whatever elegance may be credited to ELIZA lies in the fact that ELIZA maintains the illusion of understanding with so little machinery" (Weizenbaum, 1966, p. 43).

Indeed, much of the apparent intelligence of ELIZA is a contribution of the human participant in the conversation, who assumes that ELIZA understands its inputs and that even strange comments made by ELIZA are made for an intelligent reason.

> The 'sense' and the continuity the person conversing with ELIZA perceives is supplied largely by the person himself. He assigns meanings and interpretations to what ELIZA 'says' that confirm his initial hypothesis that the system does understand, just as he might do with what a fortune-teller says to him.
> (Weizenbaum, 1976, p. 190)

Weizenbaum believed that natural language understanding was beyond the capability of computers, and also believed that ELIZA illustrated this belief. However, ELIZA was received in a fashion that Weizenbaum did not anticipate, and which was opposite to his intent. He was so dismayed that he wrote a book that served as a scathing critique of artificial intelligence research (Weizenbaum, 1976, p. 2): "My own shock was administered not by any important political figure in establishing his philosophy of science, but by some people who insisted on misinterpreting a piece of work I had done."

The ease with which ELIZA was misinterpreted—that is, the ease with which it passed a striking form of Turing's test—caused Weizenbaum (1976) to question most research on the computer simulation of intelligence. Much of Weizenbaum's concern was rooted in AI's adoption of Turing's (1950) test as a measure of intelligence.

> An entirely too simplistic notion of intelligence has dominated both popular and scientific thought, and this notion is, in part, responsible for permitting artificial intelligence's perverse grand fantasy to grow. (Weizenbaum, 1976, p. 203)

However, perhaps a more reasoned response would be to adopt a stricter means of evaluating cognitive simulations. While the Turing test has had more than fifty years of extreme influence, researchers are aware of its limitations and have proposed a number of ways to make it more sensitive (French, 2000).

For instance, the Total Turing Test (French, 2000) removes the teletype and requires that a simulation of cognition be not only conversationally indistinguishable from a human, but also physically indistinguishable. Only a humanoid robot could pass such a test, and only do so by not only speaking but also behaving (in very great detail) in ways indistinguishable from a human. A fictional version of the Total Turing Test is the Voight-Kampff scale described in Dick's (1968) novel *Do Androids Dream of Electric Sheep?* This scale used behavioural measures of empathy, including pupil dilation, to distinguish humans from androids.

3.12 Towards Strong Equivalence

The Turing test has had a long, influential history (French, 2000). However, many would agree that it is flawed, perhaps because it is too easily passed. As a consequence, some have argued that artificial intelligence research is very limited (Weizenbaum, 1976). Others have argued for more stringent versions of the Turing test, such as the Total Turing Test.

Classical cognitive science recognizes that the Turing test provides a necessary, but not a sufficient, measure of a model's validity. This is because it really only establishes weak equivalence, by collecting evidence that two systems are computationally equivalent. It accomplishes this by only examining the two devices at the level of the input-output relationship. This can only establish weak equivalence, because systems that use very different algorithms and architectures can still compute the same function.

Classical cognitive science has the goal of going beyond weak equivalence. It attempts to do so by establishing additional relationships between models and subjects, identities between both algorithms and architectures. This is an attempt to establish what is known as strong equivalence (Pylyshyn, 1984). Two systems are said to be strongly equivalent if they compute the same input-output function (i.e., if they are weakly equivalent), accomplish this with the same algorithm, and bring this algorithm to life with the same architecture. Cognitive scientists are in the business of making observations that establish the strong equivalence of their models to human thinkers.

Classical cognitive science collects these observations by measuring particular behaviours that are unintended consequences of information processing, and which can therefore reveal the nature of the algorithm that is being employed. Newell and Simon (1972) named these behaviours second-order effects; in Chapter 2 these behaviours were called artifacts, to distinguish them from the primary or intended responses of an information processor. In Chapter 2, I discussed three general classes of evidence related to artifactual behaviour: intermediate state evidence, relative complexity evidence, and error evidence.

Note that although similar in spirit, the use of these three different types of evidence to determine the relationship between the algorithms used by model and subject is not the same as something like the Total Turing Test. Classical cognitive science does not require physical correspondence between model and subject. However, algorithmic correspondences established by examining behavioural artifacts put much stronger constraints on theory validation than simply looking for stimulus-response correspondences. To illustrate this, let us consider some examples of how intermediate state evidence, relative complexity evidence, and error evidence can be used to validate models.

One important source of information that can be used to validate a model is intermediate state evidence (Pylyshyn, 1984). Intermediate state evidence involves determining the intermediate steps that a symbol manipulator takes to solve a problem, and then collecting evidence to determine whether a modelled subject goes through the same intermediate steps. Intermediate state evidence is notoriously difficult to collect, because human information processors are black boxes—we cannot directly observe internal cognitive processing. However, clever experimental paradigms can be developed to permit intermediate states to be inferred.

A famous example of evaluating a model using intermediate state evidence is found in some classic and pioneering research on human problem solving (Newell & Simon, 1972). Newell and Simon collected data from human subjects as they solved problems; their method of data collection is known as protocol analysis (Ericsson & Simon, 1984). In protocol analysis, subjects are trained to think out loud as they work. A recording of what is said by the subject becomes the primary data of interest.

The logic of collecting verbal protocols is that the thought processes involved in active problem solving are likely to be stored in a person's short-term memory (STM), or working memory. Cognitive psychologists have established that items stored in such a memory are stored as an articulatory code that permits verbalization to maintain the items in memory (Baddeley, 1986, 1990; Conrad, 1964a, 1964b; Waugh & Norman, 1965). As a result, asking subjects to verbalize their thinking steps is presumed to provide accurate access to current cognitive processing, and to do so with minimal disruption. "Verbalization will not interfere with ongoing processes if the information stored in STM is encoded orally, so that an articulatory code can readily be activated" " (Ericsson & Simon, 1984, p. 68).

In order to study problem solving, Newell and Simon (1972) collected verbal protocols for problems that were difficult enough to engage subjects and generate interesting behaviour, but simple enough to be solved. For instance, when a subject was asked to decode the cryptarithmetic problem DONALD + GERALD = ROBERT after being told that D = 5, they solved the problem in twenty minutes and produced a protocol that was 2,186 words in length.

The next step in the study was to create a problem behaviour graph from a subject's protocol. A problem behaviour graph is a network of linked nodes. Each node represents a state of knowledge. For instance, in the cryptarithmetic problem such a state might be the observation that "R is odd." A horizontal link from a node to a node on its right represents the application of an operation that changed the state of knowledge. An example operation might be "Find a column that contains a letter of interest and process that column." A vertical link from a node to a node below represents backtracking. In many instances, a subject would reach a dead end in a line of thought and return to a previous state of knowledge in order to explore a different approach. The 2,186-word protocol produced a problem behaviour graph that consisted of 238 different nodes.

The initial node in a problem behaviour graph represents a subject's starting state of knowledge when given a problem. A node near the end of the problem behaviour graph represents the state of knowledge when a solution has been achieved. All of the other nodes represent intermediate states of knowledge. Furthermore, in Newell and Simon's (1972) research, these intermediate states represent very detailed elements of knowledge about the problem as it is being solved.

The goal of the simulation component of Newell and Simon's (1972) research was to create a computer model that would generate its own problem behaviour graph. The model was intended to produce a very detailed mimicry of the subject's behaviour—it was validated by examining the degree to which the simulation's problem behaviour graph matched the graph created for the subject. The meticulous nature of such intermediate state evidence provided additional confidence for the use of verbal protocols as scientific data. "For the more information conveyed in their responses, the more difficult it becomes to construct a model that will produce precisely those responses adventitiously—hence the more confidence we can place in a model that does predict them" (Ericsson & Simon, 1984, p. 7).

Newell and Simon (1972) created a computer simulation by examining a subject's problem behaviour graph, identifying the basic processes that it revealed in its links between nodes, and coding each of these processes as a production in a production system. Their model developed from the protocol for the DONALD + GERALD = ROBERT problem consisted of only 14 productions. The behaviour of this fairly small program was able to account for 75 to 80 percent of the human subject's problem behaviour graph. "All of this analysis shows how a verbal thinking-aloud protocol can be used as the raw material for generating and testing a theory of problem solving behavior" (Newell & Simon, 1972, p. 227).

The contribution of Newell and Simon's (1972) research to classical cognitive science is impossible to overstate. One of their central contributions was to demonstrate that human problem solving could be characterized as searching through a problem space. A problem space consists of a set of knowledge states—starting state, one or more goal states, and a potentially large number of intermediate states—that each represent current knowledge about a problem. A link between two knowledge states shows how the application of a single rule can transform the first state into the second. A problem behaviour graph is an example of a problem space. Searching the problem space involves finding a route—a sequence of operations—that will transform the initial state into a goal state. From this perspective, problem solving becomes the domain of control: finding as efficiently as possible an acceptable sequence of problem-solving operations. An enormous number of different search strategies exist (Knuth, 1997; Nilsson, 1980); establishing the strong equivalence of a problem-solving model requires collecting evidence (e.g., using protocol analysis) to ensure that the same search or control strategy is used by both model and agent.

A second kind of evidence that is used to investigate the validity of a model is relative complexity evidence (Pylyshyn, 1984). Relative complexity evidence generally involves examining the relative difficulty of problems, to see whether the problems that are hard (or easy) for a model are the same problems that are hard (or easy) for a modelled subject. The most common kind of relative complexity evidence collected by cognitive scientists is response latency (Luce, 1986; Posner, 1978). It is assumed that the time taken for a system to generate a response is an artifactual behaviour that can reveal properties of an underlying algorithm and be used to examine the algorithmic relationship between model and subject.

One domain in which measures of response latency have played an important role is the study of visual cognition (Kosslyn & Osherson, 1995; Pinker, 1985). Visual cognition involves solving information processing problems that involve spatial relationships or the spatial layout of information. It is a rich domain of study because it seems to involve qualitatively different kinds of information processing: the data-driven or preattentive detection of visual features (Marr, 1976;

Richards, 1988; Treisman, 1985), top-down or high-level cognition to link combinations of visual features to semantic interpretations or labels (Jackendoff, 1983, 1987; Treisman, 1986, 1988), and processing involving visual attention or visual routines that include both data-driven and top-down characteristics, and which serve as an intermediary between feature detection and object recognition (Cooper & Shepard, 1973a, 1973b; Ullman, 1984; Wright, 1998).

Visual search tasks are frequently used to study visual cognition. In such a task, a subject is usually presented with a visual display consisting of a number of objects. In the odd-man-out version of this task, in one half of the trials one of the objects (the target) is different from all of the other objects (the distracters). In the other half of the trials, the only objects present are distracters. Subjects have to decide as quickly and accurately as possible whether a target is present in each display. The dependent measures in such tasks are search latency functions, which represent the time required to detect the presence or absence of a target as a function of the total number of display elements.

Pioneering work on visual search discovered the so-called pop-out effect: the time required to detect the presence of a target that is characterized by one of a small number of unique features (e.g., colour, orientation, contrast, motion) is largely independent of the number of distractor elements in a display, producing a search latency function that is essentially flat (Treisman & Gelade, 1980). This is because, regardless of the number of elements in the display, when the target is present it seems to pop out of the display, bringing itself immediately to attention. Notice how the target pops out of the display illustrated in Figure 3-11.

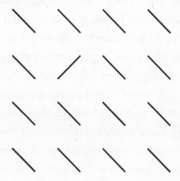

Figure 3-11. Unique features pop out of displays, regardless of display size.

In contrast, the time to detect a target defined by a unique combination of features generally increases with the number of distractor items, producing search latency functions with positive slopes. Figure 3-12 illustrates visual search in objects that are either connected or unconnected (Dawson & Thibodeau, 1998); connectedness is a property that is not local, but is only defined by relations between multiple features (Minsky & Papert, 1988). The larger the number of display items, the longer it takes to find the target when it is present in the display. Is there a target in Figure 3-12? If so, is it harder to find than the one that was present in Figure 3-11?

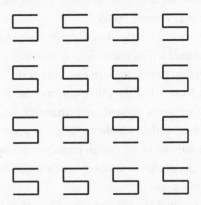

Figure 3-12. Unique combinations of features do not pop out.

Search latency results as those described above, which revealed that some objects pop out but others do not, formed the basis for feature integration theory (Treisman, 1985, 1986, 1988; Treisman & Gelade, 1980; Treisman & Gormican, 1988; Treisman, Sykes, & Gelade, 1977). Feature integration theory is a multistage account of visual cognition. In the first state, preattentive processors register the locations of a small set of primitive visual features on independent feature maps. These maps represent a small number of properties (e.g., orientation, colour, contrast movement) that also appear to be transduced by early neural visual detectors (Livingstone & Hubel, 1988). If such a feature is unique to a display, then it will be the only active location in its feature map. This permits pop out to occur, because the location of the unique, primitive feature is preattentively available.

Unique combinations of features do not produce unique activity in a single feature map and therefore cannot pop out. Instead, they require additional processing in order to be detected. First, attentional resources must be used to bring the various independent feature maps into register with respect to a master map of locations. This master map of locations will indicate what combinations of features coexist at each location in the map. Second, a "spotlight" of attention is used to scan the master map of locations in search of a unique object. Because this attentional spotlight can only process a portion of the master map at any given time, and because it must be scanned from location to location on the master map, it takes longer for unique combinations of features to be found. Furthermore, the search of the master map will become longer and longer as more of its locations are filled, explaining why the latency to detect unique feature combinations is affected by the number of distractors present.

Relative complexity evidence can also be used to explore some of the components of feature integration theory. For example, several researchers have proposed models of the how the attentional spotlight is shifted to detect targets in a visual search task (Fukushima, 1986; Gerrissen, 1991; Grossberg, 1980; Koch & Ullman, 1985; LaBerge, Carter, & Brown, 1992; Sandon, 1992). While the specific details of these models differ, their general structure is quite similar. First, these models represent the display being searched as an array of processors whose activities encode the visual distinctiveness of the location that each processor represents (i.e., how different it is in appearance relative to its neighbours). Second, these processors engage in a winner-take-all (WTA) competition (Feldman & Ballard, 1982) to identify the most distinctive location. This competition is defined by lateral inhibition: each processor uses its activity as an inhibitory signal in an attempt to reduce the activity of its neighbours. Third, the display element at the winning location is examined to see whether or not it is the target. If it is, the search stops. If it is not, activity at this location either decays or is inhibited (Klein, 1988), and a new WTA competition is used to find the next most distinctive location in the display.

This type of model provides a straightforward account of search latency functions obtained for targets defined by unique conjunctions of features. They also lead to a unique prediction: if inhibitory processes are responsible for directing the shift of the attentional spotlight, then search latency functions should be affected by the overall adapting luminance of the display. This is because there is a greater degree of inhibition during the processing of bright visual displays than there is for dimmer displays (Barlow, Fitzhugh, & Kuffler, 1957; Derrington & Lennie, 1982; Ransom-Hogg & Spillmann, 1980; Rohaly & Buchsbaum, 1989).

A visual search study was conducted to test this prediction (Dawson & Thibodeau, 1998). Modifying a paradigm used to study the effect of adaptive luminance on motion perception (Dawson & Di Lollo, 1990), Dawson and Thibodeau (1998) had subjects perform a visual search task while viewing the displays through neutral density filters that modified display luminance while not affecting the relative contrast of elements. There were two major findings that supported the kinds of models of attentional shift described above. First, when targets pop out, the response latency of subjects was not affected by adaptive luminance. This is consistent with feature integration theory, in the sense that a shifting attentional spotlight is not required for pop out to occur. Second, for targets that did not pop out, search latency functions were affected by the level of adaptive luminance. For darker displays, both the intercept and the slope of the search latency functions increased significantly. This is consistent with the hypothesis that this manipulation interferes with the inhibitory processes that guide shifts of attention.

A third approach to validating a model involves the use of error evidence. This approach assumes that errors are artifacts, in the sense that they are a natural consequence of an agent's information processing, and that they are not a deliberate or intended product of this processing.

One source of artifactual errors is the way information processing can be constrained by limits on internal resources (memory or attention) or by external demands (the need for real time responses). These restrictions on processing produce bounded rationality (Simon, 1982). Another reason for artifactual errors lies in the restrictions imposed by the particular structure-process pairing employed by an information processor. "A tool too gains its power from the fact that it permits certain actions and not others. For example, a hammer has to be rigid. It can therefore not be used as a rope" (Weizenbaum, 1976, p. 37). Like a tool, a particular structure-process pairing may not be suited for some tasks and therefore produces errors when faced with them.

One example of the importance of error evidence is found in the large literature on human, animal, and robot navigation (Cheng, 2005; Cheng & Newcombe, 2005; Healy, 1998; Jonsson, 2002; Milford, 2008). How do organisms find their place in the world? One approach to answering this question is to set up small, manageable indoor environments. These "arenas" can provide a variety of cues to animals that learn to navigate within them. If an agent is reinforced for visiting a particular location, what cues does it use to return to this place?

One paradigm for addressing this question is the reorientation task invented by Ken Cheng (1986). In the reorientation task, an agent is typically placed within a rectangular arena. Reinforcement is typically provided at one of the corner locations in the arena. That is, the agent is free to explore the arena, and eventually finds a reward at a location of interest—it learns that this is the "goal location." The agent is then removed from the arena, disoriented, and returned to an (often different) arena, with the task of using the available cues to relocate the goal. Of particular interest are experimental conditions in which the arena has been altered from the one in which the agent was originally trained.

An arena that is used in the reorientation task can provide two different kinds of navigational information: geometric cues and feature cues (Cheng & Newcombe, 2005). Geometric cues are relational, while feature cues are not.

> A geometric property of a surface, line, or point is a property it possesses by virtue of its position relative to other surfaces, lines, and points within the same space. A non-geometric property is any property that cannot be described by relative position alone. (Gallistel, 1990, p. 212)

In a rectangular arena, metric properties (e.g., wall lengths, angles between walls) combined with an agent's distinction between left and right (e.g., the long wall is to the left of the short wall) provide geometric cues. Non-geometric cues or feature cues can be added as well. For instance, one arena wall can have a different colour than the others (Cheng, 1986), or different coloured patterns can be placed at each corner of the arena (Kelly, Spetch, & Heth, 1998).

One question of interest concerns the relative contributions of these different cues for reorientation. This is studied by seeing how the agent reorients after it has been returned to an arena in which cues have been altered. For example, the feature cues might have been moved to new locations. This places feature cues in conflict with geometric cues. Will the agent move to a location defined by geometric information, or will it move to a different location indicated by feature information? Extensive use of the reorientation task has revealed some striking regularities.

Some of the most interesting regularities found in the reorientation task pertain to a particular error in reorientation. In an arena with no unique feature cues (no unique wall colour, no unique pattern at each corner), geometric cues are the only information available for reorienting. However, geometric cues cannot uniquely specify a goal location in a rectangular arena. This is because the geometric cues at the goal location (e.g., 90° angle, shorter wall to the left and longer wall to the right) are identical to the geometric cues present at the diagonally opposite corner (often called the rotational location). Under these conditions, the agent will produce rotational error (Cheng, 1986, 2005). When rotational error occurs, the trained agent goes to the goal location at above-chance levels; however, the animal goes to the rotational location equally often. Rotational error is usually taken as evidence that the agent is relying upon the geometric properties of the environment.

When feature cues are present in a rectangular arena, a goal location can be uniquely specified. In fact, when cues are present, an agent should not even need to pay attention to geometric cues, because these cues are not relevant. However, evidence suggests that geometric cues still influence behaviour even when such cues are not required to solve the task.

First, in some cases subjects continue to make some rotational errors even when feature cues specify the goal location (Cheng, 1986; Hermer & Spelke, 1994). Second, when feature cues present during training are removed from the arena in which reorientation occurs, subjects typically revert to generating rotational error (Kelly, Spetch, and Heth, 1998; Sovrano, Bisazza, & Vallortigara, 2003). Third, in studies in which local features are moved to new locations in the new arena, there is a conflict between geometric and feature cues. In this case, reorientation appears to be affected by both types of cues. The animals will not only increase their tendency to visit the corner marked by the feature cues that previously signaled the goal, but also produce rotational error for two other locations in the arena (Brown, Spetch, & Hurd, 2007; Kelly, Spetch, and Heth, 1998).

Rotational error is an important phenomenon in the reorientation literature, and it is affected by a complex interaction between geometric and feature cues. A growing variety of models of reorientation are appearing in the literature, including models consistent with the symbol-manipulating fundamental of classical cognitive science (Cheng, 1986; Gallistel, 1990), neural network models that are part of connectionist cognitive science (Dawson et al., 2010), and behaviour-based robots that are the domain of embodied cognitive science (Dawson, Dupuis, & Wilson, 2010; Nolfi, 2002). All of these models have two things in common. First, they can produce rotational error and many of its nuances. Second, this error is produced as a natural byproduct of a reorientation algorithm; the errors produced by the models are used in aid of their validation.

3.13 The Impenetrable Architecture

Classical cognitive scientists often develop theories in the form of working computer simulations. These models are validated by collecting evidence that shows they are strongly equivalent to the subjects or phenomena being modelled. This begins by first demonstrating weak equivalence, that both model and subject are computing the same input-output function. The quest for strong equivalence is furthered by using intermediate state evidence, relative complexity evidence, and error evidence to demonstrate, in striking detail, that both model and subject are employing the same algorithm.

However, strong equivalence can only be established by demonstrating an additional relationship between model and subject. Not only must model and subject be employing the same algorithm, but both must also be employing the same primitive processes. Strong equivalence requires architectural equivalence.

The primitives of a computer simulation are readily identifiable. A computer simulation should be a collection of primitives that are designed to generate a behaviour of interest (Dawson, 2004). In order to create a model of cognition, one must define the basic properties of a symbolic structure, the nature of the processes that can manipulate these expressions, and the control system that chooses when to apply a particular rule, operation, or process. A model makes these primitive characteristics explicit. When the model is run, its behaviour shows what these primitives can produce.

While identifying a model's primitives should be straightforward, determining the architecture employed by a modelled subject is far from easy. To illustrate this, let us consider research on mental imagery.

Mental imagery is a cognitive phenomenon in which we experience or imagine mental pictures. Mental imagery is often involved in solving spatial problems (Kosslyn, 1980). For instance, imagine being asked how many windows there are on the front wall of the building in which you live. A common approach to answering this question would be to imagine the image of this wall and to inspect the image, mentally counting the number of windows that are displayed in it. Mental imagery is also crucially important for human memory (Paivio, 1969, 1971, 1986; Yates, 1966): we are better at remembering items if we can create a mental image of them. Indeed, the construction of bizarre mental images, or of images that link two or more items together, is a standard tool of the mnemonic trade (Lorayne, 1985, 1998, 2007; Lorayne & Lucas, 1974).

An early achievement of the cognitive revolution in psychology (Miller, 2003; Vauclair & Perret, 2003) was a rekindled interest in studying mental imagery, an area that had been neglected during the reign of behaviourism (Paivio, 1971, 1986). In the early stages of renewed imagery research, traditional paradigms were modified to solidly establish that concept imageability was a key predictor of verbal behaviour and associative learning (Paivio, 1969). In later stages, new paradigms were invented to permit researchers to investigate the underlying nature of mental images (Kosslyn, 1980; Shepard & Cooper, 1982).

For example, consider the relative complexity evidence obtained using the mental rotation task (Cooper & Shepard, 1973a, 1973b; Shepard & Metzler, 1971). In this task, subjects are presented with a pair of images. In some instances, the two images are of the same object. In other instances, the two images are different (e.g., one is a mirror image of the other). The orientation of the images can also be varied—for instance, they can be rotated to different degrees in the plane of view. The angular disparity between the two images is the key independent variable. A subject's task is to judge whether the images are the same or not; the key dependent measure is the amount of time required to respond.

In order to perform the mental rotation task, subjects first construct a mental image of one of the objects, and then imagine rotating it to the correct orientation to enable them to judge whether it is the same as the other object. The standard finding in this task is that there is a linear relationship between response latency and the amount of mental rotation that is required. From these results it has been concluded that "the process of mental rotation is an analog one in that intermediate states in the process have a one-to-one correspondence with intermediate stages in the external rotation of an object" (Shepard & Cooper, 1982, p. 185). That is, mental processes rotate mental images in a holistic fashion, through intermediate orientations, just as physical processes can rotate real objects.

Another source of relative complexity evidence concerning mental imagery is the image scanning task (Kosslyn, 1980; Kosslyn, Ball, & Reisler, 1978). In the most famous version of this task, subjects are first trained to create an accurate mental image of an island map on which seven different locations are marked. Then subjects are asked to construct this mental image, focusing their attention at one of the locations. They are then provided with a name, which may or may not be one of the other map locations. If the name is of another map location, then subjects are instructed to scan across the image to it, pressing a button when they arrive at the second location.

In the map-scanning version of the image-scanning task, the dependent variable was the amount of time from the naming of the second location to a subject's button press, and the independent variable was the distance on the map between the first and second locations. The key finding was that there was nearly a perfectly linear relationship between latency and distance (Kosslyn Ball, & Reisler, 1978): an increased distance led to an increased response latency, suggesting that the image had spatial extent, and that it was scanned at a constant rate.

> The scanning experiments support the claim that portions of images depict corresponding portions of the represented objects, and that the spatial relations between portions of the image index the spatial relations between the corresponding portions of the imaged objects. (Kosslyn, 1980, p. 51)

The relative complexity evidence obtained from tasks like mental rotation and image scanning provided the basis for a prominent account of mental imagery known as the depictive theory (Kosslyn, 1980, 1994; Kosslyn, Thompson, & Ganis, 2006). This theory is based on the claim that mental images are not merely internal representations that *describe* visuospatial information (as would be the case with words or with logical propositions), but instead *depict* this information because the format of an image is quasi-pictorial. That is, while a mental image is not claimed to literally be a picture in the head, it nevertheless represents content by resemblance.

> There is a correspondence between parts and spatial relations of the representation and those of the object; this structural mapping, which confers a type of resemblance, underlies the way images convey specific content. In this respect images are like pictures. Unlike words and symbols, depictions are not arbitrarily paired with what they represent. (Kosslyn, Thompson, & Ganis, 2006, p. 44)

The depictive theory specifies primitive properties of mental images, which have sometimes been called privileged properties (Kosslyn, 1980). What are these primitives? One is that images occur in a spatial medium that is functionally equivalent to a coordinate space. A second is that images are patterns that are produced by activating local regions of this space to produce an "abstract spatial isomorphism" (Kosslyn, 1980, p. 33) between the image and what it represents. This isomorphism is a correspondence between an image and a represented object in terms of their parts as well as spatial relations amongst these parts. A third is that images not only depict spatial extent, they also depict properties of visible surfaces such as colour and texture.

These privileged properties are characteristic of the format mental images—the *structure* of images as symbolic expressions. When such a structure is paired with particular primitive *processes*, certain types of questions are easily answered. These processes are visual in nature: for instance, mental images can be scanned, inspected at different apparent sizes, or rotated. The coupling of such processes with the depictive structure of images is well-suited to solving visuospatial problems. Other structure-process pairings—in particular, logical operations on propositional expressions that describe spatial properties (Pylyshyn, 1973)—do not make spatial information explicit and arguably will not be as adept at solving visuospatial problems. Kosslyn (1980, p. 35) called the structural properties of images privileged because their possession "[distinguishes] an image from other forms of representation."

That the depictive theory makes claims about the primitive properties of mental images indicates quite clearly that it is an account of cognitive architecture. That it is a theory about architecture is further supported by the fact that the latest phase of imagery research has involved the supplementing behavioural data with evidence concerning the cognitive neuroscience of imagery (Kosslyn, 1994; Kosslyn et al., 1995; Kosslyn et al., 1999; Kosslyn, Thompson, & Alpert, 1997; Kosslyn, Thompson, & Ganis, 2006). This research has attempted to ground the architectural properties of images into topographically organized regions of the cortex.

Computer simulation has proven to be a key medium for evaluating the depictive theory of mental imagery. Beginning with work in the late 1970s (Kosslyn & Shwartz, 1977), the privileged properties of mental images have been converted into a working computer model (Kosslyn, 1980, 1987, 1994; Kosslyn et al., 1984; Kosslyn et al., 1985). In general terms, over time these models represent an elaboration of a general theoretical structure: long-term memory uses propositional structures to store spatial information. Image construction processes convert this propositional information into depictive representations on a spatial medium that enforces the primitive structural properties of images. Separate from this medium are primitive processes that operate on the depicted information (e.g., scan, inspect, interpret). This form of model has shown that the privileged properties of images that define the depictive theory are sufficient for simulating a wide variety of the regularities that govern mental imagery.

The last few paragraphs have introduced Kosslyn's (e.g., 1980) depictive theory, its proposals about the privileged properties of mental images, and the success that computer simulations derived from this theory have had at modelling behavioural results. All of these topics concern statements about primitives in the domain of a theory or model about mental imagery. Let us now turn to one issue that has not yet been addressed: the nature of the primitives employed by the modelled subject, the human imager.

The status of privileged properties espoused by the depictive theory has been the subject of a decades-long imagery debate (Block, 1981; Tye, 1991). At the heart of the imagery debate is a basic question: are the privileged properties parts of the architecture or not? The imagery debate began with the publication of a seminal paper (Pylyshyn, 1973), which proposed that the primitive properties of images were not depictive, but were instead descriptive properties based on a logical or propositional representation. This position represents the basic claim of the propositional theory, which stands as a critical alternative to the depictive theory. The imagery debate continues to the present day; propositional theory's criticism of the depictive position has been prolific and influential (Pylyshyn, 1981a, 1981b, 1984, 2003a, 2003b, 2003c, 2007).

The imagery debate has been contentious, has involved a number of different subtle theoretical arguments about the relationship between theory and data, and has shown no signs of being clearly resolved. Indeed, some have argued that it is a debate that is cannot be resolved, because it is impossible to identify data that is appropriate to differentiate the depictive and propositional theories (Anderson, 1978). In this section, the overall status of the imagery debate is not of concern. We are instead interested in a particular type of evidence that has played an important role in the debate: evidence concerning cognitive penetrability (Pylyshyn, 1980, 1984, 1999).

Recall from the earlier discussion of algorithms and architecture that Newell (1990) proposed that the rate of change of various parts of a physical symbol system would differ radically depending upon which component was being examined. Newell observed that data should change rapidly, stored programs should be more enduring, and the architecture that interprets stored programs should be even more stable. This is because the architecture is wired in. It may change slowly (e.g., in human cognition because of biological development), but it should be the most stable information processing component. When someone claims that they have changed their mind, we interpret this as meaning that they have updated their facts, or that they have used a new approach or strategy to arrive at a conclusion. We don't interpret this as a claim that they have altered their basic mental machinery—when we change our mind, we don't change our cognitive architecture!

The cognitive penetrability criterion (Pylyshyn, 1980, 1984, 1999) is an experimental paradigm that takes advantage of the persistent "wired in" nature of the architecture. If some function is part of the architecture, then it should not be affected by changes in cognitive content—changing beliefs should not result in a changing architecture. The architecture is cognitively impenetrable. In contrast, if some function changes because of a change in content that is semantically related to the function, then this is evidence that it is not part of the architecture.

> If a system is cognitively penetrable then the function it computes is sensitive,
> in a semantically coherent way, to the organism's goals and beliefs, that is, it can
> be altered in a way that bears some logical relation to what the person knows.
> (Pylyshyn, 1999, p. 343)

The architecture is not cognitively penetrable.

Cognitive penetrability provides a paradigm for testing whether a function of interest is part of the architecture or not. First, some function is measured as part of a pre-test. For example, consider Figure 3-13, which presents the Müller-Lyer illusion, which was discovered in 1889 (Gregory, 1978). In a pre-test, it would be determined whether you experience this illusion. Some measurement would be made to determine whether you judge the horizontal line segment of the top arrow to be longer than the horizontal line segment of the bottom arrow.

Second, a strong manipulation of a belief related to the function that produces the Müller-Lyer illusion would be performed. You, as a subject, might be told that the two horizontal line segments were equal in length. You might be given a ruler, and asked to measure the two line segments, in order to convince yourself that your experience was incorrect and that the two lines were of the same length.

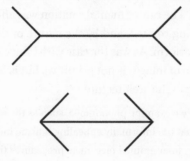

Figure 3-13. The Müller-Lyer illusion.

Third, a post-test would determine whether you still experienced the illusion. Do the line segments still appear to be of different length, even though you are armed with the knowledge that this appearance is false? This illusion has had such a long history because its appearance is not affected by such cognitive content. The mechanism that is responsible for the Müller-Lyer illusion is cognitively impenetrable.

This paradigm has been applied to some of the standard mental imagery tasks in order to show that some of the privileged properties of images are cognitively penetrable and therefore cannot be part of the architecture. For instance, in his 1981 dissertation, Liam Bannon examined the map scanning task for cognitive penetrability (for methodological details, see Pylyshyn, 1981a). Bannon reasoned that the instructions given to subjects in the standard map scanning study (Kosslyn, Ball, & Reiser, 1978) instilled a belief that image scanning was like scanning a picture. Bannon was able to replicate the Kosslyn, Ball, & Reiser results in one condition. However, in other conditions the instructions were changed so that the images had to be scanned to answer a question, but no beliefs about scanning were instilled. In one study, Bannon had subjects shift attention from the first map location to the second (named) location, and then judge the compass direction from the second location to the first. In this condition, the linearly increasing relationship between distance and time disappeared. Image scanning appears to be cognitively penetrable, challenging some of the architectural claims of depictive theory. "Images can be examined without the putative constraints of the surface display postulated by Kosslyn and others" (Pylyshyn, 1981a, p. 40).

The cognitive penetrability paradigm has also been applied to the mental rotation task (Pylyshyn, 1979b). Pylyshyn reasoned that if mental rotation is accomplished by primitive mechanisms, then it must be cognitively impenetrable. One prediction that follows from this reasoning is that the rate of mental rotation should be independent of the content being rotated—an image depicting simple content should, by virtue of its putative architectural nature, be rotated at the same rate as a different image depicting more complex content.

Pylyshyn (1979b) tested this hypothesis in two experiments and found evidence of cognitive penetration. The rate of mental rotation was affected by practice, by the content of the image being rotated, and by the nature of the comparison task that subjects were asked to perform. As was the case with image scanning, it would seem that the "analog" rotation of images is not primitive, but is instead based on simpler processes that do belong to the architecture.

> The more carefully we examine phenomena, such as the mental rotation findings, the more we find that the informally appealing holistic image-manipulation views must be replaced by finer grained piecemeal procedures that operate upon an analyzed and structured stimulus using largely serial, resource-limited mechanisms. (Pylyshyn, 1979b, p. 27)

Cognitive penetrability has played an important role in domains other than mental imagery. For instance, in the literature concerned with social perception and prediction, there is debate between a classical theory called theory-theory (Gopnik & Meltzoff, 1997; Gopnik & Wellman, 1992) and a newer approach called simulation theory (Gordon, 1986, 2005b), which is nicely situated in the embodied cognitive science that is the topic of Chapter 5. There is a growing discussion about whether cognitive penetrability can be used to discriminate between these two theories (Greenwood, 1999; Heal, 1996; Kuhberger et al., 2006; Perner et al., 1999; Stich & Nichols, 1997). Cognitive penetrability has also been applied to various topics in visual perception (Raftopoulos, 2001), including face perception (Bentin & Golland, 2002) and the perception of illusory motion (Dawson, 1991; Dawson & Wright, 1989; Wright & Dawson, 1994).

While cognitive penetrability is an important tool when faced with the challenge of examining the architectural equivalence between model and subject, it is not without its problems. For instance, in spite of it being applied to the study of mental imagery, the imager debate rages on, suggesting that penetrability evidence is not as compelling or powerful as its proponents might hope. Perhaps one reason for this is that it seeks a null result—the absence of an effect of cognitive content on cognitive function. While cognitive penetrability can provide architectural evidence for strong equivalence, other sources of evidence are likely required. One source of such additional evidence is cognitive neuroscience.

3.14 Modularity of Mind

Classical cognitive science assumes that cognition is computation, and endorses the physical symbol system hypothesis. As a result, it merges two theoretical positions that in the seventeenth century were thought to be in conflict. The first is Cartesian rationalism, the notion that the products of thought were rational conclusions drawn from the rule-governed manipulation of pre-existing ideas. The second is anti-Cartesian materialism, the notion that the processes of thought are carried out by physical mechanisms.

The merging of rationalism and materialism has resulted in the modification of a third idea, innateness, which is central to both Cartesian philosophy and classical cognitive science. According to Descartes, the contents of some mental states were innate, and served as mental axioms that permitted the derivation of new content (Descartes, 1996, 2006). Variations of this claim can be found in classical cognitive science (Fodor, 1975). However, it is much more typical for classical cognitive science to claim innateness for the mechanisms that manipulate content, instead of claiming it for the content itself. According to classical cognitive science, it is the architecture that is innate.

Innateness is but one property that can serve to constrain theories about the nature of the architecture (Newell, 1990). It is a powerful assumption that leads to particular predictions. If the architecture is innate, then it should be universal (i.e., shared by all humans), and it should develop in a systematic pattern that can be linked to biological development. These implications have guided a tremendous amount of research in linguistics over the last several decades (Jackendoff, 2002). However, innateness is but one constraint, and many radically different architectural proposals might all be consistent with it. What other constraints might be applied to narrow the field of potential architectures?

Another constraining property is modularity (Fodor, 1983). Modularity is the claim that an information processor is not just one homogeneous system used to handle every information processing problem, but is instead a collection of special-purpose processors, each of which is especially suited to deal with a narrower range of more specific problems. Modularity offers a general solution to what is known as the packing problem (Ballard, 1986).

The packing problem is concerned with maximizing the computational power of a physical device with limited resources, such as a brain with a finite number of neurons and synapses. How does one pack the maximal computing power into a finite brain? Ballard (1986) argued that many different subsystems, each designed to deal with a limited range of computations, will be easier to fit into a finite package than will be a single general-purpose device that serves the same purpose as all of the subsystems.

Of course, in order to enable a resource-limited system to solve the same class of problems as a universal machine, a compromise solution to the packing problem may be required. This is exactly the stance adopted by Fodor in his influential 1983 monograph *The Modularity of Mind*. Fodor imagined an information processor that used general central processing, which he called isotropic processes, operating on representations delivered by a set of special-purpose input systems that are now known as modules.

> If, therefore, we are to start with anything like Turing machines as models in cognitive psychology, we must think of them as embedded in a matrix of subsidiary systems which affect their computations in ways that are responsive to the flow of environmental events. The function of these subsidiary systems is to provide the central machine with information about the world. (Fodor, 1983, p. 39)

According to Fodor (1983), a module is a neural substrate that is specialized for solving a particular information processing problem. It takes input from transducers, preprocesses this input in a particular way (e.g., computing three-dimensional structure from transduced motion signals [Hildreth, 1983; Ullman, 1979]), and passes the result of this preprocessing on to central processes. Because modules are specialized processors, they are domain specific. Because the task of modules is to inform central processing about the dynamic world, modules operate in a fast, mandatory fashion. In order for modules to be fast, domain-specific, and mandatory devices, they will be "wired in," meaning that a module will be associated with fixed neural architecture. A further consequence of this is that a module will exhibit characteristic breakdown patterns when its specialized neural circuitry fails. All of these properties entail that a module will exhibit informational encapsulation: it will be unaffected by other models or by higher-level results of isotropic processes. In other words, modules are cognitively impenetrable (Pylyshyn, 1984). Clearly any function that can be shown to be modular in Fodor's sense must be a component of the architecture.

Fodor (1983) argued that modules should exist for all perceptual modalities, and that there should also be modular processing for language. There is a great deal of evidence in support of this position.

For example, consider visual perception. Evidence from anatomy, physiology, and clinical neuroscience has led many researchers to suggest that there exist two distinct pathways in the human visual system (Livingstone & Hubel, 1988; Maunsell & Newsome, 1987; Ungerleider & Mishkin, 1982). One is specialised for processing visual form, i.e., detecting an object's appearance: the "what pathway." The other is specialised for processing visual motion, i.e., detecting an object's changing location: the "where pathway." This evidence suggests that object appearance and object motion are processed by distinct modules. Furthermore, these modules are likely hierarchical, comprising systems of smaller modules. More than 30 distinct visual processing modules, each responsible for processing a very specific kind of information, have been identified (van Essen, Anderson, & Felleman, 1992).

A similar case can be made for the modularity of language. Indeed, the first biological evidence for the localization of brain function was Paul Broca's presentation of the aphasic patient Tan's brain to the Paris Société d'Anthropologie in 1861 (Gross, 1998). This patient had profound agrammatism; his brain exhibited clear abnormalities in a region of the frontal lobe now known as Broca's area. The Chomskyan tradition in linguistics has long argued for the distinct biological existence of a language faculty (Chomsky, 1957, 1965, 1966). The hierarchical nature of this faculty—the notion that it is a system of independent submodules—has been a fruitful avenue of research (Garfield, 1987); the biological nature of this system, and theories about how it evolved, are receiving considerable contemporary attention (Fitch, Hauser, & Chomsky, 2005; Hauser, Chomsky, & Fitch, 2002). Current accounts of neural processing of auditory signals suggest that there are two pathways analogous to the what-where streams in vision, although the distinction between the two is more complex because both are sensitive to speech (Rauschecker & Scott, 2009).

From both Fodor's (1983) definition of modularity and the vision and language examples briefly mentioned above, it is clear that neuroscience is a key source of evidence about modularity. "The intimate association of modular systems with neural hardwiring is pretty much what you would expect given the assumption that the key to modularity is informational encapsulation" (p. 98). This is why modularity is an important complement to architectural equivalence: it is supported by seeking data from cognitive neuroscience that complements the cognitive penetrability criterion.

The relation between modular processing and evidence from cognitive neuroscience leads us to a controversy that has arisen from Fodor's (1983) version of modularity. We have listed a number of properties that Fodor argues are true of modules. However, Fodor also argues that these same properties cannot be true of central or isotropic processing. Isotropic processes are not informationally encapsulated, domain specific, fast, mandatory, associated with fixed neural architecture, or cognitively impenetrable. Fodor proceeds to conclude that because isotropic processes do not have these properties, cognitive science will not be able to explain them.

I should like to propose a generalization; one which I fondly hope will someday come to be known as 'Fodor's First Law of the Nonexistence of Cognitive Science.' It goes like this: the more global (e.g., the more isotropic) a cognitive process is, the less anybody understands it. (Fodor, 1983, p. 107)

Fodor's (1983) position that explanations of isotropic processes are impossible poses a strong challenge to a different field of study, called evolutionary psychology (Barkow, Cosmides, & Tooby, 1992), which is controversial in its own right (Stanovich, 2004). Evolutionary psychology attempts to explain how psychological processes arose via evolution. This requires the assumption that these processes provide some survival advantage and are associated with a biological substrate, so that they are subject to natural selection. However, many of the processes of particular interest to evolutionary psychologists involve reasoning, and so would be classified by Fodor as being isotropic. If they are isotropic, and if Fodor's first law of the nonexistence of cognitive science is true, then evolutionary psychology is not possible.

Evolutionary psychologists have responded to this situation by proposing the massive modularity hypothesis (Carruthers, 2006; Pinker, 1994, 1997), an alternative to Fodor (1983). According to the massive modularity hypothesis, most cognitive processes—including high-level reasoning—are modular. For instance, Pinker (1994, p. 420) has proposed that modular processing underlies intuitive mechanics, intuitive biology, intuitive psychology, and the self-concept. The mind is "a collection of instincts adapted for solving evolutionarily significant problems—the mind as a Swiss Army knife" (p. 420). The massive modularity hypothesis proposes to eliminate isotropic processing from cognition, spawning modern discussions about how modules should be defined and about what kinds of processing are modular or not (Barrett & Kurzban, 2006; Bennett, 1990; Fodor, 2000; Samuels, 1998).

The modern debate about massive modularity indicates that the concept of *module* is firmly entrenched in cognitive science. The issue in the debate is not the existence of modularity, but is rather modularity's extent. With this in mind, let us return to the methodological issue at hand, investigating the nature of the architecture. To briefly introduce the types of evidence that can be employed to support claims about modularity, let us consider another topic made controversial by proponents of massive modularity: the modularity of musical cognition.

As we have seen, massive modularity theorists see a pervasive degree of specialization and localization in the cognitive architecture. However, one content area that these theorists have resisted to classify as modular is musical cognition. One reason for this is that evolutionary psychologists are hard pressed to explain how music benefits survival. "As far as biological cause and effect are concerned, music is useless. It shows no signs of design for attaining a goal such as long life, grandchildren, or accurate perception and prediction of the world" (Pinker, 1997, p. 528). As a result, musical processing is instead portrayed as a tangential, nonmodular function that is inconsequentially related to other modular processes. "Music is auditory cheesecake, an exquisite confection crafted to tickle the sensitive spots of at least six of our mental faculties" (p. 534).

Not surprisingly, researchers interested in studying music have reacted strongly against this position. There is currently a growing literature that provides support for the notion that musical processing—in particular the perception of rhythm and of tonal profile—is indeed modular (Alossa & Castelli, 2009; Peretz, 2009; Peretz & Coltheart, 2003; Peretz & Hyde, 2003; Peretz & Zatorre, 2003, 2005). The types of evidence reported in this literature are good examples of the ways in which cognitive neuroscience can defend claims about modularity.

One class of evidence concerns dissociations that are observed in patients who have had some type of brain injury. In a dissociation, an injury to one region of the brain disrupts one kind of processing but leaves another unaffected, suggesting that the two kinds of processing are separate and are associated with different brain areas. Those who do not believe in the modularity of music tend to see music as being strongly related to language. However, musical processing and language processing have been shown to be dissociated. Vascular damage to the left hemisphere of the Russian composer Shebalin produced severe language deficits but did not affect his ability to continue composing some of his best works (Luria, Tsvetkova, & Futer, 1965). Reciprocal evidence indicates that there is in fact a double dissociation between language and music: bilateral damage to the brain of another patient produced severe problems in music memory and perception but did not affect her language (Peretz et al., 1994).

Another class of evidence is to seek dissociations involving music that are related to congenital brain disorders. Musical savants demonstrate such a dissociation: they exhibit low general intelligence but at the same time demonstrate exceptional musical abilities (Miller, 1989; Pring, Woolf, & Tadic, 2008). Again, the dissociation is double. Approximately 4 percent of the population is tone deaf, suffering from what is called congenital amusia (Ayotte, Peretz, & Hyde, 2002; Peretz et al., 2002). Congenital amusics are musically impaired, but they are of normal intelligence and have normal language abilities. For instance, they have normal spatial abilities (Tillmann et al., 2010), and while they have short-term memory problems for musical stimuli, they have normal short-term memory for verbal materials (Tillmann, Schulze, & Foxton, 2009). Finally, there is evidence that congenital amusia is genetically inherited, which would be a plausible consequence of the modularity of musical processing (Peretz, Cummings, & Dube, 2007).

A third class of evidence that cognitive neuroscience can provide about modularity comes from a variety of techniques that noninvasively measure regional brain activity as information processing occurs (Cabeza & Kingstone, 2006; Gazzaniga, 2000). Brain imaging data can be used to seek dissociations and attempt to localize function. For instance, by seeing which regions of the brain are active during musical processing but not active when a nonmusical control task is performed, a researcher can attempt to associate musical functions with particular areas of the brain.

Brain imaging techniques have been employed by cognitive neuroscientists interested in studying musical processing (Peretz & Zatorre, 2003). Surprisingly, given the other extensive evidence concerning the dissociation of music, this kind of evidence has not provided as compelling a case for the localization of musical processing in the human brain (Warren, 2008). Instead, it appears to reveal that musical processing invokes activity in many different areas throughout the brain (Schuppert et al., 2000). "The evidence of brain imaging studies has demonstrated that music shares basic brain circuitry with other types of complex sound, and no single brain area can be regarded as exclusively dedicated to music" (Warren, 2008, p. 34). This is perhaps to be expected, under the assumption that "musical cognition" is itself a fairly broad notion, and that it is likely accomplished by a variety of subprocesses, many of which are plausibly modular. Advances in imaging studies of musical cognition may require considering finer distinctions between musical and nonmusical processing, such as studying the areas of the brain involved with singing versus those involved with speech (Peretz, 2009).

Disparities between behavioural evidence concerning dissociations and evidence from brain imaging studies do not necessarily bring the issue of modularity into question. These disparities might simply reveal the complicated relationship between the functional and the implementational nature of an architectural component. For instance, imagine that the cognitive architecture is indeed a production system. An individual production, functionally speaking, is ultra-modular. However, it is possible to create systems in which the modular functions of different productions do not map onto localized physical components, but are instead defined as a constellation of physical properties distributed over many components (Dawson et al., 2000). We consider this issue in a later chapter where the relationship between production systems and connectionist networks is investigated in more detail.

Nevertheless, the importance of using evidence from neuroscience to support claims about modularity cannot be understated. In the absence of such evidence, arguments that some function is modular can be easily undermined.

For instance, Gallistel (1990) has argued that the processing of geometric cues by animals facing the reorientation task is modular in Fodor's (1983) sense. This is because the processing of geometric cues is mandatory (as evidenced by the pervasiveness of rotational error) and not influenced by "information about surfaces other than their relative positions" (Gallistel, 1990, p. 208). However, a variety of theories that are explicitly nonmodular are capable of generating appropriate rotational error in a variety of conditions (Dawson, Dupuis, & Wilson, 2010; Dawson et al., 2010;

Miller, 2009; Miller & Shettleworth, 2007, 2008; Nolfi, 2002). As a result, the modularity of geometric cue processing is being seriously re-evaluated (Cheng, 2008).

In summary, many researchers agree that the architecture of cognition is modular. A variety of different kinds of evidence can be marshaled to support the claim that some function is modular and therefore part of the architecture. This evidence is different from, and can complement, evidence about cognitive penetrability. Establishing the nature of the architecture is nonetheless challenging and requires combining varieties of evidence from behavioural and cognitive neuroscientific studies.

3.15 Reverse Engineering

Methodologically speaking, what is classical cognitive science? The goal of classical cognitive science is to explain an agent's cognitive abilities. Given an intact, fully functioning cognitive agent, the classical cognitive scientist must construct a theory of the agent's internal processes. The working hypothesis is that this theory will take the form of a physical symbol system. Fleshing this hypothesis out will involve proposing a theory, and hopefully a working computer simulation, that will make explicit proposals about the agent's symbol structures, primitive processes, and system of control.

Given this scenario, a classical cognitive scientist will almost inevitably engage in some form of reverse engineering.

> In reverse engineering, one figures out what a machine was designed to do. Reverse engineering is what the boffins at Sony do when a new product is announced by Panasonic, or vice versa. They buy one, bring it back to the lab, take a screwdriver to it, and try to figure out what all the parts are for and how they combine to make the device work. (Pinker, 1997, p. 21)

The reverse engineering conducted by a classical cognitive science is complicated by the fact that one can't simply take cognitive agents apart with a screwdriver to learn about their design. However, the assumption that the agent is a physical symbol system provides solid guidance and an effective methodology.

The methodology employed by classical cognitive science is called functional analysis (Cummins, 1975, 1983). Functional analysis is a top-down form of reverse engineering that maps nicely onto the multiple levels of investigation that were introduced in Chapter 2.

Functional analysis begins by choosing and defining a function of interest to explain. Defining a function of interest entails an investigation at the computational level. What problem is being solved? Why do we say this problem is being solved and not some other? What constraining properties can be assumed to aid the solution to the problem? For instance, we saw earlier that a computational theory of language learning (identifying a grammar in the limit) might be used to motivate possible properties that must be true of a language or a language learner.

The next step in a functional analysis is to decompose the function of interest into a set of subcomponents that has three key properties. First, each subcomponent is defined functionally, not physically. Second, each subcomponent is simpler than the original function. Third, the organization of the subcomponents—the flow of information from one component to another—is capable of producing the input-output behaviour of the original function of interest. "Functional analysis consists in analyzing a disposition into a number of less problematic dispositions such that the programmed manifestation of these analyzing dispositions amounts to a manifestation of the analyzed disposition" (Cummins, 1983, p. 28). These properties permit the functional analysis to proceed in such a way that Ryle's regress will be avoided, and that eventually the homunculi produced by the analysis (i.e., the functional subcomponents) can be discharged, as was discussed in Chapter 2.

The analytic stage of a functional analysis belongs to the algorithmic level of analysis. This is because the organized system of subfunctions produced at this stage is identical to a program or algorithm for producing the overall input-output behaviour of the agent. However, the internal cognitive processes employed by the agent cannot be directly observed. What methods can be used to carve up the agent's behaviour into an organized set of functions? In other words, how can observations of behaviour support decisions about functional decomposition?

The answer to this question reveals why the analytic stage belongs to the algorithmic level of analysis. It is because the empirical methods of cognitive psychology are designed to motivate and validate functional decompositions.

For example, consider the invention that has become known as the modal model of memory (Baddeley, 1986), which was one of the triumphs of cognitivism in the 1960s (Shiffrin & Atkinson, 1969; Waugh & Norman, 1965). According to this model, to-be-remembered information is initially kept in primary memory, which has a small capacity and short duration, and codes items acoustically. Without additional processing, items will quickly decay from primary memory. However, maintenance rehearsal, in which an item from memory is spoken aloud and thus fed back to the memory in renewed form, will prevent this decay. With additional processing like maintenance rehearsal, some of the items in primary memory pass into secondary memory, which has large capacity and long duration, and employs a semantic code.

The modal memory model was inspired and supported by experimental data. In a standard free-recall experiment, subjects are asked to remember the items from a presented list (Glanzer & Cunitz, 1966; Postman & Phillips, 1965). The first few items presented are better remembered than the items presented in the middle— the primacy effect. Also, the last few items presented are better remembered than the middle items—the recency effect. Further experiments demonstrated a functional dissociation between the primacy and recency effects: variables that influenced one effect left the other unaffected. For example, introducing a delay before subjects recalled the list eliminated the recency effect but not the primacy effect (Glanzer & Cunitz, 1966). If a list was presented very quickly, or was constructed from low-frequency words, the primacy effect—but not the recency effect—vanished (Glanzer, 1972). To explain such functional dissociation, researchers assumed an organized system of submemories (the modal model), each with different properties.

The analytic stage of a functional analysis is iterative. That is, one can take any of the subfunctions that have resulted from one stage of analysis and decompose it into an organized system of even simpler sub-subfunctions. For instance, as experimental techniques were refined, the 1960s notion of primary memory has been decomposed into an organized set of subfunctions that together produce what is called working memory (Baddeley, 1986, 1990). Working memory is decomposed into three basic subfunctions. The central executive is responsible for operating on symbols stored in buffers, as well as for determining how attention will be allocated across simultaneously ongoing tasks. The visuospatial buffer stores visual information. The phonological loop is used to store verbal (or speech-like) information. The phonological loop has been further decomposed into subfunctions. One is a phonological store that acts as a memory by holding symbols. The other is a rehearsal process that preserves items in the phonological store.

We saw in Chapter 2 that functional decomposition cannot proceed indefinitely if the analysis is to serve as a scientific explanation. Some principles must be applied to stop the decomposition in order to exit Ryle's regress. For Cummins' (1983) functional analysis, this occurs with a final stage—causal subsumption. To causally subsume a function is to explain how physical mechanisms bring the function into being. "A functional analysis is complete when the program specifying it is explicable via instantiation—i.e., when we can show how the program is executed by the system whose capacities are being explained" (p. 35). Cummins called seeking such explanations of functions the subsumption strategy. Clearly the subsumption strategy is part of an architectural level of investigation, employing evidence involving cognitive impenetrability and modularity. It also leans heavily on evidence gathered from an implementational investigation (i.e., neuroscience).

From a methodological perspective, classical cognitive science performs reverse engineering, in the form of functional analysis, to develop a theory (and likely a simulation) of cognitive processing. This enterprise involves both formal and empirical methods as well as the multiple levels of investigation described in Chapter 2. At the same time, classical cognitive science will also be involved in collecting data to establish the strong equivalence between the theory and the agent by establishing links between the two at the different levels of analysis, as we have been discussing in the preceding pages of the current chapter.

3.16 What is Classical Cognitive Science?

The purpose of the current chapter was to introduce the foundations of classical cognitive science—the "flavour" of cognitive science that first emerged in the late 1950s—and the school of thought that still dominates modern cognitive science. The central claim of classical cognitive science is that "cognition is computation." This short slogan has been unpacked in this chapter to reveal a number of philosophical assumptions, which guide a variety of methodological practices.

The claim that cognition is computation, put in its modern form, is identical to the claim that cognition is information processing. Furthermore, classical cognitive science views such information processing in a particular way: it is processing that is identical to that carried out by a physical symbol system, a device like a modern digital computer. As a result, classical cognitive science adopts the representational theory of mind. It assumes that the mind contains internal representations (i.e., symbolic expressions) that are in turn manipulated by rules or processes that are part of a mental logic or a (programming) language of thought. Further to this, a control mechanism must be proposed to explain how the cognitive system chooses what operation to carry out at any given time.

The classical view of cognition can be described as the merging of two distinct traditions. First, many of its core ideas—appeals to rationalism, computation, innateness—are rooted in Cartesian philosophy. Second, it rejects Cartesian dualism by attempting to provide materialist explanations of representational processing. The merging of rationality and materialism is exemplified by the physical symbol system hypothesis. A consequence of this is that the theories of classical cognitive science are frequently presented in the form of working computer simulations.

In Chapter 2, we saw that the basic properties of information processing systems required that they be explained at multiple levels. Not surprisingly, classical cognitive scientists conduct their business at multiple levels of analysis, using formal methods to answer computational questions, using simulation and behavioural methods to answer algorithmic questions, and using a variety of behavioural and biological methods to answer questions about architecture and implementation.

The multidisciplinary nature of classical cognitive science is revealed in its most typical methodology, a version of reverse engineering called functional analysis. We have seen that the different stages of this type of analysis are strongly related to the multiple levels of investigations that were discussed in Chapter 2. The same relationship to these levels is revealed in the comparative nature of classical cognitive science as it attempts to establish the strong equivalence between a model and a modelled agent.

The success of classical cognitive science is revealed by its development of successful, powerful theories and models that have been applied to an incredibly broad range of phenomena, from language to problem solving to perception. This chapter has emphasized some of the foundational ideas of classical cognitive science at the expense of detailing its many empirical successes. Fortunately, a variety of excellent surveys exist to provide a more balanced account of classical cognitive science's practical success (Bechtel, Graham, & Balota, 1998; Bermúdez, 2010; Boden, 2006; Gleitman & Liberman, 1995; Green, 1996; Kosslyn & Osherson, 1995; Lepore & Pylyshyn, 1999; Posner, 1991; Smith & Osherson, 1995; Stillings, 1995; Stillings et al., 1987; Thagard, 1996; Wilson & Keil, 1999).

Nevertheless, classical cognitive science is but one perspective, and it is not without its criticisms and alternatives. Some cognitive scientists have reacted against its avoidance of the implementational (because of multiple realization), its reliance on the structure/process distinction, its hypothesis that cognitive information processing is analogous to that of a digital computer, its requirement of internal representations, and its dependence on the sense-think-act cycle. Chapter 4 turns to the foundations of a different "flavour" of cognitive science that is a reaction against the classical approach: connectionist cognitive science.

Elements of Connectionist Cognitive Science

4.0 Chapter Overview

The previous chapter introduced the elements of classical cognitive science, the school of thought that dominated cognitive science when it arose in the 1950s and which still dominates the discipline today. However, as cognitive science has matured, some researchers have questioned the classical approach. The reason for this is that in the 1950s, the only plausible definition of information processing was that provided by a relatively new invention, the electronic digital computer. Since the 1950s, alternative notions of information processing have arisen, and these new notions have formed the basis for alternative approaches to cognition.

The purpose of the current chapter is to present the core elements of one of these alternatives, connectionist cognitive science. The chapter begins with several sections (4.1 through 4.4) in which are described the core properties of connectionism and of the artificial neural networks that connectionists use to model cognitive phenomena. These elements are presented as a reaction against the foundational assumptions of classical cognitive science. Many of these elements are inspired by issues related to the implementational level of investigation. That is, connectionists aim to develop biologically plausible or neuronally inspired models of information processing.

The chapter then proceeds with an examination of connectionism at the remaining three levels of investigation. The computational level of analysis is the focus of

Sections 4.5 through 4.7. These sections investigate the kinds of tasks that artificial neural networks can accomplish and relate them to those that can be accomplished by the devices that have inspired the classical approach. The general theme of these sections is that artificial neural networks belong to the class of universal machines.

Sections 4.8 through 4.13 focus on the algorithmic level of investigation of connectionist theories. Modern artificial neural networks employ several layers of processing units that create interesting representations which are used to mediate input-output relationships. At the algorithmic level, one must explore the internal structure of these representations in an attempt to inform cognitive theory. These sections illustrate a number of different techniques for this investigation.

Architectural issues are the topics of Sections 4.14 through 4.17. In particular, these sections show that researchers must seek the simplest possible networks for solving tasks of interest, and they point out that some interesting cognitive phenomena can be captured by extremely simple networks.

The chapter ends with an examination of the properties of connectionist cognitive science, contrasting the various topics introduced in the current chapter with those that were explored in Chapter 3 on classical cognitive science.

4.1 Nurture versus Nature

The second chapter of John Locke's (1977) *An Essay Concerning Human Understanding*, originally published in 1706, begins as follows:

> It is an established opinion among some men that there are in the understanding certain innate principles; some primary notions, characters, as it were, stamped upon the mind of man, which the soul receives in its very first being, and brings into the world with it. (Locke, 1977, p. 17)

Locke's most famous work was a reaction against this view; of the "some men" being referred to, the most prominent was Descartes himself (Thilly, 1900).

Locke's *Essay* criticized Cartesian philosophy, questioning its fundamental teachings, its core principles and their necessary implications, and its arguments for innate ideas, not to mention all scholars who maintained the existence of innate ideas (Thilly, 1900). Locke's goal was to replace Cartesian rationalism with empiricism, the view that the source of ideas was experience. Locke (1977) aimed to show "how men, barely by the use of their natural faculties, may attain to all of the knowledge they have without the help of any innate impressions" (p. 17). Locke argued for experience over innateness, for nurture over nature.

The empiricism of Locke and his descendants provided a viable and popular alternative to Cartesian philosophy (Aune, 1970). It was also a primary influence on some of the psychological theories that appeared in the late nineteenth and early twentieth centuries (Warren, 1921). Thus it should be no surprise that empiricism is reflected in a different form of cognitive science, connectionism. Furthermore, just as empiricism challenged most of the key ideas of rationalism, connectionist cognitive science can be seen as challenging many of the elements of classical cognitive science.

Surprisingly, the primary concern of connectionist cognitive science is not classical cognitive science's nativism. It is instead the classical approach's excessive functionalism, due largely to its acceptance of the multiple realization argument. Logic gates, the core element of digital computers, are hardware independent because different physical mechanisms could be used to bring the two-valued logic into being (Hillis, 1998). The notion of a universal machine is an abstract, logical one (Newell, 1980), which is why physical symbol systems, computers, or universal machines can be physically realized using LEGO (Agulló et al., 2003), electric train sets (Stewart, 1994), gears (Swade, 1993), hydraulic valves (Hillis, 1998) or silicon chips (Reid, 2001). Physical constraints on computation do not seem to play an important role in classical cognitive science.

To connectionist cognitive science, the multiple realization argument is flawed because connectionists believe that the information processing responsible for human cognition depends critically on the properties of particular hardware, the brain. The characteristics of the brain place constraints on the kinds of computations that it can perform and on the manner in which they are performed (Bechtel & Abrahamsen, 2002; Churchland, Koch, & Sejnowski, 1990; Churchland & Sejnowski, 1992; Clark, 1989, 1993; Feldman & Ballard, 1982).

Brains have long been viewed as being different kinds of information processors than electronic computers because of differences in componentry (von Neumann, 1958). While electronic computers use a small number of fast components, the brain consists of a large number of very slow components, that is, neurons. As a result, the brain must be a parallel processing device that "will tend to pick up as many logical (or informational) items as possible simultaneously, and process them simultaneously" (von Neumann, 1958, p. 51).

Von Neumann (1958) argued that neural information processing would be far less precise, in terms of decimal point precision, than electronic information processing. However, this low level of neural precision would be complemented by a comparatively high level of reliability, where noise or missing information would have far less effect than it would for electronic computers. Given that the basic architecture of the brain involves many connections amongst many elementary components, and that these connections serve as a memory, the brain's memory capacity should also far exceed that of digital computers.

The differences between electronic and brain-like information processing are at the root of connectionist cognitive science's reaction against classic cognitive science. The classical approach has a long history of grand futuristic predictions that fail to materialize (Dreyfus, 1992, p. 85): "Despite predictions, press releases, films, and warnings, artificial intelligence is a promise and not an accomplished fact." Connectionist cognitive science argues that this pattern of failure is due to the fundamental assumptions of the classical approach that fail to capture the basic principles of human cognition.

Connectionists propose a very different theory of information processing— a potential paradigm shift (Schneider, 1987)—to remedy this situation. Even staunch critics of artificial intelligence research have indicated a certain sympathy with the connectionist view of information processing (Dreyfus & Dreyfus, 1988; Searle, 1992). "The fan club includes the most unlikely collection of people. . . . Almost everyone who is discontent with contemporary cognitive psychology and current 'information processing' models of the mind has rushed to embrace the 'connectionist alternative'" (Fodor & Pylyshyn, 1988, p. 4).

What are the key problems that connectionists see in classical models? Classical models invoke serial processes, which make them far too slow to run on sluggish componentry (Feldman & Ballard, 1982). They involve explicit, local, and digital representations of both rules and symbols, making these models too brittle. "If in a digital system of notations a single pulse is missing, absolute perversion of meaning, i.e., nonsense, may result" (von Neumann, 1958, p. 78). Because of this brittleness, the behaviour of classical models does not degrade gracefully when presented with noisy inputs, and such models are not damage resistant. All of these issues arise from one underlying theme: classical algorithms reflect the kind of information processing carried out by electronic computers, not the kind that characterizes the brain. In short, classical theories are not biologically plausible.

Connectionist cognitive science "offers a radically different conception of the basic processing system of the mind-brain, one inspired by our knowledge of the nervous system" (Bechtel & Abrahamsen, 2002, p. 2). The basic medium of connectionism is a type of model called an artificial neural network, or a parallel distributed processing (PDP) network (McClelland & Rumelhart, 1986; Rumelhart & McClelland, 1986c). Artificial neural networks consist of a number of simple processors that perform basic calculations and communicate the results to other processors by sending signals through weighted connections. The processors operate in parallel, permitting fast computing even when slow componentry is involved. They exploit implicit, distributed, and redundant representations, making these networks not brittle. Because networks are not brittle, their behaviour degrades gracefully when presented with noisy inputs, and such models are damage resistant. These advantages accrue because artificial neural networks are intentionally biologically plausible or neuronally inspired.

Classical cognitive science develops models that are purely symbolic and which can be described as asserting propositions or performing logic. In contrast, connectionist cognitive science develops models that are subsymbolic (Smolensky, 1988) and which can be described as statistical pattern recognizers. Networks use representations (Dawson, 2004; Horgan & Tienson, 1996), but these representations do not have the syntactic structure of those found in classical models (Waskan & Bechtel, 1997). Let us take a moment to describe in a bit more detail the basic properties of artificial neural networks.

An artificial neural network is a computer simulation of a "brain-like" system of interconnected processing units (see Figures 4-1 and 4-5 later in this chapter). In general, such a network can be viewed as a multiple-layer system that generates a desired response to an input stimulus. That is, like the devices described by cybernetics (Ashby, 1956, 1960), an artificial neural network is a machine that computes a mapping between inputs and outputs.

A network's stimulus or input pattern is provided by the environment and is encoded as a pattern of activity (i.e., a vector of numbers) in a set of input units. The response of the system, its output pattern, is represented as a pattern of activity in the network›s output units. In modern connectionism—sometimes called New Connectionism—there will be one or more intervening layers of processors in the network, called hidden units. Hidden units detect higher-order features in the input pattern, allowing the network to make a correct or appropriate response.

The behaviour of a processor in an artificial neural network, which is analogous to a neuron, can be characterized as follows. First, the processor computes the total signal (its net input) being sent to it by other processors in the network. Second, the unit uses an activation function to convert its net input into internal activity (usually a continuous number between 0 and 1) on the basis of this computed signal. Third, the unit converts its internal activity into an output signal, and sends this signal on to other processors. A network uses parallel processing because many, if not all, of its units will perform their operations simultaneously.

The signal sent by one processor to another is a number that is transmitted through a weighted connection, which is analogous to a synapse. The connection serves as a communication channel that amplifies or attenuates signals being sent through it, because these signals are multiplied by the weight associated with the connection. The weight is a number that defines the nature and strength of the connection. For example, inhibitory connections have negative weights, and excitatory connections have positive weights. Strong connections have strong weights (i.e., the absolute value of the weight is large), while weak connections have near-zero weights.

The pattern of connectivity in a PDP network (i.e., the network's entire set of connection weights) defines how signals flow between the processors. As a result, a network's connection weights are analogous to a program in a conventional computer (Smolensky, 1988). However, a network's "program" is not of the same type that defines a classical model. A network's program does not reflect the classical structure/process distinction, because networks do not employ either explicit symbols or rules. Instead, a network's program is a set of causal or associative links from signaling processors to receiving processors. The activity that is produced in the receiving units is literally caused by having an input pattern of activity modulated by an array of connection weights between units. In this sense, connectionist models seem markedly associationist in nature (Bechtel, 1985); they can be comfortably related to the old associationist psychology (Warren, 1921).

Artificial neural networks are not necessarily embodiments of empiricist philosophy. Indeed, the earliest artificial neural networks did not learn from experience; they were nativist in the sense that they had to have their connection weights "hand wired" by a designer (McCulloch & Pitts, 1943). However, their associationist characteristics resulted in a natural tendency for artificial neural networks to become the face of modern empiricism. This is because associationism has always been strongly linked to empiricism; empiricist philosophers invoked various laws of association to explain how complex ideas could be constructed from the knowledge provided by experience (Warren, 1921). By the late 1950s, when computers were being used to bring networks to life, networks were explicitly linked to empiricism (Rosenblatt, 1958). Rosenblatt's artificial neural networks were not hand wired. Instead, they learned from experience to set the values of their connection weights.

What does it mean to say that artificial neural networks are empiricist? A famous passage from Locke (1977, p. 54) highlights two key elements: "Let us then suppose the mind to be, as we say, white paper, void of all characters, without any *idea*, how comes it to be furnished? . . . To this I answer, in one word, from *experience*."

The first element in the above quote is the "white paper," often described as the *tabula rasa*, or the blank slate: the notion of a mind being blank in the absence of experience. Modern connectionist networks can be described as endorsing the notion of the blank slate (Pinker, 2002). This is because prior to learning, the pattern of connections in modern networks has no pre-existing structure. The networks either start literally as blank slates, with all connection weights being equal to zero (Anderson et al., 1977; Eich, 1982; Hinton & Anderson, 1981), or they start with all connection weights being assigned small, randomly selected values (Rumelhart, Hinton, & Williams, 1986a, 1986b).

The second element in Locke's quote is that the source of ideas or knowledge or structure is experience. Connectionist learning rules provide a modern embodiment of this notion. Artificial neural networks are exposed to environmental stimulation—activation of their input units—which results in changes to connection weights. These changes furnish a network's blank slate, resulting in a pattern of connectivity that represents knowledge and implements a particular input-output mapping.

In some systems, called self-organizing networks, experience shapes connectivity via unsupervised learning (Carpenter & Grossberg, 1992; Grossberg, 1980, 1987, 1988; Kohonen, 1977, 1984). When learning is unsupervised, networks are only provided with input patterns. They are not presented with desired outputs that are paired with each input pattern. In unsupervised learning, each presented pattern causes activity in output units; this activity is often further refined by a winner-take-all competition in which one output unit wins the competition to be paired with the current input pattern. Once the output unit is selected via internal network dynamics, its connection weights, and possibly the weights of neighbouring output units, are updated via a learning rule.

Networks whose connection weights are modified via unsupervised learning develop sensitivity to statistical regularities in the inputs and organize their output units to reflect these regularities. For instance, in a famous kind of self-organizing network called a Kohonen network (Kohonen, 1984), output units are arranged in a two-dimensional grid. Unsupervised learning causes the grid to organize itself into a map that reveals the discovered structure of the inputs, where related patterns produce neighbouring activity in the output map. For example, when such networks are presented with musical inputs, they often produce output maps that are organized according to the musical circle of fifths (Griffith & Todd, 1999; Todd & Loy, 1991).

In cognitive science, most networks reported in the literature are not self-organizing and are not structured via unsupervised learning. Instead, they are networks that are instructed to mediate a desired input-output mapping. This is accomplished via supervised learning. In supervised learning, it is assumed that the network has an external teacher. The network is presented with an input pattern and produces a response to it. The teacher compares the response generated by the network to the desired response, usually by calculating the amount of error associated with each output unit. The teacher then provides the error as feedback to the network. A learning rule uses feedback about error to modify weights in such a way that the next time this pattern is presented to the network, the amount of error that it produces will be smaller.

A variety of learning rules, including the delta rule (Rosenblatt, 1958, 1962; Stone, 1986; Widrow, 1962; Widrow & Hoff, 1960) and the generalized delta rule (Rumelhart, Hinton, & Williams, 1986b), are supervised learning rules that work by correcting network errors. (The generalized delta rule is perhaps the most popular learning rule in modern connectionism, and is discussed in more detail in Section 4.9.) This kind of learning involves the repeated presentation of a number of input-output pattern pairs, called a training set. Ideally, with enough presentations of a training set, the amount of error produced to each member of the training set will be negligible, and it can be said that the network has learned the desired input-output mapping. Because these techniques require many presentations of a set of patterns for learning to be completed, they have sometimes been criticized as being examples of "slow learning" (Carpenter, 1989).

Connectionism's empiricist and associationist nature cast it close to the very position that classical cognitivism reacted against: psychological behaviourism (Miller, 2003). Modern classical arguments against connectionist cognitive science (Fodor & Pylyshyn, 1988) cover much of the same ground as arguments against behaviourist and associationist accounts of language (Bever, Fodor, & Garrett, 1968; Chomsky, 1957, 1959a, 1959b, 1965). That is, classical cognitive scientists argue that artificial neural networks, like their associationist cousins, do not have the computational power to capture the kind of regularities modelled with recursive rule systems.

However, these arguments against connectionism are flawed. We see in later sections that computational analyses of artificial neural networks have proven that they too belong to the class "universal machine." As a result, the kinds of input-output mappings that have been realized in artificial neural networks are both vast and diverse. One can find connectionist models in every research domain that has also been explored by classical cognitive scientists. Even critics of connectionism admit that "the study of connectionist machines has led to a number of striking and unanticipated findings; it's surprising how much computing can be done with a uniform network of simple interconnected elements" (Fodor & Pylyshyn, 1988, p. 6).

That connectionist models can produce unanticipated results is a direct result of their empiricist nature. Unlike their classical counterparts, connectionist researchers do not require a fully specified theory of how a task is accomplished before modelling begins (Hillis, 1988). Instead, they can let a learning rule discover how to mediate a desired input-output mapping. Connectionist learning rules serve as powerful methods for developing new algorithms of interest to cognitive science. Hillis (1988, p. 176) has noted that artificial neural networks allow "for the possibility of constructing intelligence without first understanding it."

One problem with connectionist cognitive science is that the algorithms that learning rules discover are extremely difficult to retrieve from a trained network (Dawson, 1998, 2004, 2009; Dawson & Shamanski, 1994; McCloskey, 1991; Mozer & Smolensky, 1989; Seidenberg, 1993). This is because these algorithms involve distributed, parallel interactions amongst highly nonlinear elements. "One thing that connectionist networks have in common with brains is that if you open them up and peer inside, all you can see is a big pile of goo" (Mozer & Smolensky, 1989, p. 3).

In the early days of modern connectionist cognitive science, this was not a concern. This was a period of what has been called "gee whiz" connectionism (Dawson, 2009), in which connectionists modelled phenomena that were typically described in terms of rule-governed symbol manipulation. In the mid-1980s it was sufficiently interesting to show that such phenomena might be accounted for by parallel distributed processing systems that did not propose explicit rules or symbols. However, as connectionism matured, it was necessary for its researchers to spell out the details of the alternative algorithms embodied in their networks (Dawson, 2004). If these algorithms could not be extracted from networks, then "connectionist networks should not be viewed as theories of human cognitive functions, or as simulations of theories, or even as demonstrations of specific theoretical points" (McCloskey, 1991, p. 387). In response to such criticisms, connectionist cognitive scientists have developed a number of techniques for recovering algorithms from their networks (Berkeley et al., 1995; Dawson, 2004, 2005; Gallant, 1993; Hanson & Burr, 1990; Hinton, 1986; Moorhead, Haig, & Clement, 1989; Omlin & Giles, 1996).

What are the elements of connectionism, and how do they relate to cognitive science in general and to classical cognitive science in particular? The purpose of the remainder of this chapter is to explore the ideas of connectionist cognitive science in more detail.

4.2 Associations

Classical cognitive science has been profoundly influenced by seventeenth-century Cartesian philosophy (Descartes, 1996, 2006). The Cartesian view that thinking is equivalent to performing mental logic—that it is a mental discourse of computation or calculation (Hobbes, 1967)—has inspired the logicism that serves as the foundation of the classical approach. Fundamental classical notions, such as the assumption that cognition is the result of rule-governed symbol manipulation (Craik, 1943) or that innate knowledge is required to solve problems of underdetermination (Chomsky, 1965, 1966), have resulted in the classical being viewed as a newer variant of Cartesian rationalism (Paivio, 1986). One key classical departure from Descartes is its rejection of dualism. Classical cognitive science has appealed to recursive rules to permit finite devices to generate an infinite variety of potential behaviour.

Classical cognitive science is the modern rationalism, and one of the key ideas that it employs is recursion. Connectionist cognitive science has very different philosophical roots. Connectionism is the modern form of empiricist philosophy (Berkeley, 1710; Hume, 1952; Locke, 1977), where knowledge is not innate, but is instead provided by sensing the world. "No man's knowledge here can go beyond his experience" (Locke, 1977, p. 83). If recursion is fundamental to the classical approach's rationalism, then what notion is fundamental to connectionism's empiricism? The key idea is association: different ideas can be linked together, so that if one arises, then the association between them causes the other to arise as well.

For centuries, philosophers and psychologists have studied associations empirically, through introspection (Warren, 1921). These introspections have revealed the existence of sequences of thought that occur during thinking. Associationism attempted to determine the laws that would account for these sequences of thought.

The earliest detailed introspective account of such sequences of thought can be found in the 350 BC writings of Aristotle (Sorabji, 2006, p. 54): "Acts of recollection happen because one change is of a nature to occur after another." For Aristotle, ideas were images (Cummins, 1989). He argued that a particular sequence of images occurs either because this sequence is a natural consequence of the images, or because the sequence has been learned by habit. Recall of a particular memory, then, is achieved by cuing that memory with the appropriate prior images, which initiate the desired sequence of images. "Whenever we recollect, then, we undergo one of the earlier changes, until we undergo the one after which the change in question habitually occurs" (Sorabji, 2006, p. 54).

Aristotle's analysis of sequences of thought is central to modern mnemonic techniques for remembering ordered lists (Lorayne, 2007; Lorayne & Lucas, 1974). Aristotle noted that recollection via initiating a sequence of mental images could be a deliberate and systematic process. This was because the first image in the sequence could be selected so that it would be recollected fairly easily. Recall of the sequence, or of the target image at the end of the sequence, was then dictated by lawful relationships between adjacent ideas. Thus Aristotle invented laws of association.

Aristotle considered three different kinds of relationships between the starting image and its successor: similarity, opposition, and (temporal) contiguity:

> And this is exactly why we hunt for the successor, starting in our thoughts from the present or from something else, and from something similar, or opposite, or neighbouring. By this means recollection occurs. (Sorabji, 2006, p. 54)

In more modern associationist theories, Aristotle's laws would be called the law of similarity, the law of contrast, and the law of contiguity or the law of habit.

Aristotle's theory of memory was essentially ignored for many centuries (Warren, 1921). Instead, pre-Renaissance and Renaissance Europe were more interested in the artificial memory—mnemonics—that was the foundation of Greek oratory. These techniques were rediscovered during the Middle Ages in the form of *Ad Herennium*, a circa 86 BC text on rhetoric that included a section on enhancing the artificial memory (Yates, 1966). *Ad Herennium* described the mnemonic techniques invented by Simonides circa 500 BC. While the practice of mnemonics flourished during the Middle Ages, it was not until the seventeenth century that advances in associationist theories of memory and thought began to flourish.

The rise of modern associationism begins with Thomas Hobbes (Warren, 1921). Hobbes' (1967) notion of thought as mental discourse was based on his observation that thinking involved an orderly sequence of ideas. Hobbes was interested in explaining how such sequences occurred. While Hobbes' own work was very preliminary, it inspired more detailed analyses carried out by the British empiricists who followed him.

Empiricist philosopher John Locke coined the phrase *association of ideas*, which first appeared as a chapter title in the fourth edition of *An Essay Concerning Human Understanding* (Locke, 1977). Locke's work was an explicit reaction against Cartesian philosophy (Thilly, 1900); his goal was to establish experience as the foundation of all thought. He noted that connections between simple ideas might not reflect a natural order. Locke explained this by appealing to experience:

> Ideas that in themselves are not at all of kin, come to be so united in some men's minds that it is very hard to separate them, they always keep in company, and the one no sooner at any time comes into the understanding but its associate appears with it. (Locke, 1977, p. 122)

Eighteenth-century British empiricists expanded Locke's approach by exploring and debating possible laws of association. George Berkeley (1710) reiterated Aristotle's law of contiguity and extended it to account for associations involving different modes of sensation. David Hume (1852) proposed three different laws of association: resemblance, contiguity in time or place, and cause or effect. David Hartley, one of the first philosophers to link associative laws to brain function, saw contiguity as the primary source of associations and ignored Hume's law of resemblance (Warren, 1921).

Debates about the laws of association continued into the nineteenth century. James Mill (1829) only endorsed the law of contiguity, and explicitly denied Hume's laws of cause and effect or resemblance. Mill's ideas were challenged and modified by his son, John Stuart Mill. In his revised version of his father's book (Mill & Mill, 1869), Mill posited a completely different set of associative laws, which included a reintroduction of Hume's law of similarity. He also replaced his father's linear, mechanistic account of complex ideas with a "mental chemistry" that endorsed nonlinear emergence. This is because in this mental chemistry, when complex ideas were created via association, the resulting whole was more than just the sum of its parts. Alexander Bain (1855) refined the associationism of John Stuart Mill, proposing four different laws of association and attempting to reduce all intellectual processes to these laws. Two of these were the familiar laws of contiguity and of similarity.

Bain was the bridge between philosophical and psychological associationism (Boring, 1950). He stood,

> exactly at a corner in the development of psychology, with philosophical psychology stretching out behind, and experimental physiological psychology lying ahead, in a new direction. The psychologists of the twentieth century can read much of Bain with hearty approval; perhaps John Locke could have done the same. (Boring, 1950, p. 240)

One psychologist who approved of Bain was William James; he frequently cited Bain in his *Principles of Psychology* (James, 1890a). Chapter 14 of this work provided

James' own treatment of associationism. James criticized philosophical association-ism's emphasis on associations between mental contents. James proposed a mecha-nistic, biological theory of associationism instead, claiming that associations were made between brain states:

> We ought to talk of the association of *objects*, not of the association of *ideas*. And so far as association stands for a *cause*, it is between *processes in the brain*—it is these which, by being associated in certain ways, determine what successive objects shall be thought. (James, 1890a, p. 554, original italics)

James (1890a) attempted to reduce other laws of association to the law of contiguity, which he called the law of habit and expressed as follows: "When two elementary brain-processes have been active together or in immediate succession, one of them, on reoccurring, tends to propagate its excitement into the other" (p. 566). He illus-trated the action of this law with a figure (James, 1890a, p. 570, Figure 40), a version of which is presented as Figure 4-1.

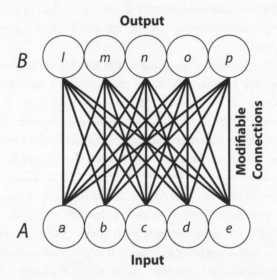

Figure 4-1. A distributed memory, initially described by James (1890a) but also part of modern connectionism.

Figure 4-1 illustrates two ideas, *A* and *B*, each represented as a pattern of activity in its own set of neurons. *A* is represented by activity in neurons *a, b, c, d,* and *e; B* is represented by activity in neurons *l, m, n, o,* and *p.* The assumption is that *A* repre-sents an experience that occurred immediately before *B.* When *B* occurs, activating its neurons, residual activity in the neurons representing *A* permits the two patterns to be associated by the law of habit. That is, the "tracts" connecting the neurons (the "modifiable connections" in Figure 4-1) have their strengths modified.

The ability of A's later activity to reproduce B is due to these modified connections between the two sets of neurons.

> The thought of A must awaken that of B, because a, b, c, d, e, will each and all discharge into l through the paths by which their original discharge took place. Similarly they will discharge into m, n, o, and p; and these latter tracts will also each reinforce the other's action because, in the experience B, they have already vibrated in unison. (James, 1890a, p. 569)

James' (1890a) biological account of association reveals three properties that are common to modern connectionist networks. First, his system is parallel: more than one neuron can be operating at the same time. Second, his system is convergent: the activity of one of the output neurons depends upon receiving or summing the signals sent by multiple input neurons. Third, his system is distributed: the association between A and B is the *set* of states of the many "tracts" illustrated in Figure 4-1; there is not just a single associative link.

James's (1890a) law of habit was central to the basic mechanism proposed by neuroscientist Donald Hebb (1949) for the development of cell assemblies. Hebb provided a famous modern statement of James' law of habit:

> When an axon of cell A is near enough to excite a cell B and repeatedly or persistently takes part in firing it, some growth process or metabolic change takes place in one or both cells such that A's efficiency, as one of the cells firing B, is increased. (Hebb, 1949, p. 62)

This makes explicit the modern connectionist idea that learning is modifying the strength of connections between processors. Hebb's theory inspired the earliest computer simulations of memory systems akin to the one proposed by James (Milner, 1957; Rochester et al., 1956). These simulations revealed a critical role for inhibition that led Hebb (1959) to revise his theory. Modern neuroscience has discovered a phenomenon called long-term potentiation that is often cited as a biologically plausible instantiation of Hebb's theory (Brown, 1990; Gerstner & Kistler, 2002; Martinez & Derrick, 1996; van Hemmen & Senn, 2002).

The journey from James through Hebb to the first simulations of memory (Milner, 1957; Rochester et al., 1956) produced a modern associative memory system called the standard pattern associator (McClelland, 1986). The standard pattern associator, which is structurally identical to Figure 4-1, is a memory capable of learning associations between pairs of input patterns (Steinbuch, 1961; Taylor, 1956) or learning to associate an input pattern with a categorizing response (Rosenblatt, 1962; Selfridge, 1956; Widrow & Hoff, 1960).

The standard pattern associator is empiricist in the sense that its knowledge is acquired by experience. Usually the memory begins as a blank slate: all of the connections between processors start with weights equal to zero. During a learning phase, pairs of to-be-associated patterns simultaneously activate the input and output units in Figure 4-1. With each presented pair, all of the connection weights—the strength of each connection between an input and an output processor—are modified by adding a value to them. This value is determined in accordance with some version of Hebb's (1949) learning rule. Usually, the value added to a weight is equal to the activity of the processor at the input end of the connection, multiplied by the activity of the processor at the output end of the connection, and multiplied by some fractional value called a learning rate. The mathematical details of such learning are provided in Chapter 9 of Dawson (2004).

The standard pattern associator is called a distributed memory because its knowledge is stored throughout all the connections in the network, and because this one set of connections can store several different associations. During a recall phase, a cue pattern is used to activate the input units. This causes signals to be sent through the connections in the network. These signals are equal to the activation value of an input unit multiplied by the weight of the connection through which the activity is being transmitted. Signals received by the output processors are used to compute net input, which is simply the sum of all of the incoming signals. In the standard pattern associator, an output unit's activity is equal to its net input. If the memory is functioning properly, then the pattern of activation in the output units will be the pattern that was originally associated with the cue pattern.

The standard pattern associator is the cornerstone of many models of memory created after the cognitive revolution (Anderson, 1972; Anderson et al., 1977; Eich, 1982; Hinton & Anderson, 1981; Murdock, 1982; Pike, 1984; Steinbuch, 1961; Taylor, 1956). These models are important, because they use a simple principle—James' (1890a, 1890b) law of habit—to model many subtle regularities of human memory, including errors in recall. In other words, the standard pattern associator is a kind of memory that has been evaluated with the different kinds of evidence cited in Chapters 2 and 3, in an attempt to establish strong equivalence.

The standard pattern associator also demonstrates another property crucial to modern connectionism, graceful degradation. How does this distributed model behave if it is presented with a noisy cue, or with some other cue that was never tested during training? It generates a response that has the same degree of noise as its input (Dawson, 1998, Table 3-1). That is, there is a match between the quality of the memory's input and the quality of its output.

The graceful degradation of the standard pattern associator reveals that it is sensitive to the similarity of noisy cues to other cues that were presented during training. Thus modern pattern associators provide some evidence for James' (1890a) attempt to reduce other associative laws, such as the law of similarity, to the basic law of habit or contiguity.

In spite of the popularity and success of distributed associative memories as models of human learning and recall (Hinton & Anderson, 1981), they are extremely limited in power. When networks learn via the Hebb rule, they produce errors when they are overtrained, are easily confused by correlated training patterns, and do not learn from their errors (Dawson, 2004). An error-correcting rule called the delta rule (Dawson, 2004; Rosenblatt, 1962; Stone, 1986; Widrow & Hoff, 1960) can alleviate some of these problems, but it does not eliminate them. While association is a fundamental notion in connectionist models, other notions are required by modern connectionist cognitive science. One of these additional ideas is nonlinear processing.

4.3 Nonlinear Transformations

John Stuart Mill modified his father's theory of associationism (Mill & Mill, 1869; Mill, 1848) in many ways, including proposing a mental chemistry "in which it is proper to say that the simple ideas generate, rather than . . . compose, the complex ones" (Mill, 1848, p. 533). Mill's mental chemistry is an early example of emergence, where the properties of a whole (i.e., a complex idea) are more than the sum of the properties of the parts (i.e., a set of associated simple ideas).

> The generation of one class of mental phenomena from another, whenever it can be made out, is a highly interesting fact in mental chemistry; but it no more supersedes the necessity of an experimental study of the generated phenomenon than a knowledge of the properties of oxygen and sulphur enables us to deduce those of sulphuric acid without specific observation and experiment. (Mill, 1848, p. 534)

Mathematically, emergence results from nonlinearity (Luce, 1999). If a system is linear, then its whole behaviour is exactly equal to the sum of the behaviours of its parts. The standard pattern associator that was illustrated in Figure 4-1 is an example of such a system. Each output unit in the standard pattern associator computes a net input, which is the sum of all of the individual signals that it receives from the input units. Output unit activity is exactly equal to net input. In other words, output activity is exactly equal to the sum of input signals in the standard pattern associator. In order to increase the power of this type of pattern associator—in order to facilitate emergence—a nonlinear relationship between input and output must be introduced.

Neurons demonstrate one powerful type of nonlinear processing. The inputs to a neuron are weak electrical signals, called graded potentials, which stimulate and travel through the dendrites of the receiving neuron. If enough of these weak graded potentials arrive at the neuron's soma at roughly the same time, then their cumulative effect disrupts the neuron's resting electrical state. This results in a massive depolarization of the membrane of the neuron's axon, called an action potential, which is a signal of constant intensity that travels along the axon to eventually stimulate some other neuron.

A crucial property of the action potential is that it is an all-or-none phenomenon, representing a nonlinear transformation of the summed graded potentials. The neuron converts continuously varying inputs into a response that is either on (action potential generated) or off (action potential not generated). This has been called the all-or-none law (Levitan & Kaczmarek, 1991, p. 43): "The all-or-none law guarantees that once an action potential is generated it is always full size, minimizing the possibility that information will be lost along the way." The all-or-none output of neurons is a nonlinear transformation of summed, continuously varying input, and it is the reason that the brain can be described as digital in nature (von Neumann, 1958).

The all-or-none behaviour of a neuron makes it logically equivalent to the relays or switches that were discussed in Chapter 2. This logical interpretation was exploited in an early mathematical account of the neural information processing (McCulloch & Pitts, 1943). McCulloch and Pitts used the all-or-none law to justify describing neurons very abstractly as devices that made true or false logical assertions about input information:

> The all-or-none law of nervous activity is sufficient to insure that the activity of any neuron may be represented as a proposition. Physiological relations existing among nervous activities correspond, of course, to relations among the propositions; and the utility of the representation depends upon the identity of these relations with those of the logical propositions. To each reaction of any neuron there is a corresponding assertion of a simple proposition. (McCulloch & Pitts, 1943, p. 117)

McCulloch and Pitts (1943) invented a connectionist processor, now known as the McCulloch-Pitts neuron (Quinlan, 1991), that used the all-or-none law. Like the output units in the standard pattern associator (Figure 4-1), a McCulloch-Pitts neuron first computes its net input by summing all of its incoming signals. However, it then uses a nonlinear activation function to transform net input into internal activity. The activation function used by McCulloch and Pitts was the Heaviside step function, named after nineteenth-century electrical engineer Oliver Heaviside. This function compares the net input to a threshold. If the net input is less than the threshold, the unit's activity is equal to 0. Otherwise, the unit's activity is equal to 1. (In other artificial neural networks [Rosenblatt, 1958, 1962], below-threshold net inputs produced activity of –1.)

The output units in the standard pattern associator (Figure 4-1) can be described as using the linear identity function to convert net input into activity, because output unit activity is equal to net input. If one replaced the identity function with the Heaviside step function in the standard pattern associator, it would then become a different kind of network, called a perceptron (Dawson, 2004), which was invented by Frank Rosenblatt during the era in which cognitive science was born (Rosenblatt, 1958, 1962).

Perceptrons (Rosenblatt, 1958, 1962) were artificial neural networks that could be trained to be pattern classifiers: given an input pattern, they would use their non-linear outputs to decide whether or not the pattern belonged to a particular class. In other words, the nonlinear activation function used by perceptrons allowed them to assign perceptual predicates; standard pattern associators do not have this ability. The nature of the perceptual predicates that a perceptron could learn to assign was a central issue in an early debate between classical and connectionist cognitive science (Minsky & Papert, 1969; Papert, 1988).

The Heaviside step function is nonlinear, but it is also discontinuous. This was problematic when modern researchers sought methods to train more complex networks. Both the standard pattern associator and the perceptron are one-layer networks, meaning that they have only one layer of connections, the direct connections between input and output units (Figure 4-1). More powerful networks arise if intermediate processors, called hidden units, are used to preprocess input signals before sending them on to the output layer. However, it was not until the mid-1980s that learning rules capable of training such networks were invented (Ackley, Hinton, & Sejnowski, 1985; Rumelhart, Hinton, & Williams, 1986b). The use of calculus to derive these new learning rules became possible when the discontinuous Heaviside step function was replaced by a continuous approximation of the all-or-none law (Rumelhart, Hinton, & Williams, 1986b).

One continuous approximation of the Heaviside step function is the sigmoid-shaped logistic function. It asymptotes to a value of 0 as its net input approaches negative infinity, and asymptotes to a value of 1 as its net input approaches positive infinity. When the net input is equal to the threshold (or bias) of the logistic, activity is equal to 0.5. Because the logistic function is continuous, its derivative can be calculated, and calculus can be used as a tool to derive new learning rules (Rumelhart, Hinton, & Williams, 1986b). However, it is still nonlinear, so logistic activities can still be interpreted as truth values assigned to propositions.

Modern connectionist networks employ many different nonlinear activation functions. Processing units that employ the logistic activation function have been called integration devices (Ballard, 1986) because they convert a sum (net input) and "squash" it into the range between 0 and 1. Other processing units might be tuned to generate maximum responses to a narrow range of net inputs. Ballard (1986) called such processors value units. A different nonlinear continuous function, the Gaussian equation, can be used to mathematically define a value unit, and calculus can be used to derive a learning rule for this type of artificial neural network (Dawson, 1998, 2004; Dawson & Schopflocher, 1992b).

Many other activation functions exist. One review paper has identified 640 different activation functions employed in connectionist networks (Duch & Jankowski, 1999). One characteristic of the vast majority of all of these activation functions is their nonlinearity. Connectionist cognitive science is associationist, but it is also nonlinear.

4.4 The Connectionist Sandwich

Both the McCulloch-Pitts neuron (McCulloch & Pitts, 1943) and the perceptron (Rosenblatt, 1958, 1962) used the Heaviside step function to implement the all-or-none law. As a result, both of these architectures generated a "true" or "false" judgment about each input pattern. Thus both of these architectures are digital, and their basic function is pattern recognition or pattern classification.

The two-valued logic that was introduced in Chapter 2 can be cast in the context of such digital pattern recognition. In the two-valued logic, functions are computed over two input propositions, p and q, which themselves can either be true or false. As a result, there are only four possible combinations of p and q, which are given in the first two columns of Table 4-1. Logical functions in the two-valued logic are themselves judgments of true or false that depend on combinations of the truth values of the input propositions p and q. As a result, there are 16 different logical operations that can be defined in the two-valued logic; these were provided in Table 2-2.

The truth tables for two of the sixteen possible operations in the two-valued logic are provided in the last two columns of Table 4-1. One is the AND operation ($p \cdot q$), which is only true when both propositions are true. The other is the XOR operation ($p \wedge q$), which is only true when one or the other of the propositions is true.

p	q	$p \cdot q$	$p \wedge q$
1	1	1	0
1	0	0	1
0	1	0	1
0	0	0	0

Table 4-1. Truth tables for the logical operations AND ($p \cdot q$) and XOR ($p \wedge q$), where the truth value of each operation is given as a function of the truth of each of two propositions, p and q. '1' indicates "true" and '0' indicates "false." The logical notation is taken from McCulloch (1988b).

That AND or XOR are examples of digital pattern recognition can be made more explicit by representing their truth tables graphically as pattern spaces. In a pattern space, an entire row of a truth table is represented as a point on a graph. The coordinates of a point in a pattern space are determined by the truth values of the input propositions. The colour of the point represents the truth value of the operation computed over the inputs.

Figure 4-2A illustrates the pattern space for the AND operation of Table 4-1. Note that it has four graphed points, one for each row of the truth table. The coordinates of each graphed point—(1,1), (1,0), (0,1), and (0,0)—indicate the truth values of the propositions p and q. The AND operation is only true when both of these propositions are true. This is represented by colouring the point at coordinate (1,1) black. The other three points are coloured white, indicating that the logical operator returns a "false" value for each of them.

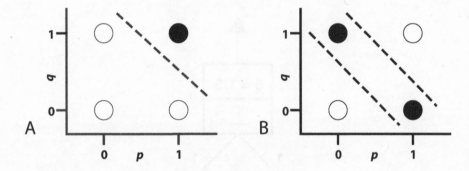

Figure 4-2. (A) Pattern space for AND; (B) Pattern space for XOR.

Pattern spaces are used for digital pattern recognition by carving them into decision regions. If a point that represents a pattern falls in one decision region, then it is classified in one way. If that point falls in a different decision region, then it is classified in a different way. Learning how to classify a set of patterns involves learning how to correctly carve the pattern space up into the desired decision regions.

The AND problem is an example of a linearly separable problem. This is because a single straight cut through the pattern space divides it into two decision regions that generate the correct pattern classifications. The dashed line in Figure 4-2A indicates the location of this straight cut for the AND problem. Note that the one "true" pattern falls on one side of this cut, and that the three "false" patterns fall on the other side of this cut.

Not all problems are linearly separable. A linearly nonseparable problem is one in which a single straight cut is not sufficient to separate all of the patterns of one type from all of the patterns of another type. An example of a linearly nonseparable problem is the XOR problem, whose pattern space is illustrated in Figure 4-2B. Note that the positions of the four patterns in Figure 4-2B are identical to the positions in Figure 4-2A, because both pattern spaces involve the same propositions.

The only difference is the colouring of the points, indicating that XOR involves making a different judgment than AND. However, this difference between graphs is important, because now it is impossible to separate all of the black points from all of the white points with a single straight cut. Instead, two different cuts are required, as shown by the two dashed lines in Figure 4-2B. This means that XOR is not linearly separable.

Linear separability defines the limits of what can be computed by a Rosenblatt perceptron (Rosenblatt, 1958, 1962) or by a McCulloch-Pitts neuron (McCulloch & Pitts, 1943). That is, if some pattern recognition problem is linearly separable, then either of these architectures is capable of representing a solution to that problem. For instance, because AND is linearly separable, it can be computed by a perceptron, such as the one illustrated in Figure 4-3.

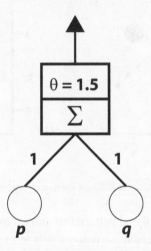

Figure 4-3. A Rosenblatt perceptron that can compute the AND operation.

This perceptron consists of two input units whose activities respectively represent the state (i.e., either 0 or 1) of the propositions p and q. Each of these input units sends a signal through a connection to an output unit; the figure indicates that the weight of each connection is 1. The output unit performs two operations. First, it computes its net input by summing the two signals that it receives (the Σ component of the output unit). Second, it transforms the net input into activity by applying the Heaviside step function. The figure indicates in the second component of the output unit that the threshold for this activation function (θ) is 1.5. This means that output unit activity will only be 1 if net input is greater than or equal to 1.5; otherwise, output unit activity will be equal to 0.

If one considers the four different combinations of input unit activities that would be presented to this device—(1,1), (1,0), (0,1), and (0,0)—then it is clear that the only time that output unit activity will equal 1 is when both input units are activated with 1 (i.e., when p and q are both true). This is because this situation will produce a net input of 2, which exceeds the threshold. In all other cases, the net input will either be 1 or 0, which will be less than the threshold, and which will therefore produce output unit activity of 0.

The ability of the Figure 4-3 perceptron to compute AND can be described in terms of the pattern space in Figure 4-2A. The threshold and the connection weights of the perceptron provide the location and orientation of the single straight cut that carves the pattern space into decision regions (the dashed line in Figure 4-2A). Activating the input units with some pattern presents a pattern space location to the perceptron. The perceptron examines this location to decide on which side of the cut the location lies, and responds accordingly.

This pattern space account of the Figure 4-3 perceptron also points to a limitation. When the Heaviside step function is used as an activation function, the perceptron only defines a single straight cut through the pattern space and therefore can only deal with linearly separable problems. A perceptron akin to the one illustrated in Figure 4-3 would not be able to compute XOR (Figure 4-2B) because the output unit is incapable of making the *two* required cuts in the pattern space.

How does one extend computational power beyond the perceptron? One approach is to add additional processing units, called hidden units, which are intermediaries between input and output units. Hidden units can detect additional features that transform the problem by increasing the dimensionality of the pattern space. As a result, the use of hidden units can convert a linearly nonseparable problem into a linearly separable one, permitting a single binary output unit to generate the correct responses.

Figure 4-4 shows how the AND circuit illustrated in Figure 4-3 can be added as a hidden unit to create a multilayer perceptron that can compute the linearly nonseparable XOR operation (Rumelhart, Hinton, & Williams, 1986a). This perceptron also has two input units whose activities respectively represent the state of the propositions p and q. Each of these input units sends a signal through a connection to an output unit; the figure indicates that the weight of each connection is 1. The threshold of the output's activation function (θ) is 0.5. If we were to ignore the hidden unit in this network, the output unit would be computing OR, turning on when one or both of the input propositions are true.

However, this network does not compute OR, because the input units are also connected to a hidden unit, which in turn sends a third signal to be added into the output unit's net input. The hidden unit is identical to the AND circuit from Figure 4-3. The signal that it sends to the output unit is strongly inhibitory; the weight of the connection between the two units is –2.

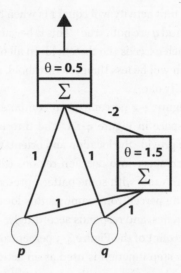

Figure 4-4. A multilayer perceptron that can compute XOR.

The action of the hidden unit is crucial to the behaviour of the system. When neither or only one of the input units activates, the hidden unit does not respond, so it sends a signal of 0 to the output unit. As a result, in these three situations the output unit turns on when either of the inputs is on (because the net input is over the threshold) and turns off when neither input unit is on. When both input units are on, they send an excitatory signal to the output unit. However, they also send a signal that turns on the hidden unit, causing it to send inhibition to the output unit. In this situation, the net input of the output unit is 1 + 1 – 2 = 0 which is below threshold, producing zero output unit activity. The entire circuit therefore performs the XOR operation.

The behaviour of the Figure 4-4 multilayer perceptron can also be related to the pattern space of Figure 4-2B. The lower cut in that pattern space is provided by the output unit. The upper cut in that pattern space is provided by the hidden unit. The coordination of the two units permits the circuit to solve this linearly nonseparable problem.

Interpreting networks in terms of the manner in which they carve a pattern space into decision regions suggests that learning can be described as determining where cuts in a pattern space should be made. Any hidden or output unit that uses a nonlinear, monotonic function like the Heaviside or the logistic can be viewed as making a single cut in a space. The position and orientation of this cut is determined by the weights of the connections feeding into the unit, as well as the threshold or bias (θ) of the unit. A learning rule modifies all of these components. (The bias of a unit can be trained as if it were just another connection weight by assuming that it is the signal coming from a special, extra input unit that is always turned on [Dawson, 2004, 2005].)

The multilayer network illustrated in Figure 4-4 is atypical because it directly connects input and output units. Most modern networks eliminate such direct connections by using at least one layer of hidden units to isolate the input units from the output units, as shown in Figure 4-5. In such a network, the hidden units can still be described as carving a pattern space, with point coordinates provided by the input units, into a decision region. However, because the output units do not have direct access to input signals, they do not carve the pattern space. Instead, they divide an alternate space, the hidden unit space, into decision regions. The hidden unit space is similar to the pattern space, with the exception that the coordinates of the points that are placed within it are provided by hidden unit activities.

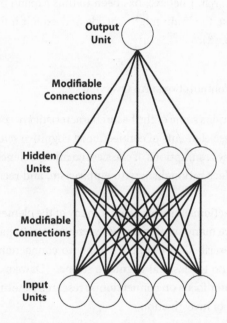

Figure 4-5. A typical multilayer perceptron has no direct connections between input and output units.

When there are no direct connections between input and output units, the hidden units provide output units with an internal representation of input unit activity. Thus it is proper to describe a network like the one illustrated in Figure 4-5 as being just as representational (Horgan & Tienson, 1996) as a classical model. That connectionist representations can be described as a nonlinear transformation of the input unit representation, permitting higher-order nonlinear features to be detected, is why a network like the one in Figure 4-5 is far more powerful than one in which no hidden units appear (e.g., Figure 4-3).

When there are no direct connections between input and output units, the representations held by hidden units conform to the classical sandwich that characterized classical models (Hurley, 2001)—a connectionist sandwich (Calvo & Gomila, 2008, p. 5): "Cognitive sandwiches need not be Fodorian. A feed forward connectionist network conforms equally to the sandwich metaphor. The input layer is identified with a perception module, the output layer with an action one, and hidden space serves to identify metrically, in terms of the distance relations among patterns of activation, the structural relations that obtain among concepts. The hidden layer this time contains the meat of the connectionist sandwich."

A difference between classical and connectionist cognitive science is not that the former is representational and the latter is not. Both are representational, but they disagree about the nature of mental representations. "The major lesson of neural network research, I believe, has been to thus expand our vision of the ways a physical system like the brain might encode and exploit information and knowledge" (Clark, 1997, p. 58).

4.5 Connectionist Computations: An Overview

In the preceding sections some of the basic characteristics of connectionist networks were presented. These elements of connectionist cognitive science have emerged as a reaction against key assumptions of classical cognitive science. Connectionist cognitive scientists replace rationalism with empiricism, and recursion with chains of associations.

Although connectionism reacts against many of the elements of classical cognitive science, there are many similarities between the two. In particular, the multiple levels of analysis described in Chapter 2 apply to connectionist cognitive science just as well as they do to classical cognitive science (Dawson, 1998). The next two sections of this chapter focus on connectionist research in terms of one of these, the computational level of investigation.

Connectionism's emphasis on both empiricism and associationism has raised the spectre, at least in the eyes of many classical cognitive scientists, of a return to the behaviourism that cognitivism itself revolted against. When cognitivism arose, some of its early successes involved formal proofs that behaviourist and associationist theories were incapable of accounting for fundamental properties of human languages (Bever, Fodor, & Garrett, 1968; Chomsky, 1957, 1959b, 1965, 1966). With the rise of modern connectionism, similar computational arguments have been made against artificial neural networks, essentially claiming that they are not sophisticated enough to belong to the class of universal machines (Fodor & Pylyshyn, 1988).

In Section 4.6, "Beyond the Terminal Meta-postulate," we consider the in-principle power of connectionist networks, beginning with two different types of tasks that networks can be used to accomplish. One is pattern classification: assigning an input pattern in an all-or-none fashion to a particular category. A second is function approximation: generating a continuous response to a set of input values.

Section 4.6 then proceeds to computational analyses of how capable networks are of accomplishing these tasks. These analyses prove that networks are as powerful as need be, provided that they include hidden units. They can serve as arbitrary pattern classifiers, meaning that they can solve any pattern classification problem with which they are faced. They can also serve as universal function approximators, meaning that they can fit any continuous function to an arbitrary degree of precision. This computational power suggests that artificial neural networks belong to the class of universal machines. The section ends with a brief review of computational analyses, which conclude that connectionist networks indeed can serve as universal Turing machines and are therefore computationally sophisticated enough to serve as plausible models for cognitive science.

Computational analyses need not limit themselves to considering the general power of artificial neural networks. Computational analyses can be used to explore more specific questions about networks. This is illustrated in Section 4.7, "What Do Output Unit Activities Represent?" in which we use formal methods to answer the question that serves as the section's title. The section begins with a general discussion of theories that view biological agents as intuitive statisticians who infer the probability that certain events may occur in the world (Peterson & Beach, 1967; Rescorla, 1967, 1968). An empirical result is reviewed that suggests artificial neural networks are also intuitive statisticians, in the sense that the activity of an output unit matches the probability that a network will be "rewarded" (i.e., trained to turn on) when presented with a particular set of cues (Dawson et al., 2009).

The section then ends by providing an example computational analysis: a formal proof that output unit activity can indeed literally be interpreted as a conditional probability. This proof takes advantage of known formal relations between neural networks and the Rescorla-Wagner learning rule (Dawson, 2008; Gluck & Bower, 1988; Sutton & Barto, 1981), as well as known formal relations between the Rescorla-Wagner learning rule and contingency theory (Chapman & Robbins, 1990).

4.6 Beyond the Terminal Meta-postulate

Connectionist networks are associationist devices that map inputs to outputs, systems that convert stimuli into responses. However, we saw in Chapter 3 that classical cognitive scientists had established that the stimulus-response theories of behaviourist psychology could not adequately deal with the recursive structure of natural language (Chomsky, 1957, 1959b, 1965, 1966). In the terminal meta-postulate argument (Bever, Fodor, and Garrett, 1968), it was noted that the rules of associative theory defined a "terminal vocabulary of a theory, i.e., over the vocabulary in which behavior is described" (p. 583). Bever, Fodor, and Garrett then proceeded to prove that the terminal vocabulary of associationism is not powerful enough to accept or reject languages that have recursive clausal structure.

If connectionist cognitive science is another instance of associative or behaviourist theory, then it stands to reason that it too is subject to these same problems and therefore lacks the computational power required of cognitive theory. One of the most influential criticisms of connectionism has essentially made this point, arguing against the computational power of artificial neural networks because they lack the componentiality and systematicity associated with recursive rules that operate on components of symbolic expressions (Fodor & Pylyshyn, 1988). If artificial neural networks do not belong to the class of universal machines, then they cannot compete against the physical symbol systems that define classical cognitive science (Newell, 1980; Newell & Simon, 1976).

What tasks can artificial neural networks perform, and how well can they perform them? To begin, let us consider the most frequent kind of problem that artificial neural networks are used to solve: pattern recognition (Pao, 1989; Ripley, 1996). Pattern recognition is a process by which varying input patterns, defined by sets of features which may have continuous values, are assigned to discrete categories in an all-or-none fashion (Harnad, 1987). In other words, it requires that a system perform a mapping from continuous inputs to discrete outputs. Artificial neural networks are clearly capable of performing this kind of mapping, provided either that their output units use a binary activation function like the Heaviside, or that their continuous output is extreme enough to be given a binary interpretation. In this context, the pattern of "on" and "off" responses in a set of output units represents the digital name of the class to which an input pattern has been assigned.

We saw earlier that pattern recognition problems can be represented using pattern spaces (Figure 4-2). To classify patterns, a system carves a pattern space into decision regions that separate all of the patterns belonging to one class from the patterns that belong to others. An arbitrary pattern classifier would be a system that could, in principle, solve any pattern recognition problem with which it was faced. In order to have such ability, such a system must have complete flexibility in carving a pattern space into decision regions: it must be able to slice the space into regions of any required shape or number.

Artificial neural networks can categorize patterns. How well can they do so? It has been shown that a multilayer perceptron with three layers of connections—two layers of hidden units intervening between the input and output layers—is indeed an arbitrary pattern classifier (Lippmann, 1987, 1989). This is because the two layers of hidden units provided the required flexibility in carving pattern spaces into decision regions, assuming that the hidden units use a sigmoid-shaped activation function such as the logistic. "No more than three layers are required in perceptron-like feed-forward nets" (Lippmann, 1987, p. 16).

When output unit activity is interpreted digitally—as delivering "true" or "false" judgments—artificial neural networks can be interpreted as performing one kind of task, pattern classification. However, modern networks use continuous activation functions that do not need to be interpreted digitally. If one applies an analog interpretation to output unit activity, then networks can be interpreted as performing a second kind of input-output mapping task, function approximation.

In function approximation, an input is a set of numbers that represents the values of variables passed into a function, i.e., the values of the set $x_1, x_2, x_3, \ldots x_N$. The output is a single value y that is the result of computing some function of those variables, i.e., $y = f(x_1, x_2, x_3, \ldots x_N)$. Many artificial neural networks have been trained to approximate functions (Girosi & Poggio, 1990; Hartman, Keeler, & Kowalski, 1989; Moody & Darken, 1989; Poggio & Girosi, 1990; Renals, 1989). In these networks, the value of each input variable is represented by the activity of an input unit, and the continuous value of an output unit's activity represents the computed value of the function of those input variables.

A system that is most powerful at approximating functions is called a universal function approximator. Consider taking any continuous function and examining a region of this function from a particular starting point (e.g., one set of input values) to a particular ending point (e.g., a different set of input values). A universal function approximator is capable of approximating the shape of the function between these bounds to an arbitrary degree of accuracy.

Artificial neural networks can approximate functions. How well can they do so? A number of proofs have shown that a multilayer perceptron with two layers of connections—in other words, a single layer of hidden units intervening between the input and output layers—is capable of universal function approximation (Cotter, 1990; Cybenko, 1989; Funahashi, 1989; Hartman, Keeler, & Kowalski, 1989; Hornik, Stinchcombe, & White, 1989). "If we have the right connections from the input units to a large enough set of hidden units, we can always find a representation that will perform any mapping from input to output" (Rumelhart, Hinton, & Williams, 1986a, p. 319).

That multilayered networks have the in-principle power to be arbitrary pattern classifiers or universal function approximators suggests that they belong to the class "universal machine," the same class to which physical symbol systems belong (Newell, 1980). Newell (1980) proved that physical symbol systems belonged to this class by showing how a universal Turing machine could be simulated by a physical symbol system. Similar proofs exist for artificial neural networks, firmly establishing their computational power.

The Turing equivalence of connectionist networks has long been established. McCulloch and Pitts (1943) proved that a network of McCulloch-Pitts neurons could be used to build the machine head of a universal Turing machine; universal power was then achieved by providing this system with an external memory. "To psychology, however defined, specification of the net would contribute all that could be achieved in that field" (p. 131). More modern results have used the analog nature of modern processors to internalize the memory, indicating that an artificial neural network can simulate the entire Turing machine (Siegelmann, 1999; Siegelmann & Sontag, 1991, 1995).

Modern associationist psychologists have been concerned about the implications of the terminal meta-postulate and have argued against it in an attempt to free their theories from its computational shackles (Anderson & Bower, 1973; Paivio, 1986). The hidden units of modern artificial neural networks break these shackles by capturing higher-order associations—associations between associations—that are not defined in a vocabulary restricted to input and output activities. The presence of hidden units provides enough power to modern networks to firmly plant them in the class "universal machine" and to make them viable alternatives to classical simulations.

4.7 What Do Output Unit Activities Represent?

When McCulloch and Pitts (1943) formalized the information processing of neurons, they did so by exploiting the all-or-none law. As a result, whether a neuron responded could be interpreted as assigning a "true" or "false" value to some proposition computed over the neuron's outputs. McCulloch and Pitts were able to design artificial neurons capable of acting as 14 of the 16 possible primitive functions on the two-valued logic that was described in Chapter 2.

McCulloch and Pitts (1943) formalized the all-or-none law by using the Heaviside step equation as the activation function for their artificial neurons. Modern activation functions such as the logistic equation provide a continuous approximation of the step function. It is also quite common to interpret the logistic function in digital, step function terms. This is done by interpreting a modern unit as being "on" or "off" if its activity is sufficiently extreme. For instance, in simulations conducted with my laboratory software (Dawson, 2005) it is typical to view a unit as being "on" if its activity is 0.9 or higher, or "off" if its activity is 0.1 or lower.

Digital activation functions, or digital interpretations of continuous activation functions, mean that pattern recognition is a primary task for artificial neural networks (Pao, 1989; Ripley, 1996). When a network performs pattern recognition, it is trained to generate a digital or binary response to an input pattern, where this response is interpreted as representing a class to which the input pattern is unambiguously assigned.

What does the activity of a unit in a connectionist network mean? Under the strict digital interpretation described above, activity is interpreted as the truth value of some proposition represented by the unit. However, modern activation functions such as the logistic or Gaussian equations have continuous values, which permit more flexible kinds of interpretation. Continuous activity might model the frequency with which a real unit (i.e., a neuron) generates action potentials. It could represent a degree of confidence in asserting that a detected feature is present, or it could represent the amount of a feature that is present (Waskan & Bechtel, 1997).

In this section, a computational-level analysis is used to prove that, in the context of modern learning theory, continuous unit activity can be unambiguously interpreted as a candidate measure of degree of confidence with conditional probability (Waskan & Bechtel, 1997).

In experimental psychology, some learning theories are motivated by the ambiguous or noisy nature of the world. Cues in the real world do not signal outcomes with complete certainty (Dewey, 1929). It has been argued that adaptive systems deal with worldly uncertainty by becoming "intuitive statisticians," whether these systems are humans (Peterson & Beach, 1967) or animals (Gallistel, 1990; Shanks, 1995). An agent that behaves like an intuitive statistician detects contingency in the world, because cues signal the likelihood (and *not* the certainty) that certain events (such as being rewarded) will occur (Rescorla, 1967, 1968).

Evidence indicates that a variety of organisms are intuitive statisticians. For example, the matching law is a mathematical formalism that was originally used to explain variations in response frequency. It states that the rate of a response reflects the rate of its obtained reinforcement. For instance, if response A is reinforced twice as frequently as response B, then A will appear twice as frequently as B (Herrnstein, 1961). The matching law also predicts how response strength varies with reinforcement frequency (de Villiers & Herrnstein, 1976). Many results show that the matching law governs numerous tasks in psychology and economics (Davison & McCarthy, 1988; de Villiers, 1977; Herrnstein, 1997).

Another phenomenon that is formally related (Herrnstein & Loveland, 1975) to the matching law is probability matching, which concerns choices made by agents faced with competing alternatives. Under probability matching, the likelihood that an agent makes a choice amongst different alternatives mirrors the probability associated with the outcome or reward of that choice (Vulkan, 2000). Probability matching has been demonstrated in a variety of organisms, including insects (Fischer, Couvillon, & Bitterman, 1993; Keasar et al., 2002; Longo, 1964; Niv et al., 2002), fish (Behrend & Bitterman, 1961), turtles (Kirk & Bitterman, 1965), pigeons (Graf, Bullock, & Bitterman, 1964), and humans (Estes & Straughan, 1954).

Perceptrons, too, can match probabilities (Dawson et al., 2009). Dawson et al. used four different cues, or discriminative stimuli (DSs), but did not "reward" them 100 percent of the time. Instead, they rewarded one DS 20 percent of the time, another 40 percent, a third 60 percent, and a fourth 80 percent. After 300 epochs, where each epoch involved presenting each cue alone 10 different times in random order, these contingencies were inverted (i.e., subtracted from 100). The dependent measure was perceptron activity when a cue was presented; the activation function employed was the logistic. Some results of this experiment are presented in Figure 4-6. It shows that after a small number of epochs, the output unit activity becomes equal to the probability that a presented cue was rewarded. It also shows that perceptron responses quickly readjust when contingencies are suddenly modified, as shown by the change in Figure 4-6 around epoch 300. In short, perceptrons are capable of probability matching.

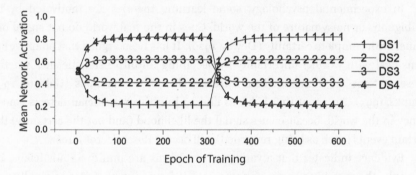

Figure 4-6. Probability matching by perceptrons. Each line shows the perceptron activation when a different cue (or discriminative stimulus, DS) is presented. Activity levels quickly become equal to the probability that each cue was reinforced (Dawson et al., 2009).

That perceptrons match probabilities relates them to contingency theory. Formal statements of this theory formalize contingency as a contrast between conditional probabilities (Allan, 1980; Cheng, 1997; Cheng & Holyoak, 1995; Cheng & Novick, 1990, 1992; Rescorla, 1967, 1968).

For instance, consider the simple situation in which a cue can either be presented, C, or not, $\sim C$. Associated with either of these states is an outcome (e.g., a reward) that can either occur, O, or not, $\sim O$. In this simple situation, involving a single cue and a single outcome, the contingency between the cue and the outcome is formally defined as the difference in conditional probabilities, ΔP, where $\Delta P = P(O|C) - P(O|\sim C)$ (Allan, 1980). More sophisticated models, such as the probabilistic contrast model (e.g., Cheng & Novick, 1990) or the power PC theory (Cheng, 1997), define more complex probabilistic contrasts that are possible when multiple cues occur and can be affected by the context in which they are presented.

Empirically, the probability matching of perceptrons, illustrated in Figure 4-6, suggests that their behaviour can represent ΔP. When a cue is presented, activity is equal to the probability that the cue signals reinforcement—that is, $P(O|C)$. This implies that the difference between a perceptron's activity when a cue is presented and its activity when a cue is absent must be equal to ΔP. Let us now turn to a computational analysis to prove this claim.

What is the formal relationship between formal contingency theories and theories of associative learning (Shanks, 2007)? Researchers have compared the predictions of an influential account of associative learning, the Rescorla-Wagner model (Rescorla & Wagner, 1972), to formal theories of contingency (Chapman & Robbins, 1990; Cheng, 1997; Cheng & Holyoak, 1995). It has been shown that while in some instances the Rescorla-Wagner model predicts the conditional contrasts defined by a formal contingency theory, in other situations it fails to generate these predictions (Cheng, 1997).

Comparisons between contingency learning and Rescorla-Wagner learning typically involve determining equilibria of the Rescorla-Wagner model. An equilibrium of the Rescorla-Wagner model is a set of associative strengths defined by the model, at the point where the asymptote of changes in error defined by Rescorla-Wagner learning approaches zero (Danks, 2003). In the simple case described earlier, involving a single cue and a single outcome, the Rescorla-Wagner model is identical to contingency theory. This is because at equilibrium, the associative strength between cue and outcome is exactly equal to ΔP (Chapman & Robbins, 1990).

There is also an established formal relationship between the Rescorla-Wagner model and the delta rule learning of a perceptron (Dawson, 2008; Gluck & Bower, 1988; Sutton & Barto, 1981). Thus by examining the equilibrium state of a perceptron facing a simple contingency problem, we can formally relate this kind of network to contingency theory and arrive at a formal understanding of what output unit activity represents.

When a continuous activation function is used in a perceptron, calculus can be used to determine the equilibrium of the perceptron. Let us do so for a single cue situation in which some cue, C, when presented, is rewarded a frequency of a times, and is not rewarded a frequency of b times. Similarly, when the cue is not presented, the perceptron is rewarded a frequency of c times and is not rewarded a frequency of d times. Note that to reward a perceptron is to train it to generate a desired response of 1, and that to not reward a perceptron is to train it to generate a desired response of 0, because the desired response indicates the presence or absence of the unconditioned stimulus (Dawson, 2008).

Assume that when the cue is present, the logistic activation function computes an activation value that we designate as o_c, and that when the cue is absent it returns the activation value designated as $o_{\sim c}$. We can now define the total error of responding for the perceptron, that is, its total error for the $(a + b + c + d)$ number of patterns that represent a single epoch, in which each instance of the contingency problem is presented once. For instance, on a trial in which C is presented and the perceptron is reinforced, the perceptron's error for that trial is the squared difference between the reward, 1, and o_c. As there are a of these trials, the total contribution of this type of trial to overall error is $a(1 - o_c)^2$. Applying this logic to the other three pairings of cue and outcome, total error E can be defined as follows:

$$E = a(1 - o_c)^2 + b(0 - o_c)^2 + c(1 - o_{\sim c})^2 + d(0 - o_{\sim c})^2$$
$$= a(1 - o_c)^2 + b(o_c)^2 + c(1 - o_{\sim c})^2 + d(o_{\sim c})^2$$

For a perceptron to be at equilibrium, it must have reached a state in which total error has been optimized, so that the error can no longer be decreased by using the delta rule to alter the perceptron's weight. To determine the equilibrium of the perceptron for the single cue contingency problem, we begin by taking the derivative of the error equation with respect to the activity of the perceptron when the cue is present, o_c:

$$\frac{\partial E}{\partial o_c} = 2(a(o_c - 1) + bo_c)$$

One condition of the perceptron at equilibrium is that o_c is a value that causes this derivative to be equal to 0. The equation below sets the derivative to 0 and solves for o_c. The result is $a/(a + b)$, which is equal to the conditional probability $P(O|C)$ if the single cue experiment is represented with a traditional contingency table:

$$0 = 2(a(o_c - 1) + bo_c)$$
$$= a(o_c - 1) + bo_c$$
$$= ao_c - a + bo_c$$
$$a = o_c(a + b)$$
$$\frac{a}{a + b} = o_c$$
$$P(O|C) = o_c$$

Similarly, we can take the derivative of the error equation with respect to the activity of the perceptron when the cue is *not* present, $o_{\sim c}$:

$$\frac{\partial E}{\partial o_{\sim c}} = 2(c(o_{\sim c} - 1) + do_{\sim c})$$

A second condition of the perceptron at equilibrium is that $o_{~c}$ is a value that causes the derivative above to be equal to 0. As before, we can set the derivative to 0 and solve for the value of $o_{~c}$. This time the result is $c/(c + d)$, which in a traditional contingency table is equal to the conditional probability $P(O|{\sim}C)$:

$$0 = 2(c(o_{~c} - 1) + do_{~c})$$
$$= c(o_{~c} - 1) + do_{~c}$$
$$= co_{~c} - c + do_{~c}$$
$$c = o_{~c}(c + d)$$
$$\frac{c}{c + d} = o_{~c}$$
$$P(O|{\sim}C) = o_{~c}$$

The main implication of the above equations is that they show that perceptron activity is literally a conditional probability. This provides a computational proof for the empirical hypothesis about perceptron activity that was generated from examining Figure 4-6.

A second implication of the proof is that when faced with the same contingency problem, a perceptron's equilibrium is not the same as that for the Rescorla-Wagner model. At equilibrium, the associative strength for the cue C that is determined by Rescorla-Wagner training is literally ΔP (Chapman & Robbins, 1990). This is *not* the case for the perceptron. For the perceptron, ΔP must be computed by taking the difference between its output when the cue is present and its output when the cue is absent. That is, ΔP is not directly represented as a connection weight, but instead is the difference between perceptron behaviours under different cue situations— that is, the difference between the conditional probability output by the perceptron when a cue is present and the conditional probability output by the perceptron when the cue is absent.

Importantly, even though the perceptron and the Rescorla-Wagner model achieve different equilibria for the same problem, it is clear that both are sensitive to contingency when it is formally defined as ΔP. Differences between the two reflect an issue that was raised in Chapter 2, that there exist many different possible algorithms for computing the same function. Key differences between the perceptron and the Rescorla-Wagner model—in particular, the fact that the former performs a nonlinear transformation on internal signals, while the latter does not—cause them to adopt very different structures, as indicated by different equilibria. Nonetheless, these very different systems are equally sensitive to exactly the same contingency.

This last observation has implications for the debate between contingency theory and associative learning (Cheng, 1997; Cheng & Holyoak, 1995; Shanks, 2007). In the current phase of this debate, modern contingency theories have been proposed as alternatives to Rescorla-Wagner learning. While in some instances equilibria for the Rescorla-Wagner model predict the conditional contrasts defined by a formal contingency theory like the power PC model, in other situations this is not the case (Cheng, 1997). However, the result above indicates that differences in equilibria do not necessarily reflect differences in system abilities. Clearly equilibrium differences cannot be used as the sole measure when different theories of contingency are compared.

4.8 Connectionist Algorithms: An Overview

In the last several sections we have explored connectionist cognitive science at the computational level of analysis. Claims about linear separability, the in-principle power of multilayer networks, and the interpretation of output unit activity have all been established using formal analyses.

In the next few sections we consider connectionist cognitive science from another perspective that it shares with classical cognitive science: the use of algorithmic-level investigations. The sections that follow explore how modern networks, which develop internal representations with hidden units, are trained, and also describe how one might interpret the internal representations of a network after it has learned to accomplish a task of interest. Such interpretations answer the question *How does a network convert an input pattern into an output response?* — and thus provide information about network algorithms.

The need for algorithmic-level investigations is introduced by noting in Section 4.9 that most modern connectionist networks are multilayered, meaning that they have at least one layer of hidden units lying between the input units and the output units. This section introduces a general technique for training such networks, called the generalized delta rule. This rule extends empiricism to systems that can have powerful internal representations.

Section 4.10 provides one example of how the internal representations created by the generalized delta rule can be interpreted. It describes the analysis of a multilayered network that has learned to classify different types of musical chords. An examination of the connection weights between the input units and the hidden units reveals a number of interesting ways in which this network represents musical regularities. An examination of the network's hidden unit space shows how these musical regularities permit the network to rearrange different types of chord types so that they may then be carved into appropriate decision regions by the output units.

In section 4.11 a biologically inspired approach to discovering network algorithms is introduced. This approach involves wiretapping the responses of hidden units when the network is presented with various stimuli, and then using these responses to determine the trigger features that the hidden units detect. It is also shown that changing the activation function of a hidden unit can lead to interesting complexities in defining the notion of a trigger feature, because some kinds of hidden units capture families of trigger features that require further analysis.

In Section 4.12 we describe how interpreting the internal structure of a network begins to shed light on the relationship between algorithms and architectures. Also described is a network that, as a result of training, translates a classical model of a task into a connectionist one. This illustrates an intertheoretic reduction between classical and connectionist theories, raising the possibility that both types of theories can be described in the same architecture.

4.9 Empiricism and Internal Representations

The ability of hidden units to increase the computational power of artificial neural networks was well known to Old Connectionism (McCulloch & Pitts, 1943). Its problem was that while a learning rule could be used to train networks with no hidden units (Rosenblatt, 1958, 1962), no such rule existed for multilayered networks. The reason that a learning rule did not exist for multilayered networks was because learning was defined in terms of minimizing the error of unit responses. While it was straightforward to define output unit error, no parallel definition existed for hidden unit error. A hidden unit's error could not be defined because it was not related to any directly observable outcome (e.g., external behaviour). If a hidden unit's error could not be defined, then Old Connectionist rules could not be used to modify its connections.

The need to define and compute hidden unit error is an example of the credit assignment problem:

> In playing a complex game such as chess or checkers, or in writing a computer program, one has a definite success criterion—the game is won or lost. But in the course of play, each ultimate success (or failure) is associated with a vast number of internal decisions. If the run is successful, how can we assign credit for the success among the multitude of decisions? (Minsky, 1963, p. 432)

The credit assignment problem that faced Old Connectionism was the inability to assign the appropriate credit—or more to the point, the appropriate blame—to each hidden unit for its contribution to output unit error. Failure to solve this problem prevented Old Connectionism from discovering methods to make their most powerful networks belong to the domain of empiricism and led to its demise (Papert, 1988).

The rebirth of connectionist cognitive science in the 1980s (McClelland & Rumelhart, 1986; Rumelhart & McClelland, 1986c) was caused by the discovery of a solution to Old Connectionism's credit assignment problem. By employing a nonlinear but continuous activation function, calculus could be used to explore changes in network behaviour (Rumelhart, Hinton, & Williams, 1986b). In particular, calculus could reveal how an overall network error was altered, by changing a component deep within the network, such as a single connection between an input unit and a hidden unit. This led to the discovery of the "backpropagation of error" learning rule, sometimes known as the generalized delta rule (Rumelhart Hinton, & Williams, 1986b). The calculus underlying the generalized delta rule revealed that hidden unit error could be defined as the sum of weighted errors being sent backwards through the network from output units to hidden units.

The generalized delta rule is an error-correcting method for training multi-layered networks that shares many characteristics with the original delta rule for perceptrons (Rosenblatt, 1958, 1962; Widrow, 1962; Widrow & Hoff, 1960). A more detailed mathematical treatment of this rule, and its relationship to other connectionist learning rules, is provided by Dawson (2004). A less technical account of the rule is given below.

The generalized delta rule is used to train a multilayer perceptron to mediate a desired input-output mapping. It is a form of supervised learning, in which a finite set of input-output pairs is presented iteratively, in random order, during training. Prior to training, a network is a "pretty blank" slate; all of its connection weights, and all of the biases of its activation functions, are initialized as small, random numbers. The generalized delta rule involves repeatedly presenting input-output pairs and then modifying weights. The purpose of weight modification is to reduce overall network error.

A single presentation of an input-output pair proceeds as follows. First, the input pattern is presented, which causes signals to be sent to hidden units, which in turn activate and send signals to the output units, which finally activate to represent the network's response to the input pattern. Second, the output unit responses are compared to the desired responses, and an error term is computed for each output unit. Third, an output unit's error is used to modify the weights of its connections. This is accomplished by adding a weight change to the existing weight. The weight change is computed by multiplying four different numbers together: a learning rate, the derivative of the unit's activation function, the output unit's error, and the current activity at the input end of the connection. Up to this point, learning is functionally the same as performing gradient descent training on a perceptron (Dawson, 2004).

The fourth step differentiates the generalized delta rule from older rules: each hidden unit computes its error. This is done by treating an output unit's error as if it were activity and sending it backwards as a signal through a connection to a hidden unit. As this signal is sent, it is multiplied by the weight of the connection. Each hidden unit computes its error by summing together all of the error signals that it receives from the output units to which it is connected. Fifth, once the hidden unit error has been computed, the weights of the hidden units can be modified using the same equation that was used to alter the weights of each of the output units.

This procedure can be repeated iteratively if there is more than one layer of hidden units. That is, the error of each hidden unit in one layer can be propagated backwards to an adjacent layer as an error signal once the hidden unit weights have been modified. Learning about this pattern stops once all of the connections have been modified. Then the next training pattern can be presented to the input units, and the learning process occurs again.

There are a variety of different ways in which the generic algorithm given above can be realized. For instance, in stochastic training, connection weights are updated after each pattern is presented (Dawson, 2004). This approach is called stochastic because each pattern is presented once per epoch of training, but the order of presentation is randomized for each epoch. Another approach, batch training, is to accumulate error over an epoch and to only update weights once at the end of the epoch, using accumulated error (Rumelhart, Hinton, & Williams, 1986a). As well, variations of the algorithm exist for different continuous activation functions. For instance, an elaborated error term is required to train units that have Gaussian activation functions, but when this is done, the underlying mathematics are essentially the same as in the original generalized delta rule (Dawson & Schopflocher, 1992b).

New Connectionism was born when the generalized delta rule was invented. Interestingly, the precise date of its birth and the names of its parents are not completely established. The algorithm was independently discovered more than once. Rumelhart, Hinton, and Williams (1986a, 1986b) are its most famous discoverers and popularizers. It was also discovered by David Parker in 1985 and by Yann LeCun in 1986 (Anderson, 1995). More than a decade earlier, the algorithm was reported in Paul Werbos' (1974) doctoral thesis. The mathematical foundations of the generalized delta rule can be traced to an earlier decade, in a publication by Shun-Ichi Amari (1967).

In an interview (Anderson & Rosenfeld, 1998), neural network pioneer Stephen Grossberg stated that "Paul Werbos, David Parker, and Shun-Ichi Amari should have gotten credit for the backpropagation model, instead of Rumelhart, Hinton, and Williams" (pp. 179–180). Regardless of the credit assignment problem associated with the scientific history of this algorithm, it transformed cognitive science in the mid-1980s, demonstrating "how the lowly concepts of *feedback* and *derivatives* are the essential building blocks needed to understand *and replicate* higher-order phenomena like learning, emotion and intelligence at all levels of the human mind" (Werbos, 1994, p. 1).

4.10 Chord Classification by a Multilayer Perceptron

Artificial neural networks provide a medium in which to explore empiricism, for they acquire knowledge via experience. This knowledge is used to mediate an input-output mapping and usually takes the form of a distributed representation. Distributed representations provide some of the putative connectionist advantages over classical cognitive science: damage resistance, graceful degradation, and so on. Unfortunately, distributed representations are also tricky to interpret, making it difficult for them to provide new theories for cognitive science.

However, interpreting the internal structures of multilayered networks, though difficult, is not impossible. To illustrate this, let us consider a multilayer perceptron trained to classify different types of musical chords. The purpose of this section is to discuss the role of hidden units, to demonstrate that networks that use hidden units can also be interpreted, and to introduce a decidedly connectionist notion called the coarse code.

Chords are combinations of notes that are related to musical scales, where a scale is a sequence of notes that is subject to certain constraints. A chromatic scale is one in which every note played is one semitone higher than the previous note. If one were to play the first thirteen numbered piano keys of Figure 4-7 in order, then the result would be a chromatic scale that begins on a low C and ends on another C an octave higher.

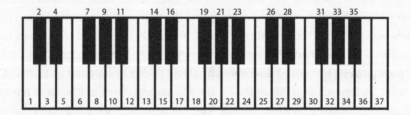

Figure 4-7. A small piano keyboard with numbered keys. Key 1 is C.

A major scale results by constraining a chromatic scale such that some of its notes are *not* played. For instance, the C major scale is produced if *only* the white keys numbered from 1 to 13 in Figure 4-7 are played in sequence (i.e., if the black keys numbered 2, 4, 7, 9, and 11 are *not* played).

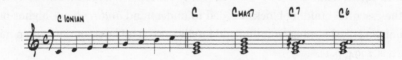

Figure 4-8. The C major scale and some of its added note chords.

The musical notation for the C major scale is provided in the sequence of notes illustrated in the first part of Figure 4-8. The Greeks defined a variety of modes for each scale; different modes were used to provoke different aesthetic experiences (Hanslick, 1957). The C major scale in the first staff of Figure 4-8 is in the Ionian mode because it begins on the note C, which is the root note, designated *I*, for the C major key.

One can define various musical chords in the context of C major in two different senses. First, the key signature of each chord is the same as C major (i.e., no sharps or flats). Second, each of these chords is built on the root of the C major scale (the note C). For instance, one basic chord is the major triad. In the key of C major, the root of this chord—the chord's lowest note—is C (e.g., piano key #1 in Figure 4-7). The major triad for this key is completed by adding two other notes to this root. The second note in the triad is 4 semitones higher than C, which is the note E (the third note in the major scale in Figure 4-8). The third note in the triad is 3 semitones higher than the second note, which in this case is G (the fifth note in the major scale in Figure 4-8). Thus the notes C-E-G define the major triad for the key of C; this is the first chord illustrated in Figure 4-8.

A fourth note can added on to any major triad to create an "added note" tetrachord (Baker, 1982). The type of added note chord that is created depends upon the relationship between the added note and the third note of the major triad. If the added note is 4 semitones higher than the third note, the result is a major 7th chord, such as the Cmaj7 illustrated in Figure 4-8. If the added note is 3 semitones higher than the third note, the result is a dominant 7th chord such as the C7 chord presented in Figure 4-8. If the added note is 2 semitones higher than the third note, then the result is a 6th chord, such as the C6 chord illustrated in Figure 4-8.

The preceding paragraphs described the major triad and some added note chords for the key of C major. In Western music, C major is one of twelve possible major keys. The set of all possible major keys is provided in Figure 4-9, which organizes them in an important cyclic structure, called the circle of fifths.

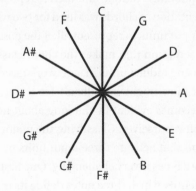

Figure 4-9. The circle of fifths.

The circle of fifths includes all 12 notes in a chromatic scale, but arranges them so that adjacent notes in the circle are a musical interval of a perfect fifth (i.e., 7 semitones) apart. The circle of fifths is a standard topic for music students, and it is foundational to many concepts in music theory. It is provided here, though, to be contrasted later with "strange circles" that are revealed in the internal structure of a network trained to identify musical chords.

Any one of the notes in the circle of fifths can be used to define a musical key and therefore can serve as the root note of a major scale. Similarly, any one of these notes can be the root of a major triad created using the pattern of root + 4 semitones + 3 semitones that was described earlier for the key of C major (Baker, 1982). Furthermore, the rules described earlier can also be applied to produce added note chords for any of the 12 major key signatures. These possible major triads and added note chords were used as inputs for training a network to correctly classify different types of chords, ignoring musical key.

A training set of 48 chords was created by building the major triad, as well as the major 7th, dominant 7th, and 6th chord for each of the 12 possible major key signatures (i.e., using each of the notes in Figure 4-9 as a root). When presented with a chord, the network was trained to classify it into one of the four types of interest: major triad, major 7th, dominant 7th, or 6th. To do so, the network had 4 output units, one for each type of chord. For any input, the network learned to turn the correct output unit on and to turn the other three output units off.

The input chords were encoded with a pitch class representation (Laden & Keefe, 1989; Yaremchuk & Dawson, 2008). In a pitch class representation, only 12 input units are employed, one for each of the 12 different notes that can appear in a scale. Different versions of the same note (i.e., the same note played at differ-. ent octaves) are all mapped onto the same input representation. For instance, notes 1, 13, 25, and 37 in Figure 4-7 all correspond to different pitches but belong to the same pitch *class*—they are all C notes, played at different octaves of the keyboard. In a pitch class representation, the playing of any of these input notes would be encoded by turning on a single input unit—the one unit used to represent the pitch class of C.

A pitch class representation of chords was used for two reasons. First, it requires a very small number of input units to represent all of the possible stimuli. Second, it is a fairly abstract representation that makes the chord classification task difficult, which in turn requires using hidden units in a network faced with this task.

Why chord classification might be difficult for a network when pitch class encoding is employed becomes evident by thinking about how we might approach the problem if faced with it ourselves. Classifying the major chords is simple: they are the only input stimuli that activate three input units instead of four. However, classifying the other chord types is very challenging. One first has to determine what key the stimulus is in, identify which three notes define its major chord component, and then determine the relationship between the third note of the major chord component and the fourth "added" note. This is particularly difficult because of the pitch class representation, which throws away note-order information that might be useful in identifying chord type.

It was decided that the network that would be trained on the chord classification task would be a network of value units (Dawson & Schopflocher, 1992b). The hidden units and output units in a network of value units use a Gaussian activation function, which means that they behave as if they carve two parallel planes through a pattern space. Such networks can be trained with a variation of the generalized delta rule. This type of network was chosen for this problem for two reasons. First, networks of value units have emergent properties that make them easier to interpret than other types of networks trained on similar problems (Dawson, 2004; Dawson et al., 1994). One reason for this is because value units behave as if they are "tuned" to respond to very particular input signals. Second, previous research on different versions of chord classification problems had produced networks that revealed elegant internal structure (Yaremchuk & Dawson, 2005, 2008).

The simplest network of value units that could learn to solve the chord classification problem required three hidden units. At the start of training, the value of m for each unit was initialized as 0. (The value of m for a value unit is analogous to a threshold in other types of units [Dawson, Kremer, & Gannon, 1994; Dawson & Schopflocher, 1992b]; if a value unit's net input is equal to m then the unit generates a maximum activity of 1.00.) All connection weights were set to values randomly selected from the range between -0.1 and 0.1. The network was trained with a learning rate of 0.01 until it produced a "hit" for every output unit on every pattern. Because of the continuous nature of the activation function, a hit was defined as follows: a value of 0.9 or higher when the desired output was 1, and a value of 0.1 or lower when the desired output was 0. The network that is interpreted below learned the chord classification task after 299 presentations of the training set.

What is the role of a layer of hidden units? In a perceptron, which has no hidden units, input patterns can only be represented in a pattern space. Recall from the discussion of Figure 4-2 that a pattern space represents each pattern as a point in space. The dimensionality of this space is equal to the number of input units. The coordinates of each pattern's point in this space are given by the activities of the input units. For some networks, the positioning of the points in the pattern space prevents some patterns from being correctly classified, because the output units are unable to adequately carve the pattern space into the appropriate decision regions.

In a multilayer perceptron, the hidden units serve to solve this problem. They do so by transforming the pattern space into a hidden unit space (Dawson, 2004). The dimensionality of a hidden unit space is equal to the number of hidden units in the layer. Patterns are again represented as points in this space; however, in this space their coordinates are determined by the activities they produce in each hidden unit. The hidden unit space is a transformation of the pattern space that involves detecting higher-order features. This usually produces a change in dimensionality—the hidden unit space often has a different number of dimensions than does the pattern space—and a repositioning of the points in the new space. As a result, the output units are able to carve the hidden unit space into a set of decision regions that permit all of the patterns, repositioned in the hidden unit space, to be correctly classified.

This account of the role of hidden units indicates that the interpretation of the internal structure of a multilayer perceptron involves answering two different questions. First, what kinds of features are the hidden units detecting in order to map patterns from the pattern space into the hidden unit space? Second, how do the output units process the hidden unit space to solve the problem of interest? The chord classification network can be used to illustrate how both questions can be addressed.

First, when mapping the input patterns into the hidden unit space, the hidden units must be detecting some sorts of musical regularities. One clue as to what these regularities may be is provided by simply examining the connection weights that feed into them, provided in Table 4-2.

Input Note	Hidden 1	Hidden 1 Class	Hidden 2	Hidden 2 Class	Hidden 3	Hidden 3 Class
B	0.53	Circle of Major Thirds 1	0.12	Circle of Major Thirds 1	0.75	Circle of Major Seconds 1
D#	0.53		0.12		0.75	
G	0.53		0.12		0.75	
A	−0.53	Circle of Major Thirds 2	−0.12	Circle of Major Thirds 2	0.75	
C#	−0.53		−0.12		0.75	
F	−0.53		−0.12		0.75	
C	0.12	Circle of Major Thirds 3	−0.53	Circle of Major Thirds 3	−0.77	Circle of Major Seconds 2
G#	0.12		−0.53		−0.77	
E	0.12		−0.53		−0.77	
F#	−0.12	Circle of Major Thirds 4	0.53	Circle of Major Thirds 4	−0.77	
A#	−0.12		0.53		−0.77	
D	−0.12		0.53		−0.77	

Table 4-2. Connection weights from the 12 input units to each of the three hidden units. Note that the first two hidden units adopt weights that assign input notes to the four circles of major thirds. The third hidden unit adopts weights that assign input notes to the two circles of major seconds.

In the pitch class representation used for this network, each input unit stands for a distinct musical note. As far as the hidden units are concerned, the "name" of each note is provided by the connection weight between the input unit and the hidden unit. Interestingly, Table 4-2 reveals that all three hidden units take input notes that we would take as being different (because they have different names, as in the circle of fifths in Figure 4-9) and treat them as being identical. That is, the hidden units assign the same "name," or connection weight, to input notes that we would give different names to.

Furthermore, assigning the same "name" to different notes by the hidden units is not done randomly. Notes are assigned according to strange circles, that is, circles of major thirds and circles of major seconds. Let us briefly describe these circles, and then return to an analysis of Table 4-2.

The circle of fifths (Figure 4-9) is not the only way in which notes can be arranged geometrically. One can produce other circular arrangements by exploiting other musical intervals. These are strange circles in the sense that they would very rarely be taught to music students as part of a music theory curriculum. However, these strange circles are formal devices that can be as easily defined as can be the circle of fifths.

For instance, if one starts with the note C and moves up a major second (2 semitones) then one arrives at the note D. From here, moving up another major second arrives at the note E. This can continue until one circles back to C but an octave higher than the original, which is a major second higher than A#. This circle of major seconds captures half of the notes in the chromatic scale, as is shown in the top part of Figure 4-10. A complementary circle of major seconds can also be constructed (bottom circle of Figure 4-10); this circle contains all the remaining notes that are not part of the first circle.

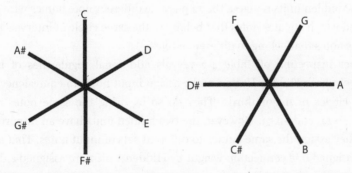

Figure 4-10. The two circles of major seconds.

An alternative set of musical circles can be defined by exploiting a different musical interval. In each circle depicted in Figure 4-11, adjacent notes are a major third (4 semitones) apart. As shown in Figure 4-11 four such circles are possible.

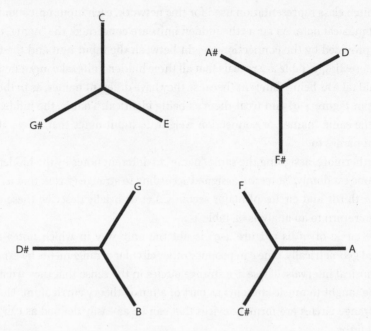

Figure 4-11. The four circles of major thirds.

What do these strange circles have to do with the internal structure of the network trained to classify the different types of chords? A close examination of Table 4-2 indicates that these strange circles are reflected in the connection weights that feed into the network's hidden units. For Hidden Units 1 and 2, if notes belong to the same circle of major thirds (Figure 4-11), then they are assigned the same connection weight. For Hidden Unit 3, if notes belong to the same circle of major seconds (Figure 4-10), then they are assigned the same connection weight. In short, each of the hidden units replaces the 12 possible different note names with a much smaller set, which equates notes that belong to the same circle of intervals and differentiates notes that belong to different circles.

Further inspection of Table 4-2 reveals additional regularities of interest. Qualitatively, both Hidden Units 1 and 2 assign input notes to equivalence classes based on circles of major thirds. They do so by using the same note "names": 0.53, 0.12, −0.12, and −0.53. However, the two hidden units have an important difference: they assign the same names to different sets of input notes. That is, notes that are assigned one connection weight by Hidden Unit 1 are assigned a different connection weight by Hidden Unit 2.

The reason that the difference in weight assignment between the two hidden units is important is that the behaviour of each hidden unit is not governed by a single incoming signal, but is instead governed by a combination of three or four input signals coming from all of the units. The connection weights used by the hidden units place meaningful constraints on how these signals are combined.

Let us consider the role of the particular connection weights used by the hidden units. Given the binary nature of the input encoding, the net input of any hidden unit is simply the sum of the weights associated with each of the activated input units. For a value unit, if the net input is equal to the value of the unit's m then the output generates a maximum value of 1.00. As the net input moves away from m in either a positive or negative direction, activity quickly decreases. At the end of training, the values of m for the three hidden units were 0.00, 0.00, and −0.03 for Hidden Units 1, 2, and 3, respectively. Thus for each hidden unit, if the incoming signals are essentially zero—that is if all the incoming signals cancel each other out—then high activity will be produced.

Why then do Hidden Units 1 and 2 use the same set of four connection weights but assign these weights to different sets of input notes? The answer is that these hidden units capture similar chord relationships but do so using notes from different strange circles.

This is shown by examining the responses of each hidden unit to each input chord after training. Table 4-3 summarizes these responses, and shows that each hidden unit generated identical responses to different subsets of input chords.

Chord	Input Chord Chord Root	Activation Hid1	Hid2	Hid3
Major	C, D, A, F#, G#, A#	0.16	0.06	0.16
	C#, D#, F, G, A, B	0.06	0.16	0.16
Major7	C, D, A, F#, G#, A#	0.01	0.12	1.00
	C#, D#, F, G, A, B	0.12	0.01	1.00
Dom7	C, D, A, F#, G#, A#	0.27	0.59	0.00
	C#, D#, F, G, A, B	0.59	0.27	0.00
6th	C, D, A, F#, G#, A#	0.84	0.03	1.00
	C#, D#, F, G, A, B	0.03	0.84	1.00

Table 4-3. The activations produced in each hidden unit by different subsets of input chords.

From Table 4-3, one can see that the activity of Hidden Unit 3 is simplest to describe: when presented with a dominant 7th chord, it produces an activation of 0 and a weak activation to a major triad. When presented with either a major 7th or a 6th chord, it produces maximum activity. This pattern of activation is easily explained by considering the weights that feed into Hidden Unit 3 (Table 4-2). Any major 7th or 6th chord is created out of two notes from one circle of major seconds and two notes from the other circle. The sums of pairs of weights from different circles cancel each other out, producing near-zero net input and causing maximum activation.

In contrast, the dominant 7th chords use three notes from one circle of major seconds and only one from the other circle. As a result, the signals do not cancel out completely, given the weights in Table 4-2. Instead, a strong non-zero net input is produced, and the result is zero activity.

Finally, any major triad involves only three notes: two from one circle of major seconds and one from the other. Because of the odd number of input signals, cancellation to zero is not possible. However, the weights have been selected so that the net input produced by a major triad is close enough to m to produce weak activity.

The activation patterns for Hidden Units 1 and 2 are more complex. It is possible to explain all of them in terms of balancing (or failing to balance) signals associated with different circles of major thirds. However, it is more enlightening to consider these two units at a more general level, focusing on the relationship between their activations.

In general terms, Hidden Units 1 and 2 generate activations of different intensities to different classes of chords. In general, they produce the highest activity to 6th chords and the lowest activity to major 7th chords. Importantly, they do not generate the same activity to all chords of the same type. For instance, for the 12 possible 6th chords, Hidden Unit 1 generates activity of 0.84 to 6 of them but activity of only 0.03 to the other 6 chords. An inspection of Table 4-3 indicates that for every chord type, both Hidden Units 1 and 2 generate one level of activity with half of them, but produce another level of activity with the other half.

The varied responses of these two hidden units to different chords of the same type are related to the circle of major seconds (Figure 4-10). For example, Hidden Unit 1 generates a response of 0.84 to 6th chords whose root note belongs to the top circle of Figure 4-10, and a response of 0.03 to 6th chords whose root note belongs to the bottom circle of Figure 4-10. Indeed, for all of the chord types, both of these hidden units generate one response if the root note belongs to one circle of major seconds and a different response if the root note belongs to the other circle.

Furthermore, the responses of Hidden Units 1 and 2 complement one another: for any chord type, those chords that produce low activity in Hidden Unit 1 produce higher activity in Hidden Unit 2. As well, those chords that produce low activity in Hidden Unit 2 produce higher activity in Hidden Unit 1. This complementing is again related to the circles of major seconds: Hidden Unit 1 generates higher responses to chords whose root belongs to one circle, while Hidden Unit 2 generates higher responses to chords whose roots belong to the other. Which circle is "preferred" by a hidden unit depends on chord type.

Clearly each of the three hidden units is sensitive to musical properties. However, it is not clear how these properties support the network's ability to classify chords. For instance, none of the hidden units by themselves pick out a set of properties that uniquely define a particular type of chord. Instead, hidden units generate some activity to different chord types, suggesting the existence of a coarse code.

In order to see how the activities of the hidden units serve as a distributed representation that mediates chord classification, we must examine the hidden unit space. The hidden unit space plots each input pattern as a point in a space whose dimensionality is determined by the number of hidden units. The coordinates of the point in the hidden unit space are the activities produced by an input pattern in each hidden unit. The three-dimensional hidden unit space for the chord classification network is illustrated in Figure 4-12.

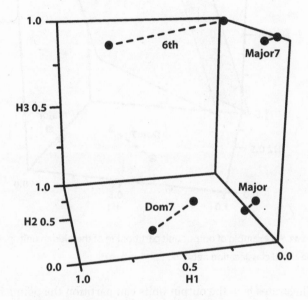

Figure 4-12. The hidden unit space for the chord classification network. H1, H2, and H3 provide the activity of hidden units 1, 2, and 3 respectively.

Because the hidden units generate identical responses to many of the chords, instead of 48 different visible points in this graph (one for each input pattern), there are only 8. Each point represents 6 different chords that fall in exactly the same location in the hidden unit space.

The hidden unit space reveals that each chord type is represented by two different points. That these points capture the same class is represented in Figure 4-12 by joining a chord type's points with a dashed line. Two points are involved in defining a chord class in this space because, as already discussed, each hidden unit is sensitive to the organization of notes according to the two circles of major seconds. For each chord type, chords whose root belongs to one of these circles are mapped to one point, and chords whose root belongs to the other are mapped to the other point. Interestingly, there is no systematic relationship in the graph that maps onto the two circles. For instance, it is not the case that the four points toward the back of the Figure 4-12 cube all map onto the same circle of major seconds.

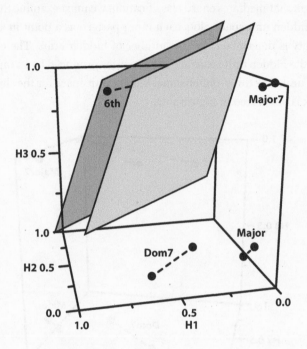

Figure 4-13. An example of output unit partitioning of the hidden unit space for the chord classification network.

Figure 4-13 illustrates how the output units can partition the points in the hidden unit space in order to classify chords. Each output unit in this network is a value unit, which carves two parallel hyperplanes through a pattern space. To solve the chord classification problem, the connection weights and the bias of each output unit must take on values that permit these two planes to isolate the two points associated with one chord type from all of the other points in the space. Figure 4-13 shows how this would be accomplished by the output unit that signals that a 6th chord has been detected.

4.11 Trigger Features

For more than half a century, neuroscientists have studied vision by mapping the receptive fields of individual neurons (Hubel & Wiesel, 1959; Lettvin, Maturana, McCulloch, & Pitts, 1959). To do this, they use a method called microelectrode recording or wiretapping (Calvin & Ojemann, 1994), in which the responses of single neurons are measured while stimuli are being presented to an animal. With this technique, it is possible to describe a neuron as being sensitive to a trigger feature, a specific pattern that when detected produces maximum activity in the cell.

That individual neurons may be described as detecting trigger features has led some to endorse a neuron doctrine for perceptual psychology. This doctrine has the goal of discovering the trigger features for all neurons (Barlow, 1972, 1995). This is because,

> a description of that activity of a single nerve cell which is transmitted to and influences other nerve cells, and of a nerve cell's response to such influences from other cells, is a complete enough description for functional understanding of the nervous system. (Barlow, 1972, p. 380)

The validity of the neuron doctrine is a controversial issue (Bowers, 2009; Gross, 2002). Regardless, there is a possibility that identifying trigger features can help to interpret the internal workings of artificial neural networks.

For some types of hidden units, trigger features can be identified analytically, without requiring any wiretapping of hidden unit activities (Dawson, 2004). For instance, the activation function for an integration device (e.g., the logistic equation) is monotonic, which means that increases in net input always produce increases in activity. As a result, if one knows the maximum and minimum possible values for input signals, then one can define an integration device's trigger feature simply by inspecting the connection weights that feed into it (Dawson, Kremer, & Gannon, 1994). The trigger feature is that pattern which sends the minimum signal through every inhibitory connection and the maximum signal through every excitatory connection. The monotonicity of an integration device's activation function ensures that it will have only one trigger feature.

The notion of a trigger feature for other kinds of hidden units is more complex. Consider a value unit whose bias, m, in its Gaussian activation function is equal to 0. The trigger feature for this unit will be the feature that causes it to produce maximum activation. For this value unit, this will occur when the net input to the unit is equal to 0 (i.e., equal to the value of μ) (Dawson & Schopflocher, 1992b). The net input of a value unit is defined by a particular linear algebra operation, called the inner product, between a vector that represents a stimulus and a vector that represents the connection weights that fan into the unit (Dawson, 2004). So, when net input equals 0, this means that the inner product is equal to 0.

However, when an inner product is equal to 0, this indicates that the two vectors being combined are orthogonal to one another (that is, there is an angle of 90° between the two vectors). Geometrically speaking, then, the trigger feature for a value unit is an input pattern represented by a vector of activities that is at a right angle to the vector of connection weights.

This geometric observation raises complications, because it implies that a hidden value unit will *not* have a single trigger feature. This is because there are many input patterns that are orthogonal to a vector of connection weights. *Any* input vector that lies in the hyperplane that is perpendicular to the vector of connection weights will serve as a trigger feature for the hidden value unit (Dawson, 2004); this is illustrated in Figure 4-14.

Another consequence of the geometric account provided above is that there should be families of other input patterns that share the property of producing the same hidden unit activity, but one that is lower than the maximum activity produced by one of the trigger features. These will be patterns that all fall into the same hyperplane, but this hyperplane is *not* orthogonal to the vector of connection weights.

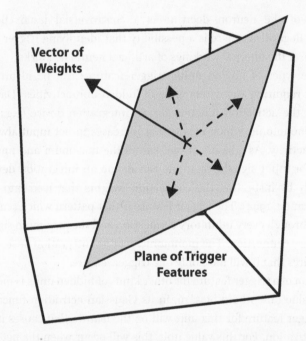

Figure 4-14. Any input pattern (dashed lines) whose vector falls in the plane orthogonal to the vector of connection weights (solid line) will be a trigger feature for a hidden value unit.

The upshot of all of this is that if one trains a network of value units and then wiretaps its hidden units, the resulting hidden unit activities should be highly organized. Instead of having a rectangular distribution of activation values, there should be regular groups of activations, where each group is related to a different family of input patterns (i.e., families related to different hyperplanes of input patterns).

Empirical support for this analysis was provided by the discovery of activity banding when a hidden unit's activities were plotted using a jittered density plot (Berkeley et al., 1995). A jittered density plot is a two-dimensional scatterplot of points; one such plot can be created for each hidden unit in a network. Each plotted point represents one of the patterns presented to the hidden unit during wiretapping. The x-value of the point's position in the graph is the activity produced in that hidden unit by the pattern. The y-value of the point's position in the scatterplot is a random value that is assigned to reduce overlap between points.

An example of a jittered density plot for a hidden value unit is provided in Figure 4-15. Note that the points in this plot are organized into distinct bands, which is consistent with the geometric analysis. This particular unit belongs to a network of value units trained on a logic problem discussed in slightly more detail below (Bechtel & Abrahamsen, 1991), and was part of a study that examined some of the implications of activity banding (Dawson & Piercey, 2001).

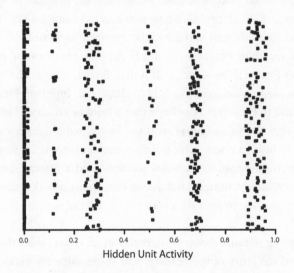

Figure 4-15. An example of banding in a jittered density plot of a hidden value unit in a network that was trained on a logic problem.

Bands in jittered density plots of hidden value units can be used to reveal the kinds of features that are being detected by these units. For instance, Berkeley et al. (1995) reported that all of the patterns that fell into the same band on a single jittered density plot in the networks did so because they shared certain local properties or features, which are called definite features.

There are two types of definite features. The first is called a definite unary feature. When a definite unary feature exists, it means that a single feature has the same value for every pattern in the band. The second is called a definite binary feature. With this kind of definite feature, an individual feature is not constant within a band. However, its *relationship* to some other feature is constant—variations in one feature are perfectly correlated with variations in another. Berkeley et al. (1995) showed how definite features could be both objectively defined and easily discovered using simple descriptive statistics (see also Dawson, 2005).

Definite features are always expressed in terms of the values of input unit activities. As a result, they can be assigned meanings using knowledge of a network's input unit encoding scheme.

One example of using this approach was presented in Berkeley et al.'s (1995) analysis of a network on the Bechtel and Abrahamsen (1991) logic task. This task consists of a set of 576 logical syllogisms, each of which can be expressed as a pattern of binary activities using 14 input units. Each problem is represented as a first sentence that uses two variables, a connective or a second sentence that states a variable, and a conclusion that states a variable. Four different problem types were created in this format: *modus ponens*, *modus tollens*, disjunctive syllogism, and alternative syllogism. Each problem type was created using one of three different connectives and four different variables: the connectives were *If...then*, *Or*, or *Not Both...And*; the variables were *A*, *B*, *C*, and *D*. An example of a valid modus ponens argument in this format is "Sentence 1: 'If A then B'; Sentence 2: 'A'; Conclusion: 'B'."

For this problem, a network's task is to classify an input problem into one of the four types and to classify it as being either a valid or an invalid example of that problem type. Berkeley et al. (1995) successfully trained a network of value units that employed 10 hidden units. After training, each of these units were wiretapped using the entire training set as stimulus patterns, and a jittered density plot was produced for each hidden unit. All but one of these plots revealed distinct banding. Berkeley et al. were able to provide a very detailed set of definite features for each of the bands.

After assigning definite features, Berkeley et al. (1995) used them to explore how the internal structure of the network was responsible for making the correct logical judgments. They expressed input logic problems in terms of which band of activity they belonged to for each jittered density plot. They then described each pattern as the combination of definite features from each of these bands, and they found that the internal structure of the network represented rules that were very classical in nature.

For example, Berkeley et al. (1995) found that every valid modus ponens problem was represented as the following features: having the connective *If...then*, having the first variable in Sentence 1 identical to Sentence 2, and having the second variable in Sentence 1 identical to the Conclusion. This is essentially the rule for valid modus ponens that could be taught in an introductory logic class (Bergmann, Moor, & Nelson, 1990). Berkeley et al. found several such rules; they also found a number that were not so traditional, but which could still be expressed in a classical form. This result suggests that artificial neural networks might be more symbolic in nature than connectionist cognitive scientists care to admit (Dawson, Medler, & Berkeley, 1997).

Importantly, the Berkeley et al. (1995) analysis was successful because the definite features that they identified were local. That is, by examining a single band in a single jittered density plot, one could determine a semantically interpretable set of features. However, activity bands are not always local. In some instances hidden value units produce nicely banded jittered density plots that possess definite features, but these features are difficult to interpret semantically (Dawson & Piercey, 2001). This occurs when the semantic interpretation is itself distributed across different bands for different hidden units; an interpretation of such a network requires definite features from multiple bands to be considered in concert.

While the geometric argument provided earlier motivated a search for the existence of bands in the hidden units of value unit networks, banding has been observed in networks of integration devices as well (Berkeley & Gunay, 2004). That being said, banding is not seen in every value unit network either. The existence of banding is likely an interaction between network architecture and problem representation; banding is useful when discovered, but it is only one tool available for network interpretation.

The important point is that practical tools exist for interpreting the internal structure of connectionist networks. Many of the technical issues concerning the relationship between classical and connectionist cognitive science may hinge upon network interpretations: "In our view, questions like 'What is a classical rule?' and 'Can connectionist networks be classical in nature?' are also hopelessly unconstrained. Detailed analyses of the internal structure of particular connectionist networks provide a specific framework in which these questions can be fruitfully pursued" (Dawson, Medler, & Berkeley, 1997, p. 39).

4.12 A Parallel Distributed Production System

One of the prototypical architectures for classical cognitive science is the production system (Anderson, 1983; Kieras & Meyer, 1997; Meyer et al., 2001; Meyer & Kieras, 1997a, 1997b; Newell, 1973, 1990; Newell & Simon, 1972). A production system is a set of condition-action pairs. Each production works in parallel, scanning working memory for a pattern that matches its condition. If a production finds such a match, then it takes control, momentarily disabling the other productions, and performs its action, which typically involves adding, deleting, copying, or moving symbols in the working memory.

Production systems have been proposed as a lingua franca for cognitive science, capable of describing any connectionist or embodied cognitive science theory and therefore of subsuming such theories under the umbrella of classical cognitive science (Vera & Simon, 1993). This is because Vera and Simon (1993) argued that *any* situation-action pairing can be represented either as a single production in a production system or, for complicated situations, as a set of productions. "Productions provide an essentially neutral language for describing the linkages between information and action at any desired (sufficiently high) level of aggregation" (p. 42). Other philosophers of cognitive science have endorsed similar positions. For instance, von Eckardt (1995) suggested that if one considers distributed representations in artificial neural networks as being "higher-level" representations, then connectionist networks can be viewed as being analogous to classical architectures. This is because when examined at this level, connectionist networks have the capacity to input and output represented information, to store represented information, and to manipulate represented information. In other words, the symbolic properties of classical architectures may emerge from what are known as the subsymbolic properties of networks (Smolensky, 1988).

However, the view that artificial neural networks are classical in general or examples of production systems in particular is not accepted by all connectionists. It has been claimed that connectionism represents a Kuhnian paradigm shift away from classical cognitive science (Schneider, 1987). With respect to Vera and Simon's (1993) particular analysis, their definition of symbol has been deemed too liberal by some neural network researchers (Touretzky & Pomerleau, 1994). Touretzky and Pomerlau (1994) claimed of a particular neural network discussed by Vera and Simon, ALVINN (Pomerleau, 1991), that its hidden unit "patterns are not arbitrarily shaped symbols, and they are not combinatorial. Its hidden unit feature detectors are tuned filters" (Touretzky & Pomerleau, 1994, p. 348). Others have viewed ALVINN from a position of compromise, noting that "some of the processes are symbolic and some are not" (Greeno & Moore, 1993, p. 54).

Are artificial neural networks equivalent to production systems? In the philosophy of science, if two apparently different theories are in fact identical, then one theory can be translated into the other. This is called intertheoretic reduction (Churchland, 1985, 1988; Hooker, 1979, 1981). The widely accepted view that classical and connectionist cognitive science are fundamentally different (Schneider, 1987) amounts to the claim that intertheoretic reduction between a symbolic model and a connectionist network is impossible. One research project (Dawson et al., 2000) directly examined this issue by investigating whether a production system model could be translated into an artificial neural network.

Dawson et al. (2000) investigated intertheoretic reduction using a benchmark problem in the machine learning literature, classifying a very large number (8,124) of mushrooms as being either edible or poisonous on the basis of 21 different features (Schlimmer, 1987). Dawson et al. (2000) used a standard machine learning technique, the ID3 algorithm (Quinlan, 1986) to induce a decision tree for the mushroom problem. A decision tree is a set of tests that are performed in sequence to classify patterns. After performing a test, one either reaches a terminal branch of the tree, at which point the pattern being tested can be classified, or a node of the decision tree, which is to say another test that must be performed. The decision tree is complete for a pattern set if every pattern eventually leads the user to a terminal branch. Dawson et al. (2000) discovered that a decision tree consisting of only five different tests could solve the Schlimmer mushroom classification task. Their decision tree is provided in Table 4-4.

Step	Tests and Decision Points
1	*What is the mushroom's odour?* If it is almond or anise then it is edible. **(Rule 1 Edible)** If it is creosote or fishy or foul or musty or pungent or spicy then it is poisonous. **(Rule 1 Poisonous)** If it has no odour then proceed to Step 2.
2	*Obtain the spore print of the mushroom.* If the spore print is black or brown or buff or chocolate or orange or yellow then it is edible. **(Rule 2 Edible)** If the spore print is green or purple then it is poisonous. **(Rule 2 Poisonous)** If the spore print is white then proceed to Step 3.
3	*Examine the gill size of the mushroom.* If the gill size is broad, then it is edible. **(Rule 3 Edible)** If the gill size is narrow, then proceed to Step 4.
4	*Examine the stalk surface above the mushroom's ring.* If the surface is fibrous then it is edible. **(Rule 4 Edible)** If the surface is silky or scaly then it is poisonous. **(Rule 4 Poisonous)** If the surface is smooth then proceed to Step 5.
5	*Examine the mushroom for bruises.* If it has no bruises then it is edible. **(Rule 5 Edible)** If it has bruises then it is poisonous. **(Rule 5 Poisonous)**

Table 4-4. Dawson et al.'s (2000) step decision tree for classifying mushrooms. Decision points in this tree where mushrooms are classified (e.g., *Rule 1 Edible*) are given in bold.

The decision tree provided in Table 4-4 is a classical theory of how mushrooms can be classified. It is not surprising, then, that one can translate this decision tree into the lingua franca: Dawson et al. (2000) rewrote the decision tree as an equivalent set of production rules. They did so by using the features of mushrooms that must be true at each terminal branch of the decision tree as the conditions for a production. The action of this production is to classify the mushroom (i.e., to assert that a mushroom is either edible or poisonous). For instance, at the Rule 1 Edible decision point in Table 4-4, one could create the following production rule: "If the odour is anise or almond, then the mushroom is edible." Similar productions can be created for later decision points in the algorithm; these productions will involve a longer list of mushroom features. The complete set of productions that were created for the decision tree algorithm is provided in Table 4-5.

Dawson et al. (2000) trained a network of value units to solve the mushroom classification problem and to determine whether a classical model (such as the decision tree from Table 4-4 or the production system from Table 4-5) could be translated into a network. To encode mushroom features, their network used 21 input units, 5 hidden value units, and 10 output value units. One output unit encoded the edible/poisonous classification—if a mushroom was edible, this unit was trained to turn on; otherwise this unit was trained to turn off.

Decision Point From Table 4-4	Equivalent Production	Network Cluster
Rule 1 Edible	P1: if (odor = anise) ∨ (odor = almond) → edible	2 or 3
Rule 1 Poisonous	P2: if (odor ≠ anise) ∧ (odor ≠ almond) ∧ (odor ≠ none) → not edible	1
Rule 2 Edible	P3: if (odor = none) ∧ (spore print colour ≠ green) ∧ (spore print colour ≠ purple) ∧ (spore print colour [1] white) → edible	9
Rule 2 Poisonous	P4: if (odor = none) ∧ ((spore print colour = green) ∨ (spore print colour = purple)) → not edible	6
Rule 3 Edible	P5: if (odor = none) ∧ (spore print colour = white) ∧ (gill size = broad) → edible	4
Rule 4 Edible	P6: if (odor = none) ∧ (spore print colour = white) ∧ (gill size = narrow) ∧ (stalk surface above ring = fibrous) → edible	7 or 11

Decision Point From Table 4-4	Equivalent Production	Network Cluster
Rule 4 Poisonous	P7: if (odor = none) ∧ (spore print colour = white) ∧ (gill size = narrow) ∧ ((stalk surface above ring = silky) ∨ (stalk surface above ring = scaly)) → not edible	5
Rule 5 Edible	P8: if (odor = none) ∧ (spore print colour = white) ∧ (gill size = narrow) ∧ (stalk surface above ring = smooth) ∧ (bruises = no) → edible	8 or 12
Rule 5 Poisonous	P9: if (odor = none) ∧ (spore print colour = white) ∧ (gill size = narrow) ∧ (stalk surface above ring = smooth) ∧ (bruises = yes) → not edible	10

Table 4-5. Dawson et al.'s (2000) production system translation of Table 4-4. Conditions are given as sets of features. The *Network Cluster* column pertains to their artificial neural network trained on the mushroom problem and is described later in the text.

The other nine output units were used to provide extra output learning, which was the technique employed to insert a classical theory into the network. Normally, a pattern classification system is only provided with information about what correct pattern labels to assign. For instance, in the mushroom problem, the system would typically only be taught to generate the label *edible* or the label *poisonous*. However, more information about the pattern classification task is frequently available. In particular, it is often known *why* an input pattern belongs to one class or another. It is possible to incorporate this information to the pattern classification problem by teaching the system not only to assign a pattern to a class (e.g., "edible", "poisonous") but to also generate a reason for making this classification (e.g., "passed Rule 1", "failed Rule 4"). Elaborating a classification task along such lines is called the injection of hints or extra output learning (Abu-Mostafa, 1990; Suddarth & Kergosien, 1990).

Dawson et al. (2000) hypothesized that extra output learning could be used to insert the decision tree from Table 4-4 into a network. Table 4-4 provides nine different terminal branches of the decision tree at which mushrooms are assigned to categories ("Rule 1 edible", "Rule 1 poisonous", "Rule 2 edible", etc.). The network learned to "explain" why it classified an input pattern in a particular way by turning on one of the nine extra output units to indicate which terminal branch of the decision tree was involved. In other words, the network (which required 8,699 epochs of training on the 8,124 different input patterns!) classified networks "for the same reasons" as would the decision tree. This is why Dawson et al. hoped that this classical theory would literally be translated into the network.

Apart from the output unit behaviour, how could one support the claim that a classical theory had been translated into a connectionist network? Dawson et al. (2000) interpreted the internal structure of the network in an attempt to see whether such a network analysis would reveal an internal representation of the classical algorithm. If this were the case, then standard training practices would have succeeded in translating the classical algorithm into a PDP network.

One method that Dawson et al. (2000) used to interpret the trained network was a multivariate analysis of the network's hidden unit space. They represented each mushroom as the vector of five hidden unit activation values that it produced when presented to the network. They then performed a k-means clustering of this data. The k-means clustering is an iterative procedure that assigns data points to k different clusters in such a way that each member of a cluster is closer to the centroid of that cluster than to the centroid of any other cluster to which other data points have been assigned.

However, whenever cluster analysis is performed, one question that must be answered is *How many clusters should be used?*—in other words, what should the value of k be?. An answer to this question is called a stopping rule. Unfortunately, no single stopping rule has been agreed upon (Aldenderfer & Blashfield, 1984; Everitt, 1980). As a result, there exist many different types of methods for determining k (Milligan & Cooper, 1985).

While no general method exists for determining the optimal number of clusters, one can take advantage of heuristic information concerning the domain being clustered in order to come up with a satisfactory stopping rule for this domain. Dawson et al. (2000) argued that when the hidden unit activities of a trained network are being clustered, there must be a correct mapping from these activities to output responses, because one trained network itself has discovered one such mapping. They used this position to create the following stopping rule: "Extract the smallest number of clusters such that every hidden unit activity vector assigned to the same cluster produces the same output response in the network." They used this rule to determine that the k-means analysis of the network's hidden unit activity patterns required the use of 12 different clusters.

Dawson et al. (2000) then proceeded to examine the mushroom patterns that belonged to each cluster in order to determine what they had in common. For each cluster, they determined the set of descriptive features that each mushroom shared. They realized that each set of shared features they identified could be thought of as a condition, represented internally by the network as a vector of hidden unit activities, which results in the network producing a particular action, in particular, the edible/poisonous judgement represented by the first output unit.

For example, mushrooms that were assigned to Cluster 2 had an odour that was either almond or anise, which is represented by the network's five hidden units adopting a particular vector of activities. These activities serve as a condition that causes the network to assert that the mushroom is edible.

By interpreting a hidden unit vector in terms of condition features that are prerequisites to network responses, Dawson et al. (2000) discovered an amazing relationship between the clusters and the set of productions in Table 4-5. They determined that each distinct class of hidden unit activities (i.e., each cluster) corresponded to one, and only one, of the productions listed in the table. This mapping is provided in the last column of Table 4-5. In other words, when one describes the network as generating a response because its hidden units are in one state of activity, one can translate this into the claim that the network is executing a particular production. This shows that the extra output learning translated the classical algorithm into a network model.

The translation of a network into a production system, or vice versa, is an example of new wave reductionism (Bickle, 1996; Endicott, 1998). In new wave reductionism, one does not reduce a secondary theory directly to a primary theory. Instead, one takes the primary theory and constructs from it a structure that is analogous to the secondary theory, but which is created in the vocabulary of the primary theory. Theory reduction involves constructing a mapping between the secondary theory and its image constructed from the primary theory. "The older theory, accordingly, is never deduced; it is just the target of a relevantly adequate mimicry" (Churchland, 1985, p. 10).

Dawson et al.'s (2000) interpretation is a new wave intertheoretic reduction because the production system of Table 4-5 represents the intermediate structure that is analogous to the decision tree of Table 4-4. "Adequate mimicry" was established by mapping different classes of hidden unit states to the execution of particular productions. In turn, there is a direct mapping from any of the productions back to the decision tree algorithm. Dawson et al. concluded that they had provided an exact translation of a classical algorithm into a network of value units.

The relationship between hidden unit activities and productions in Dawson et al.'s (2000) mushroom network is in essence an example of equivalence between symbolic and subsymbolic accounts. This implies that one cannot assume that classical models and connectionist networks are fundamentally different at the algorithmic level, because one type of model can be translated into the other. It is possible to have a classical model that is exactly equivalent to a PDP network.

This result provides very strong support for the position proposed by Vera and Simon (1993). The detailed analysis provided by Dawson et al. (2000) permitted them to make claims of the type "Network State x is equivalent to Production y." Of course, this one result cannot by itself validate Vera and Simon's argument. For instance, can any classical theory be translated into a network? This is one type of algorithmic-level issue that requires a great deal of additional research. As well, the translation works both ways: perhaps artificial neural networks provide a biologically plausible lingua franca for classical architectures!

4.13 Of Coarse Codes

The notion of representation in classical cognitive science is tightly linked to the structure/process distinction that is itself inspired by the digital computer. An explicit set of rules is proposed to operate on a set of symbols that permits its components to be identified, digitally, as tokens that belong to particular symbol types.

In contrast, artificial neural networks dispense (at first glance) with the sharp distinction between structure and process that characterizes classical cognitive science. Instead, networks themselves take the form of dynamic symbols that represent information at the same time as they transform it. The dynamic, distributed nature of artificial neural networks appears to make them more likely to be explained using statistical mechanics than using propositional logic.

One of the putative advantages of connectionist cognitive science is that it can inspire alternative notions of representation. The blurring of the structure/process distinction, the seemingly amorphous nature of the internal structure that characterizes many multilayer networks, leads to one such proposal, called coarse coding.

A coarse code is one in which an individual unit is very broadly tuned, sensitive to either a wide range of features or at least to a wide range of values for an individual feature (Churchland & Sejnowski, 1992; Hinton, McClelland, & Rumelhart, 1986). In other words, individual processors are themselves very inaccurate devices for measuring or detecting a feature. The accurate representation of a feature can become possible, though, by pooling or combining the responses of many such inaccurate detectors, particularly if their perspectives are slightly different (e.g., if they are sensitive to different ranges of features, or if they detect features from different input locations).

A familiar example of coarse coding is provided by the nineteenth trichromatic theory of colour perception (Helmholtz, 1968; Wasserman, 1978). According to this theory, colour perception is mediated by three types of retinal cone receptors. One is maximally sensitive to short (blue) wavelengths of light, another is maximally sensitive to medium (green) wavelengths, and the third is maximally sensitive to long (red) wavelengths. Thus none of these types of receptors are capable of representing, by themselves, the rich rainbow of perceptible hues.

> However, these receptors are broadly tuned and have overlapping sensitivities. As a result, most light will activate all three channels simultaneously, but to different degrees. Actual colored light does not produce sensations of absolutely pure

color; that red, for instance, even when completely freed from all admixture of white light, still does not excite those nervous fibers which alone are sensitive to impressions of red, but also, to a very slight degree, those which are sensitive to green, and perhaps to a still smaller extent those which are sensitive to violet rays. (Helmholtz, 1968, p. 97)

The pooling of different activities of the three channels permits a much greater variety of colours to be represented and perceived.

We have already seen examples of coarse coding in some of the network analyses that were presented earlier in this chapter. For instance, consider the chord recognition network. It was shown in Table 4-3 that none of its hidden units were accurate chord detectors. Hidden Units 1 and 2 did not achieve maximum activity when presented with any chord. When Hidden Unit 3 achieved maximum activity, this did not distinguish a 6th chord from a major 7th chord. However, when patterns were represented as points in a three-dimensional space, where the coordinates of each point were defined by a pattern's activity in each of the three hidden units (Figures 4-12 and 4-13), perfect chord classification was possible.

Other connectionist examples of coarse coding are found in studies of networks trained to accomplish navigational tasks, such as making judgments about the distance or direction between pairs of cities on a map (Dawson & Boechler, 2007; Dawson, Boechler, & Orsten, 2005; Dawson, Boechler, & Valsangkar-Smyth, 2000). For instance, Dawson and Boechler (2007) trained a network to judge the heading from one city on a map of Alberta to another. Seven hidden value units were required to accomplish this task. Each of these hidden units could be described as being sensitive to heading. However, this sensitivity was extremely coarse—some hidden units could resolve directions only to the nearest 180°. Nevertheless, a linear combination of the activities of all seven hidden units represented the desired direction between cities with a high degree of accuracy.

Similarly, Dawson, Boechler, and Valsangkar-Smyth (2000) trained a network of value units to make distance judgments between all possible pairs of 13 Albertan cities. This network required six hidden units to accomplish this task. Again, these units provided a coarse coding solution to the problem. Each hidden unit could be described as occupying a location on the map of Alberta through which a line was drawn at a particular orientation. This oriented line provided a one-dimensional map of the cities: connection weights encoded the projections of the cities from the two-dimensional map onto each hidden unit's one-dimensional representation. However, because the hidden units provided maps of reduced dimensionality, they were wildly inaccurate. Depending on the position of the oriented line, two cities that were far apart in the actual map could lie close together on a hidden unit's representation. Fortunately, because each of these inaccurate hidden unit maps encoded projections from different perspectives, the combination of their activities was able to represent the actual distance between all city pairs with a high degree of accuracy.

The discovery of coarse coding in navigational networks has important theoretical implications. Since the discovery of place cells in the hippocampus (O'Keefe & Dostrovsky, 1971), it has been thought that one function of the hippocampus is to instantiate a cognitive map (O'Keefe & Nadel, 1978). One analogy used to explain cognitive maps is that they are like graphical maps (Kitchin, 1994). From this, one might predict that the cognitive map is a metric, topographically organized, two-dimensional array in which each location in the map (i.e., each place in the external world) is associated with the firing of a particular place cell, and neighbouring place cells represent neighbouring places in the external world.

However, this prediction is not supported by anatomical evidence. First, place cells do not appear to be topographically organized (Burgess, Recce, & O'Keefe, 1995; McNaughton et al., 1996). Second, the receptive fields of place cells are at best locally metric, because one cannot measure the distance between points that are more than about a dozen body lengths apart because of a lack of receptive field overlap (Touretzky, Wan, & Redish, 1994). Some researchers now propose that the cognitive map doesn't really exit, but that map-like properties emerge when place cells are coordinated with other types of cells, such as head direction cells, which fire when an animal's head is pointed in a particular direction, regardless of the animal's location in space (McNaughton et al., 1996; Redish, 1999; Redish & Touretzky, 1999; Touretzky, Wan, & Redish, 1994).

Dawson et al. (2000) observed that their navigational network is also subject to the same criticisms that have been levelled against the notion of a topographically organized cognitive map. The hidden units did not exhibit topographic organization, and their inaccurate responses suggest that they are at best locally metric.

Nevertheless, the behaviour of the Dawson et al. (2000) network indicated that it represented information about a metric space. That such behaviour can be supported by the type of coarse coding discovered in this network suggests that metric, spatial information can be encoded in a representational scheme that is not isomorphic to a graphical map. This raises the possibility that place cells represent spatial information using a coarse code which, when its individual components are inspected, is not very map-like at all. O'Keefe and Nadel (1978, p. 78) were explicitly aware of this kind of possibility: "The cognitive map is *not* a picture or image which 'looks like' what it represents; rather, it is an information structure from which map-like images can be reconstructed and from which behaviour dependent upon place information can be generated."

What are the implications of the ability to interpret the internal structure of artificial neural networks to the practice of connectionist cognitive science?

When New Connectionism arose in the 1980s, interest in it was fuelled by two complementary perspectives (Medler, 1998). First, there was growing dissatisfaction with the progress being made in classical cognitive science and symbolic artificial intelligence (Dreyfus, 1992; Dreyfus & Dreyfus, 1988). Second, seminal introductions to artificial neural networks (McClelland & Rumelhart, 1986; Rumelhart & McClelland, 1986c) gave the sense that the connectionist architecture was a radical alternative to its classical counterpart (Schneider, 1987).

The apparent differences between artificial neural networks and classical models led to an early period of research in which networks were trained to accomplish tasks that had typically been viewed as prototypical examples of classical cognitive science (Bechtel, 1994; Rumelhart & McClelland, 1986a; Seidenberg & McClelland, 1989; Sejnowski & Rosenberg, 1988). These networks were then used as "existence proofs" to support the claim that non-classical models of classical phenomena are possible. However, detailed analyses of these networks were not provided, which meant that, apart from intuitions that connectionism is not classical, there was no evidence to support claims about the non-classical nature of the networks' solutions to the classical problems. Because of this, this research perspective has been called gee whiz connectionism (Dawson, 2004, 2009).

Of course, at around the same time, prominent classical researchers were criticizing the computational power of connectionist networks (Fodor & Pylyshyn, 1988), arguing that connectionism was a throwback to less powerful notions of associationism that classical cognitive science had already vanquished (Bever, Fodor, & Garrett, 1968; Chomsky, 1957, 1959b, 1965). Thus gee whiz connectionism served an important purpose: providing empirical demonstrations that connectionism might be a plausible medium in which cognitive science can be fruitfully pursued.

However, it was noted earlier that there exists a great deal of research on the computational power of artificial neural networks (Girosi & Poggio, 1990; Hartman, Keeler, & Kowalski, 1989; Lippmann, 1989; McCulloch & Pitts, 1943; Moody & Darken, 1989; Poggio & Girosi, 1990; Renals, 1989; Siegelmann, 1999; Siegelmann & Sontag, 1991); the conclusion from this research is that multilayered networks have the same in-principle power as any universal machine. This leads, though, to the demise of gee whiz connectionism, because if connectionist systems belong to the class of universal machines, "it is neither interesting nor surprising to demonstrate that a network can learn a task of interest" (Dawson, 2004, p. 118). If a network's ability to learn to perform a task is not of interest, then what is?

> It can be extremely interesting, surprising, and informative to determine what regularities the network exploits. What kinds of regularities in the input patterns has the network discovered? How does it represent these regularities? How are these regularities combined to govern the response of the network? (Dawson, 2004, p. 118)

By uncovering the properties of representations that networks have discovered for mediating an input-output relationship, connectionist cognitive scientists can discover new properties of cognitive phenomena.

4.14 Architectural Connectionism: An Overview

In the last several sections, we have been concerned with interpreting the internal structure of multilayered artificial neural networks. While some have claimed that all that can be found within brains and networks is goo (Mozer & Smolensky, 1989), the preceding examples have shown that detailed interpretations of internal network structure are both possible and informative. These interpretations reveal algorithmic-level details about how artificial neural networks use their hidden units to mediate mappings from inputs to outputs.

If the goal of connectionist cognitive science is to make new representational discoveries, then this suggests that it be practised as a form of synthetic psychology (Braitenberg, 1984; Dawson, 2004) that incorporates both synthesis and analysis, and that involves both forward engineering and reverse engineering.

The analytic aspect of connectionist cognitive science involves peering inside a network in order to determine how its internal structure represents solutions to problems. The preceding pages of this chapter have provided several examples of this approach, which seems identical to the reverse engineering practised by classical cognitive scientists.

The reverse engineering phase of connectionist cognitive science is also linked to classical cognitive science, in the sense that the results of these analyses are likely to provide the questions that drive algorithmic-level investigations. Once a novel representational format is discovered in a network, a key issue is to determine whether it also characterizes human or animal cognition. One would expect that when connectionist cognitive scientists evaluate their representational discoveries, they should do so by gathering the same kind of relative complexity, intermediate state, and error evidence that classical cognitive scientists gather when seeking strong equivalence.

Before one can reverse engineer a network, one must create it. And if the goal of such a network is to discover surprising representational regularities, then it should be created by minimizing representational assumptions as much as possible. One takes the building blocks available in a particular connectionist architecture, creates a network from them, encodes a problem for this network in some way, and attempts to train the network to map inputs to outputs.

This synthetic phase of research involves exploring different network structures (e.g., different design decisions about numbers of hidden units, or types of activation functions) and different approaches to encoding inputs and outputs. The idea is to give the network as many degrees of freedom as possible to discover representational regularities that have not been imposed or predicted by the researcher. These decisions all involve the architectural level of investigation.

One issue, though, is that networks are greedy, in the sense that they will exploit whatever resources are available to them. As a result, fairly idiosyncratic and specialized detectors are likely to be found if too many hidden units are provided to the network, and the network's performance may not transfer well when presented with novel stimuli. To deal with this, one must impose constraints by looking for the simplest network that will reliably learn the mapping of interest. The idea here is that such a network might be the one most likely to discover a representation general enough to transfer the network's ability to new patterns.

Importantly, sometimes when one makes architectural decisions to seek the simplest network capable of solving a problem, one discovers that the required network is merely a perceptron that does not employ any hidden units. In the remaining sections of this chapter I provide some examples of simple networks that are capable of performing interesting tasks. In section 4.15 the relevance of perceptrons to modern theories of associative learning is described. In section 4.16 I present a perceptron model of the reorientation task. In section 4.17 an interpretation is given for the structure of a perceptron that learns a seemingly complicated progression of musical chords.

4.15 New Powers of Old Networks

The history of artificial neural networks can be divided into two periods, Old Connectionism and New Connectionism (Medler, 1998). New Connectionism studies powerful networks consisting of multiple layers of units, and connections are trained to perform complex tasks. Old Connectionism studied networks that belonged to one of two classes. One was powerful multilayer networks that were hand wired, not trained (McCulloch & Pitts, 1943). The other was less powerful networks that did not have hidden units but were trained (Rosenblatt, 1958, 1962; Widrow, 1962; Widrow & Hoff, 1960).

Perceptrons (Rosenblatt, 1958, 1962) belong to Old Connectionism. A perceptron is a standard pattern associator whose output units employ a nonlinear activation function. Rosenblatt's perceptrons used the Heaviside step function to convert net input into output unit activity. Modern perceptrons use continuous nonlinear activation functions, such as the logistic or the Gaussian (Dawson, 2004, 2005, 2008; Dawson et al., 2009; Dawson et al., 2010).

Perceptrons are trained using an error-correcting variant of Hebb-style learning (Dawson, 2004). Perceptron training associates input activity with output unit error as follows. First, a pattern is presented to the input units, producing output unit activity via the existing connection weights. Second, output unit error is computed by taking the difference between actual output unit activity and desired output unit activity for each output unit in the network. This kind of training is called supervised learning, because it requires an external trainer to provide the desired output unit activities. Third, Hebb-style learning is used to associate input unit activity with output unit error: weight change is equal to a learning rate times input unit activity times output unit error. (In modern perceptrons, this triple product can also be multiplied by the derivative of the output unit's activation function, resulting in gradient descent learning [Dawson, 2004]).

The supervised learning of a perceptron is designed to reduce output unit errors as training proceeds. Weight changes are proportional to the amount of generated error. If no errors occur, then weights are not changed. If a task's solution can be represented by a perceptron, then repeated training using pairs of input-output stimuli is guaranteed to eventually produce zero error, as proven in Rosenblatt's perceptron convergence theorem (Rosenblatt, 1962).

Being a product of Old Connectionism, there are limits to the range of input-output mappings that can be mediated by perceptrons. In their famous computational analyses of what perceptrons could and could not learn to compute, Minsky and Papert (1969) demonstrated that perceptrons could not learn to distinguish some basic topological properties easily discriminated by humans, such as the difference between connected and unconnected figures. As a result, interest in and funding for Old Connectionist research decreased dramatically (Medler, 1998; Papert, 1988).

However, perceptrons are still capable of providing new insights into phenomena of interest to cognitive science. The remainder of this section illustrates this by exploring the relationship between perceptron learning and classical conditioning.

The primary reason that connectionist cognitive science is related to empiricism is that the knowledge of an artificial neural network is typically acquired via experience. For instance, in supervised learning a network is presented with pairs of patterns that define an input-output mapping of interest, and a learning rule is used to adjust connection weights until the network generates the desired response to a given input pattern.

In the twentieth century, prior to the birth of artificial neural networks (McCulloch & Pitts, 1943), empiricism was the province of experimental psychology. A detailed study of classical conditioning (Pavlov, 1927) explored the subtle regularities of the law of contiguity. Pavlovian, or classical, conditioning begins with an unconditioned stimulus (US) that is capable, without training, of producing an unconditioned response (UR). Also of interest is a conditioned stimulus (CS) that when presented will not produce the UR. In classical conditioning, the CS is paired with the US for a number of trials. As a result of this pairing, which places the CS in contiguity with the UR, the CS becomes capable of eliciting the UR on its own. When this occurs, the UR is then known as the conditioned response (CR).

Classical conditioning is a very basic kind of learning, but experiments revealed that the mechanisms underlying it were more complex than the simple law of contiguity. For example, one phenomenon found in classical conditioning is blocking (Kamin, 1968). Blocking involves two conditioned stimuli, CS_A and CS_B. Either stimulus is capable of being conditioned to produce the CR. However, if training begins with a phase in which only CS_A is paired with the US and is then followed by a phase in which both CS_A and CS_B are paired with the US, then CS_B fails to produce the CR. The prior conditioning involving CS_A blocks the conditioning of CS_B, even though in the second phase of training CS_B is contiguous with the UR.

The explanation of phenomena such as blocking required a new model of associative learning. Such a model was proposed in the early 1970s by Robert Rescorla and Allen Wagner (Rescorla & Wagner, 1972). This mathematical model of learning has been described as being cognitive, because it defines associative learning in terms of expectation. Its basic idea is that a CS is a signal about the likelihood that a US will soon occur. Thus the CS sets up expectations of future events. If these expectations are met, then no learning will occur. However, if these expectations are not met, then associations between stimuli and responses will be modified. "Certain expectations are built up about the events following a stimulus complex; expectations initiated by that complex and its component stimuli are then only modified when consequent events disagree with the composite expectation" (p. 75).

The expectation-driven learning that was formalized in the Rescorla-Wagner model explained phenomena such as blocking. In the second phase of learning in the blocking paradigm, the coming US was already signaled by CS_A. Because there was no surprise, no conditioning of CS_B occurred. The Rescorla-Wagner model has had many other successes; though it is far from perfect (Miller, Barnet, & Grahame, 1995; Walkenbach & Haddad, 1980), it remains an extremely influential, if not the most influential, mathematical model of learning.

The Rescorla-Wagner proposal that learning depends on the amount of surprise parallels the notion in supervised training of networks that learning depends on the amount of error. What is the relationship between Rescorla-Wagner learning and perceptron learning?

Proofs of the equivalence between the mathematics of Rescorla-Wagner learning and the mathematics of perceptron learning have a long history. Early proofs demonstrated that one learning rule could be translated into the other (Gluck & Bower, 1988; Sutton & Barto, 1981). However, these proofs assumed that the networks had linear activation functions. Recently, it has been proven that if when it is more properly assumed that networks employ a nonlinear activation function, one can still translate Rescorla-Wagner learning into perceptron learning, and vice versa (Dawson, 2008).

One would imagine that the existence of proofs of the computational equivalence between Rescorla-Wagner learning and perceptron learning would mean that perceptrons would not be able to provide any new insights into classical conditioning. However, this is not correct. Dawson (2008) has shown that if one puts aside the formal comparison of the two types of learning and uses perceptrons to simulate a wide variety of different classical conditioning paradigms, then some puzzling results occur. On the one hand, perceptrons generate the same results as the Rescorla-Wagner model for many different paradigms. Given the formal equivalence between the two types of learning, this is not surprising. On the other hand, for some paradigms, perceptrons generate different results than those predicted from the Rescorla-Wagner model (Dawson, 2008, Chapter 7). Furthermore, in many cases these differences represent improvements over Rescorla-Wagner learning. If the two types of learning are formally equivalent, then how is it possible for such differences to occur?

Dawson (2008) used this perceptron paradox to motivate a more detailed comparison between Rescorla-Wagner learning and perceptron learning. He found that while these two models of learning were equivalent at the computational level of investigation, there were crucial differences between them at the algorithmic level. In order to train a perceptron, the network must first behave (i.e., respond to an input pattern) in order for error to be computed to determine weight changes. In contrast, Dawson showed that the Rescorla-Wagner model defines learning in such a way that behaviour is not required!

Dawson's (2008) algorithmic analysis of Rescorla-Wagner learning is consistent with Rescorla and Wagner's (1972) own understanding of their model: "Independent assumptions will necessarily have to be made about the mapping of associative strengths into responding in any particular situation" (p. 75). Later, they make this same point much more explicitly:

> We need to provide some mapping of [associative] values into behavior. We are
> not prepared to make detailed assumptions in this instance. In fact, we would
> assume that any such mapping would necessarily be peculiar to each experi-
> mental situation, and depend upon a large number of 'performance' variables.
> (Rescorla & Wagner, 1972, p. 77)

Some knowledge is tacit: we can know more than we can tell (Polanyi, 1966). Dawson (2008) noted that the Rescorla-Wagner model presents an interesting variant of this theme, where if there is no explicit need for a behavioural theory, then there is no need to specify it explicitly. Instead, researchers can ignore Rescorla and Wagner's (1972) call for explicit models to convert associative strengths into behaviour and instead assume unstated, tacit theories such as "strong associations produce stronger, or more intense, or faster behavior." Researchers evaluate the Rescorla-Wagner model (Miller, Barnet, & Grahame, 1995; Walkenbach & Haddad, 1980) by agreeing that associations will eventually lead to behaviour, without actually stating how this is done. In the Rescorla-Wagner model, learning comes first and behaviour comes later—maybe.

Using perceptrons to study classical conditioning paradigms contributes to the psychological understanding of such learning in three ways. First, at the computational level, it demonstrates equivalences between independent work on learning conducted in computer science, electrical engineering, and psychology (Dawson, 2008; Gluck & Bower, 1988; Sutton & Barto, 1981).

Second, the results of training perceptrons in these paradigms raise issues that lead to a more sophisticated understanding of learning theories. For instance, the perceptron paradox led to the realization that when the Rescorla-Wagner model is typically used, accounts of converting associations into behaviour are unspecified. Recall that one of the advantages of computer simulation research is exposing tacit assumptions (Lewandowsky, 1993).

Third, the activation functions that are a required property of a perceptron serve as explicit theories of behaviour to be incorporated into the Rescorla-Wagner model. More precisely, changes in activation function result in changes to how the perceptron responds to stimuli, indicating the importance of choosing a particular architecture (Dawson & Spetch, 2005). The wide variety of activation functions that are available for artificial neural networks (Duch & Jankowski, 1999) offers a great opportunity to explore how changing theories of behaviour—or altering architectures—affect the nature of associative learning.

The preceding paragraphs have shown how the perceptron can be used to inform theories of a very old psychological phenomenon, classical conditioning. We now consider how perceptrons can play a role in exploring a more modern topic, reorientation, which was described from a classical perspective in Chapter 3 (Section 3.12).

4.16 Connectionist Reorientation

In the reorientation task, an agent learns that a particular place—usually a corner of a rectangular arena—is a goal location. The agent is then removed from the arena, disoriented, and returned to an arena. Its task is to use the available cues to relocate the goal. Theories of reorientation assume that there are two types of cues available for reorienting: local feature cues and relational geometric cues. Studies indicate that both types of cues are used for reorienting, even in cases where geometric cues are irrelevant (Cheng & Newcombe, 2005). As a result, some theories have proposed that a geometric module guides reorienting behaviour (Cheng, 1986; Gallistel, 1990).

The existence of a geometric module has been proposed because different kinds of results indicate that the processing of geometric cues is mandatory. First, in some cases agents continue to make rotational errors (i.e., the agent does not go to the goal location, but goes instead to an incorrect location that is geometrically identical to the goal location) even when a feature disambiguates the correct corner (Cheng, 1986; Hermer & Spelke, 1994). Second, when features are removed following training, agents typically revert to choosing both of the geometrically correct locations (Kelly et al., 1998; Sovrano et al., 2003). Third, when features are moved, agents generate behaviours that indicate that both types of cues were processed (Brown, Spetch, & Hurd, 2007; Kelly, Spetch, & Heth, 1998).

Recently, some researchers have begun to question the existence of geometric modules. One reason for this is that the most compelling evidence for claims of modularity comes from neuroscience (Dawson, 1998; Fodor, 1983), but such evidence about the modularity of geometry in the reorientation task is admittedly sparse (Cheng & Newcombe, 2005). This has led some researchers to propose alternative notions of modularity when explaining reorientation task regularities (Cheng, 2005, 2008; Cheng & Newcombe, 2005).

Still other researchers have explored how to abandon the notion of the geometric module altogether. They have proceeded by creating models that produce the main findings from the reorientation task, but they do so without using a geometric module. A modern perceptron that uses the logistic activation function has been shown to provide just such a model (Dawson et al., 2010).

The perceptrons used by Dawson et al. (2010) used a single output unit that, when the perceptron was "placed" in the original arena, was trained to turn on to the goal location and turn off to all of the other locations. A set of input units was used to represent the various cues—featural and geometric—available at each location. Both feature cues and geometric cues were treated in an identical fashion by the network; no geometric module was built into it.

After training, the perceptron was "placed" into a new arena; this approach was used to simulate the standard variations of the reorientation task in which geometric cues and feature cues could be placed in conflict. In the new arena, the perceptron was "shown" all of the possible goal locations by activating its input units with the features available at each location. The resulting output unit activity was interpreted as representing the likelihood that there was a reward at any of the locations in the new arena.

The results of the Dawson et al. (2010) simulations replicated the standard reorientation task findings that have been used to argue for the existence of a geometric module. However, this was accomplished without using such a module. These simulations also revealed new phenomena that have typically not been explored in the reorientation task that relate to the difference between excitatory cues, which indicate the presence of a reward, and inhibitory cues, which indicate the absence of a reward. In short, perceptrons have been used to create an associative, nonmodular theory of reorientation.

4.17 Perceptrons and Jazz Progressions

We have seen that a particular type of network from Old Connectionism, the perceptron, can be usefully applied in the studies of classical conditioning and reorientation. In the current section we see that it can also be used to explore musical regularities. Also illustrated is the interpretation of the internal structure of such a network, which demonstrates that even simple networks can reveal some interesting algorithmic properties.

Jazz progressions are sequences of chords. Consider the C major scale presented earlier, in Figure 4-8. If one takes the first note of the scale, C, as the root and adds every second note in the scale—E, G, and B)—the result is a four-note chord—a tetrachord—called the C major 7th chord (Cmaj7). Because the root of this chord is the first note of the scale, this is identified as the I chord for C major. Other tetrachords can also be built for this key. Starting with the second note in the scale, D, and adding the notes F, A, and C produces D minor 7th (Dm7). Because its root is the second note of the scale, this is identified as the II chord for the key of C major. Using G as the root and adding the notes B, D, and F creates the G dominant 7th chord (G7). It is the V chord of the key of C major because its root is the fifth note of the C major scale.

The I, II, and V chords are the three most commonly played jazz chords, and in jazz they often appear in the context of the II-V-I progression (Levine, 1989). This chord progression involves playing these chords in a sequence that begins with the II chord, moves to the V chord, and ends on the I chord. The II-V-I progression is important for several reasons.

First, chord progressions are used to establish tonality, that is, to specify to the listener the musical key in which a piece is being played. They do so by setting up expectancies about what is to be played next. For any major key, the most stable tones are notes I, IV, and V (Krumhansl, 1990), and the most stable chords are the ones built on those three notes.

Second, in the perception of chord sequences there are definite preferences for the IV chord to resolve into the V chord and for the V chord to resolve into the I chord, producing the IV-V-I progression that is common in cadences in classical music (Bharucha, 1984; Jarvinen, 1995; Katz, 1995; Krumhansl, Bharucha, & Kessler, 1982; Rosner & Narmour, 1992). There is a similar relationship between the IV chord and the II chord if the latter is minor (Steedman, 1984). Thus the II-V-I progression is a powerful tool for establishing the tonality of a musical piece.

Third, the II-V-I progression lends itself to a further set of chord progressions that move from key to key, providing variety but also establishing tonality. After playing the Cmaj7 chord to end the II-V-I progression for C major, one can change two notes to transform Cmaj7 into Cm7, which is the II chord of a different musical key, A♯ major. As a result, one can move from performing the II-V-I progression in C major to performing the same progression in a major key one tone lower. This process can be repeated; the full set of chord changes is provided in Table 4-6. Note that this progression eventually returns to the starting key of C major, providing another powerful cue of tonality.

Key	Chord Progression For Key		
	II	V	I
C	Dm7	G7	Cmaj7
A♯	Cm7	F7	A♯maj7
G♯	A♯m7	D♯7	G♯maj7
F♯	G♯m7	C♯7	F♯maj7
E	F♯m7	B7	Emaj7
D	Em7	A7	Dmaj7
C	Dm7	G7	Cmaj7

Table 4-6. A progression of II-V-I progressions, descending from the key of C major. The chords in each row are played in sequence, and after playing one row, the next row is played.

A connectionist network can be taught the II-V-I chord progression. During training, one presents, in pitch class format, a chord belonging to the progression. The network learns to output the next chord to be played in the progression, again using pitch class format. Surprisingly, this problem is very simple: it is linearly separable and can be solved by a perceptron!

How does a perceptron represent this jazz progression? Because a perceptron has no hidden units, its representation must be stored in the set of connection weights between the input and output units. However, this matrix of connection weights is too complex to reveal its musical representations simply by inspecting it. Instead, multivariate statistics must be used.

First, one can convert the raw connection weights into a correlation matrix. That is, one can compute the similarity of each pair of output units by computing the correlation between the connection weights that feed into them. Once the weights have been converted into correlations, further analyses are then available to interpret network representations. Multidimensional scaling (MDS) can summarize the relationships within a correlation matrix made visible by creating a map (Kruskal & Wish, 1978; Romney, Shepard, & Nerlove, 1972; Shepard, Romney, & Nerlove, 1972). Items are positioned in the map in such a way that the more similar items are, the closer together they are in the map.

The MDS of the jazz progression network's correlations produced a one-dimensional map that provided a striking representation of musical relationships amongst the notes. In a one-dimensional MDS solution, each data point is assigned a single number, which is its coordinate on the single axis that is the map. The coordinate for each note is presented in a bar chart in Figure 4-16.

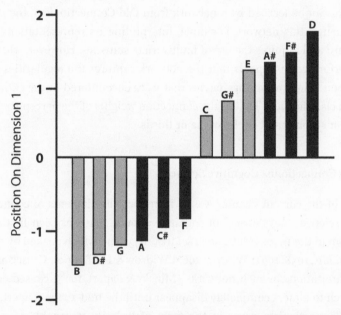

Figure 4-16. Coordinates associated with each output note, taken from an MDS of the Table 4-8 correlations. Shading reflects groupings of notes as circles of major thirds.

The first regularity evident from Figure 4-16 is that half of the notes have negative coordinates, while the other half have positive coordinates. That is, the perceptron's connection weights separate musical notes into two equal-sized classes. These classes reflect a basic property of the chord progressions learned by the network: all of the notes that have positive coordinates were also used as major keys in which the II-V-I progression was defined, while none of the notes with negative coordinates were used in this fashion.

Another way to view the two classes of notes revealed by this analysis is in terms of the two circles of major seconds that were presented in Figure 4-10. The first circle of major seconds contains only those notes that have positive coordinates in Figure 4-16. The other circle of major seconds captures the set of notes that have negative coordinates in Figure 4-16. In other words, the jazz progression network acts as if it has classified notes in terms of the circles of major seconds!

The order in which the notes are arranged in the one-dimensional map is also related to the four circles of major thirds that were presented in Figure 4-11. The bars in Figure 4-16 have been coloured to reveal four sets of three notes each. Each of these sets of notes defines a circle of major thirds. The MDS map places notes in such a way that the notes of one such circle are listed in order, followed by the notes of another circle of major thirds.

To summarize, one musical formalism is the II-V-I jazz progression. Interestingly, this formalism can be learned by a network from Old Connectionism, the perceptron. Even though this network is simple, interpreting its representations is not straightforward and requires the use of multivariate statistics. However, when such analysis is performed, it appears that the network captures the regularities of this jazz progression using the strange circles that were encountered in the earlier section on chord classification. That is, the connection weights of the perceptron reveal circles of major seconds and circles of major thirds.

4.18 What Is Connectionist Cognitive Science?

The purpose of the current chapter was to introduce the elements of connectionist cognitive science, the "flavour" of cognitive science that was seen first as Old Connectionism in the 1940s (McCulloch & Pitts, 1943) and which peaked by the late 1950s (Rosenblatt, 1958, 1962; Widrow, 1962; Widrow & Hoff, 1960). Criticisms concerning the limitations of such networks (Minsky & Papert, 1969) caused connectionist research to almost completely disappear until the mid-1980s (Papert, 1988), when New Connectionism arose in the form of techniques capable of training powerful multilayered networks (McClelland & Rumelhart, 1986; Rumelhart & McClelland, 1986c).

Connectionism is now well established as part of mainstream cognitive science, although its relationship to classical cognitive science is far from clear. Artificial neural networks have been used to model a dizzying variety of phenomena including animal learning (Enquist & Ghirlanda, 2005; Schmajuk, 1997), cognitive development (Elman et al., 1996), expert systems (Gallant, 1993), language (Mammone, 1993; Sharkey, 1992), pattern recognition and perception (Pao, 1989; Ripley, 1996; Wechsler, 1992), and musical cognition (Griffith & Todd, 1999; Todd & Loy, 1991).

Given the breadth of connectionist cognitive science, only a selection of its elements have been introduced in this chapter; capturing all of the important contributions of connectionism in a single chapter is not possible. A proper treatment of connectionism requires a great deal of further reading; fortunately connectionism is described in a rich and growing literature (Amit, 1989; Anderson, 1995; Anderson & Rosenfeld, 1998; Bechtel & Abrahamsen, 2002; Carpenter & Grossberg, 1992; Caudill & Butler, 1992a, 1992b; Churchland, 1986; Churchland & Sejnowski, 1992; Clark, 1989, 1993; Dawson, 2004, 2005; Grossberg, 1988; Horgan & Tienson, 1996; Quinlan, 1991; Ramsey, Stich, & Rumelhart, 1991; Ripley, 1996; Rojas, 1996).

Connectionist cognitive science is frequently described as a reaction against the foundational assumptions of classical cognitive science. The roots of classical cognitive science draw inspiration from the rationalist philosophy of Descartes, with an emphasis on nativism and logicism (Chomsky, 1966; Devlin, 1996). In contrast, the foundations of connectionist cognitive science are the empiricist philosophy of Locke and the associationist psychology that can be traced from the early British empiricists to the more modern American behaviourists. Connectionist networks acquire structure or knowledge via experience; they often begin as blank slates (Pinker, 2002) and acquire structure as they learn about their environments (Bechtel, 1985; Clark, 1989, 1993; Hillis, 1988).

Classical cognitive science departed from Cartesian philosophy by seeking materialist accounts of mentality. This view was inspired by the digital computer and the fact that electronic switches could be assigned abstract logical interpretations (Shannon, 1938).

Connectionism is materialist as well, but arguably in a more restricted sense than classical cognitive science. The classical approach appeals to the multiple realization argument when it notes that under the proper interpretation, almost any physical substrate could instantiate information processing or symbol manipulation (Hillis, 1998). In contrast, connectionism views the digital computer metaphor as mistaken. Connectionists claim that the operations of such a device—regardless of its material nature—are too slow, brittle, and inflexible to be appropriate for modelling cognition. Connectionism posits instead that the brain is the only appropriate material for realizing the mind and researchers attempt to frame its theories in terms of information processing that is biologically plausible or neuronally inspired (Amit, 1989; Burnod, 1990; Gluck & Myers, 2001).

In adopting the digital computer metaphor and the accompanying logicist view that cognition is the result of rule-governed symbol manipulation, classical cognitive science is characterized by a marked structure/process distinction. That is, classical models—typified by Turing machines (Turing, 1936) or production systems (Newell & Simon, 1972)—distinguish between the symbols being manipulated and the explicit rules doing the manipulating. This distinction is usually marked in models by having separate locations for structure and process, such as a memory that holds symbols and a central controller that holds the processes.

In abandoning the digital computer metaphor and adopting a notion of information processing that is biologically inspired, connectionist cognitive science abandons or blurs the structure/process distinction. Neural networks can be viewed as both structure and process; they have been called active data structures (Hillis, 1985). This has led to an extensive debate about whether theories of cognition require explicit rules (Ramsey, Stich, & Rumelhart, 1991).

The digital computer metaphor adopted by classical cognitive science leads it to also adopt a particular notion of control. In particular, classical models invoke a notion of serial control in which representations can only be manipulated one rule at a time. When classical problem solvers search a problem space in order to solve a problem (Newell & Simon, 1972), they do so to discover a *sequence* of operations to perform.

In contrast, when connectionist cognitive science abandons the digital computer metaphor, it abandons with it the assumption of centralized serial control. It does so because it views this as a fatal flaw in classical models, generating a "von Neumann bottleneck" that makes classical theories too slow to be useful in real time (Feldman & Ballard, 1982; Hillis, 1985). In the stead of centralized serial control, connectionists propose decentralized control in which many simple processes can be operating in parallel (see Dawson & Schopflocher, 1992a).

Clearly, from one perspective, there are obvious and important differences between connectionist and classical cognitive science. However, a shift in perspective can reveal a view in which striking similarities between these two approaches are evident. We saw earlier that classical cognitive science is performed at multiple levels of analysis, using formal methods to explore the computational level, behavioural methods to investigate the algorithmic level, and a variety of behavioural and biological techniques to elaborate the architectural and implementational levels. It is when connectionist cognitive science is examined from this same multiple-levels viewpoint that its relationship to classical cognitive science is made apparent (Dawson, 1998).

Analyses at the computational level involve using some formal language to make proofs about cognitive systems. Usually these proofs concern statements about what kind of computation is being performed or what the general capabilities of a system are. Computational-level analyses have had a long and important history in connectionist cognitive science, and they have been responsible, for example, for proofs that particular learning rules will converge to desired least-energy or low-error states (Ackley, Hinton, & Sejnowski, 1985; Hopfield, 1982; Rosenblatt, 1962; Rumelhart, Hinton, & Williams, 1986b). Other examples of computational analyses were provided earlier in this chapter, in the discussion of carving pattern spaces into decision regions and the determination that output unit activities could be interpreted as being conditional probabilities.

That computational analysis is possible for both connectionist and classical cognitive science highlights one similarity between these two approaches. The results of some computational analyses, though, reveal a more striking similarity. One debate in the literature has concerned whether the associationist nature of artificial neural networks limits their computational power, to the extent that they are not appropriate for cognitive science. For instance, there has been considerable debate about whether PDP networks demonstrate appropriate systematicity and componentiality (Fodor & McLaughlin, 1990; Fodor & Pylyshyn, 1988; Hadley, 1994a, 1994b, 1997; Hadley & Hayward, 1997), two characteristics important for the use of recursion in classical models. However, beginning with the mathematical analyses of Warren McCulloch (McCulloch & Pitts, 1943) and continuing with modern computational analyses (Girosi & Poggio, 1990; Hartman, Keeler, & Kowalski, 1989; Lippmann, 1989; McCulloch & Pitts, 1943; Moody & Darken, 1989; Poggio & Girosi, 1990; Renals, 1989; Siegelmann, 1999; Siegelmann & Sontag, 1991), we have seen that artificial neural networks belong to the class of universal machines. Classical and connectionist cognitive science are not distinguishable at the computational level of analysis (Dawson, 1998, 2009).

Let us now turn to the next level of analysis, the algorithmic level. For classical cognitive science, the algorithmic level involves detailing the specific information processing steps that are involved in solving a problem. In general, this almost always involves analyzing behaving systems in order to determine how representations are being manipulated, an approach typified by examining human problem solving with the use of protocol analysis (Ericsson & Simon, 1984; Newell & Simon, 1972). Algorithmic-level analyses for connectionists also involve analyzing the internal structure of intact systems—trained networks—in order to determine how they mediate stimulus-response regularities. We have seen examples of a variety of techniques that can and have been used to uncover the representations that are hidden within network structures, and which permit networks to perform desired input-output mappings. Some of these representations, such as coarse codes, look like alternatives to classical representations. Thus one of classical cognitive science's contributions may be to permit new kinds of representations to be discovered and explored.

Nevertheless, algorithmic-level analyses also reveal further similarities between connectionist and classical cognitive science. While these two approaches may propose different kinds of representations, they still are both representational. There is no principled difference between the classical sandwich and the connectionist sandwich (Calvo & Gomila, 2008). Furthermore, it is not even guaranteed that the contents of these two types of sandwiches will differ. One can peer inside an artificial neural network and find classical rules for logic (Berkeley et al., 1995) or even an entire production system (Dawson et al., 2000).

At the architectural level of analysis, stronger differences between connectionist and classical cognitive science can be established. Indeed, the debate between these two approaches is in essence a debate about architecture. This is because many of the dichotomies introduced earlier—rationalism vs. empiricism, digital computer vs. analog brain, structure/process vs. dynamic data, serialism vs. parallelism—are differences in opinion about cognitive architecture.

In spite of these differences, and in spite of connectionism's search for biologically plausible information processing, there is a key similarity at the architectural level between connectionist and classical cognitive science: at this level, both propose architectures that are *functional*, not physical. The connectionist architecture consists of a set of building blocks: units and their activation functions, modifiable connections, learning rules. But these building blocks are functional accounts of the information processing properties of neurons; other brain-like properties are ignored. Consider one response (Churchland & Churchland, 1990) to the claim that the mind is the product of the causal powers of the brain (Searle, 1990):

> We presume that Searle is not claiming that a successful artificial mind must have all the causal powers of the brain, such as the power to smell bad when rotting, to harbor slow viruses such as kuru, to stain yellow with horseradish peroxidase and so forth. Requiring perfect parity would be like requiring that an artificial flying device lay eggs. (Churchland & Churchland, 1990, p. 37)

It is the functional nature of the connectionist architecture that enables it to be almost always studied by simulating it—on a digital computer!

The functional nature of the connectionist architecture raises some complications when the implementational level of analysis is considered. On the one hand, many researchers view connectionism as providing implementational-level theories of cognitive phenomena. At this level, one finds researchers exploring relationships between biological receptive fields and patterns of connectivity and similar properties of artificial networks (Ballard, 1986; Bankes & Margoliash, 1993; Bowers, 2009; Guzik, Eaton, & Mathis, 1999; Keith, Blohm, & Crawford, 2010; Moorhead, Haig, & Clement, 1989; Poggio, Torre, & Koch, 1985; Zipser & Andersen, 1988). One also encounters researchers finding biological mechanisms that map onto architectural properties such as learning rules. For example, there is a great deal of interest in relating the actions of certain neurotransmitters to Hebb learning (Brown, 1990; Gerstner & Kistler, 2002; van Hemmen & Senn, 2002). Similarly, it has been argued that connectionist networks provide an implementational account of associative learning (Shanks, 1995), a position that ignores its potential contributions at other levels of analysis (Dawson, 2008).

On the other hand, the functional nature of the connectionist architecture has resulted in its biological status being questioned or challenged. There are many important differences between biological and artificial neural networks (Crick & Asanuma, 1986; Douglas & Martin, 1991; McCloskey, 1991). There is very little biological evidence in support of important connectionist learning rules such as backpropagation of error (Mazzoni, Andersen, & Jordan, 1991; O'Reilly, 1996; Shimansky, 2009). Douglas and Martin (1991, p. 292) dismissed artificial neural networks as merely being "stick and ball models." Thus whether connectionist cognitive science is a biologically plausible alternative to classical cognitive science remains an open issue.

That connectionist cognitive science has established itself as a reaction against classical cognitive science cannot be denied. However, as we have seen in this section, it is not completely clear that connectionism represents a radical alternative to the classical approach (Schneider, 1987), or that it is rather much more closely related to classical cognitive science than a brief glance at some of the literature might suggest (Dawson, 1998). It is certainly the case that connectionist cognitive science has provided important criticisms of the classical approach and has therefore been an important contributor to theory of mind.

Interestingly, many of the criticisms that have been highlighted by connectionist cognitive science—slowness, brittleness, biological implausibility, overemphasis of logicism and disembodiment—have been echoed by a third school, embodied cognitive science. Furthermore, related criticisms have been applied by embodied cognitive scientists against connectionist cognitive science. Not surprisingly, then, embodied cognitive science has generated a very different approach to deal with these issues than has connectionist cognitive science.

In Chapter 5 we turn to the elements of this third "flavour" of cognitive science. As has been noted in this final section of Chapter 4, there appears to be ample room for finding relationships between connectionism and classicism such that the umbrella *cognitive science* can be aptly applied to both. We see that embodied cognitive science poses some interesting and radical challenges, and that its existence calls many of the core features shared by connectionism and classicism into question.

Elements of Embodied Cognitive Science

5.0 Chapter Overview

One of the key reactions against classical cognitive science was connectionism. A second reaction against the classical approach has also emerged. This second reaction is called embodied cognitive science, and the purpose of this chapter is to introduce its key elements.

Embodied cognitive science explicitly abandons the disembodied mind that serves as the core of classical cognitive science. It views the purpose of cognition not as building representations of the world, but instead as directing actions upon the world. As a result, the structure of an agent's body and how this body can sense and act upon the world become core elements. Embodied cognitive science emphasizes the embodiment and situatedness of agents.

Embodied cognitive science's emphasis on embodiment, situatedness, and action upon the world is detailed in the early sections of the chapter. This emphasis leads to a number of related elements: feedback between agents and environments, stigmergic control of behaviour, affordances and enactive perception, and cognitive scaffolding. In the first half of this chapter these notions are explained, showing how they too can be traced back to some of the fundamental assumptions of cybernetics. Also illustrated is how such ideas are radical departures from the ideas emphasized by classical cognitive scientists.

Not surprisingly, such differences in fundamental ideas lead to embodied cognitive science adopting methodologies that are atypical of classical cognitive science. Reverse engineering is replaced with forward engineering, as typified by behaviour-based robotics. These methodologies use an agent's environment to increase or leverage its abilities, and in turn they have led to novel accounts of complex human activities. For instance, embodied cognitive science can construe social interactions either as sense-act cycles in a social environment or as mediated by simulations that use our own brains or bodies as physical stand-ins for other agents.

In spite of such differences, it is still the case that there are structural similarities between embodied cognitive science and the other two approaches that have been introduced in the preceding chapters. The current chapter ends with a consideration of embodied cognitive science in light of Chapter 2's multiple levels of investigation, which were earlier used as a context in which to consider the research of both classical and of connectionist cognitive science.

5.1 Abandoning Methodological Solipsism

The goal of Cartesian philosophy was to provide a core of incontestable truths to serve as an anchor for knowledge (Descartes, 1960, 1996). Descartes believed that he had achieved this goal. However, the cost of this accomplishment was a fundamental separation between mind and body. Cartesian dualism disembodied the mind, because Descartes held that the mind's existence was independent of the existence of the body.

> I am not that structure of limbs which is called a human body, I am not even some thin vapor which permeates the limbs—a wind, fire, air, breath, or whatever I depict in my imagination, for these are things which I have supposed to be nothing. (Descartes, 1996, p. 18)

Cartesian dualism permeates a great deal of theorizing about the nature of mind and self, particularly in our current age of information technology. One such theory is posthumanism (Dewdney, 1998; Hayles, 1999). Posthumanism results when the content of information is more important than the physical medium in which it is represented, when consciousness is considered to be epiphenomenal, and when the human body is simply a prosthetic. Posthumanism is rooted in the pioneering work of cybernetics (Ashby, 1956, 1960; MacKay, 1969; Wiener, 1948), and is sympathetic to such futuristic views as uploading our minds into silicon bodies (Kurzweil, 1999, 2005; Moravec, 1988, 1999), because, in this view, the nature of the body is irrelevant to the nature of the mind. Hayles uncomfortably notes that a major implication of posthumanism is its "systematic devaluation of materiality and embodiment" (Hayles, 1999, p. 48); "because we are essentially information, we can do away with the body" (Hayles, 1999, p. 12).

Some would argue that similar ideas pervade classical cognitive science. American psychologist Sylvia Scribner wrote that cognitive science "is haunted by a metaphysical spectre. The spectre goes by the familiar name of Cartesian dualism, which, in spite of its age, continues to cast a shadow over inquiries into the nature of human nature" (Scribner & Tobach, 1997, p. 308).

In Chapter 3 we observed that classical cognitive science departed from the Cartesian approach by seeking materialist explanations of cognition. Why then should it be haunted by dualism?

To answer this question, we examine how classical cognitive science explains, for instance, how a single agent produces different behaviours. Because classical cognitive science appeals to the representational theory of mind (Pylyshyn, 1984), it must claim that different behaviours must ultimately be rooted in different mental representations.

If different behaviours are caused by differences between representations, then classical cognitive science must be able to distinguish or individuate representational states. How is this done? The typical position adopted by classical cognitive science is called methodological solipsism (Fodor, 1980). Methodological solipsism individuates representational states only in terms of their relations to other representational states. Relations of the states to the external world—the agent's environment—are not considered. "Methodological solipsism in psychology is the view that psychological states should be construed without reference to anything beyond the boundary of the individual who has those states" (Wilson, 2004, p. 77).

The methodological solipsism that accompanies the representational theory of mind is an example of the classical sandwich (Hurley, 2001). The classical sandwich is the view that links between a cognitive agent's perceptions and a cognitive agent's actions must be mediated by internal thinking or planning. In the classical sandwich, models of cognition take the form of sense-think-act cycles (Brooks, 1999; Clark, 1997; Pfeifer & Scheier, 1999). Furthermore, these theories tend to place a strong emphasis on the purely mental part of cognition—the thinking—and at the same time strongly de-emphasize the physical—the action. In the classical sandwich, perception, thinking, and action are separate and unequal.

> On this traditional view, the mind passively receives sensory input from its environment, structures that input in cognition, and then marries the products of cognition to action in a peculiar sort of shotgun wedding. Action is a by-product of genuinely mental activity. (Hurley, 2001, p. 11)

Although connectionist cognitive science is a reaction against classical cognitivism, this reaction does not include a rejection of the separation of perception and action via internal representation. Artificial neural networks typically have undeveloped models of perception (i.e., input unit encodings) and action (i.e., output unit encodings), and in modern networks communication between the two must be moderated by representational layers of hidden units.

> Highly artificial choices of input and output representations and poor choices of problem domains have, I believe, robbed the neural network revolution of some of its initial momentum. . . . The worry is, in essence, that a good deal of the research on artificial neural networks leaned too heavily on a rather classical conception of the nature of the problems: (Clark, 1997, p. 58)

The purpose of this chapter is to introduce embodied cognitive science, a fairly modern reaction against classical cognitive science. This approach is an explicit rejection of methodological solipsism. Embodied cognitive scientists argue that a cognitive theory must include an agent's environment as well as the agent's experience of that environment (Agre, 1997; Chemero, 2009; Clancey, 1997; Clark, 1997; Dawson, Dupuis, & Wilson, 2010; Dourish, 2001; Gibbs, 2006; Johnson, 2007; Menary, 2008; Pfeifer & Scheier, 1999; Shapiro, 2011; Varela, Thompson, & Rosch, 1991). They recognize that this experience depends on how the environment is sensed, which is situation; that an agent's situation depends upon its physical nature, which is embodiment; and that an embodied agent can act upon and change its environment (Webb & Consi, 2001). The embodied approach replaces the notion that cognition is representation with the notion that cognition is the control of actions upon the environment. As such, it can also be viewed as a reaction against a great deal of connectionist cognitive science.

In embodied cognitive science, the environment contributes in such a significant way to cognitive processing that some would argue that an agent's mind has leaked into the world (Clark, 1997; Hutchins, 1995; Menary, 2008, 2010; Noë, 2009; Wilson, 2004). For example, research in behaviour-based robotics eliminates resource-consuming representations of the world by letting the world serve as its own representation, one that can be accessed by a situated agent (Brooks, 1999). This robotics tradition has also shown that nonlinear interactions between an embodied agent and its environment can produce surprisingly complex behaviour, even when the internal components of an agent are exceedingly simple (Braitenberg, 1984; Grey Walter, 1950a, 1950b, 1951, 1963; Webb & Consi, 2001).

In short, embodied cognitive scientists argue that classical cognitive science's reliance on methodological solipsism—its Cartesian view of the disembodied mind—is a deep-seated error. "Classical rule-and-symbol-based AI may have made a fundamental error, mistaking the cognitive profile of the agent plus the environment for the cognitive profile of the naked brain" (Clark, 1997, p. 61).

In reacting against classical cognitive science, the embodied approach takes seriously the idea that Simon's (1969) parable of the ant might also be applicable to human cognition: "A man, viewed as a behaving system, is quite simple. The apparent complexity of his behavior over time is largely a reflection of the complexity of the environment in which he finds himself" (p. 25). However, when it comes to specifics about applying such insight, embodied cognitive science is frustratingly fractured. "Embodied cognition, at this stage in its very brief history, is better considered a *research program* than a well-defined theory" (Shapiro, 2011, p. 2). Shapiro (2011) went on to note that this is because embodied cognitive science "exhibits much greater latitude in its subject matter, ontological commitment, and methodology than does standard cognitive science" (p. 2).

Shapiro (2011) distinguished three key themes that are present, often to differing degrees, in a variety of theories that belong to embodied cognitive science. The first of Shapiro's themes is *conceptualization*. According to this theme, the concepts that an agent requires to interact with its environment depend on the form of the agent's body. If different agents have different bodies, then their understanding or engagement with the world will differ as well. We explore the theme of conceptualization later in this chapter, in the discussion of concepts such as *umwelten*, affordances, and enactive perception.

Shapiro's (2011) second theme of embodied cognitive science is *replacement*: "An organism's body in interaction with its environment replaces the need for representational processes thought to have been at the core of cognition" (p. 4). The theme of replacement is central to the idea of cognitive scaffolding, in which agents exploit environmental resources for problem representation and solution.

> The biological brain takes all the help it can get. This help includes the use of external physical structures (both natural and artifactual), the use of language and cultural institutions, and the extensive use of other agents. (Clark, 1997, p. 80)

Shapiro's (2011) third theme of embodied cognitive science is *constitution*. According to this theme, the body or the world has more than a causal role in cognition—they are literally constituents of cognitive processing. The constitution hypothesis leads to one of the more interesting and radical proposals from embodied cognitive science, the extended mind. According to this hypothesis, which flies in the face of the Cartesian mind, the boundary of the mind is not the skin or the skull (Clark, 1997, p. 53): "Mind is a leaky organ, forever escaping its 'natural' confines and mingling shamelessly with body and with world."

One reason that Shapiro (2011) argued that embodied cognitive science is not a well-defined theory, but is instead a more ambiguous research program, is because these different themes are endorsed to different degrees by different embodied cognitive scientists. For example, consider the replacement hypothesis. On the one hand, some researchers, such as behaviour-based roboticists (Brooks, 1999) or radical embodied cognitive scientists (Chemero, 2009), are strongly anti-representational; their aim is to use embodied insights to expunge representational issues from cognitive science. On the other hand, some other researchers, such as philosopher Andy Clark (1997), have a more moderate view in which both representational and non-representational forms of cognition might be present in the same agent.

Shapiro's (2011) three themes of conceptualization, replacement, and constitution characterize important principles that are the concern of the embodied approach. These principles also have important effects on the practice of embodied cognitive science. Because of their concern with environmental contributions to behavioural complexity, embodied cognitive scientists are much more likely to practise forward engineering or synthetic psychology (Braitenberg, 1984; Dawson, 2004; Dawson, Dupuis, & Wilson, 2010; Pfeifer & Scheier, 1999). In this approach, devices are first constructed and placed in an environment, to examine what complicated or surprising behaviours might emerge. Thus while in reverse engineering behavioural observations are the source of models, in forward engineering models are the source of behaviour to observe. Because of their concern about how engagement with the world is dependent upon the physical nature and abilities of agents, embodied cognitive scientists actively explore the role that embodiment plays in cognition. For instance, their growing interest in humanoid robots is motivated by the realization that human intelligence and development require human form (Breazeal, 2002; Brooks et al., 1999).

In the current chapter we introduce some of the key elements that characterize embodied cognitive science. These ideas are presented in the context of reactions against classical cognitive science in order to highlight their innovative nature. However, it is important to keep potential similarities between embodied cognitive science and the other two approaches in mind; while they are not emphasized here, the possibility of such similarities is a central theme of Part II of this book.

5.2 Societal Computing

The travelling salesman problem is a vital optimization problem (Gutin & Punnen, 2002; Lawler, 1985). It involves determining the order in which a salesman should visit a sequence of cities, stopping at each city only once, such that the shortest total distance is travelled. The problem is tremendously important: a modern bibliography cites 500 studies on how to solve it (Laporte & Osman, 1995).

One reason for the tremendous amount of research on the travelling salesman problem is that its solution can be applied to a dizzying array of real-world problems and situations (Punnen, 2002), including scheduling tasks, minimizing interference amongst a network of transmitters, data analysis in psychology, X-ray crystallography, overhauling gas turbine engines, warehouse order-picking problems, and wallpaper cutting. It has also attracted so much attention because it is difficult. The travelling salesman problem is an NP-complete problem (Kirkpatrick, Gelatt, & Vecchi, 1983), which means that as the number of cities involved in the salesman's tour increases linearly, the computational effort for finding the shortest route increases exponentially.

Because of its importance and difficulty, a number of different approaches to solving the travelling salesman problem have been explored. These include a variety of numerical optimization algorithms (Bellmore & Nemhauser, 1968). Some other algorithms, such as simulated annealing, are derived from physical metaphors (Kirkpatrick, Gelatt, & Vecchi, 1983). Still other approaches are biologically inspired and include neural networks (Hopfield & Tank, 1985; Siqueira, Steiner, & Scheer, 2007), genetic algorithms (Braun, 1991; Fogel, 1988), and molecular computers built using DNA molecules (Lee et al., 2004).

Given the difficulty of the travelling salesman problem, it might seem foolish to suppose that cognitively simple agents are capable of solving it. However, evidence shows that a colony of ants is capable of solving a version of this problem, which has inspired new algorithms for solving the travelling salesman problem (Dorigo & Gambardella, 1997)!

One study of the Argentine ant *Iridomyrmex humilis* used a system of bridges to link the colony's nest to a food supply (Goss et al., 1989). The ants had to choose between two different routes at two different locations in the network of bridges; some of these routes were shorter than others. When food was initially discovered, ants traversed all of the routes with equal likelihood. However, shortly afterwards, a strong preference emerged: almost all of the ants chose the path that produced the shortest journey between the nest and the food.

The ants' solution to the travelling salesmen problem involved an interaction between the world and a basic behaviour: as *Iridomyrmex humilis* moves, it deposits a pheromone trail; the potency of this trail fades over time. An ant that by chance chooses the shortest path will add to the pheromone trail at the decision points sooner than will an ant that has taken a longer route. This means that as other ants arrive at a decision point they will find a stronger pheromone trail in the shorter direction, they will be more likely to choose this direction, and they will also add to the pheromone signal.

> Each ant that passes the choice point modifies the following ant's probability of choosing left or right by adding to the pheromone on the chosen path. This positive feedback system, after initial fluctuation, rapidly leads to one branch being 'selected.' (Goss et al., 1989, p. 581)

The ability of ants to choose shortest routes does not require a great deal of individual computational power. The solution to the travelling salesman problem emerges from the actions of the ant colony as a whole.

> The selection of the shortest branch is not the result of individual ants comparing the different lengths of each branch, but is instead a collective and self-organizing

process, resulting from the interactions between the ants marking in both directions. (Goss et al., 1989, p. 581)

5.3 Stigmergy and Superorganisms

To compute solutions to the travelling salesman problem, ants from a colony interact with and alter their environment in a fairly minimal way: they deposit a pheromone trail that can be later detected by other colony members. However, impressive examples of richer interactions between social insects and their world are easily found.

For example, wasps are social insects that house their colonies in nests of intricate structure that exhibit, across species, tremendous variability in size, shape, and location (Downing & Jeanne, 1986). The size of nests ranges from a mere dozen to nearly a million cells or combs (Theraulaz, Bonabeau, & Deneubourg, 1998). The construction of some nests requires that specialized labour be coordinated (Jeanne, 1996, p. 473): "In the complexity and regularity of their nests and the diversity of their construction techniques, wasps equal or surpass many of the ants and bees."

More impressive nests are constructed by other kinds of insect colonies, such as termites, whose vast mounds are built over many years by millions of individual insects. A typical termite mound has a height of 2 metres, while some as high as 7 metres have been observed (von Frisch, 1974). Termite mounds adopt a variety of structural innovations to control their internal temperature, including ventilation shafts or shape and orientation to minimize the effects of sun or rain. Such nests,

> seem [to be] evidence of a master plan which controls the activities of the builders and is based on the requirements of the community. How this can come to pass within the enormous complex of millions of blind workers is something we do not know. (von Frisch, 1974, p. 150)

How do colonies of simple insects, such as wasps or termites, coordinate the actions of individuals to create their impressive, intricate nests? "One of the challenges of insect sociobiology is to explain how such colony-level behavior emerges from the individual decisions of members of the colony" (Jeanne, 1996, p. 473).

One theoretical approach to this problem is found in the pioneering work of entomologist William Morton Wheeler, who argued that biology had to explain how organisms cope with complex and unstable environments. With respect to social insects, Wheeler (1911) proposed that a colony of ants, considered as a whole, is actually an organism, calling the colony-as-organism the superorganism: "The animal colony is a true organism and not merely the analogue of the person" (p. 310).

Wheeler (1926) agreed that the characteristics of a superorganism must emerge from the actions of its parts, that is, its individual colony members. However, Wheeler also argued that higher-order properties could not be reduced to properties of the superorganism's components. He endorsed ideas that were later popularized by Gestalt psychology, such as the notion that the whole is not merely the sum of its parts (Koffka, 1935; Köhler, 1947).

> The unique qualitative character of organic wholes is due to the peculiar non-additive relations or interactions among their parts. In other words, the whole is not merely a sum, or resultant, but also an emergent novelty, or creative synthesis. (Wheeler, 1926, p. 433)

Wheeler's theory is an example of holism (Sawyer, 2002), in which the regularities governing a whole system cannot be easily reduced to a theory that appeals to the properties of the system's parts. Holistic theories have often been criticized as being nonscientific (Wilson & Lumsden, 1991). The problem with these theories is that in many instances they resist traditional, reductionist approaches to defining the laws responsible for emerging regularities. "Holism is an idea that has haunted biology and philosophy for nearly a century, without coming into clear focus" (Wilson & Lumsden, 1991, p. 401).

Theorists who rejected Wheeler's proposal of the superorganism proposed alternative theories that reduced colonial intelligence to the actions of individual colony members. A pioneer of this alternative was a contemporary of Wheeler, French biologist Etienne Rabaud. "His entire work on insect societies was an attempt to demonstrate that each individual insect in a society behaves as if it were alone" (Theraulaz & Bonabeau, 1999). Wilson and Lumsden adopted a similar position:

> It is tempting to postulate some very complex force distinct from individual repertories and operating at the level of the colony. But a closer look shows that the superorganismic order is actually a straightforward summation of often surprisingly simple individual responses. (Wilson & Lumsden, 1991, p. 402)

Of interest to embodied cognitive science are theories which propose that dynamic environmental control guides the construction of the elaborate nests.

The first concern of such a theory is the general account that it provides of the behaviour of each individual. For example, consider one influential theory of wasp behaviour (Evans, 1966; Evans & West-Eberhard, 1970), in which a hierarchy of internal drives serves to release behaviours. For instance, high-level drives might include mating, feeding, and brood-rearing. Such drives set in motion lower-level sequences of behaviour, which in turn might activate even lower-level behavioural sequences. In short, Evans views wasp behaviour as being rooted in innate programs, where a program is a set of behaviours that are produced in a particular sequence, and where the sequence is dictated by the control of a hierarchical arrangement of drives. For example, a brood-rearing drive might activate a drive for capturing prey, which in turn activates a set of behaviours that produces a hunting flight.

Critically, though, Evans' programs are also controlled by releasing stimuli that are *external* to the wasp. In particular, one behaviour in the sequence is presumed to produce an environmental signal that serves to initiate the next behaviour in the sequence. For instance, in Evans' (1966) model of the construction of a burrow by a solitary digger wasp, the digging behaviour of a wasp produces loosened soil, which serves as a signal for the wasp to initiate scraping behaviour. This behaviour in turn causes the burrow to be clogged, which serves as a signal for clearing behaviour. Having a sequence of behaviours under the control of both internal drives and external releasers provides a balance between rigidity and flexibility; the internal drives serve to provide a general behavioural goal, while variations in external releasers can produce variations in behaviours: e.g., resulting in an atypical nest structure when nest damage elicits a varied behavioural sequence. "Each element in the 'reaction chain' is dependent upon that preceding it as well as upon certain factors in the environment (often gestalts), and each act is capable a certain latitude of execution" (p. 144).

If an individual's behaviour is a program whose actions are under some environmental control (Evans, 1966; Evans & West-Eberhard, 1970), then it is a small step to imagine how the actions of one member of a colony can affect the later actions of other members, even in the extreme case where there is absolutely no direct communication amongst colony members; an individual in the colony simply changes the environment in such a way that new behaviours are triggered by other colony members.

This kind of theorizing is prominent in modern accounts of nest construction by social paper wasps (Theraulaz & Bonabeau, 1999). A nest for such wasps consists of a lattice of cells, where each cell is essentially a comb created from a hexagonal arrangement of walls. When a large nest is under construction, where will new cells be added?

Theraulaz and Bonabeau (1999) answered this question by assuming that the addition of new cells was under environmental control. They hypothesized that an individual wasp's decision about where to build a new cell wall was driven by its perception of existing walls. Their theory consisted of two simple rules. First, if there is a location on the nest in which three walls of a cell already existed, then this was proposed as a stimulus to cause a wasp to add another wall here with high probability. Second, if only two walls already existed as part of a cell, this was also a stimulus to add a wall, but this stimulus produced this action with a much lower probability.

The crucial characteristic of this approach is that behaviour is controlled, and the activities of the members of a colony are coordinated, by a dynamic environment. That is, when an individual is triggered to add a cell wall to the nest, then the nest structure changes. Such changes in nest appearance in turn affect the behaviour of other wasps, affecting choices about the locations where walls will be added next. Theraulaz and Bonabeau (1999) created a nest building simulation that only used these two rules, and demonstrated that it created simulated nests that were very similar in structure to real wasp nests.

In addition to adding cells laterally to the nest, wasps must also lengthen existing walls to accommodate the growth of larvae that live inside the cells. Karsai (1999) proposed another environmentally controlled model of this aspect of nest building. His theory is that wasps perceive the relative difference between the longest and the shortest wall of a cell. If this difference was below a threshold value, then the cell was untouched. However, if this difference exceeded a certain threshold, then this would cause a wasp to lengthen the shortest wall. Karsai used a computer simulation to demonstrate that this simple model provided an accurate account of the three-dimensional growth of a wasp nest over time.

The externalization of control illustrated in theories of wasp nest construction is called stigmergy (Grasse, 1959). The term comes from the Greek *stigma*, meaning "sting," and *ergon*, meaning "work," capturing the notion that the environment is a stimulus that causes particular work, or behaviour, to occur. It was first used in theories of termite mound construction proposed by French zoologist Pierre-Paul Grassé (Theraulaz & Bonabeau, 1999). Grassé demonstrated that the termites themselves do not coordinate or regulate their building behaviour, but that this is instead controlled by the mound structure itself.

Stigmergy is appealing because it can explain how very simple agents create extremely complex products, particularly in the case where the final product, such as a termite mound, is extended in space and time far beyond the life expectancy of the organisms that create it. As well, it accounts for the building of large, sophisticated nests without the need for a complete blueprint and without the need for direct communication amongst colony members (Bonabeau et al., 1998; Downing & Jeanne, 1988; Grasse, 1959; Karsai, 1999; Karsai & Penzes, 1998; Karsai & Wenzel, 2000; Theraulaz & Bonabeau, 1995).

Stigmergy places an emphasis on the importance of the environment that is typically absent in the classical sandwich that characterizes theories in both classical and connectionist cognitive science. However, early classical theories were sympathetic to the role of stigmergy (Simon, 1969). In Simon's famous parable of the ant, observers recorded the path travelled by an ant along a beach. How might we account for the complicated twists and turns of the ant's route? Cognitive scientists tend to explain complex behaviours by invoking complicated representational mechanisms (Braitenberg, 1984). In contrast, Simon (1969) noted that the path might result from simple internal processes reacting to complex external forces—the various obstacles along the natural terrain of the beach: "Viewed as a geometric figure, the ant's path is irregular, complex, hard to describe. But its complexity is really a complexity in the surface of the beach, not a complexity in the ant" (p. 24).

Similarly, Braitenberg (1984) argued that when researchers explain behaviour by appealing to internal processes, they ignore the environment: "When we analyze a mechanism, we tend to overestimate its complexity" (p. 20). He suggested an alternative approach, synthetic psychology, in which simple agents (such as robots) are built and then observed in environments of varying complexity. This approach can provide cognitive science with more powerful, and much simpler, theories by taking advantage of the fact that not all of the intelligence must be placed inside an agent.

Embodied cognitive scientists recognize that the external world can be used to scaffold cognition and that working memory—and other components of a classical architecture—have leaked into the world (Brooks, 1999; Chemero, 2009; Clark, 1997, 2003; Hutchins, 1995; Pfeifer & Scheier, 1999). In many respect, embodied cognitive science is primarily a reaction against the overemphasis of internal processing that is imposed by the classical sandwich.

5.4 Embodiment, Situatedness, and Feedback

Theories that incorporate stigmergy demonstrate the plausibility of removing central cognitive control; perhaps embodied cognitive science could replace the classical sandwich's sense-think-act cycle with sense-act reflexes.

> The realization was that the so-called central systems of intelligence—or core AI as it has been referred to more recently—was perhaps an unnecessary illusion, and that all the power of intelligence arose from the coupling of perception and actuation systems. (Brooks, 1999, p. viii)

For a stigmergic theory to have any power at all, agents must exhibit two critical abilities. First, they must be able to sense their world. Second, they must be able to physically act upon the world. For instance, stigmergic control of nest construction would be impossible if wasps could neither sense local attributes of nest structure nor act upon the nest to change its appearance.

In embodied cognitive science, an agent's ability to sense its world is called situatedness. For the time being, we will simply equate situatedness with the ability to sense. However, situatedness is more complicated than this, because it depends critically upon the physical nature of an agent, including its sensory apparatus and its bodily structure. These issues will be considered in more detail in the next section.

In embodied cognitive science, an agent's ability to act upon and alter its world depends upon its embodiment. In the most general sense, to say that an agent is embodied is to say that it is an artifact, that it has physical existence. Thus while neither a thought experiment (Braitenberg, 1984) nor a computer simulation (Wilhelms & Skinner, 1990) for exploring a Braitenberg vehicle are embodied, a physical robot that acts like a Braitenberg vehicle (Dawson, Dupuis, & Wilson, 2010) *is* embodied. The physical structure of the robot itself is important in the sense that it is a source of behavioural complexity. Computer simulations of Braitenberg vehicles are idealizations in which all motors and sensors work perfectly. This is impossible in a physically realized robot. In an embodied agent, one motor will be less powerful than another, or one sensor may be less effective than another. Such differences will alter robot behaviour. These imperfections are another important source of behavioural complexity, but are absent when such vehicles are created in simulated and idealized worlds.

However, embodiment is more complicated than mere physical existence. Physically existing agents can be embodied to different degrees (Fong, Nourbakhsh, & Dautenhahn, 2003). This is because some definitions of embodiment relate to the extent to which an agent can alter its environment. For instance, Fong, Nourbakhsh, & Dautenhahn (2003, p. 149) argued that "embodiment is grounded in the relationship between a system and its environment. The more a robot can perturb an environment, and be perturbed by it, the more it is embodied." As a result, not all robots are equally embodied (Dawson, Dupuis, & Wilson, 2010). A robot that is more strongly embodied than another is a robot that is more capable of affecting, and being affected by, its environment.

The power of embodied cognitive science emerges from agents that are both situated and embodied. This is because these two characteristics provide a critical source of nonlinearity called feedback (Ashby, 1956; Wiener, 1948). Feedback occurs when information about an action's effect on the world is used to inform the progress of that action. As Ashby (1956, p. 53) noted, "'feedback' exists between two parts when each affects the other," when "circularity of action exists between the parts of a dynamic system."

Wiener (1948) realized that feedback was central to a core of problems involving communication, control, and statistical mechanics, and that it was crucial to both biological agents and artificial systems. He provided a mathematical framework for studying communication and control, defining the discipline that he called cybernetics. The term *cybernetics* was derived from the Greek word for "steersman" or "governor." "In choosing this term, we wish to recognize that the first significant paper on feedback mechanisms is an article on governors, which was published by Clerk Maxwell in 1868" (Wiener, 1948, p. 11). Interestingly, engine governors make frequent appearances in formal discussions of the embodied approach (Clark, 1997; Port & van Gelder, 1995b; Shapiro, 2011).

The problem with the nonlinearity produced by feedback is that it makes computational analyses extraordinarily difficult. This is because the mathematics of feedback relationships between even small numbers of components is essentially intractable. For instance, Ashby (1956) realized that feedback amongst a machine that only consisted of four simple components could not analyzed:

> When there are only two parts joined so that each affects the other, the properties
> of the feedback give important and useful information about the properties of the
> whole. But when the parts rise to even as few as four, if everyone affects the other
> three, then twenty circuits can be traced through them; and knowing the proper-
> ties of all the twenty circuits does *not* give complete information about the system.
> (Ashby, 1956, p. 54)

For this reason, embodied cognitive science is often practised using forward engi-neering, which is a kind of synthetic methodology (Braitenberg, 1984; Dawson, 2004; Pfeifer & Scheier, 1999). That is, researchers do not take a complete agent and reverse engineer it into its components. Instead, they take a small number of simple components, compose them into an intact system, set the components in motion in an environment of interest, and observe the resulting behaviours.

For instance, Ashby (1960) investigated the complexities of his four-compo-nent machine not by dealing with intractable mathematics, but by building and observing a working device, the Homeostat. It comprised four identical machines (electrical input-output devices), incorporated mutual feedback, and permitted him to observe the behaviour, which was the movement of indicators for each machine. Ashby discovered that the Homeostat could learn; he reinforced its responses by physically manipulating the dial of one component to "punish" an incorrect response (e.g., for moving one of its needles in the incorrect direction). Ashby also found that the Homeostat could adapt to two different environments that were alternated from trial to trial. This knowledge was unattainable from mathematical analyses. "A better demonstration can be given by a machine, built so that we know its nature exactly and on which we can observe what will happen in various conditions" (p. 99).

Braitenberg (1984) has argued that an advantage of forward engineering is that it will produce theories that are simpler than those that will be attained by reverse engineering. This is because when complex or surprising behaviours emerge, pre-existing knowledge of the components—which were constructed by the researcher—can be used to generate simpler explanations of the behaviour.

> Analysis is more difficult than invention in the sense in which, generally, induc-
> tion takes more time to perform than deduction: in induction one has to
> search for the way, whereas in deduction one follows a straightforward path.
> (Braitenberg, 1984, p. 20)

Braitenberg called this the law of uphill analysis and downhill synthesis.

Another way in which to consider the law of uphill analysis and downhill syn-thesis is to apply Simon's (1969) parable of the ant. If the environment is taken seriously as a contributor to the complexity of the behaviour of a situated and embodied agent, then one can take advantage of the agent's world and propose less complex internal mechanisms that still produce the desired intricate results. This idea is central to the replacement hypothesis that Shapiro (2011) has argued is a fundamental characteristic of embodied cognitive science.

5.5 *Umwelten*, Affordances, and Enactive Perception

The situatedness of an agent is not merely perception; the nature of an agent's perceptual apparatus is a critical component of situatedness. Clearly agents can only experience the world in particular ways because of limits, or specializations, in their sensory apparatus (Uexküll, 2001). Ethologist Jakob von Uexküll coined the term *umwelt* to denote the "island of the senses" produced by the unique way in which an organism is perceptually engaged with its world. Uexküll realized that because different organisms experience the world in different ways, they can live in the same world but at the same time exist in different *umwelten*. Similarly, the ecological theory of perception (Gibson, 1966, 1979) recognized that one could not separate the characteristics of an organism from the characteristics of its environment. "It is often neglected that the words *animal* and *environment* make an inseparable pair" (Gibson, 1979, p. 8).

The inseparability of animal and environment can at times even be rooted in the structure of an agent's body. For instance, bats provide a prototypical example of an active-sensing system (MacIver, 2008) because they emit a high-frequency sound and detect the location of targets by processing the echo. The horizontal position of a target (e.g., a prey insect) is uniquely determined by the difference in time between the echo's arrival to the left and right ears. However, this information is not sufficient to specify the vertical position of the target. The physical nature of bat ears solves this problem. The visible external structure (the pinna and the tragus) of the bat's ear has an extremely intricate shape. As a result, returning echoes strike the ear at different angles of entry. This provides additional auditory cues that vary systematically with the vertical position of the target (Wotton, Haresign, & Simmons, 1995; Wotton & Simmons, 2000). In other words, the bat's body—in particular, the shape of its ears—is critical to its *umwelt*.

Passive and active characteristics of an agent's body are central to theories of perception that are most consistent with embodied cognitive science (Gibson, 1966, 1979; Noë, 2004). This is because embodied cognitive science has arisen as part of a reaction against the Cartesian view of mind that inspired classical cognitive science. In particular, classical cognitive science inherited Descartes' notion (Descartes, 1960, 1996) of the disembodied mind that had descended from Descartes' claim of *Cogito ergo sum*. Embodied cognitive scientists have been strongly influenced by philosophical positions which arose as reactions against Descartes, such as Martin Heidegger's *Being and Time* (Heidegger, 1962), originally published in 1927. Heidegger criticized Descartes for adopting many of the terms of older philosophies but failing to recognize a critical element, their interactive relationship to the world: "The ancient way of interpreting the Being of entities is oriented towards the 'world' or 'Nature' in the widest sense" (Heidegger, 1962, p. 47). Heidegger argued instead for Being-in-the-world as a primary mode of existence. Being-in-the-world is not just being spatially located in an environment, but is a mode of existence in which an agent is actively engaged with entities in the world.

Dawson, Dupuis, and Wilson (2010) used a passive dynamic walker to illustrate this inseparability of agent and environment. A passive dynamic walker is an agent that walks without requiring active control: its walking gait is completely due to gravity and inertia (McGeer, 1990). Their simplicity and low energy requirements have made them very important models for the development of walking robots (Alexander, 2005; Collins et al., 2005; Kurz et al., 2008; Ohta, Yamakita, & Furuta, 2001; Safa, Saadat, & Naraghi, 2007; Wisse, Schwab, & van der Helm, 2004). Dawson, Dupuis, and Wilson constructed a version of McGeer's (1990) original walker from LEGO. The walker itself was essentially a straight-legged hinge that would walk down an inclined ramp. However, the ramp had to be of a particular slope and had to have properly spaced platforms with gaps in between to permit the agent's legs to swing. Thus the LEGO hinge that Dawson, Dupuis, and Wilson (2010) built had the disposition to walk, but it required a specialized environment to have this disposition realized. The LEGO passive dynamic walker is only a walker when it interacts with the special properties of its ramp. Passive dynamic walking is not a characteristic of a device, but is instead a characteristic of a device being in a particular world.

Being-in-the-world is related to the concept of affordances developed by psychologist James J. Gibson (Gibson, 1979). In general terms, the affordances of an object are the possibilities for action that a particular object permits a particular agent. "The *affordances* of the environment are what it *offers* the animal, what it *provides* or *furnishes*, either for good or ill" (p. 127). Again, affordances emerge from an integral relationship between an object's properties and an agent's abilities to act.

> Note that the four properties listed—horizontal, flat, extended, and rigid—would be *physical* properties of a surface if they were measured with the scales and standard units used in physics. As an affordance of support for a species of animal, however, they have to be measured *relative to the animal*. They are unique for that animal. They are not just abstract physical properties. (p. 127)

Given that affordances are defined in terms of an organism's potential actions, it is not surprising that action is central to Gibson's (1966, 1979) ecological approach to perception. Gibson (1966, p. 49) noted that "when the 'senses' are considered as active systems they are classified by modes of activity not by modes of conscious quality." Gibson's emphasis on action and the world caused his theory to be criticized by classical cognitive science (Fodor & Pylyshyn, 1981). Perhaps it is not surprising that the embodied reaction to classical cognitive science has been accompanied by a modern theory of perception that has descended from Gibson's work: the enactive approach to perception (Noë, 2004).

Enactive perception reacts against the traditional view that perception is constructing internal representations of the external world. Enactive perception argues instead that the role of perception is to access information in the world when it is needed. That is, perception is not a representational process, but is instead a sensorimotor skill (Noë, 2004). "Perceiving is a way of acting. Perception is not something that happens to us, or in us. It is something we do" (p. 1).

Action plays multiple central roles in the theory of enactive perception (Noë, 2004). First, the purpose of perception is not viewed as building internal representations of the world, but instead as controlling action on the world. Second, and related to the importance of controlling action, our perceptual understanding of objects is sensorimotor, much like Gibson's (1979) notion of affordance. That is, we obtain an understanding of the external world that is related to its changes in appearance that would result by changing our position—by acting on an object, or by moving to a new position. Third, perception is to be an intrinsically exploratory process. As a result, we do not construct complete visual representations of the world. Instead, perceptual objects are virtual—we have access to properties in the world when needed, and only through action.

> Our sense of the perceptual presence of the cat as a whole now does not require
> us to be committed to the idea that we represent the whole cat in consciousness at
> once. What it requires, rather, is that we take ourselves to have *access*, now, to the
> whole cat. The cat, the tomato, the bottle, the detailed scene, all are present percep-
> tually in the sense that they are perceptually accessible to us. (Noë, 2004, p. 63)

Empirical support for the virtual presence of objects is provided by the phenomenon of change blindness. Change blindness occurs when a visual change occurs in plain sight of a viewer, but the viewer does not notice the change. For instance, in one experiment (O'Regan et al., 2000), subjects inspect an image of a Paris street scene. During this inspection, the colour of a car in the foreground of the image changes, but a subject does not notice this change! Change blindness supports the view that representations of the world are not constructed. "The upshot of this is that *all* detail is present in experience not as represented, but rather as accessible" (Noë, 2004, p. 193). Accessibility depends on action, and action also depends on embodiment. "To perceive like us, it follows, you must have a body like ours" (p. 25).

5.6 Horizontal Layers of Control

Classical cognitive science usually assumes that the primary purpose of cognition is planning (Anderson, 1983; Newell, 1990); this planning is used to mediate perception and action. As a result, classical theories take the form of the sense-think-act cycle (Pfeifer & Scheier, 1999). Furthermore, the "thinking" component of this cycle is emphasized far more than either the "sensing" or the "acting." "One problem with psychology's attempt at cognitive theory has been our persistence in thinking about cognition without bringing in perceptual and motor processes" (Newell, 1990, p. 15).

Embodied cognitive science (Agre, 1997; Brooks, 1999, 2002; Chemero, 2009; Clancey, 1997; Clark, 1997, 2003, 2008; Pfeifer & Scheier, 1999; Robbins & Aydede, 2009; Shapiro, 2011; Varela, Thompson, & Rosch, 1991) recognizes the importance of sensing and acting, and reacts against central cognitive control. Its more radical proponents strive to completely replace the sense-think-act cycle with sense-act mechanisms.

This reaction is consistent with several themes in the current chapter: the importance of the environment, degrees of embodiment, feedback between the world and the agent, and the integral relationship between an agent's body and its *umwelt*. Given these themes, it becomes quite plausible to reject the proposal that cognition is used to plan, and to posit instead that the purpose of cognition is to guide action:

> The brain should not be seen as primarily a locus of inner *descriptions* of external states of affairs; rather, it should be seen as a locus of internal *structures* that act as operators upon the world via their role in determining actions. (Clark, 1997, 47)

Importantly, these structures do not stand between sensing and acting, but instead provide direct links between them.

The action-based reaction against classical cognitivism is typified by pioneering work in behaviour-based robotics (Brooks, 1989, 1991, 1999, 2002; Brooks & Flynn, 1989). Roboticist Rodney Brooks construes the classical sandwich as a set of vertical processing layers that separate perception and action. His alternative is a hierarchical arrangement of horizontal processing layers that directly connect perception and action.

Brooks' action-based approach to behaviour is called the subsumption architecture (Brooks, 1999). The subsumption architecture is a set of modules. However, these modules are somewhat different in nature than those that were discussed in Chapter 3 (see also Fodor, 1983). This is because each module in the subsumption architecture can be described as a sense-act mechanism. That is, every module can have access to sensed information, as well as to actuators. This means that modules in the subsumption architecture do not separate perception from action. Instead, each module is used to control some action on the basis of sensed information.

The subsumption architecture arranges modules hierarchically. Lower-level modules provide basic, general-purpose, sense-act functions. Higher-level modules provide more complex and more specific sense-act functions that can exploit the operations of lower-level operations. For instance, in an autonomous robot the lowest-level module might simply activate motors to move a robot forward (e.g., Dawson, Dupuis. & Wilson, 2010, Chapter 7). The next level might activate a steering mechanism. This second level causes the robot to wander by taking advantage of the movement provided by the lower level. If the lower level were not operating, then wandering would not occur: because although the steering mechanism was operating, the vehicle would not be moving forward.

Vertical sense-act modules, which are the foundation of the subsumption architecture, also appear to exist in the human brain (Goodale, 1988, 1990, 1995; Goodale & Humphrey, 1998; Goodale, Milner, Jakobson, & Carey, 1991; Jakobson et al., 1991).

There is a long-established view that two distinct physiological pathways exist in the human visual system (Livingstone & Hubel, 1988; Maunsell & Newsome, 1987; Ungerleider & Mishkin, 1982): one, the ventral stream, for processing the appearance of objects; the other, the dorsal stream, for processing their locations. In short, in object perception the ventral stream delivers the "what," while the dorsal stream delivers the "where." This view is supported by double dissociation evidence observed in clinical patients: brain injuries can cause severe problems in seeing motion but leave form perception unaffected, or vice versa (Botez, 1975; Hess, Baker, & Zihl, 1989; Zihl, von Cramon, & Mai, 1983).

There has been a more recent reconceptualization of this classic distinction: the duplex approach to vision (Goodale & Humphrey, 1998), which maintains the physiological distinction between the ventral and dorsal streams but reinterprets their functions. In the duplex theory, the ventral stream creates perceptual representations, while the dorsal stream mediates the visual control of action.

> The functional distinction is not between 'what' and 'where,' but between the way in which the visual information about a broad range of object parameters are transformed either for perceptual purposes or for the control of goal-directed actions.
> (Goodale & Humphrey, 1998, p. 187)

The duplex theory can be seen as representational theory that is elaborated in such a way that fundamental characteristics of the subsumption architecture are present. These results can be used to argue that the human brain is not completely structured as a "classical sandwich." On the one hand, in the duplex theory the purpose of the ventral stream is to create a representation of the perceived world (Goodale & Humphrey, 1998). On the other hand, in the duplex theory the purpose of the dorsal stream is the control of action, because it functions to convert visual information directly into motor commands. In the duplex theory, the ventral stream is strikingly similar to the vertical layers of the subsumption architecture.

Double dissociation evidence from cognitive neuroscience has been used to support the duplex theory. The study of one brain-injured subject (Goodale et al., 1991) revealed normal basic sensation. However, the patient could not describe the orientation or shape of any visual contour, no matter what visual information was used to create it. While this information could not be consciously reported, it was available, and could control actions. The patient could grasp objects, or insert objects through oriented slots, in a fashion indistinguishable from control subjects, even to the fine details that are observed when such actions are initiated and then carried out. This pattern of evidence suggests that the patient's ventral stream was damaged, but that the dorsal stream was unaffected and controlled visual actions. "At some level in normal brains the visual processing underlying 'conscious' perceptual judgments must operate separately from that underlying the 'automatic' visuomotor guidance of skilled actions of the hand and limb" (p. 155).

Other kinds of brain injuries produce a very different pattern of abnormalities, establishing the double dissociation that supports the duplex theory. For instance, damage to the posterior parietal cortex—part of the dorsal stream—can cause optic ataxia, in which visual information cannot be used to control actions towards objects presented in the part of the visual field affected by the brain injury (Jakobson et al., 1991). Optic ataxia, however, does not impair the ability to perceive the orientation and shapes of visual contours.

Healthy subjects can also provide support for the duplex theory. For instance, in one study subjects reached toward an object whose position changed during a saccadic eye movement (Pelisson et al., 1986). As a result, subjects were not conscious of the target's change in location. Nevertheless, they compensated to the object's new position when they reached towards it. "No perceptual change occurred, while the hand pointing response was shifted systematically, showing that different mechanisms were involved in visual perception and in the control of the motor response" (p. 309). This supports the existence of "horizontal" sense-act modules in the human brain.

5.7 Mind in Action

Shakey was a 1960s robot that used a variety of sensors and motors to navigate through a controlled indoor environment (Nilsson, 1984). It did so by uploading its sensor readings to a central computer that stored, updated, and manipulated a model of Shakey's world. This representation was used to develop plans of action to be put into effect, providing the important filling for Shakey's classical sandwich.

Shakey impressed in its ability to navigate around obstacles and move objects to desired locations. However, it also demonstrated some key limitations of the classical sandwich. In particular, Shakey was extremely slow. Shakey typically required several hours to complete a task (Moravec, 1999), because the internal model of its world was computationally expensive to create and update. The problem with the sense-think-act cycle in robots like Shakey is that by the time the (slow) thinking is finished, the resulting plan may fail because the world has changed in the meantime.

The subsumption architecture of behaviour-based robotics (Brooks, 1999, 2002) attempted to solve such problems by removing the classical sandwich; it was explicitly anti-representational. The logic of this radical move was that the world was its own best representation (Clark, 1997).

Behaviour-based robotics took advantage of Simon's (1969) parable of the ant, reducing costly and complex internal representations by recognizing that the external world is a critical contributor to behaviour. Why expend computational resources on the creation and maintenance of an internal model of the world, when externally the world was already present, open to being sensed and to being acted upon? Classical cognitive science's emphasis on internal representations and planning was a failure to take this parable to heart.

Interestingly, action was more important to earlier cognitive theories. Take, for example, Piaget's theory of cognitive development (Inhelder & Piaget, 1958, 1964; Piaget, 1970a, 1970b, 1972; Piaget & Inhelder, 1969). According to this theory, in their early teens children achieve the stage of formal operations. Formal operations describe adult-level cognitive abilities that are classical in the sense that they involve logical operations on symbolic representations. Formal operations involve completely abstract thinking, where relationships between propositions are considered.

However, Piagetian theory departs from classical cognitive science by including actions in the world. The development of formal operations begins with the sensorimotor stage, which involves direct interactions with objects in the world. In the next preoperational stage these objects are internalized as symbols. The preoperational stage is followed by concrete operations. When the child is in the stage of concrete operations, symbols are manipulated, but not in the abstract: concrete operations are applied to "manipulable objects (effective or immediately imaginable manipulations), in contrast to operations bearing on propositions or simple verbal statements (logic of propositions)" (Piaget, 1972, p. 56). In short, Piaget rooted fully representational or symbolic thought (i.e., formal operations) in the child's physical manipulation of his or her world. "The starting-point for the understanding, even of verbal concepts, is still the actions and operations of the subject" (Inhelder & Piaget, 1964, p. 284).

For example, classification and seriation (i.e., grouping and ordering entities) are operations that can be formally specified using logic or mathematics. One goal of Piagetian theory is to explain the development of such abstract competence. It does so by appealing to basic actions on the world experienced prior to the stage of formal operations, "actions which are quite elementary: putting things in piles, separating piles into lots, making alignments, and so on" (Inhelder & Piaget, 1964, p. 291).

Other theories of cognitive development share the Piagetian emphasis on the role of the world, but elaborate the notion of what aspects of the world are involved (Vygotsky, 1986). Vygotsky (1986), for example, highlighted the role of social systems—a different conceptualization of the external world—in assisting cognitive development. Vygotsky used the term *zone of proximal development* to define the difference between a child's ability to solve problems without aid and their ability to solve problems when provided support or assistance. Vygotsky was strongly critical of instructional approaches that did not provide help to children as they solved problems.

Vygotsky (1986) recognized that sources of support for development were not limited to the physical world. He expanded the notion of worldly support to include social and cultural factors: "The true direction of the development of thinking is not from the individual to the social, but from the social to the individual" (p. 36). For example, to Vygotsky language was a tool for supporting cognition:

> Real concepts are impossible without words, and thinking in concepts does not exist beyond verbal thinking. That is why the central moment in concept formation, and its generative cause, is a specific use of words as functional 'tools.'
> (Vygotsky, 1986, p. 107)

Clark (1997, p. 45) wrote: "We may often solve problems by 'piggy-backing' on reliable environmental properties. This exploitation of external structure is what I mean by the term scaffolding." Cognitive scaffolding—the use of the world to support or extend thinking—is characteristic of theories in embodied cognitive science. Clark views scaffolding in the broad sense of a world or structure that descends from Vygotsky's theory:

> Advanced cognition depends crucially on our abilities to dissipate reasoning: to diffuse knowledge and practical wisdom through complex social structures, and to reduce the loads on individual brains by locating those brains in complex webs of linguistic, social, political, and institutional constraints. (Clark, 1997, p. 180)

While the developmental theories of Piaget and Vygotsky are departures from typical classical cognitive science in their emphasis on action and scaffolding, they are very traditional in other respects. American psychologist Sylvia Scribner pointed out that these two theorists, along with Newell and Simon, shared Aristotle's "preoccupation with modes of thought central to theoretical inquiry—with logical operations, scientific concepts, and problem solving in symbolic domains," maintaining "Aristotle's high esteem for theoretical thought and disregard for the practical" (Scribner & Tobach, 1997, p. 338).

Scribner's own work (Scribner & Tobach, 1997) was inspired by Vygotskian theory but aimed to extend its scope by examining practical cognition. Scribner described her research as the study of mind in action, because she viewed cognitive processes as being embedded with human action in the world. Scribner's studies analyzed "the characteristics of memory and thought as they function in the larger, purposive activities which cultures organize and in which individuals engage" (p. 384). In other words, the everyday cognition studied by Scribner and her colleagues provided ample evidence of cognitive scaffolding: "Practical problem solving is an open system that includes components lying outside the formal problem—objects and information in the environment and goals and interests of the problem solver" (pp. 334–335).

One example of Scribner's work on mind in action was the observation of problem-solving strategies exhibited by different types of workers at a dairy (Scribner & Tobach, 1997). It was discovered that a reliable difference between expert and novice dairy workers was that the former were more versatile in finding solutions to problems, largely because expert workers were much more able to exploit environmental resources. "The physical environment did not determine the problem-solving process but . . . was drawn into the process through worker initiative" (p. 377).

For example, one necessary job in the dairy was assembling orders. This involved using a computer printout of a wholesale truck driver's order for products to deliver the next day, to fetch from different areas in the dairy the required number of cases and partial cases of various products to be loaded onto the driver's truck. However, while the driver's order was placed in terms of individual units (e.g., particular numbers of quarts of skim milk, of half-pints of chocolate milk, and so on), the computer printout converted these individual units into "case equivalents." For example, one driver might require 20 quarts of skim milk. However, one case contains only 16 quarts. The computer printout for this part of the order would be 1 + 4, indicating one full case plus 4 additional units.

Scribner found differences between novice and expert product assemblers in the way in which these mixed numbers from the computer printout were converted into gathered products. Novice workers would take a purely mental arithmetic approach. As an example, consider the following protocol obtained from a novice worker:

> It was one case minus six, so there's two, four, six, eight, ten, sixteen (determines how many in a case, points finger as she counts). So there should be ten in here. Two, four, six, ten (counts units as she moves them from full to empty). One case minus six would be ten. (Scribner & Tobach, 1997, p. 302)

In contrast, expert workers were much more likely to scaffold this problem solving by working directly from the visual appearance of cases, as illustrated in a very different protocol:

> I walked over and I visualized. I knew the case I was looking at had ten out of it, and I only wanted eight, so I just added two to it. I don't never count when I'm making the order, I do it visual, a visual thing you know. (Scribner & Tobach, 1997, p. 303)

It was also found that expert workers flexibly alternated the distribution of scaffolded and mental arithmetic, but did so in a systematic way: when more mental arithmetic was employed, it was done to decrease the amount of physical exertion required to complete the order. This led to Scribner postulating a law of mental effort: "In product assembly, mental work will be expended to save physical work" (Scribner & Tobach, 1997, p. 348).

The law of mental effort was the result of Scribner's observation that expert workers in the dairy demonstrated marked diversity and flexibility in their solutions to work-related problems. Intelligent agents may be flexible in the manner in which they allocate resources between sense-act and sense-think-act processing. Both types of processes may be in play simultaneously, but they may be applied in different amounts when the same problem is encountered at different times and under different task demands (Hutchins, 1995).

Such flexible information processing is an example of *bricolage* (Lévi-Strauss, 1966). A *bricoleur* is an "odd job man" in France.

> The '*bricoleur*' is adept at performing a large number of diverse tasks; but, unlike the engineer, he does not subordinate each of them to the availability of raw materials and tools conceived and procured for the purpose of the project. His universe of instruments is closed and the rules of his game are always to make do with 'whatever is at hand.' (Lévi-Strauss, 1966, p. 17)

Bricolage seems well suited to account for the flexible thinking of the sort described by Scribner. Lévi-Strauss (1966) proposed *bricolage* as an alternative to formal, theoretical thinking, but cast it in a negative light: "The '*bricoleur*' is still someone who works with his hands and uses devious means compared to those of a craftsman" (pp. 16–17). Devious means are required because the *bricoleur* is limited to using only those components or tools that are at hand. "The engineer is always trying to make his way out of and go beyond the constraints imposed by a particular state of civilization while the '*bricoleur*' by inclination or necessity always remains within them" (p. 19).

Recently, researchers have renewed interest in *bricolage* and presented it in a more positive light than did Lévi-Strauss (Papert, 1980; Turkle, 1995). To Turkle (1995), *bricolage* was a sort of intuition, a mental tinkering, a dialogue mediated by a virtual interface that was increasingly important with the visual GUIs of modern computing devices.

> As the computer culture's center of gravity has shifted from programming to dealing with screen simulations, the intellectual values of *bricolage* have become far more important. . . . Playing with simulation encourages people to develop the skills of the more informal soft mastery because it is so easy to run 'What if?' scenarios and tinker with the outcome. (Turkle, 1995, p. 52)

Papert (1980) argued that *bricolage* demands greater respect because it may serve as "a model for how scientifically legitimate theories are built" (p. 173).

The *bricolage* observed by Scribner and her colleagues when studying mind in action at the dairy revealed that practical cognition is flexibly and creatively scaffolded by an agent's environment. However, many of the examples reported by Scribner suggest that this scaffolding involves using the environment as an external representation or memory of a problem. That the environment can be used in this fashion, as an externalized extension of memory, is not surprising. Our entire print culture—the use of handwritten notes, the writing of books—has arisen from a technology that serves as an extension of memory (McLuhan, 1994, p. 189): "Print provided a vast new memory for past writings that made a personal memory inadequate."

However, the environment can also provide a more intricate kind of scaffolding. In addition to serving as an external store of information, it can also be exploited to manipulate its data. For instance, consider a naval navigation task in which a ship's speed is to be computed by measuring of how far the ship has travelled over a recent interval of time (Hutchins, 1995). An internal, representational approach to performing this computation would be to calculate speed based on internalized knowledge of algebra, arithmetic, and conversions between yards and nautical miles. However, an easier external solution is possible. A navigator is much more likely to draw a line on a three-scale representation called a nomogram. The top scale of this tool indicates duration, the middle scale indicates distance, and the bottom scale indicates speed. The user marks the measured time and distance on the first two scales, joins them with a straight line, and reads the speed from the intersection of this line with the bottom scale. Thus the answer to the problem isn't as much computed as it is inspected. "Much of the computation was done by the tool, or by its designer. The person somehow could succeed by doing less because the tool did more" (Hutchins, 1995, p. 151).

Classical cognitive science, in its championing of the representational theory of mind, demonstrates a modern persistence of the Cartesian distinction between mind and body. Its reliance on mental representation occurs at the expense of ignoring potential contributions of both an agent's body and world. Early representational theories were strongly criticized because of their immaterial nature.

For example, consider the work of Edward Tolman (1932, 1948). Tolman appealed to representational concepts to explain behaviour, such as his proposal that rats navigate and locate reinforcers by creating and manipulating a cognitive map. The mentalistic nature of Tolman's theories was a source of harsh criticism:

> Signs, in Tolman's theory, occasion in the rat realization, or cognition, or judgment, or hypotheses, or abstraction, but they do not occasion action. In his concern with what goes on in the rat's mind, Tolman has neglected to predict what the rat will do. So far as the theory is concerned the rat is left buried in thought; if he gets to the food-box at the end that is his concern, not the concern of the theory. (Guthrie, 1935, p. 172)

The later successes, and current dominance, of cognitive theory make such criticisms appear quaint. But classical theories are nonetheless being rigorously reformulated by embodied cognitive science.

Embodied cognitive scientists argue that classical cognitive science, with its emphasis on the disembodied mind, has failed to capture important aspects of thinking. For example, Hutchins (1995, p. 171) noted that "by failing to understand the source of the computational power in our interactions with simple 'unintelligent' physical devices, we position ourselves well to squander opportunities with so-called intelligent computers." Embodied cognitive science proposes that the modern form of dualism exhibited by classical cognitive science is a mistake. For instance, Scribner hoped that her studies of mind in action conveyed "a conception of mind which is not hostage to the traditional cleavage between the mind and the hand, the mental and the manual" (Scribner & Tobach, 1997, p. 307).

5.8 The Extended Mind

In preceding pages of this chapter, a number of interrelated topics that are central to embodied cognitive science have been introduced: situation and embodiment, feedback between agents and environments, stigmergic control of behaviour, affordances and enactive perception, and cognitive scaffolding. These topics show that embodied cognitive science places much more emphasis on body and world, and on sense and action, than do other "flavours" of cognitive science.

This change in emphasis can have profound effects on our definitions of *mind* or *self* (Bateson, 1972). For example, consider this famous passage from anthropologist Gregory Bateson:

> But what about 'me'? Suppose I am a blind man, and I use a stick. I go tap, tap, tap. Where do *I* start? Is my mental system bounded at the handle of the stick? Is it bounded by my skin? (Bateson, 1972, p. 465)

The embodied approach's emphasis on agents embedded in their environments leads to a radical and controversial answer to Bateson's questions, in the form of the extended mind (Clark, 1997, 1999, 2003, 2008; Clark & Chalmers, 1998; Menary, 2008, 2010; Noë, 2009; Rupert, 2009; Wilson, 2004, 2005). According to the extended mind hypothesis, the mind and its information processing are not separated from the world by the skull. Instead, the mind interacts with the world in such a way that information processing is both part of the brain and part of the world—the boundary between the mind and the world is blurred, or has disappeared.

Where is the mind located? The traditional view—typified by the classical approach introduced in Chapter 3—is that thinking is inside the individual, and that sensing and acting involve the world outside. However, if cognition is scaffolded, then some thinking has moved from inside the head to outside in the world. "It is the human brain *plus* these chunks of external scaffolding that finally constitutes the smart, rational inference engine we call mind" (Clark, 1997, p. 180). As a result, Clark (1997) described the mind as a leaky organ, because it has spread from inside our head to include whatever is used as external scaffolding.

The extended mind hypothesis has enormous implications for the cognitive sciences. The debate between classical and connectionist cognitive science does not turn on this issue, because both approaches are essentially representational. That is, both approaches tacitly endorse the classical sandwich; while they have strong disagreements about the nature of representational processes in the filling of the sandwich, neither of these approaches views the mind as being extended. Embodied cognitive scientists who endorse the extended mind hypothesis thus appear to be moving in a direction that strongly separates the embodied approach from the other two. It is small comfort to know that all cognitive scientists might agree that they are in the business of studying the mind, when they can't agree upon what minds are.

For this reason, the extended mind hypothesis has increasingly been a source of intense philosophical analysis and criticism (Adams & Aizawa, 2008; Menary, 2010; Robbins & Aydede, 2009). Adams and Aizawa (2008) are strongly critical of the extended mind hypothesis because they believe that it makes no serious attempt to define the "mark of the cognitive," that is, the principled differences between cognitive and non-cognitive processing:

> If just any sort of information processing is cognitive processing, then it is not
> hard to find cognitive processing in notebooks, computers and other tools. The
> problem is that this theory of the cognitive is wildly implausible and evidently not
> what cognitive psychologists intend. A wristwatch is an information processor,
> but not a cognitive agent. What the advocates of extended cognition need, but, we
> argue, do not have, is a plausible theory of the difference between the cognitive and

the non-cognitive that does justice to the subject matter of cognitive psychology. (Adams & Aizawa, 2008, p. 11)

A variety of other critiques can be found in various contributions to Robbins and Aydede's (2009) *Cambridge Handbook of Situated Cognition*. Prinz made a pointed argument that the extended mind has nothing to contribute to the study of consciousness. Rupert noted how the notion of innateness poses numerous problems for the extended mind. Warneken and Tomasello examined cultural scaffolding, but they eventually adopted a position where these cultural tools have been internalized by agents. Finally, Bechtel presented a coherent argument from the philosophy of biology that there is good reason for the skull to serve as the boundary between the world and the mind. Clearly, the degree to which extendedness is adopted by situated researchers is far from universal.

In spite of the currently unresolved debate about the plausibility of the extended mind, the extended mind hypothesis is an idea that is growing in popularity in embodied cognitive science. Let us briefly turn to another implication that this hypothesis has for the practice of cognitive science.

The extended mind hypothesis is frequently applied to single cognitive agents. However, this hypothesis also opens the door to co-operative or public cognition in which a group of agents are embedded in a shared environment (Hutchins, 1995). In this situation, more than one cognitive agent can manipulate the world that is being used to support the information processing of other group members.

Hutchins (1995) provided one example of public cognition in his description of how a team of individuals is responsible for navigating a ship. He argued that "organized groups may have cognitive properties that differ from those of the individuals who constitute the group" (p. 228). For instance, in many cases it is very difficult to translate the heuristics used by a solo navigator into a procedure that can be implemented by a navigation team.

Collective intelligence—also called swarm intelligence or co-operative computing—is also of growing importance in robotics. Entomologists used the concept of the superorganism (Wheeler, 1911) to explain how entire colonies could produce more complex results (such as elaborate nests) than one would predict from knowing the capabilities of individual colony members. Swarm intelligence is an interesting evolution of the idea of the superorganism; it involves a collective of agents operating in a shared environment. Importantly, a swarm's components are only involved in local interactions with each other, resulting in many advantages (Balch & Parker, 2002; Sharkey, 2006).

For instance, a computing swarm is scalable—it may comprise varying numbers of agents, because the same control structure (i.e., local interactions) is used regardless of how many agents are in the swarm. For the same reason, a computing swarm is flexible: agents can be added or removed from the swarm without reorganizing the entire system. The scalability and flexibility of a swarm make it robust, as it can continue to compute when some of its component agents no longer function properly. Notice how these advantages of a swarm of agents are analogous to the advantages of connectionist networks over classical models, as discussed in Chapter 4.

Nonlinearity is also a key ingredient of swarm intelligence. For a swarm to be considered intelligent, the whole must be greater than the sum of its parts. This idea has been used to identify the presence of swarm intelligence by relating the amount of work done by a collective to the number of agents in the collection (Beni & Wang, 1991). If the relationship between work accomplished and number of agents is linear, then the swarm is not considered to be intelligent. However, if the relationship is nonlinear—for instance, exponentially increasing—then swarm intelligence is present. The nonlinear relationship between work and numbers may itself be mediated by other nonlinear relationships. For example, Dawson, Dupuis, and Wilson (2010) found that in collections of simple LEGO robots, the presence of additional robots influenced robot paths in an arena in such a way that a sorting task was accomplished far more efficiently.

While early studies of robot collectives concerned small groups of homogenous robots (Gerkey & Mataric, 2004), researchers are now more interested in complex collectives consisting of different types of machines for performing diverse tasks at varying locations or times (Balch & Parker, 2002; Schultz & Parker, 2002). This leads to the problem of coordinating the varying actions of diverse collective members (Gerkey & Mataric, 2002, 2004; Mataric, 1998). One general approach to solving this coordination problem is intentional co-operation (Balch & Parker, 2002; Parker, 1998, 2001), which uses direct communication amongst robots to prevent unnecessary duplication (or competition) between robot actions. However, intentional co-operation comes with its own set of problems. For instance, communication between robots is costly, particularly as more robots are added to a communicating team (Kube & Zhang, 1994). As well, as communication makes the functions carried out by individual team members more specialized, the robustness of the robot collective is jeopardized (Kube & Bonabeau, 2000). Is it possible for a robot collective to coordinate its component activities, and solve interesting problems, in the absence of direction communication?

The embodied approach has generated a plausible answer to this question via stigmergy (Kube & Bonabeau, 2000). Kube and Bonabeau (2000) demonstrated that the actions of a large collective of robots could be stigmergically coordinated so that the collective could push a box to a goal location in an arena. Robots used a variety of sensors to detect (and avoid) other robots, locate the box, and locate the goal location. A subsumption architecture was employed to instantiate a fairly simple set of sense-act reflexes. For instance, if a robot detected that is was in contact with the box and could see the goal, then box-pushing behaviour was initiated. If it was in contact with the box but could not see the goal, then other movements were triggered, resulting in the robot finding contact with the box at a different position.

This subsumption architecture caused robots to seek the box, push it towards the goal, and do so co-operatively by avoiding other robots. Furthermore, when robot activities altered the environment, this produced corresponding changes in behaviour of other robots. For instance, a robot pushing the box might lose sight of the goal because of box movement, and it would therefore leave the box and use its other exploratory behaviours to come back to the box and push it from a different location. "Cooperation in some tasks is possible without direct communication" (Kube & Bonabeau, 2000, p. 100). Importantly, the solution to the box-pushing problem required such co-operation, because the box being manipulated was too heavy to be moved by a small number of robots!

The box-pushing research of Kube and Bonabeau (2000) is an example of stigmergic processing that occurs when two or more individuals collaborate on a task using a shared environment. Hutchins (1995) brought attention to less obvious examples of public cognition that exploit specialized environmental tools. Such scaffolding devices cannot be dissociated from culture or history. For example, Hutchins noted that navigation depends upon centuries-old mathematics of chart projections, not to mention millennia-old number systems.

These observations caused Hutchins (1995) to propose an extension of Simon's (1969) parable of the ant. Hutchins argued that rather than watching an individual ant on the beach, we should arrive at a beach after a storm and watch generations of ants at work. As the ant colony matures, the ants will appear smarter, because their behaviours are more efficient. But this is because,

> the environment is not the same. Generations of ants have left their marks
> on the beach, and now a dumb ant has been made to appear smart through
> its simple interaction with the residua of the history of its ancestor's actions.
> (Hutchins, 1995, p. 169)

Hutchins' (1995) suggestion mirrored concerns raised by Scribner's studies of mind in action. She observed that the diversity of problem solutions generated by dairy workers, for example, was due in part to social scaffolding.

> We need a greater understanding of the ways in which the institutional setting,
> norms and values of the work group and, more broadly, cultural understandings
> of labor contribute to the reorganization of work tasks in a given community.
> (Scribner & Tobach, 1997, p. 373)

Furthermore, Scribner pointed out that the traditional methods used by classical researchers to study cognition were not suited for increasing this kind of understanding. The extended mind hypothesis leads not only to questions about the nature of mind, but also to the questions about the methods used to study mentality.

5.9 The Roots of Forward Engineering

The most typical methodology to be found in classical cognitive science is reverse engineering. Reverse engineering involves observing the behaviour of an intact system in order to infer the nature and organization of the system's internal processes. Most cognitive theories are produced by using a methodology called functional analysis (Cummins, 1975, 1983), which uses experimental results to iteratively carve a system into a hierarchy of functional components until a basic level of subfunctions, the cognitive architecture, is reached.

A practical problem with functional analysis or reverse engineering is the frame of reference problem (Pfeifer & Scheier, 1999). This problem arises during the distribution of responsibility for the complexity of behaviour between the internal processes of an agent and the external influences of its environment. Classical cognitive science, a major practitioner of functional analysis, endorses the classical sandwich; its functional analyses tend to attribute behavioural complexity to the internal processes of an agent, while at the same time ignoring potential contributions of the environment. In other words, the frame of reference problem is to ignore Simon's (1969) parable of the ant.

Embodied cognitive scientists frequently adopt a different methodology, forward engineering. In forward engineering, a system is constructed from a set of primitive functions of interest. The system is then observed to determine whether it generates surprising or complicated behaviour. "Only about 1 in 20 'gets it'—that is, the idea of thinking about psychological problems by inventing mechanisms for them and then trying to see what they can and cannot do" (Minsky, personal communication, 1995). This approach has also been called synthetic psychology (Braitenberg, 1984). Reverse engineers collect data to create their models; in contrast, forward engineers build their models first and use them as primary sources of data (Dawson, 2004).

We noted in Chapter 3 that classical cognitive science has descended from the seventeenth-century rationalist philosophy of René Descartes (1960, 1996). It was observed in Chapter 4 that connectionist cognitive science descended from the early eighteenth-century empiricism of John Locke (1977), which was itself a reaction against Cartesian rationalism. The synthetic approach seeks "understanding by building" (Pfeifer & Scheier, 1999), and as such permits us to link embodied cognitive science to another eighteenth-century reaction against Descartes, the philosophy of Giambattista Vico (Vico, 1990, 1988, 2002).

Vico based his philosophy on the analysis of word meanings. He argued that the Latin term for truth, *verum*, had the same meaning as the Latin term *factum*, and therefore concluded that "it is reasonable to assume that the ancient sages of Italy entertained the following beliefs about the true: 'the true is precisely what is made'" (Vico, 1988, p. 46). This conclusion led Vico to his argument that humans could only understand the things that they made, which is why he studied societal artifacts, such as the law.

Vico's work provides an early motivation for forward engineering: "To know (*scire*) is to put together the elements of things" (Vico, 1988, p. 46). Vico's account of the mind was a radical departure from Cartesian disembodiment. To Vico, the Latins "thought every work of the mind was sense; that is, whatever the mind does or undergoes derives from contact with bodies" (p. 95). Indeed, Vico's *verum-factum* principle is based upon embodied mentality. Because the mind is "immersed and buried in the body, it naturally inclines to take notice of bodily things" (p. 97).

While the philosophical roots of forward engineering can be traced to Vico's eighteenth-century philosophy, its actual practice—as far as cognitive science is concerned—did not emerge until cybernetics arose in the 1940s. One of the earliest examples of synthetic psychology was the Homeostat (Ashby, 1956, 1960), which was built by cyberneticist William Ross Ashby in 1948. The Homeostat was a system that changed its internal states to maximize stability amongst the interactions between its internal components and the environment. William Grey Walter (1963, p. 123) noted that it was "like a fireside cat or dog which only stirs when disturbed, and then methodically finds a comfortable position and goes to sleep again."

Ashby's (1956, 1960) Homeostat illustrated the promise of synthetic psychology. The feedback that Ashby was interested in could not be analyzed mathematically; it was successfully studied synthetically with Ashby's device. Remember, too, that when the Homeostat was created, computer simulations of feedback were still in the future.

As well, it was easier to produce interesting behaviour in the Homeostat than it was to analyze it. This is because the secret to its success was a large number of potential internal states, which provided many degrees of freedom for producing stability. At the same time, this internal variability was an obstacle to traditional analysis. "Although the machine is man-made, the experimenter cannot tell at any moment exactly what the machine's circuit is without 'killing' it and dissecting out the 'nervous system'" (Grey Walter, 1963, p. 124).

Concerns about this characteristic of the Homeostat inspired the study of the first autonomous robots, created by cyberneticist William Grey Walter (1950a, 1950b, 1951, 1963). The first two of these machines were constructed in 1948 (de Latil, 1956); comprising surplus war materials, their creation was clearly an act of *bricolage*. "The first model of this species was furnished with pinions from old clocks and gas meters" (Grey Walter, 1963, p. 244). By 1951, these two had been replaced by six improved machines (Holland, 2003a), two of which are currently displayed in museums.

The robots came to be called Tortoises because of their appearance: they seemed to be toy tractors surrounded by a tortoise-like shell. Grey Walter viewed them as an artificial life form that he classified as *Machina speculatrix*. *Machina speculatrix* was a reaction against the internal variability in Ashby's Homeostat. The goal of Grey Walter's robotics research was to explore the degree to which one could produce complex behaviour from such very simple devices (Boden, 2006). When Grey Walter modelled behaviour he "was determined to wield Occam's razor. That is, he aimed to posit as simple a mechanism as possible to explain apparently complex behaviour. And simple, here, meant simple" (Boden, 2006, p. 224). Grey Walter restricted a Tortoise's internal components to "two functional elements: two miniature radio tubes, two sense organs, one for light and the other for touch, and two effectors or motors, one for crawling and the other for steering" (Grey Walter, 1950b, p. 43).

The interesting behaviour of the Tortoises was a product of simple reflexes that used detected light (via a light sensor mounted on the robot's steering column) and obstacles (via movement of the robot's shell) to control the actions of the robot's two motors. Light controlled motor activity as follows. In dim light, the Tortoise's drive motor would move the robot forward, while the steering motor slowly turned the front wheel. Thus in dim light the Tortoise "explored." In moderate light, the drive motor continued to run, but the steering motor stopped. Thus in moderate light the Tortoise "approached." In bright light, the drive motor continued to run, but the steering motor ran at twice the normal speed, causing marked oscillatory movements. Thus in bright light the Tortoise "avoided."

The motors were affected by the shell's sense of touch as follows. When the Tortoise's shell was moved by an obstacle, an oscillating signal was generated that first caused the robot to drive fast while slowly turning, and then to drive slowly while quickly turning. The alternation of these behaviours permitted the Tortoise to escape from obstacles. Interestingly, when movement of the Tortoise shell triggered such behaviour, signals from the photoelectric cell were rendered inoperative for a few moments. Thus Grey Walter employed a simple version of what later would be known as Brooks' (1999) subsumption architecture: a higher layer of touch processing could inhibit a lower layer of light processing.

In accordance with forward engineering, after Grey Walter constructed his robots, he observed their behaviour by recording the paths that they took in a number of simple environments. He preserved a visual record of their movement by using time-lapse photography; because of lights mounted on the robots, their paths were literally traced on each photograph (Holland, 2003b). Like the paths on the beach traced in Simon's (1969) parable of the ant, the photographs recorded Tortoise behaviour that was "remarkably unpredictable" (Grey Walter, 1950b, p. 44).

Grey Walter observed the behaviours of his robots in a number of different environments. For example, in one study the robot was placed in a room where a light was hidden from view by an obstacle. The Tortoise began to explore the room, bumped into the obstacle, and engaged in its avoidance behaviour. This in turn permitted the robot to detect the light, which it approached. However, it didn't collide with the light. Instead the robot circled it cautiously, veering away when it came too close. "Thus the machine can avoid the fate of the moth in the candle" (Grey Walter, 1963, p. 128).

When the environment became more complicated, so too did the behaviours produced by the Tortoise. If the robot was confronted with two stimulus lights instead of one, it would first be attracted to one, which it circled, only to move away and circle the other, demonstrating an ability to choose: it solved the problem "of Buridan's ass, which starved to death, as some animals acting trophically in fact do, because two exactly equal piles of hay were precisely the same distance away" (Grey Walter, 1963, p. 128). If a mirror was placed in its environment, the mirror served as an obstacle, but it reflected the light mounted on the robot, which was an attractant. The resulting dynamics produced the so-called "mirror dance" in which the robot,

> lingers before a mirror, flickering, twittering and jigging like a clumsy Narcissus. The behaviour of a creature thus engaged with its own reflection is quite specific, and on a purely empirical basis, if it were observed in an animal, might be accepted as evidence of some degree of self-awareness. (Grey Walter, 1963, pp. 128–129)

In less controlled or open-ended environments, the behaviour that was produced was lifelike in its complexity. The Tortoises produced "the exploratory, speculative behaviour that is so characteristic of most animals" (Grey Walter, 1950b, p. 43). Examples of such behaviour were recounted by cyberneticist Pierre de Latil (1956):

> Elsie moved to and fro just like a real animal. A kind of head at the end of a long neck towered over the shell, like a lighthouse on a promontory and, like a lighthouse; it veered round and round continuously. (de Latil, 1956, p. 209)

The *Daily Mail* reported that,

> the toys possess the senses of sight, hunger, touch, and memory. They can walk about the room avoiding obstacles, stroll round the garden, climb stairs, and feed themselves by automatically recharging six-volt accumulators from the light in the room. And they can dance a jig, go to sleep when tired, and give an electric shock if disturbed when they are not playful. (Holland, 2003a, p. 2090)

Grey Walter released the Tortoises to mingle with the audience at a 1955 meeting of the British Association (Hayward, 2001): "The tortoises, with their in-built attraction towards light, moved towards the pale stockings of the female delegates whilst avoiding the darker legs of the betrousered males" (p. 624).

Grey Walter was masterfully able to promote his work to the general public (Hayward, 2001; Holland, 2003a). However, he worried that public reception of his machines would decrease their scientific importance. History has put such concerns to rest; Grey Walter's pioneering research has influenced many modern researchers (Reeve & Webb, 2003). Grey Walter's,

> ingenious devices were seriously intended as working models for understanding biology: a 'mirror for the brain' that could both generally enrich our understanding of principles of behavior (such as the complex outcome of combining simple tropisms) and be used to test specific hypotheses (such as Hebbian learning). (Reeve & Webb, 2003, p. 2245)

5.10 Reorientation without Representation

The robotics work of Grey Walter has been accurately described as an inspiration to modern studies of autonomous systems (Reeve & Webb, 2003). Indeed, the kind of research conducted by Grey Walter seems remarkably similar to the "new wave" of behaviour-based or biologically inspired robotics (Arkin, 1998; Breazeal, 2002; Sharkey, 1997; Webb & Consi, 2001).

In many respects, this represents an important renaissance of Grey Walter's search for "mimicry of life" (Grey Walter, 1963, p. 114). Although the Tortoises were described in his very popular 1963 book *The Living Brain*, they essentially disappeared from the scientific picture for about a quarter of a century. Grey Walter was involved in a 1970 motorcycle accident that ended his career; after this accident, the whereabouts of most of the Tortoises was lost. One remained in the possession of his son after Grey Walter's death in 1977; it was located in 1995 after an extensive search by Owen Holland. This discovery renewed interest in Grey Walter's work (Hayward, 2001; Holland, 2003a, 2003b), and has re-established its important place in modern research.

The purpose of the current section is to briefly introduce one small segment of robotics research that has descended from Grey Walter's pioneering work. In Chapter 3, we introduced the reorientation task that is frequently used to study how geometric and feature cues are used by an agent to navigate through its world. We also described a classical theory, the geometric module (Cheng, 1986; Gallistel, 1990), which has been used to explain some of the basic findings concerning this task. In Chapter 4, we noted that the reorientation task has also been approached from the perspective of connectionist cognitive science. A simple artificial neural network, the perceptron, has been offered as a viable alternative to classical theory (Dawson et al., 2010). In this section we briefly describe a third approach to the reorientation task, because embodied cognitive science has studied it in the context of behaviour-based robotics.

Classical and connectionist cognitive science provide very different accounts of the co-operative and competitive interactions between geometric and featural cues when an agent attempts to relocate the target location in a reorientation arena. However, these different accounts are both representational. One of the themes pervading embodied cognitive science is a reaction against representational explanations of intelligent behaviour (Shapiro, 2011). One field that has been a test bed for abandoning internal representations is known as new wave robotics (Sharkey, 1997).

New wave roboticists strive to replace representation with reaction (Brooks, 1999), to use sense-act cycles in the place of representational sense-think-act processing. This is because "embodied and situated systems can solve rather complicated tasks without requiring internal states or internal representations" (Nolfi & Floreano, 2000, p. 93). One skill that has been successfully demonstrated in new wave robotics is navigation in the context of the reorientation task (Lund & Miglino, 1998).

The Khepera robot (Bellmore & Nemhauser, 1968; Boogaarts, 2007) is a standard platform for the practice of new wave robotics. It has the appearance of a motorized hockey puck, uses two motor-driven wheels to move about, and has eight sensors distributed around its chassis that allow it to detect the proximity of obstacles. Roboticists have the goal of combining the proximity detector signals to control motor speed in order to produce desired dynamic behaviour. One approach to achieving this goal is to employ evolutionary robotics (Nolfi & Floreano, 2000). Evolutionary robotics involves using a genetic algorithm (Holland, 1992; Mitchell, 1996) to find a set of weights between each proximity detector and each motor.

In general, evolutionary robotics proceeds as follows (Nolfi & Floreano, 2000). First, a fitness function is defined, to evaluate the quality of robot performance. Evolution begins with an initial population of different control systems, such as different sets of sensor-to-motor weights. The fitness function is used to assess each of these control systems, and those that produce higher fitness values "survive." Survivors are used to create the next generation of control systems via prescribed methods of "mutation." The whole process of evaluate-survive-mutate is iterated; average fitness is expected to improve with each new generation. The evolutionary process ends when improvements in fitness stabilize. When evolution stops, the result is a control system that should be quite capable of performing the task that was evaluated by the fitness function.

Lund and Miglino (1998) used this procedure to evolve a control system that enabled Khepera robots to perform the reorientation task in a rectangular arena without feature cues. Their goal was to see whether a standard result—rotational error—could be produced in an agent that did not employ the geometric module, and indeed which did not represent arena properties at all. Lund and Miglino's fitness function simply measured a robot's closeness to the goal location. After 30 generations of evolution, they produced a system that would navigate a robot to the goal location from any of 8 different starting locations with a 41 percent success rate. Their robots also produced rotational error, for they incorrectly navigated to the corner 180° from the goal in another 41 percent of the test trials. These results were strikingly similar to those observed when rats perform reorientation in featureless rectangular arenas (e.g., Gallistel, 1990).

Importantly, the control system that was evolved by Lund and Miglino (1998) was simply a set of weighted connections between proximity detectors and motors, and not an encoding of arena shape.

> The geometrical properties of the environment can be assimilated in the sensory-motor schema of the robot behavior without any explicit representation. In general, our work, in contrast with traditional cognitive models, shows how environmental knowledge can be reached without any form of direct representation. (Lund and Miglino, 1998, p. 198)

If arena shape is not explicitly represented, then how does the control system developed by Lund and Miglino (1998) produce reorientation task behaviour? When the robot is far enough from the arena walls that none of the sensors are detecting an obstacle, the controller weights are such that the robot moves in a gentle curve to the left. As a result, it never encounters a short wall when it leaves from any of its eight starting locations! When a long wall is (inevitably) encountered, the robot turns left and follows the wall until it stops in a corner. The result is that the robot will be at either the target location or its rotational equivalent.

The control system evolved by Lund and Miglino (1998) is restricted to rectangular arenas of a set size. If one of their robots is placed in an arena of even a slightly different size, its performance suffers (Nolfi, 2002). Nolfi used a much longer evolutionary process (500 generations), and also placed robots in different sized arenas, to successfully produce devices that would generate typical results not only in a featureless rectangular arena, but also in arenas of different dimensions. Again, these robots did so without representing arena shape or geometry.

Nolfi's (2002) more general control system worked as follows. His robots would begin by moving forwards and avoiding walls, which would eventually lead them into a corner. When facing a corner, signals from the corner's two walls caused the robot to first turn to orient itself at an angle of 45° from one of the corner's walls. Then the robot would make an additional turn that was either clockwise or counterclockwise, depending upon whether the sensed wall was to the robot's left or the right.

The final turn away from the corner necessarily pointed the robot in a direction that would cause it to follow a long wall, because sensing a wall at 45° is an indirect measurement of wall length:

> If the robot finds a wall at about 45° on its left side and it previously left a corner, it means that the actual wall is one of the two longer walls. Conversely, if it encounters a wall at 45° on its right side, the actual wall is necessarily one of the two shorter walls. What is interesting is that the robot "measures" the relative length of the walls through action (i.e., by exploiting sensory–motor coordination) and it does not need any internal state to do so. (Nolfi, 2002, p. 141)

As a result, the robot sensed the long wall in a rectangular arena without representing wall length. It followed the long wall, which necessarily led the robot to either the goal corner or the corner that results in a rotational error, regardless of the actual dimensions of the rectangular arena.

Robots simpler than the Khepera can also perform the reorientation task, and they can at the same time generate some of its core results. The subsumption architecture has been used to design a simple LEGO robot, antiSLAM (Dawson, Dupuis, & Wilson, 2010), that demonstrates rotational error and illustrates how a new wave robot can combine geometric and featural cues, an ability not included in the evolved robots that have been discussed above.

The ability of autonomous robots to navigate is fundamental to their success. In contrast to the robots described in the preceding paragraphs, one of the major approaches to providing such navigation is called SLAM, which is an acronym for a representational approach named "simultaneous localization and mapping" (Jefferies & Yeap, 2008). Representationalists assumed that agents navigate their environment by sensing their current location and referencing it on some internal map. How is such navigation to proceed if an agent is placed in a novel environment for which no such map exists? SLAM is an attempt to answer this question. It proposes methods that enable an agent to build a new map of a novel environment and at the same time use this map to determine the agent's current location.

The representational assumptions that underlie approaches such as SLAM have recently raised concerns in some researchers who study animal navigation (Alerstam, 2006). To what extent might a completely reactive, sense-act robot be capable of demonstrating interesting navigational behaviour? The purpose of anti-SLAM (Dawson, Dupuis, & Wilson, 2010) was to explore this question in an incredibly simple platform—the robot's name provides some sense of the motivation for its construction.

AntiSLAM is an example of a Braitenberg Vehicle 3 (Braitenberg, 1984), because it uses six different sensors, each of which contributes to the speed of two motors that propel and steer it. Two are ultrasonic sensors that are used as sonar to detect obstacles, two are rotation detectors that are used to determine when the robot has stopped moving, and two are light sensors that are used to attract the robot to locations of bright illumination. The sense-act reflexes of antiSLAM were not evolved but were instead created using the subsumption architecture.

The lowest level of processing in antiSLAM is "drive," which essentially uses the outputs of the ultrasonic sensors to control motor speed. The closer to an obstacle a sensor gets, the slower is the speed of the one motor that the sensor helps to control. The next level is "escape." When both rotation sensors are signaling that the robot is stationary (i.e., stopped by an obstacle detected by both sensors), the robot executes a turn to point itself in a different direction. The next level up is "wall following": motor speed is manipulated in such a way that the robot has a strong bias to keep closer to a wall on the right than to a wall on the left. The highest level is "feature," which uses two light sensors to contribute to motor speed in such a way that it approaches areas of brighter light.

AntiSLAM performs complex, lifelike exploratory behaviour when placed in general environments. It follows walls, steers itself around obstacles, explores regions of brighter light, and turns around and escapes when it finds itself stopped in a corner or in front of a large obstacle.

When placed in a reorientation task arena, antiSLAM generates behaviours that give it the illusion of representing geometric and feature cues (Dawson, Dupuis, & Wilson, 2010). It follows walls in a rectangular arena, slowing to a halt when enters a corner. It then initiates a turning routine to exit the corner and continue exploring. Its light sensors permit it to reliably find a target location that is associated with particular geometric and local features. When local features are removed, it navigates the arena using geometric cues only, and it produces rotational errors. When local features are moved (i.e., an incorrect corner is illuminated), its choice of locations from a variety of starting points mimics the same combination of geometric and feature cues demonstrated in experiments with animals. In short, it produces some of the key features of the reorientation task—however, it does so without creating a cognitive map, and even without representing a goal. Furthermore, observations of antiSLAM's reorientation task behaviour indicated that a crucial behavioural measure, the path taken by an agent as it moves through the arena, is critical. Such paths are rarely reported in studies of reorientation.

The reorienting robots discussed above are fairly recent descendants of Grey Walter's (1963) Tortoises, but their more ancient ancestors are the eighteenth-century life-mimicking, clockwork automata (Wood, 2002). These devices brought into sharp focus the philosophical issues concerning the comparison of man and machine that was central to Cartesian philosophy (Grenville, 2001; Wood, 2002). Religious tensions concerning the mechanistic nature of man, and the spiritual nature of clockwork automata, were soothed by dualism: automata and animals were machines. Men too were machines, but unlike automata, they also had souls. It was the appearance of clockwork automata that led to their popularity, as well as to their conflicts with the church. "Until the scientific era, what seemed most alive to people was what most *looked* like a living being. The vitality accorded to an object was a function primarily of its form" (Grey Walter, 1963, p. 115).

In contrast, Grey Walter's Tortoises were not attempts to reproduce appearances, but were instead simulations of more general and more abstract abilities central to biological agents,

> exploration, curiosity, free-will in the sense of unpredictability, goal-seeking, self-regulation, avoidance of dilemmas, foresight, memory, learning, forgetting, association of ideas, form recognition, and the elements of social accommodation. Such is life. (Grey Walter, 1963, p. 120)

By situating and embodying his machines, Grey Walter invented a new kind of scientific tool that produced behaviours that were creative and unpredictable, governed by nonlinear relationships between internal mechanisms and the surrounding, dynamic world.

Modern machines that mimic lifelike behaviour still raise serious questions about what it is to be human. To Wood (2002, p. xxvii) all automata were presumptions "that life can be simulated by art or science or magic. And embodied in each invention is a riddle, a fundamental challenge to our perception of what makes us human." The challenge is that if the lifelike behaviours of the Tortoises and their descendants are merely feedback loops between simple mechanisms and their environments, then might the same be true of human intelligence?

This challenge is reflected in some of roboticist Rodney Brooks' remarks in Errol Morris' 1997 documentary *Fast, Cheap & Out of Control*. Brooks begins by describing one of his early robots: "To an observer it appears that the robot has intentions and it has goals and it is following people and chasing prey. But it's just the interaction of lots and lots of much simpler processes." Brooks then considers extending this view to human cognition: "Maybe that's all there is. Maybe a lot of what humans are doing could be explained this way."

But as the segment in the documentary proceeds, Brooks, the pioneer of behaviour-based robotics, is reluctant to believe that humans are similar types of devices:

> When I think about it, I can almost see myself as being made up of thousands and thousands of little agents doing stuff almost independently. But at the same time I fall back into believing the things about humans that we all believe about humans and living life that way. Otherwise I analyze it too much; life becomes almost meaningless. (Morris, 1997)

Conflicts like those voiced by Brooks are brought to the forefront when embodied cognitive science ventures to study humanoid robots that are designed to exploit social environments and interactions (Breazeal, 2002; Turkle, 2011).

5.11 Robotic Moments in Social Environments

The embodied approach has long recognized that an agent's environment is much more that a static array of stimuli (Gibson, 1979; Neisser, 1976; Scribner & Tobach, 1997; Vygotsky, 1986). "The richest and most elaborate affordances of the environment are provided by other animals and, for us, other people" (Gibson, 1979, p. 135). A social environment is a rich source of complexity and ranges from dynamic interactions with other agents to cognitive scaffolding provided by cultural conventions. "All higher mental processes are primarily social phenomena, made possible by cognitive tools and characteristic situations that have evolved in the course of history" (Neisser, 1976, p. 134).

In the most basic sense of *social*, multiple agents in a shared world produce a particularly complex source of feedback between each other's actions. "What the other animal affords the observer is not only behaviour but also social interaction. As one moves so does the other, the one sequence of action being suited to the other in a kind of behavioral loop" (Gibson, 1979, p. 42).

Grey Walter (1963) explored such behavioural loops when he placed two Tortoises in the same room. Mounted lights provided particularly complex stimuli in this case, because robot movements would change the position of the two lights, which in turn altered subsequent robot behaviours. In describing a photographic record of one such interaction, Grey Walter called the social dynamics of his machines,

> the formation of a cooperative and a competitive society. . . . When the two crea-
> tures are released at the same time in the dark, each is attracted by the other's
> headlight but each in being attracted extinguishes the source of attraction to the
> other. The result is a stately circulating movement of minuet-like character; when-
> ever the creatures touch they become obstacles and withdraw but are attracted
> again in rhythmic fashion. (Holland, 2003a, p. 2104)

Similar behavioural loops have been exploited to explain the behaviour of larger collections of interdependent agents, such as flocks of flying birds or schools of swimming fish (Nathan & Barbosa, 2008; Reynolds, 1987). Such an aggregate presents itself as another example of a superorganism, because the synchronized movements of flock members give "the strong impression of intentional, centralized control" (Reynolds, 1987, p. 25). However, this impression may be the result of local, stigmergic interactions in which an environment chiefly consists of other flock members in an agent's immediate vicinity.

In his pioneering work on simulating the flight of a flock of artificial birds, called boids, Reynolds (1987) created lifelike flocking behaviour by having each independently flying boid adapt its trajectory according to three simple rules: avoid collision with nearby flock mates, match the velocity of nearby flock mates, and stay close to nearby flock mates. A related model (Couzin et al., 2005) has been successfully used to predict movement of human crowds (Dyer et al., 2008; Dyer et al., 2009; Faria et al., 2010).

However, many human social interactions are likely more involved than the simple behavioural loops that defined the social interactions amongst Grey Walter's (1963) Tortoises or the flocking behaviour of Reynolds' (1987) boids. These interactions are possibly still behavioural loops, but they may be loops that involve processing special aspects of the social environment. This is because it appears that the human brain has a great deal of neural circuitry devoted to processing specific kinds of social information.

Social cognition is fundamentally involved with how we understand others (Lieberman, 2007). One key avenue to such understanding is our ability to use and interpret facial expressions (Cole, 1998; Etcoff & Magee, 1992). There is a long history of evidence that indicates that our brains have specialized circuitry for processing faces. Throughout the eighteenth and nineteenth centuries, there were many reports of patients whose brain injuries produced an inability to recognize faces but did not alter the patients' ability to identify other visual objects. This condition was called prosopagnosia, for "face blindness," by German neuroscientist Joachim Bodamer in a famous 1947 manuscript (Ellis & Florence, 1990). In the 1980s, recordings from single neurons in the monkey brain revealed cells that appeared to be tailored to respond to specific views of monkey faces (Perrett, Mistlin, & Chitty, 1987; Perrett, Rolls, & Caan, 1982). At that time, though, it was unclear whether analogous neurons for face processing were present in the human brain.

Modern brain imaging techniques now suggest that the human brain has an elaborate hierarchy of co-operating neural systems for processing faces and their expressions (Haxby, Hoffman, & Gobbini, 2000, 2002). Haxby, Hoffman, and Gobbini (2000, 2002) argue for the existence of multiple, bilateral brain regions involved in different face perception functions. Some of these are core systems that are responsible for processing facial invariants, such as relative positions of the eyes, nose, and mouth, which are required for recognizing faces. Others are extended systems that process dynamic aspects of faces in order to interpret, for instance, the meanings of facial expressions. These include subsystems that co-operatively account for lip reading, following gaze direction, and assigning affect to dynamic changes in expression.

Facial expressions are not the only source of social information. Gestures and actions, too, are critical social stimuli. Evidence also suggests that mirror neurons in the human brain (Gallese et al., 1996; Iacoboni, 2008; Rizzolatti & Craighero, 2004; Rizzolatti, Fogassi, & Gallese, 2006) are specialized for both the generation and interpretation of gestures and actions.

Mirror neurons were serendipitously discovered in experiments in which motor neurons in region F5 were recorded when monkeys performed various reaching actions (Di Pellegrino et al., 1992). By accident, it was discovered that many of the neurons that were active when a monkey performed an action also responded when similar actions were *observed* being performed by another:

> After the initial recording experiments, we incidentally observed that some experi-
> menter's actions, such as picking up the food or placing it inside the testing box,
> activated a relatively large proportion of F5 neurons in the absence of any overt
> movement of the monkey. (Di Pellegrino et al., 1992, p. 176)

The chance discovery of mirror neurons has led to an explosion of research into their behaviour (Iacoboni, 2008). It has been discovered that when the neurons fire, they do so for the entire duration of the observed action, not just at its onset. They are grasp specific: some respond to actions involving precision grips, while others respond to actions involving larger objects. Some are broadly tuned, in the sense that they will be triggered when a variety of actions are observed, while others are narrowly tuned to specific actions. All seem to be tuned to object-oriented action: a mirror neuron will respond to a particular action on an object, but it will fail to respond to the identical action if no object is present.

While most of the results described above were obtained from studies of the monkey brain, there is a steadily growing literature indicating that the human brain also has a mirror system (Buccino et al., 2001; Iacoboni, 2008).

Mirror neurons are not solely concerned with hand and arm movements. For instance, some monkey mirror neurons respond to mouth movements, such as lip smacking (Ferrari et al., 2003). Similarly, the human brain has a mirror system for the act of touching (Keysers et al., 2004). Likewise, another part of the human brain, the insula, may be a mirror system for emotion (Wicker et al., 2003). For example, it generates activity when a subject experiences disgust, and also when a subject observes the facial expressions of someone else having a similar experience.

Two decades after its discovery, extensive research on the mirror neuron system has led some researchers to claim that it provides the neural substrate for social cognition and imitative learning (Gallese & Goldman, 1998; Gallese, Keysers, & Rizzolatti, 2004; Iacoboni, 2008), and that disruptions of this system may be responsible for autism (Williams et al., 2001). The growing understanding of the mirror system and advances in knowledge about the neuroscience of face percep- tion have heralded a new interdisciplinary research program, called social cogni- tive neuroscience (Blakemore, Winston, & Frith, 2004; Lieberman, 2007; Ochsner & Lieberman, 2001).

It may once have seemed foolhardy to work out connections between fundamental neurophysiological mechanisms and highly complex social behaviour, let alone to decide whether the mechanisms are specific to social processes. However ... neuroimaging studies have provided some encouraging examples. (Blakemore, Winston, & Frith, 2004, p. 216)

The existence of social cognitive neuroscience is a consequence of humans evolving, embodied and situated, in a social environment that includes other humans and their facial expressions, gestures, and actions. The modern field of sociable robotics (Breazeal, 2002) attempts to develop humanoid robots that are also socially embodied and situated. One purpose of such robots is to provide a medium for studying human social cognition via forward engineering.

A second, applied purpose of sociable robotics is to design robots to work co-operatively with humans by taking advantage of a shared social environment. Breazeal (2002) argued that because the human brain has evolved to be expert in social interaction, "if a technology behaves in a socially competent manner, we evoke our evolved social machinery to interact with it" (p. 15). This is particularly true if a robot's socially competent behaviour is mediated by its humanoid embodiment, permitting it to gesture or to generate facial expressions. "When a robot holds our gaze, the hardwiring of evolution makes us think that the robot is interested in us. When that happens, we feel a possibility for deeper connection" (Turkle, 2011, p. 110). Sociable robotics exploits the human mechanisms that offer this deeper connection so that humans won't require expert training in interacting with sociable robots.

A third purpose of sociable robotics is to explore cognitive scaffolding, which in this literature is often called leverage, in order to extend the capabilities of robots. For instance, many of the famous platforms of sociable robotics—including Cog (Brooks et al., 1999; Scassellati, 2002), Kismet (Breazeal, 2002, 2003, 2004), Domo (Edsinger-Gonzales & Weber, 2004), and Leanardo (Breazeal, Gray, & Berlin, 2009)—are humanoid in form and are social learners—their capabilities advance through imitation and through interacting with human partners. Furthermore, the success of the robot's contribution to the shared social environment leans heavily on the contributions of the human partner. "Edsinger thinks of it as getting Domo to do more 'by leveraging the people.' Domo needs the help. It understands very little about any task as a whole" (Turkle, 2011, p. 157).

The leverage exploited by a sociable robot takes advantage of behavioural loops mediated by the expressions and gestures of both robot and human partner. For example, consider the robot Kismet (Breazeal, 2002). Kismet is a sociable robotic "infant," a dynamic, mechanized head that participates in social interactions. Kismet has auditory and visual perceptual systems that are designed to perceive social cues provided by a human "caregiver." Kismet can also deliver such social cues by changing its facial expression, directing its gaze to a location in a shared environment, changing its posture, and vocalizing.

When Kismet is communicating with a human, it uses the interaction to fulfill internal drives or needs (Breazeal, 2002). Kismet has three drives: a social drive to be in the presence of and stimulated by people, a stimulation drive to be stimulated by the environment in general (e.g., by colourful toys), and a fatigue drive that causes the robot to "sleep." Kismet sends social signals to satisfy these drives. It can manipulate its facial expression, vocalization, and posture to communicate six basic emotions: anger, disgust, fear, joy, sorrow, and surprise. These expressions work to meet the drives by manipulating the social environment in such a way that the environment changes to satisfy Kismet's needs.

For example, an unfulfilled social drive causes Kismet to express sadness, which initiates social responses from a caregiver. When Kismet perceives the caregiver's face, it wiggles its ears in greeting, and initiates a playful dialog to engage the caregiver. Kismet will eventually habituate to these interactions and then seek to fulfill a stimulation drive by coaxing the caregiver to present a colourful toy. However, if this presentation is too stimulating—if the toy is presented too closely or moved too quickly—the fatigue drive will produce changes in Kismet's behaviour that attempt to decrease this stimulation. If the world does not change in the desired way, Kismet will end the interaction by "sleeping." "But even at its worst, Kismet gives the appearance of trying to relate. At its best, Kismet appears to be in continuous, expressive conversation" (Turkle, 2011, p. 118).

Kismet's behaviour leads to lengthy, dynamic interactions that are realistically social. A young girl interacting with Kismet "becomes increasingly happy and relaxed. Watching girl and robot together, it is easy to see Kismet as increasingly happy and relaxed as well. Child and robot are a happy couple" (Turkle, 2011, p. 121). Similar results occur when adults converse with Kismet. "One moment, Rich plays at a conversation with Kismet, and the next, he is swept up in something that starts to feel real" (p. 154).

Even the designer of a humanoid robot can be "swept up" by their interactions with it. Domo (Edsinger-Gonzales & Weber, 2004) is a limbed humanoid robot that is intended to be a physical helper, by performing such actions as placing objects on shelves. It learns to behave by physically interacting with a human teacher. These physical interactions give even sophisticated users—including its designer, Edsinger—a strong sense that Domo is a social creature. Edsinger finds himself vacillating back and forth between viewing Domo as a creature or as being merely a device that he has designed.

> For Edsinger, this sequence—experiencing Domo as having desires and then talking himself out of the idea—becomes familiar. For even though he is Domo's programmer, the robot's behaviour has not become dull or predictable.

Working together, Edsigner and Domo appear to be learning from each other. (Turkle, 2011, p. 156)

That sociable robots can generate such strong reactions within humans is potentially concerning. The feeling of the uncanny occurs when the familiar is presented in unfamiliar form (Freud, 1976). The uncanny results when standard categories used to classify the world disappear (Turkle, 2011). Turkle (2011) called one such instance, when a sociable robot is uncritically accepted as a creature, the robotic moment. Edsinger's reactions to Domo illustrated its occurrence: "And this is where we are in the robotic moment. One of the world's most sophisticated robot 'users' cannot resist the idea that pressure from a robot's hand implies caring" (p. 160).

At issue in the robotic moment is a radical recasting of the posthuman (Hayles, 1999). "The boundaries between people and things are shifting" (Turkle, 2011, p. 162). The designers of sociable robots scaffold their creations by taking advantage of the expert social abilities of humans. The robotic moment, though, implies a dramatic rethinking of what such human abilities entail. Might human social interactions be reduced to mere sense-act cycles of the sort employed in devices like Kismet? "To the objection that a robot can only seem to care or understand, it has become commonplace to get the reply that people, too, may only seem to care or understand" (p. 151).

In Hayles' (1999) definition of posthumanism, the body is dispensable, because the essence of humanity is information. But this is an extremely classical view. An alternative, embodied posthumanism is one in which the mind is dispensed with, because what is fundamental to humanity is the body and its engagement with reality. "From its very beginnings, artificial intelligence has worked in this space between a mechanical view of people and a psychological, even spiritual, view of machines" (Turkle, 2011, p. 109). The robotic moment leads Turkle to ask "What will love be? And what will it mean to achieve ever-greater intimacy with our machines? Are we ready to see ourselves in the mirror of the machine and to see love as our performances of love?" (p. 165).

5.12 The Architecture of Mind Reading

Social interactions involve coordinating the activities of two or more agents. Even something as basic as a conversation between two people is highly coordinated, with voices, gestures, and facial expressions used to orchestrate joint actions (Clark, 1996). Fundamental to coordinating such social interactions is our ability to predict the actions, interest, and emotions of others. Generically, the study of the ability to make such predictions is called the study of theory of mind, because many theorists argue that these predictions are rooted in our assumption that others, like us, have minds or mental states. As a result, researchers call our ability to foretell others' actions mind reading or mentalizing (Goldman, 2006). "*Having* a mental state and *representing* another individual as having such a state are entirely different matters. The latter activity, *mentalizing* or *mind reading*, is a second-order activity: It is mind thinking about minds" (p. 3).

There are three general, competing theories about how humans perform mind reading (Goldman, 2006). The first is rationality theory, a version of which was introduced in Chapter 3 in the form of the intentional stance (Dennett, 1987). According to rationality theory, mind reading is accomplished via the ascription of contents to the putative mental states of others. In addition, we assume that other agents are rational. As a result, future behaviours are predicted by inferring what future behaviours follow rationally from the ascribed contents. For instance, if we ascribe to someone the belief that piano playing can only be improved by practising daily, and we also ascribe to them the desire to improve at piano, then according to rationality theory it would be natural to predict that they would practise piano daily.

A second account of mentalizing is called theory-theory (Goldman, 2006). Theory-theory emerged from studies of the development of theory of mind (Gopnik & Wellman, 1992; Wellman, 1990) as well as from research on cognitive development in general (Gopnik & Meltzoff, 1997; Gopnik, Meltzoff, & Kuhl, 1999). Theory-theory is the position that our understanding of the world, including our understanding of other people in it, is guided by naïve theories (Goldman, 2006). These theories are similar in form to the theories employed by scientists, because a naïve theory of the world will—eventually—be revised in light of conflicting evidence.

> Babies and scientists share the same basic cognitive machinery. They have similar programs, and they reprogram themselves in the same way. They formulate theories, make and test predictions, seek explanations, do experiments, and revise what they know in the light of new evidence. (Gopnik, Meltzoff, & Kuhl, 1999, p. 161)

There is no special role for a principle of rationality in theory-theory, which distinguishes it from rationality theory (Goldman, 2006). However, it is clear that both of these approaches to mentalizing are strikingly classical in nature. This is because both rely on representations. One senses the social environment, then thinks (by applying rationality or by using a naïve theory), and then finally predicts future actions of others. A third theory of mind reading, simulation theory, has emerged as a rival to theory-theory, and some of its versions posit an embodied account of mentalizing.

Simulation theory is the view that people mind read by replicating or emulating the states of others (Goldman, 2006). In simulation theory, "mindreading includes a crucial role for putting oneself in others' shoes. It may even be part of the brain's design to generate mental states that match, or resonate with, states of people one is observing" (p. 4).

The modern origins of simulation theory rest in two philosophical papers from the 1980s, one by Gordon (1986) and one by Heal (1986). Gordon (1986) noted that the starting point for explaining how we predict the behaviour of others should be investigating our ability to predict our own actions. We can do so with exceedingly high accuracy because "our declarations of immediate intention are causally tied to some actual precursor of behavior: perhaps tapping into the brain's updated behavioral 'plans' or into 'executive commands' that are about to guide the relevant motor sequences" (p. 159).

For Gordon (1986), our ability to accurately predict our own behaviour was a kind of practical reasoning. He proceeded to argue that such reasoning could also be used in attempts to predict others. We could predict others, or predict our own future behaviour in hypothetical situations, by *simulating* practical reasoning.

> To simulate the appropriate practical reasoning I can engage in a kind of *pretend-play*: pretend that the indicated conditions *actually obtain*, with all other conditions remaining (so far as is logically possible and physically probable) as they presently stand; then continuing the make-believe try to 'make up my mind' what to do given these (modified) conditions. (Gordon, 1986, p. 160)

A key element of such "pretend play" is that behavioural output is taken offline.

Gordon's proposal causes simulation theory to depart from the other two theories of mind reading by reducing its reliance on ascribed mental contents. For Gordon (1986, p. 162), when someone simulates practical reasoning to make predictions about someone else, "they are 'putting themselves in the other's shoes' in one sense of that expression: that is, they project themselves into the other's *situation*, but without any attempt to project themselves into, as we say, the other's 'mind.'" Heal (1986) proposed a similar approach, which she called replication.

A number of different variations of simulation theory have emerged (Davies & Stone, 1995a, 1995b), making a definitive statement of its fundamental characteristics problematic (Heal, 1996). Some versions of simulation theory remain very classical in nature. For instance, simulation could proceed by setting the values of a number of variables to define a situation of interest. These values could then be provided to a classical reasoning system, which would use these represented values to make plausible predictions.

> Suppose I am interested in predicting someone's action. . . . I place myself in what I take to be his initial state by imagining the world as it would appear from his point of view and I then deliberate, reason and reflect to see what decision emerges. (Heal, 1996, p. 137)

Some critics of simulation theory argue that it is just as Cartesian as other mind reading theories (Gallagher, 2005). For instance, Heal's (1986) notion of replication exploits shared mental abilities. For her, mind reading requires only the assumption that others "are like me in being thinkers, that they possess the same fundamental cognitive capacities and propensities that I do" (p. 137).

However, other versions of simulation theory are far less Cartesian or classical in nature. Gordon (1986, pp. 17–18) illustrated such a theory with an example from Edgar Allen Poe's *The Purloined Letter*:

> When I wish to find out how wise, or how stupid, or how good, or how wicked is any one, or what are his thoughts at the moment, I fashion the expression of my face, as accurately as possible, in accordance with the expression of his, and then wait to see what thoughts or sentiments arise in my mind or heart, as if to match or correspond with the expression. (Gordon, 1986, pp. 17–18)

In Poe's example, mind reading occurs not by using our reasoning mechanisms to take another's place, but instead by exploiting the fact that we share similar bodies. Songwriter David Byrne (1980) takes a related position in *Seen and Not Seen*, in which he envisions the implications of people being able to mould their appearance according to some ideal: "they imagined that their personality would be forced to change to fit the new appearance. . . .This is why first impressions are often correct." Social cognitive neuroscience transforms such views from art into scientific theory.

> Ultimately, subjective experience is a biological data format, a highly specific mode of presenting about the world, and the Ego is merely a complex physical event—an activation pattern in your central nervous system. (Metzinger, 208, p. 208)

Philosopher Robert Gordon's version of simulation theory (Gordon, 1986, 1992, 1995, 1999, 2005a, 2005b, 2007, 2008) provides an example of a radically embodied theory of mind reading. Gordon (2008, p. 220) could "see no reason to hold on to the assumption that our psychological competence is chiefly dependent on the application of concepts of mental states." This is because his simulation theory exploited the body in exactly the same way that Brooks' (1999) behaviour-based robots exploited the world: as a replacement for representation (Gordon, 1999). "One's own behavior control system is employed as a manipulable model of other such systems. . . . Because one human behavior control system is being used to model others, general information about such systems is unnecessary" (p. 765).

What kind of evidence exists to support a more embodied or less Cartesian simulation theory? Researchers have argued that simulation theory is supported by the discovery of the brain mechanisms of interest to social cognitive neuroscience (Lieberman, 2007). In particular, it has been argued that mirror neurons provide the neural substrate that instantiates simulation theory (Gallese & Goldman, 1998): "[Mirror neuron] activity seems to be nature's way of getting the observer into the same 'mental shoes' as the target—exactly what the conjectured simulation heuristic aims to do" (p. 497–498).

Importantly, the combination of the mirror system and simulation theory implies that the "mental shoes" involved in mind reading are not symbolic representations. They are instead motor representations; they are actions-on-objects as instantiated by the mirror system. This has huge implications for theories of social interactions, minds, and selves:

> Few great social philosophers of the past would have thought that social understanding had anything to do with the pre-motor cortex, and that 'motor ideas' would play such a central role in the emergence of social understanding. Who could have expected that shared thought would depend upon shared 'motor representations'? (Metzinger, 2009, p. 171)

If motor representations are the basis of social interactions, then simulation theory becomes an account of mind reading that stands as a reaction against classical, representational theories. Mirror neuron explanations of simulation theory replace sense-think-act cycles with sense-act reflexes in much the same way as was the case in behaviour-based robotics. Such a revolutionary position is becoming commonplace for neuroscientists who study the mirror system (Metzinger, 2009).

Neuroscientist Vittorio Gallese, one of the discoverers of mirror neurons, provides an example of this radical position:

> Social cognition is not only social metacognition, that is, explicitly thinking about the contents of some else's mind by means of abstract representations. We can certainly explain the behavior of others by using our complex and sophisticated mentalizing ability. My point is that most of the time in our daily social interactions, we do not need to do this. We have a much more direct access to the experiential world of the other. This dimension of social cognition is embodied, in that it mediates between our multimodal experiential knowledge of our own lived body and the way we experience others. (Metzinger, 2009, p. 177)

Cartesian philosophy was based upon an extraordinary act of skepticism (Descartes, 1996). In his search for truth, Descartes believed that he could not rely on his knowledge of the world, or even of his own body, because such knowledge could be illusory.

> I shall think that the sky, the air, the earth, colors, shapes, sounds, and all external things are merely the delusions of dreams which he [a malicious demon] has devised to ensnare my judgment. I shall consider myself as not having hands or eyes, or flesh, or blood or senses, but as falsely believing that I have all these things. (Descartes, 1996, p. 23)

The disembodied Cartesian mind is founded on the myth of the external world.

Embodied theories of mind invert Cartesian skepticism. The body and the world are taken as fundamental; it is the mind or the holistic self that has become the myth. However, some have argued that our notion of a holistic internal self is illusory (Clark, 2003; Dennett, 1991, 2005; Metzinger, 2009; Minsky, 1985, 2006; Varela, Thompson, & Rosch, 1991). "We are, in short, in the grip of a seductive but quite untenable illusion: the illusion that the mechanisms of mind and self can ultimately unfold only on some privileged stage marked out by the good old-fashioned skin-bag" (Clark, 2003, p. 27).

5.13 Levels of Embodied Cognitive Science

Classical cognitive scientists investigate cognitive phenomena at multiple levels (Dawson, 1998; Marr, 1982; Pylyshyn, 1984). Their materialism commits them to exploring issues concerning implementation and architecture. Their view that the mind is a symbol manipulator leads them to seek the algorithms responsible for solving cognitive information problems. Their commitment to logicism and rationality has them deriving formal, mathematical, or logical proofs concerning the capabilities of cognitive systems.

Embodied cognitive science can also be characterized as adopting these same multiple levels of investigation. Of course, this is not to say that there are not also interesting technical differences between the levels of investigation that guide embodied cognitive science and those that characterize classical cognitive science.

By definition, embodied cognitive science is committed to providing implementational accounts. Embodied cognitive science is an explicit reaction against Cartesian dualism and its modern descendant, methodological solipsism. In its emphasis on environments and embodied agents, embodied cognitive science is easily as materialist as the classical approach. Some of the more radical positions in embodied cognitive science, such as the myth of the self (Metzinger, 2009) or the abandonment of representation (Chemero, 2009), imply that implementational accounts may be even more critical for the embodied approach than is the case for classical researchers.

However, even though embodied cognitive science shares the implementational level of analysis with classical cognitive science, this does not mean that it interprets implementational evidence in the same way. For instance, consider single cell recordings from visual neurons. Classical cognitive science, with its emphasis on the creation of internal models of the world, views such data as providing evidence about what kinds of visual features are detected, to be later combined into more complex representations of objects (Livingstone & Hubel, 1988). In contrast, embodied cognitive scientists see visual neurons as being involved not in modelling, but instead in controlling action. As a result, single cell recordings are more likely to be interpreted in the context of ideas such as the affordances of ecological perception (Gibson, 1966, 1979; Noë, 2004). "Our brain does not simply register a chair, a teacup, an apple; it immediately represents the seen object as what I could do with it—as an affordance, a set of possible behaviors" (Metzinger, 2009, p. 167). In short, while embodied and classical cognitive scientists seek implementational evidence, they are likely to interpret it very differently.

The materialism of embodied cognitive science leads naturally to proposals of functional architectures. An architecture is a set of primitives, a physically grounded toolbox of core processes, from which cognitive phenomena emerge. Explicit statements of primitive processes are easily found in embodied cognitive science. For example, it is common to see subsumption architectures explicitly laid out in accounts of behaviour-based robots (Breazeal, 2002; Brooks, 1999, 2002; Kube & Bonabeau, 2000; Scassellati, 2002).

Of course, the primitive components of a typical subsumption architecture are designed to mediate actions on the world, not to aid in the creation of models of it. As a result, the assumptions underlying embodied cognitive science's primitive sense-act cycles are quite different from those underlying classical cognitive science's primitive sense-think-act processing.

As well, embodied cognitive science's emphasis on the fundamental role of an agent's environment can lead to architectural specifications that can dramatically differ from those found in classical cognitive science. For instance, a core aspect of an architecture is control—the mechanisms that choose which primitive operation or operations to execute at any given time. Typical classical architectures will internalize control; for example, the central executive in models of working memory (Baddeley, 1986). In contrast, in embodied cognitive science an agent's environment is critical to control; for example, in architectures that exploit stigmergy (Downing & Jeanne, 1988; Holland & Melhuish, 1999; Karsai, 1999; Susi & Ziemke, 2001; Theraulaz & Bonabeau, 1999). This suggests that the notion of the extended mind is really one of an extended architecture; control of processing can reside outside of an agent.

When embodied cognitive scientists posit an architectural role for the environment, as is required in the notion of stigmergic control, this means that an agent's physical body must also be a critical component of an embodied architecture. One reason for this is that from the embodied perspective, an environment cannot be defined in the absence of an agent's body, as in proposing affordances (Gibson, 1979). A second reason for this is that if an embodied architecture defines sense-act primitives, then the available actions that are available are constrained by the nature of an agent's embodiment. A third reason for this is that some environments are explicitly defined, at least in part, by bodies. For instance, the social environment for a sociable robot such as Kismet (Breazeal, 2002) includes its moveable ears, eyebrows, lips, eyelids, and head, because it manipulates these bodily components to coordinate its social interactions with others.

Even though an agent's body can be part of an embodied architecture does not mean that this architecture is not functional. The key elements of Kismet's expressive features are shape and movement; the fact that Kismet is not flesh is irrelevant because its facial features are defined in terms of their function.

> In the robotic moment, what you are made of—silicon, metal, flesh—pales in comparison with how you behave. In any given circumstance, some people and some robots are competent and some not. Like people, any particular robot needs to be judged on its own merits. (Turkle, 2011, p. 94)

That an agent's body can be part of a functional architecture is an idea that is foreign to classical cognitive science. It also leads to an architectural complication that may be unique to embodied cognitive science. Humans have no trouble relating to, and accepting, sociable robots that are obviously toy creatures, such as Kismet or the robot dog Aibo (Turkle, 2011). In general, as the appearance and behaviour of such robots becomes more lifelike, their acceptance will increase.

However, as robots become closer in resemblance to humans, they produce a reaction called the uncanny valley (MacDorman & Ishiguro, 2006; Mori, 1970). The uncanny valley is seen in a graph that plots human acceptance of robots as a function of robot appearance. The uncanny valley is the part of the graph in which acceptance, which has been steadily growing as appearance grows more lifelike, suddenly plummets when a robot's appearance is "almost human"—that is, when it is realistically human, but can still be differentiated from biological humans.

The uncanny valley is illustrated in the work of roboticist Hiroshi Ishiguro, who,

> built androids that reproduced himself, his wife, and his five-year old daughter. The daughter's first reaction when she saw her android clone was to flee. She refused to go near it and would no longer visit her father's laboratory. (Turkle, 2011, p. 128)

Producing an adequate architectural component—a body that avoids the uncanny valley—is a distinctive challenge for embodied cognitive scientists who ply their trade using humanoid robots.

In embodied cognitive science, functional architectures lead to algorithmic explorations. We saw that when classical cognitive science conducts such explorations, it uses reverse engineering to attempt to infer the program that an information processor uses to solve an information processing problem. In classical cognitive science, algorithmic investigations almost always involve observing behaviour, often at a fine level of detail. Such behavioural observations are the source of relative complexity evidence, intermediate state evidence, and error evidence, which are used to place constraints on inferred algorithms.

Algorithmic investigations in classical cognitive science are almost exclusively focused on unseen, internal processes. Classical cognitive scientists use behavioural observations to uncover the algorithms hidden within the "black box" of an agent.

Embodied cognitive science does not share this exclusive focus, because it attributes some behavioural complexities to environmental influences. Apart from this important difference, though, algorithmic investigations—specifically in the form of behavioural observations—are central to the embodied approach. Descriptions of behaviour are the primary product of forward engineering; examples in behaviour-based robotics span the literature from time lapse photographs of Tortoise trajectories (Grey Walter, 1963) to modern reports of how, over time, robots sort or rearrange objects in an enclosure (Holland & Melhuish, 1999; Melhuish et al., 2006; Scholes et al., 2004; Wilson et al., 2004). At the heart of such behavioural accounts is acceptance of Simon's (1969) parable of the ant. The embodied approach cannot understand an architecture by examining its inert components. It must see what emerges when this architecture is embodied in, situated in, and interacting with an environment.

When embodied cognitive science moves beyond behaviour-based robotics, it relies on some sorts of behavioural observations that are not employed as frequently in classical cognitive science. For example, many embodied cognitive scientists exhort the phenomenological study of cognition (Gallagher, 2005; Gibbs, 2006; Thompson, 2007; Varela, Thompson, & Rosch, 1991). Phenomenology explores how people experience their world and examines how the world is meaningful to us via our experience (Brentano, 1995; Husserl, 1965; Merleau-Ponty, 1962).

Just as enactive theories of perception (Noë, 2004) can be viewed as being inspired by Gibson's (1979) ecological account of perception, phenomenological studies within embodied cognitive science (Varela, Thompson, & Rosch, 1991) are inspired by the philosophy of Maurice Merleau-Ponty (1962). Merleau-Ponty rejected the Cartesian separation between world and mind: "Truth does not 'inhabit' only 'the inner man,' or more accurately, there is no inner man, man is in the world, and only in the world does he know himself" (p. xii). Merleau-Ponty strove to replace this Cartesian view with one that relied upon embodiment. "We shall need to reawaken our experience of the world as it appears to us in so far as we are in the world through our body, and in so far as we perceive the world with our body" (p. 239).

Phenomenology with modern embodied cognitive science is a call to further pursue Merleau-Ponty's embodied approach.

> What we are suggesting is a change in the nature of reflection from an abstract, disembodied activity to an embodied (mindful), open-ended reflection. By embodied, we mean reflection in which body and mind have been brought together. (Varela, Thompson, & Rosch, 1991, p. 27)

However, seeking evidence from such reflection is not necessarily straightforward (Gallagher, 2005). For instance, while Gallagher acknowledges that the body is critical in its shaping of cognition, he also notes that many aspects of our bodily interaction with the world are not available to consciousness and are therefore difficult to study phenomenologically.

Embodied cognitive science's interest in phenomenology is an example of a reaction against the formal, disembodied view of the mind that classical cognitive science has inherited from Descartes (Devlin, 1996). Does this imply, then, that embodied cognitive scientists do not engage in the formal analyses that characterize the computational level of analysis? No. Following the tradition established by cybernetics (Ashby, 1956; Wiener, 1948), which made extensive use of mathematics to describe feedback relations between physical systems and their environments, embodied cognitive scientists too are engaged in computational investigations. Again, though, these investigations deviate from those conducted within classical cognitive science. Classical cognitive science used formal methods to develop proofs about what information processing problem was being solved by a system (Marr, 1982), with the notion of "information processing problem" placed in the context of rule-governed symbol manipulation. Embodied cognitive science operates in a very different context, because it has a different notion of information processing. In this new context, cognition is not modelling or planning, but is instead coordinating action (Clark, 1997).

When cognition is placed in the context of coordinating action, one key element that must be captured by formal analyses is that actions unfold in time. It has been argued that computational analyses conducted by classical researchers fail to incorporate the temporal element (Port & van Gelder, 1995a): "Representations are static structures of discrete symbols. Cognitive operations are transformations from one static symbol structure to the next. These transformations are discrete, effectively instantaneous, and sequential" (p. 1). As such, classical analyses are deemed by some to be inadequate. When embodied cognitive scientists explore the computational level, they do so with a different formalism, called dynamical systems theory (Clark, 1997; Port & van Gelder, 1995b; Shapiro, 2011).

Dynamical systems theory is a mathematical formalism that describes how systems change over time. In this formalism, at any given time a system is described as being in a state. A state is a set of variables to which values are assigned. The variables define all of the components of the system, and the values assigned to these variables describe the characteristics of these components (e.g., their features) at a particular time. At any moment of time, the values of its components provide the position of the system in a state space. That is, any state of a system is a point in a multidimensional space, and the values of the system's variables provide the coordinates of that point.

The temporal dynamics of a system describe how its characteristics change over time. These changes are captured as a path or trajectory through state space. Dynamical systems theory provides a mathematical description of such trajectories, usually in the form of differential equations. Its utility was illustrated in Randall Beer's (2003) analysis of an agent that learns to categorize objects, of circuits for associative learning (Phattanasri, Chiel, & Beer, 2007), and of a walking leg controlled by a neural mechanism (Beer, 2010).

While dynamical systems theory provides a medium in which embodied cognitive scientists can conduct computational analyses, it is also intimidating and difficult. "A common criticism of dynamical approaches to cognition is that they are practically intractable except in the simplest cases" (Shapiro, 2011, pp. 127–128). This was exactly the situation that led Ashby (1956, 1960) to study feedback between multiple devices synthetically, by constructing the Homeostat. This does not mean, however, that computational analyses are impossible or fruitless. On the contrary, it is possible that such analyses can co-operate with the synthetic exploration of models in an attempt to advance both formal and behavioural investigations (Dawson, 2004; Dawson, Dupuis, & Wilson, 2010).

In the preceding paragraphs we presented an argument that embodied cognitive scientists study cognition at the same multiple levels of investigation that characterize classical cognitive science. Also acknowledged is that embodied cognitive scientists are likely to view each of these levels slightly differently than their classical counterparts. Ultimately, that embodied cognitive science explores cognition at these different levels of analysis also implies that embodied cognitive scientists are also committed to the notion of validating their theories by seeking strong equivalence. It stands to reason that the validity of a theory created within embodied cognitive science would be best established by showing that this theory is supported at all of the different levels of investigation.

5.14 What Is Embodied Cognitive Science?

To review, the central claim of classical cognitive science is that cognition is computation, where computation is taken to be the manipulation of internal representations. From this perspective, classical cognitive science construes cognition as an iterative sense-think-act cycle. The "think" part of this cycle is emphasized, because it is responsible for modelling and planning. The "thinking" also stands as a required mentalistic buffer between sensing and acting, producing what is known as the classical sandwich (Hurley, 2001). The classical sandwich represents a modern form of Cartesian dualism, in the sense that the mental (thinking) is distinct from the physical (the world that is sensed, and the body that can act upon it) (Devlin, 1996).

Embodied cognitive science, like connectionist cognitive science, arises from the view that the core logicist assumptions of classical cognitive science are not adequate to explain human cognition (Dreyfus, 1992; Port & van Gelder, 1995b; Winograd & Flores, 1987b).

The lofty goals of artificial intelligence, cognitive science, and mathematical linguistics that were prevalent in the 1950s and 1960s (and even as late as the 1970s) have now given way to the realization that the 'soft' world of people and societies is almost certainly not amenable to a precise, predictive, mathematical analysis to anything like the same degree as is the 'hard' world of the physical universe. (Devlin, 1996, p. 344)

As such a reaction, the key elements of embodied cognitive science can be portrayed as an inversion of elements of the classical approach.

While classical cognitive science abandons Cartesian dualism in one sense, by seeking materialist explanations of cognition, it remains true to it in another sense, through its methodological solipsism (Fodor, 1980). Methodological solipsism attempts to characterize and differentiate mental states without appealing to properties of the body or of the world (Wilson, 2004), consistent with the Cartesian notion of the disembodied mind.

In contrast, embodied cognitive science explicitly rejects methodological solipsism and the disembodied mind. Instead, embodied cognitive science takes to heart the message of Simon's (1969) parable of the ant by recognizing that crucial contributors to behavioural complexity include an organism's environment and bodily form. Rather than creating formal theories of disembodied minds, embodied cognitive scientists build embodied and situated agents.

Classical cognitive science adopts the classical sandwich (Hurley, 2001), construing cognition as an iterative sense-think-act cycle. There are no direct links between sensing and acting from this perspective (Brooks, 1991); a planning process involving the manipulation of internal models stands as a necessary intermediary between perceiving and acting.

In contrast, embodied cognitive science strives to replace sense-think-act processing with sense-act cycles that bypass representational processing. Cognition is seen as the control of direct action upon the world rather than the reasoning about possible action. While classical cognitive science draws heavily from the symbol-manipulating examples provided by computer science, embodied cognitive science steps further back in time, taking its inspiration from the accounts of feedback and adaptation provided by cybernetics (Ashby, 1956, 1960; Wiener, 1948).

Shapiro (2011) invoked the theme of conceptualization to characterize embodied cognitive science because it saw cognition as being directed action on the world. Conceptualization is the view that the form of an agent's body determines the concepts that it requires to interact with the world. Conceptualization is also a view that draws from embodied and ecological accounts of perception (Gibson, 1966, 1979; Merleau-Ponty, 1962; Neisser, 1976); such theories construed perception as being the result of action and as directing possible actions (affordances) on the world.

As such, the perceptual world cannot exist independently of a perceiving agent; *umwelten* (Uexküll, 2001) are defined in terms of the agent as well.

The relevance of the world to embodied cognitive science leads to another of its characteristics: Shapiro's (2011) notion of replacement. Replacement is the view that an agent's direct actions on the world can replace internal models, because the world can serve as its own best representation. The replacement theme is central to behaviour-based robotics (Breazeal, 2002; Brooks, 1991, 1999, 2002; Edsinger-Gonzales & Weber, 2004; Grey Walter, 1963; Sharkey, 1997), and leads some radical embodied cognitive scientists to argue that the notion of internal representations should be completely abandoned (Chemero, 2009). Replacement also permits theories to include the co-operative interaction between and mutual support of world and agent by exploring notions of cognitive scaffolding and leverage (Clark, 1997; Hutchins, 1995; Scribner & Tobach, 1997).

The themes of conceptualization and replacement emerge from a view of cognition that is radically embodied, in the sense that it cannot construe cognition without considering the rich relationships between mind, body, and world. This also leads to embodied cognitive science being characterized by Shapiro's (2011) third theme, constitution. This theme, as it appears in embodied cognitive science, is the extended mind hypothesis (Clark, 1997, 1999, 2003, 2008; Clark & Chalmers, 1998; Menary, 2008, 2010; Noë, 2009; Rupert, 2009; Wilson, 2004, 2005). According to the extended mind hypothesis, the world and body are literally constituents of cognitive processing; they are not merely causal contributors to it, as is the case in the classical sandwich.

Clearly embodied cognitive science has a much different view of cognition than is the case for classical cognitive science. This in turn leads to differences in the way that cognition is studied.

Classical cognitive science studies cognition at multiple levels: computational, algorithmic, architectural, and implementational. It typically does so by using a top-down strategy, beginning with the computational and moving "down" towards the architectural and implementational (Marr, 1982). This top-down strategy is intrinsic to the methodology of reverse engineering or functional analysis (Cummins, 1975, 1983). In reverse engineering, the behaviour of an intact system is observed and manipulated in an attempt to decompose it into an organized system of primitive components.

We have seen that embodied cognitive science exploits the same multiple levels of investigation that characterize classical cognitive science. However, embodied cognitive science tends to replace reverse engineering with an inverse, bottom-up methodology, as in forward engineering or synthetic psychology (Braitenberg, 1984; Dawson, 2004; Dawson, Dupuis, & Wilson, 2010; Pfeifer & Scheier, 1999). In forward engineering, a set of interesting primitives is assembled into a working system.

This system is then placed in an interesting environment in order to see what it can and cannot do. In other words, forward engineering starts with implementational and architectural investigations. Forward engineering is motivated by the realization that an agent's environment is a crucial contributor to behavioural complexity, and it is an attempt to leverage this possibility. As a result, some have argued that this approach can lead to simpler theories than is the case when reverse engineering is adopted (Braitenberg, 1984).

Shapiro (2011) has noted that it is too early to characterize embodied cognitive science as a unified school of thought. The many different variations of the embodied approach, and the important differences between them, are beyond the scope of the current chapter. A more accurate account of the current state of embodied cognitive science requires exploring an extensive and growing literature, current and historical. (Agre, 1997; Arkin, 1998; Bateson, 1972; Breazeal, 2002; Chemero, 2009; Clancey, 1997; Clark, 1997, 2003, 2008; Dawson, Dupuis, & Wilson, 2010; Dourish, 2001; Gallagher, 2005; Gibbs, 2006; Gibson, 1979; Goldman, 2006; Hutchins, 1995; Johnson, 2007; Menary, 2010; Merleau-Ponty, 1962; Neisser, 1976; Noë, 2004, 2009; Pfeifer & Scheier, 1999; Port & van Gelder, 1995b; Robbins & Aydede, 2009; Rupert, 2009; Shapiro, 2011; Varela, Thompson, & Rosch, 1991; Wilson, 2004; Winograd & Flores, 1987b).

6

Classical Music and Cognitive Science

6.0 Chapter Overview

In the previous three chapters I have presented the elements of three different approaches to cognitive science: classical, connectionist, and embodied. In the current chapter I present a review of these elements in the context of a single topic: musical cognition. In general, this is done by developing an analogy: cognitive science is like classical music. This analogy serves to highlight the contrasting characteristics between the three approaches of cognitive science, because each school of thought approaches the study of music cognition in a distinctive way.

These distinctions are made evident by arguing that the analogy between cognitive science and classical music is itself composed of three different relationships: between Austro-German classical music and classical cognitive science, between musical Romanticism and connectionist cognitive science, and between modern music and embodied cognitive science. One goal of the current chapter is to develop each of these more specific analogies, and in so doing we review the core characteristics of each approach within cognitive science.

Each of these more specific analogies is also reflected in how each school of cognitive science studies musical cognition. Classical, connectionist, and embodied cognitive scientists have all been involved in research on musical cognition, and they have not surprisingly focused on different themes.

Reviewing the three approaches within cognitive science in the context of music cognition again points to distinctions between the three approaches. However, the fact that all three approaches are involved in the study of music points to possible similarities between them. The current chapter begins to set the stage for a second theme that is fundamental to the remainder of the book: that there is the possibility for a synthesis amongst the three approaches that have been introduced in the earlier chapters. For instance, the current chapter ends by considering the possibility of a hybrid theory of musical cognition, a theory that has characteristics of classical, connectionist, and embodied cognitive science.

6.1 The Classical Nature of Classical Music

There are many striking parallels between the classical mind and classical music, particularly the music composed in the Austro-German tradition of the eighteenth and nineteenth centuries. First, both rely heavily upon formal structures. Second, both emphasize that their formal structures are content laden. Third, both attribute great importance to abstract thought inside an agent (or composer) at the expense of contributions involving the agent's environment or embodiment. Fourth, both emphasize central control. Fifth, the "classical" traditions of both mind and music have faced strong challenges, and many of the challenges in one domain can be related to analogous challenges in the other.

The purpose of this section is to elaborate the parallels noted above between classical music and classical cognitive science. One reason to do so is to begin to illustrate the analogy that *classical cognitive science is like classical music*. However, a more important reason is that this analogy, at least tacitly, has a tremendous effect on how researchers approach musical cognition. The methodological implications of this analogy are considered in detail later in this chapter.

To begin, let us consider how the notions of formalism or logicism serve as links between classical cognitive science and classical music. Classical cognitive science takes thinking to be the rule-governed manipulation of mental representations. Rules are sensitive to the form of mental symbols (Haugeland, 1985). That is, a symbol's form is used to identify it as being a token of a particular type; to be so identified means that only certain rules can be applied. While the rules are sensitive to the formal nature of symbols, they act in such a way to preserve the meaning of the information that the symbols represent. This property reflects classical cognitive science's logicism: the laws of thought are equivalent to the formal rules that define a system of logic (Boole, 2003). The goal of characterizing thought purely in the form of logical rules has been called the Boolean dream (Hofstadter, 1995).

It is not implausible that the Boolean dream might also characterize conceptions of music. Music's formal nature extends far beyond musical symbols on a sheet of staff paper. Since the time of Pythagoras, scholars have understood that music reflects regularities that are intrinsically mathematical (Ferguson, 2008). There is an extensive literature on the mathematical nature of music (Assayag et al., 2002; Benson, 2007; Harkleroad, 2006). For instance, different approaches to tuning instruments reflect the extent to which tunings are deemed mathematically sensible (Isacoff, 2001).

To elaborate, some pairs of tones played simultaneously are pleasing to the ear, such as a pair of notes that are a perfect fifth apart (see Figure 4-10)—they are consonant—while other combinations are not (Krumhansl, 1990). The consonance of notes can be explained by the physics of sound waves (Helmholtz & Ellis, 1954). Such physical relationships are ultimately mathematical, because they concern ratios of frequencies of sine waves. Consonant tone pairs have frequency ratios of 2:1 (octave), 3:2 (perfect fifth), and 4:3 (perfect fourth). The most dissonant pair of tones, the tritone (an augmented fourth) is defined by a ratio that includes an irrational number ($\sqrt[3]{}2$:1), a fact that was probably known to the Pythagoreans.

The formal nature of music extends far beyond the physics of sound. There are formal descriptions of musical elements, and of entire musical compositions, that are analogous to the syntax of linguistics (Chomsky, 1965). Some researchers have employed generative grammars to express these regularities (Lerdahl & Jackendoff, 1983; Steedman, 1984).

For instance, Lerdahl and Jackendoff (1983) argued that listeners impose a hierarchical structure on music, organizing "the sound signals into units such as motives, themes, phrases, periods, theme-groups, sections and the piece itself" (p. 12). They defined a set of well-formedness rules, which are directly analogous to generative rules in linguistics, to define how this musical organization proceeds and to rule out impossible organizations.

That classical music is expected to have a hierarchically organized, well-formed structure is a long-established view amongst scholars who do not use generative grammars to capture such regularities. Composer Aaron Copland (1939, p. 113) argued that a composition's structure is "one of the principal things to listen for" because it is "the planned design that binds an entire composition together."

One important musical structure is the sonata-allegro form (Copland, 1939), which is a hierarchical organization of musical themes or ideas. At the top level of this hierarchy are three different components that are presented in sequence: an initial exposition of melodic structures called musical themes, followed by the free development of these themes, and finishing with their recapitulation. Each of these segments is itself composed of three sub-segments, which are again presented in sequence. This structure is formal in the sense that the relationship between different themes presented in different sub-segments is defined in terms of their key signatures.

For instance, the exposition uses its first sub-segment to introduce an opening theme in the tonic key, that is, the initial key signature of the piece. The exposition's second sub-segment then presents a second theme in the dominant key, a perfect fifth above the tonic. The final sub-segment of the exposition finishes with a closing theme in the dominant key. The recapitulation has a substructure that is related to that of the exposition; it uses the same three themes in the same order, but all are presented in the tonic key. The development section, which falls between the exposition and the recapitulation, explores the exposition's themes, but does so using new material written in different keys.

Sonata-allegro form foreshadowed the modern symphony and produced a market for purely instrumental music (Rosen, 1988). Importantly, it also provided a structure, shared by both composers and their audiences, which permitted instrumental music to be expressive. Rosen notes that the sonata became popular because it,

> has an identifiable climax, a point of maximum tension to which the first part of
> the work leads and which is symmetrically resolved. It is a closed form, without the
> static frame of ternary form; it has a dynamic closure analogous to the denouement
> of 18th-century drama, in which everything is resolved, all loose ends are tied up,
> and the work rounded off. (Rosen, 1988, p. 10)

In short, the sonata-allegro form provided a logical structure that permitted the music to be meaningful.

The idea that musical form is essential to communicating musical meaning brings us to the second parallel between classical music and classical cognitive science: both domains presume that their formal structures are content-bearing.

Classical cognitive science explains cognition by invoking the intentional stance (Dennett, 1987), which is equivalent to relying on a cognitive vocabulary (Pylyshyn, 1984). If one assumes that an agent has certain intentional states (e.g., beliefs, desires, goals) and that lawful regularities (such as the principle of rationality) govern relationships between the contents of these states, then one can use the contents to predict future behaviour. "This single assumption [rationality], in combination with home truths about our needs, capacities and typical circumstances, generates both an intentional interpretation of us as believers and desirers and actual predictions of behavior in great profusion" (Dennett, 1987, p. 50). Similarly, Pylyshyn (1984, pp. 20–21) noted that "the principle of rationality . . . is indispensable for giving an account of human behavior."

Is there any sense in which the intentional stance can be applied to classical music? Classical composers are certainly of the opinion that music can express ideas. Copland noted that,

> my own belief is that all music has an expressive power, some more and some less,
> but that all music has a certain meaning behind the notes and that that meaning

behind the notes constitutes, after all, what the piece is saying, what the piece is about. (Copland, 1939, p. 12)

John Cage (1961) believed that compositions had intended meanings:

> It seemed to me that composers knew what they were doing, and that the experiments that had been made had taken place prior to the finished works, just as sketches are made before paintings and rehearsals precede performances. (John Cage, 1961, p. 7)

Scholars, too, have debated the ability of music to convey meanings. One of the central questions in the philosophy of music is whether music can represent. As late as 1790, the dominant philosophical view of music was that it was incapable of conveying ideas, but by the time that E. T. A. Hoffman reviewed Beethoven's Fifth Symphony in 1810, this view was predominately rejected (Bonds, 2006), although the autonomist school of musical aesthetics—which rejected musical representation—was active in the late nineteenth century (Hanslick, 1957). Nowadays most philosophers of music agree that music is representational, and they focus their attention on *how* musical representations are possible (Kivy, 1991; Meyer, 1956; Robinson, 1994, 1997; Sparshoot, 1994; Walton, 1994).

How might composers communicate intended meanings with their music? One answer is by exploiting particular musical forms. Conventions such as sonata-allegro form provide a structure that generates expectations, expectations that are often presumed to be shared by the audience. Copland (1939) used his book about listening to music to educate audiences about musical forms so that they could better understand his compositions as well as those of others: "In helping others to hear music more intelligently, [the composer] is working toward the spread of a musical culture, which in the end will affect the understanding of his own creations" (p. vi).

The extent to which the audience's expectations are toyed with, and ultimately fulfilled, can manipulate its interpretation of a musical performance. Some scholars have argued that these manipulations can be described completely in terms of the structure of musical elements (Meyer, 1956). The formalist's motto of classical cognitive science (Haugeland, 1985) can plausibly be applied to classical music.

A third parallel between classical cognitive science, which likely follows directly from the assumption that formal structures can represent content, is an emphasis on Cartesian disembodiment. Let us now consider this characteristic in more detail.

Classical cognitive science attempts to explain cognitive phenomena by appealing to a sense-think-act cycle (Pfeifer & Scheier, 1999). In this cycle, sensing mechanisms provide information about the world, and acting mechanisms produce behaviours that might change it. Thinking, considered as the manipulation of mental representations, is the interface between sensing and acting (Wilson, 2004). However, this interface, internal thinking, receives the most emphasis in a classical theory, with an accompanying underemphasis on sensing and acting (Clark, 1997).

One can easily find evidence for the classical emphasis on representations. Autonomous robots that were developed following classical ideas devote most of their computational resources to using internal representations of the external world (Brooks, 2002; Moravec, 1999; Nilsson, 1984). Most survey books on cognitive psychology (Anderson, 1985; Best, 1995; Haberlandt, 1994; Robinson-Riegler & Robinson-Riegler, 2003; Solso, 1995; Sternberg, 1996) have multiple chapters on representational topics such as memory and reasoning and rarely mention embodiment, sensing, or acting. Classical cognitive science's sensitivity to the multiple realization argument (Fodor, 1968b, 1975), with its accompanying focus on functional (not physical) accounts of cognition (Cummins, 1983), underlines its view of thinking as a disembodied process. It was argued in Chapter 3 that the classical notion of the disembodied mind was a consequence of its being inspired by Cartesian philosophy.

Interestingly, a composer of classical music is also characterized as being similarly engaged in a process that is abstract, rational, and disembodied. Does not a composer first think of a theme or a melody and then translate this mental representation into a musical score? Mozart "carried his compositions around in his head for days before setting them down on paper" (Hildesheimer, 1983). Benson (2007, p. 25) noted that "Stravinsky speaks of a musical work as being 'the fruit of study, reasoning, and calculation that imply exactly the converse of improvisation.'" In short, abstract thinking seems to be a prerequisite for composing.

Reactions against Austro-German classical music (Nyman, 1999) were reactions against its severe rationality. John Cage pioneered this reaction (Griffiths, 1994); beginning in the 1950s, Cage increasingly used chance mechanisms to determine musical events. He advocated "that music should no longer be conceived of as rational discourse" (Nyman, 1999, p. 32). He explicitly attacked the logicism of traditional music (Ross, 2007), declaring that "any composing strategy which is wholly 'rational' is irrational in the extreme" (p. 371).

Despite opposition such as Cage's, the disembodied rationality of classical music was one of its key features. Indeed, the cognitive scaffolding of composing is frowned upon. There is a general prejudice against composers who rely on external aids (Rosen, 2002). Copland (1939, p. 22) observed that "a current idea exists that there is something shameful about writing a piece of music at the piano." Rosen traces this idea to Giovanni Maria Artusi's criticism of composers such as Monteverdi, in 1600: "It is one thing to search with voices and instruments for something pertaining to the harmonic faculty, another to arrive at the exact truth by means of reasons seconded by the ear" (p. 17). The expectation (then and now) is that composing a piece involves "mentally planning it by logic, rules, and traditional reason" (Rosen, 2002, p. 17). This expectation is completely consistent with the disembodied, classical view of thinking, which assumes that the primary purpose of cognition is not acting, but is instead planning.

Planning has been described as solving the problem of what to do next (Dawson, 1998; Stillings, 1995). A solution to this problem involves providing an account of the control system of a planning agent; such accounts are critical components of classical cognitive science. "An adequate theory of human cognitive processes must include a description of the *control system*—the mechanism that determines the sequence in which operations will be performed" (Simon, 1979, p. 370). In classical cognitive science, such control is typically central. The notion of central control is also characteristic of classical music, providing the fourth parallel between classical cognitive science and classical music.

Within the Austro-German musical tradition, a composition is a formal structure intended to express ideas. A composer uses musical notation to signify the musical events which, when realized, accomplish this expressive goal. An orchestra's purpose is to bring the score to life, in order for the performance to deliver the intended message to the audience:

> We tend to see both the score and the performance primarily as vehicles for preserving what the composer has created. We assume that musical scores provide a permanent record or embodiment in signs; in effect, a score serves to 'fix' or objectify a musical work. (Benson, 2003, p. 9)

However, a musical score is vague; it cannot determine every minute detail of a performance (Benson, 2003; Copland, 1939). As a result, during a performance the score must be interpreted in such a way that the missing details can be filled in without distorting the composer's desired effect. In the Austro-German tradition of music, an orchestra's conductor takes the role of interpreter and controls the orchestra in order to deliver the composer's message (Green & Malko, 1975, p. 7): "The conductor acts as a guide, a solver of problems, a decision maker. His guidance chart is the composer's score; his job, to animate the score, to make it come alive, to bring it into audible being."

The conductor provides another link between classical music and classical cognitive science, because the conductor is the orchestra's central control system. The individual players are expected to submit to the conductor's control.

> Our conception of the role of a classical musician is far closer to that of self-effacing servant who faithfully serves the score of the composer. Admittedly, performers are given a certain degree of leeway; but the unwritten rules of the game are such that this leeway is relatively small and must be kept in careful check. (Benson, 2003, p. 5)

It has been suggested—not necessarily validly—that professional, classically trained musicians are incapable of improvisation (Bailey, 1992)!

The conductor is not the only controller of a performance. While it is unavoidably vague, the musical score also serves to control the musical events generated by an orchestra. If the score is a content-bearing formal expression, then it is reasonable to assume that it designates the contents that the score is literally about. Benson (2003) described this aspect of a score as follows:

> The idea of being '*treu*'—which can be translated as true or a faithful—implies faithfulness to someone or something. *Werktreue*, then, is directly a kind of faithfulness to the *Werk* (work) and, indirectly, a faithfulness to the composer. Given the centrality of musical notation in the discourse of classical music, a parallel notion is that of *Texttreue*: fidelity to the written score. (Benson, 2003, p. 5)

Note Benson's emphasis on the formal notation of the score. It highlighted the idea that the written score is analogous to a logical expression, and that converting it into the musical events that the score is about (in Brentano's sense) is not only desirable, but also rational. This logicism of classical music perfectly parallels the logicism found in classical cognitive science.

The role of the score as a source of control provides a link back to another issue discussed earlier, disembodiment. We saw in Chapter 3 that the disembodiment of modern classical cognitive science is reflected in its methodological solipsism. In methodological solipsism, representational states are individuated from one another only in terms of their relations to other representational states. Relations of the states to the external world—the agent's environment—are not considered.

It is methodological solipsism that links a score's control back to disembodiment, providing another link in the analogy between the classical mind and classical music. When a piece is performed, it is brought to life with the intent of delivering a particular message to the audience. Ultimately, then, the audience is a fundamental component of a composition's environment. To what extent does this environment affect or determine the composition itself?

In traditional classical music, the audience is presumed to have absolutely *no effect* on the composition. Composer Arnold Schoenberg believed that the audience was "merely an acoustic necessity—and an annoying one at that" (Benson, 2003, p. 14). Composer Virgil Thompson defined the ideal listener as "a person who applauds vigorously" (Copland, 1939, p. 252). In short, the purpose of the audience is to passively receive the intended message. It too is under the control of the score:

> The intelligent listener must be prepared to increase his awareness of the musical material and what happens to it. He must hear the melodies, the rhythms, the harmonies, the tone colors in a more conscious fashion. But above all he must, in order to follow the line of the composer's thought, know something of the principles of musical form. (Copland, 1939, p. 17)

To see that this is analogous to methodological solipsism, consider how we differentiate compositions from one another. Traditionally, this is done by referring to a composition's score (Benson, 2003). That is, compositions are identified in terms of a particular set of symbols, a particular formal structure. The identification of a composition does not depend upon identifying which audience has heard it. A composition can exist, and be identified, in the absence of its audience-as-environment.

Another parallel between the classical mind and classical music is that there have been significant modern reactions against the Austro-German musical tradition (Griffiths, 1994, 1995). Interestingly, these reactions parallel many of the reactions of embodied cognitive science against the classical approach. In later sections of this chapter we consider some of these reactions, and explore the idea that they make plausible the claim that "non-cognitive" processes are applicable to classical music. However, before we do so, let us first turn to consider how the parallels considered above are reflected in how classical cognitive scientists study musical cognition.

6.2 The Classical Approach to Musical Cognition

In Chapter 8 on seeing and visualizing, we see that classical theories take the purpose of visual perception to be the construction of mental models of the external, visual world. To do so, these theories must deal with the problem of underdetermination. Information in the world is not sufficient, on its own, to completely determine visual experience.

Classical solutions to the problem of underdetermination (Bruner, 1973; Gregory, 1970, 1978; Rock, 1983) propose that knowledge of the world—the contents of mental representations—is also used to determine visual experience. In other words, classical theories of perception describe visual experience as arising from the interaction of stimulus information with internal representations. Seeing is a kind of thinking.

Auditory perception has also been the subject of classical theorization. Classical theories of auditory perception parallel classical theories of visual perception in two general respects. First, since the earliest psychophysical studies of audition (Helmholtz & Ellis, 1954), hearing has been viewed as a process for building internal representations of the external world.

> We have to investigate the various modes in which the nerves themselves are excited, giving rise to their various *sensations*, and finally the laws according to which these sensations result in mental images of determinate external objects, that is, in *perceptions*. (Helmholtz & Ellis, 1954, p. 4)

Second, in classical theories of hearing, physical stimulation does not by itself determine the nature of auditory percepts. Auditory stimuli are actively organized, being grouped into distinct auditory streams, according to psychological principles of organization (Bregman, 1990). "When listeners create a mental representation of auditory input, they too must employ rules about what goes with what" (p. 11).

The existence of classical theories of auditory perception, combined with the links between classical music and classical cognitive science discussed in the previous section, should make it quite unsurprising that classical theories of music perception and cognition are well represented in the literature (Deutsch, 1999; Francès, 1988; Howell, Cross, & West, 1985; Krumhansl, 1990; Lerdahl, 2001; Lerdahl & Jackendoff, 1983; Sloboda, 1985; Snyder, 2000; Temperley, 2001). This section provides some brief examples of the classical approach to musical cognition. These examples illustrate that the previously described links between classical music and cognitive science are reflected in the manner in which musical cognition is studied.

The classical approach to musical cognition assumes that listeners construct mental representations of music. Sloboda (1985) argued that,

> a person may understand the music he hears without being moved by it. If he
> *is* moved by it then he must have passed through the cognitive stage, which
> involves forming an abstract or symbolic *internal representation* of the music.
> (Sloboda, 1985, p. 3)

Similarly, "a piece of music is a mentally constructed entity, of which scores and performances are partial representations by which the piece is transmitted" (Lerdahl & Jackendoff, 1983, p. 2). A classical theory must provide an account of such mentally constructed entities. How are they represented? What processes are required to create and manipulate them?

There is a long history of attempting to use geometric relations to map the relationships between musical pitches, so that similar pitches are nearer to one another in the map (Krumhansl, 2005). Krumhansl (1990) has shown how simple judgments about tones can be used to derive a spatial, cognitive representation of musical elements.

Krumhansl's general paradigm is called the tone probe method (Krumhansl & Shepard, 1979). In this paradigm, a musical context is established, for instance by playing a partial scale or a chord. A probe note is then played, and subjects rate how well this probe note fits into the context. For instance, subjects might rate how well the probe note serves to complete a partial scale. The relatedness between pairs of tones within a musical context can also be measured using variations of this paradigm.

Extensive use of the probe tone method has revealed a hierarchical organization of musical notes. Within a given musical context—a particular musical key—the most stable tone is the tonic, the root of the key. For example, in the musical key of C major, the note C is the most stable. The next most stable tones are those in either the third or fifth positions of the key's scale. In the key of C major, these are the notes E or

G. Less stable than these two notes are any of the set of remaining notes that belong to the context's scale. In the context of C major, these are the notes D, F, A, and B. Finally, the least stable tones are the set of five notes that do not belong to the context's scale. For C major, these are the notes C#, D#, F#, G#, and A#.

This hierarchical pattern of stabilities is revealed using different kinds of contexts (e.g., partial scales, chords), and is found in subjects with widely varying degrees of musical expertise (Krumhansl, 1990). It can also be used to account for judgments about the consonance or dissonance of tones, which is one of the oldest topics in the psychology of music (Helmholtz & Ellis, 1954).

Hierarchical tonal stability relationships can also be used to quantify relationships between different musical keys. If two different keys are similar to one another, then their tonal hierarchies should be similar as well. The correlations between tonal hierarchies were calculated for every possible pair of the 12 different major and 12 different minor musical keys, and then multidimensional scaling was performed on the resulting similarity data (Krumhansl & Kessler, 1982). A four-dimensional solution was found to provide the best fit for the data. This solution arranged the tonic notes along a spiral that wrapped itself around a toroidal surface. The spiral represents two circles of fifths, one for the 12 major scales and the other for the 12 minor scales.

The spiral arrangement of notes around the torus reflects elegant spatial relationships among tonic notes (Krumhansl, 1990; Krumhansl & Kessler, 1982). For any key, the nearest neighbours moving around from the inside to the outside of the torus are the neighbouring keys in the circle of fifths. For instance, the nearest neighbours to C in this direction are the notes F and G, which are on either side of C in the circle of fifths.

In addition, the nearest neighbour to a note in the direction along the torus (i.e., orthogonal to the direction that captures the circles of fifths) reflects relationships between major and minor keys. Every major key has a complementary minor key, and vice versa; complimentary keys have the same key signature, and are musically very similar. Complimentary keys are close together on the torus. For example, the key of C major has the key of A minor as its compliment; the tonic notes for these two scales are also close together on the toroidal map.

Krumhansl's (1990) tonal hierarchy is a classical representation in two senses. First, the toroidal map derived from tonal hierarchies provides one of the many examples of spatial representations that have been used to model regularities in perception (Shepard, 1984a), reasoning (Sternberg, 1977), and language (Tourange au & Sternberg, 1981, 1982). Second, a tonal hierarchy is not a musical property per se, but instead is a psychologically imposed organization of musical elements. "The experience of music goes beyond registering the acoustic parameters of tone frequency, amplitude, duration, and timbre. Presumably, these are recoded, organized, and stored in memory in a form different from sensory codes" (Krumhansl, 1990, p. 281). The tonal hierarchy is one such mental organization of musical tones.

In music, tones are not the only elements that appear to be organized by psychological hierarchies. "When hearing a piece, the listener naturally organizes the sound signals into units such as motives, themes, phrases, periods, theme-groups, and the piece itself" (Lerdahl & Jackendoff, 1983, p. 12). In their classic work *A Generative Theory of Tonal Music*, Lerdahl and Jackendoff (1983) developed a classical model of how such a hierarchical organization is derived.

Lerdahl and Jackendoff's (1983) research program was inspired by Leonard Bernstein's (1976) Charles Eliot Norton lectures at Harvard, in which Bernstein called for the methods of Chomskyan linguistics to be applied to music. "All musical thinkers agree that there is such a thing as a musical syntax, comparable to a descriptive grammar of speech" (p. 56). There are indeed important parallels between language and music that support developing a generative grammar of music (Jackendoff, 2009). In particular, systems for both language and music must be capable of dealing with novel stimuli, which classical researchers argue requires the use of recursive rules. However, there are important differences too. Most notable for Jackendoff (2009) is that language conveys propositional thought, while music does not. This means that while a linguistic analysis can ultimately be evaluated as being true or false, the same cannot be said for a musical analysis, which has important implications for a grammatical model of music.

Lerdahl and Jackendoff's (1983) generative theory of tonal music correspondingly has components that are closely analogous to a generative grammar for language and other components that are not. The linguistic analogs assign structural descriptions to a musical piece. These structural descriptions involve four different, but interrelated, hierarchies.

The first is grouping structure, which hierarchically organizes a piece into motives, phrases, and sections. The second is metrical structure, which relates the events of a piece to hierarchically organized alternations of strong and weak beats. The third is time-span reduction, which assigns pitches to a hierarchy of structural importance that is related to grouping and metrical structures. The fourth is prolongational reduction, which is a hierarchy that "expresses harmonic and melodic tension and relaxation, continuity and progression" (Lerdahl & Jackendoff, 1983, p. 9). Prolongational reduction was inspired by Schenkerian musical analysis (Schenker, 1979), and is represented in a fashion that is very similar to a phrase marker. As a result, it is the component of the generative theory of tonal music that is most closely related to a generative syntax of language (Jackendoff, 2009).

Each of the four hierarchies is associated with a set of well-formedness rules (Lerdahl & Jackendoff, 1983). These rules describe how the different hierarchies are constructed, and they also impose constraints that prevent certain structures from being created. Importantly, the well-formedness rules provide psychological principles for organizing musical stimuli, as one would expect in a classical theory. The rules "define a class of grouping structures that can be associated with a sequence of pitch-events, but which are not specified in any direct way by the physical signal (as pitches and durations are)" (p. 39). Lerdahl and Jackendoff take care to express these rules in plain English so as not to obscure their theory. However, they presume that the well-formedness rules could be translated into a more formal notation, and indeed computer implementations of their theory are possible (Hamanaka, Hirata, & Tojo, 2006).

Lerdahl and Jackendoff's (1983) well-formedness rules are not sufficient to deliver a unique "parsing" of a musical piece. One reason for this is because, unlike language, a musical parsing cannot be deemed to be correct; it can only be described as having a certain degree of coherence or preferredness. Lerdahl and Jackendoff supplement their well-formedness rules with a set of preference rules. For instance, one preference rule for grouping structure indicates that symmetric groups are to be preferred over asymmetric ones. Once again there is a different set of preference rules for each of the four hierarchies of musical structure.

The hierarchical structures defined by the generative theory of tonal music (Lerdahl & Jackendoff, 1983) describe the properties of a particular musical event. In contrast, the hierarchical arrangement of musical tones (Krumhansl, 1990) is a general organizational principle that applies to musical pitches in general, not to an event. Interestingly, the two types of hierarchies are not mutually exclusive. The generative theory of tonal music has been extended (Lerdahl, 2001) to include tonal pitch spaces, which are spatial representations of tones and chords in which the distance between two entities in the space reflects the cognitive distance between them. Lerdahl has shown that the properties of tonal pitch space can be used to aid in the construction of the time-span reduction and the prolongational reduction, increasing the power of the original generative theory. The theory can be used to predict listeners' judgments about the attraction and tension between tones in a musical selection (Lerdahl & Krumhansl, 2007).

Lerdahl and Jackendoff's (1983) generative theory of tonal music shares another characteristic with the linguistic theories that inspired it: it provides an account of musical competence, and it is less concerned with algorithmic accounts of music perception. The goal of their theory is to provide a "formal description of the musical intuitions of a listener who is experienced in a musical idiom" (p. 1). Musical intuition is the largely unconscious knowledge that a listener uses to organize, identify, and comprehend musical stimuli. Because characterizing such knowledge is the goal of the theory, other processing is ignored.

> Instead of describing the listener's real-time mental processes, we will be concerned only with the final state of his understanding. In our view it would be fruitless to

theorize about mental processing before understanding the organization to which the processing leads. (Lerdahl & Jackendoff, 1983, pp. 3–4)

One consequence of ignoring mental processing is that the generative theory of tonal music is generally not applied to psychologically plausible representations. For instance, in spite of being a theory about an experienced listener, the various incarnations of the theory are not applied to auditory stimuli, but are instead applied to musical scores (Hamanaka, Hirata, & Tojo, 2006; Lerdahl, 2001; Lerdahl & Jackendoff, 1983).

Of course, this is not a principled limitation of the generative theory of tonal music. This theory has inspired researchers to develop models that have a more algorithmic emphasis and operate on representations that take steps towards psychological plausibility (Temperley, 2001).

Temperley's (2001) theory can be described as a variant of the original generative theory of tonal music (Lerdahl & Jackendoff, 1983). One key difference between the two is the input representation. Temperley employs a piano-roll representation, which can be described as being a two-dimensional graph of musical input. The vertical axis, or pitch axis, is a discrete representation of different musical notes. That is, each row in the vertical axis can be associated with its own piano key. The horizontal axis is a continuous representation of time. When a note is played, a horizontal line is drawn on the piano-roll representation; the height of the line indicates which note is being played. The beginning of the line represents the note's onset, the length of the line represents the note's duration, and the end of the line represents the note's offset. Temperley assumes the psychological reality of the piano-roll representation, although he admits that the evidence for this strong assumption is inconclusive.

Temperley's (2001) model applies a variety of preference rules to accomplish the hierarchical organization of different aspects of a musical piece presented as a piano-roll representation. He provides different preference rule systems for assigning metrical structure, melodic phrase structure, contrapuntal structure, pitch class representation, harmonic structure, and key structure. In many respects, these preference rule systems represent an evolution of the well-formedness and preference rules in Lerdahl and Jackendoff's (1983) theory.

For example, one of Temperley's (2001) preference rule systems assigns metrical structure (i.e., hierarchically organized sets of beats) to a musical piece. Lerdahl and Jackendoff (1983) accomplished this by applying four different well-formedness rules and ten different preference rules. Temperley accepts two of Lerdahl and Jackendoff's well-formedness rules for metre (albeit in revised form, as preference rules) and rejects two others because they do not apply to the more realistic representation that Temperley adopts. Temperley adds three other preference rules. This system of five preference rules derives metric structure to a high degree of accuracy (i.e., corresponding to a degree of 86 percent or better with Temperley's metric intuitions).

One further difference between Temperley's (2001) algorithmic emphasis and Lerdahl and Jackendoff's (1983) emphasis on competence is reflected in how the theory is refined. Because Temperley's model is realized as a working computer model, he could easily examine its performance on a variety of input pieces and therefore identify its potential weaknesses. He took advantage of this ability to propose an additional set of four preference rules for metre, as an example, to extend the applicability of his algorithm to a broader range of input materials.

To this point, the brief examples provided in this section have been used to illustrate two of the key assumptions made by classical researchers of musical cognition. First, mental representations are used to impose an organization on music that is not physically present in musical stimuli. Second, these representations are classical in nature: they involve different kinds of rules (e.g., preference rules, well-formedness rules) that can be applied to symbolic media that have musical contents (e.g., spatial maps, musical scores, piano-roll representations). A third characteristic also is frequently present in classical theories of musical cognition: the notion that the musical knowledge reflected in these representations is acquired, or can be modified, by experience.

The plasticity of musical knowledge is neither a new idea nor a concept that is exclusively classical. We saw earlier that composers wished to inform their audience about compositional conventions so the latter could better appreciate performances (Copland, 1939). More modern examples of this approach argue that ear training, specialized to deal with some of the complexities of modern music to be introduced later in this chapter, can help to bridge the gaps between composers, performers, and audiences (Friedmann, 1990). Individual differences in musical ability were thought to be a combination of innate and learned information long before the cognitive revolution occurred (Seashore, 1967): "The ear, like the eye, is an instrument, and mental development in music consists in the acquisition of skills and the enrichment of experience through this channel" (p. 3).

The classical approach views the acquisition of musical skills in terms of changes in mental representations. "We *learn* the structures that we use to represent music" (Sloboda, 1985, p. 6). Krumhansl (1990, p. 286) noted that the robust hierarchies of tonal stability revealed in her research reflect stylistic regularities in Western tonal music. From this she suggests that "it seems probable, then, that abstract tonal and harmonic relations are learned through internalizing distributional properties characteristic of the style." This view is analogous to those classical theories of perception that propose that the structure of internal representations imposes constraints on visual transformations that mirror the constraints imposed by the physics of the external world (Shepard, 1984b).

Krumhansl's (1990) internalization hypothesis is one of many classical accounts that have descended from Leonard Meyer's account of musical meaning arising from emotions manipulated by expectation (Meyer, 1956). "Styles in music are basically complex systems of probability relationships" (p. 54). Indeed, a tremendous variety of musical characteristics can be captured by applying Bayesian models, including rhythm and metre, pitch and melody, and musical style (Temperley, 2007). A great deal of evidence also suggests that expectations about what is to come next are critical determinants of human music perception (Huron, 2006). Temperley argues that classical models of music perception (Lerdahl, 2001; Lerdahl & Jackendoff, 1983; Temperley, 2001) make explicit these probabilistic relationships. "Listeners' generative models are tuned to reflect the statistical properties of the music that they encounter" (Temperley, 2007, p. 207).

It was earlier argued that there are distinct parallels between Austro-German classical music and the classical approach to cognitive science. One of the most compelling is that both appeal to abstract, formal structures. It would appear that the classical approach to musical cognition takes this parallel very literally. That is, the representational systems proposed by classical researchers of musical cognition internalize the formal properties of music, and in turn they impose this formal structure on sounds during the perception of music.

6.3 Musical Romanticism and Connectionism

The eighteenth-century Industrial Revolution produced profound changes in the nature of European life, transferring power and wealth from the nobility to the commercial class (Plantinga, 1984). Tremendous discontentment with the existing social order, culminating in the French revolution, had a profound influence on political, intellectual, and artistic pursuits. It led to a movement called Romanticism (Claudon, 1980), which roughly spanned the period from the years leading up to the 1789 French revolution through to the end of the nineteenth century.

A precise definition of Romanticism is impossible, for it developed at different times in different countries, and in different arts—first poetry, then painting, and finally music (Einstein, 1947). Romanticism was a reaction against the reason and rationality that characterized the Enlightenment period that preceded it. Romanticism emphasized the individual, the irrational, and the imaginative. Arguably music provided Romanticism's greatest expression (Einstein, 1947; Plantinga, 1984), because music expressed mystical and imaginative ideas that could not be captured by language.

It is impossible to provide a clear characterization of Romantic music (Einstein, 1947; Longyear, 1988; Plantinga, 1984; Whittall, 1987). "We seek in vain an unequivocal idea of the nature of 'musical Romanticism'" (Einstein, 1947, p. 4).

However, there is general agreement that Romantic music exhibits,

> a preference for the original rather than the normative, a pursuit of unique
> effects and extremes of expressiveness, the mobilization to that end of an
> enriched harmonic vocabulary, striking new figurations, textures, and tone colors.
> (Plantinga, 1984, p. 21)

The list of composers who were musical Romanticism's greatest practitioners begins with Beethoven, and includes Schubert, Mendelssohn, Schumann, Chopin, Berlioz, Liszt, Wagner, and Brahms.

Romantic music can be used to further develop the analogy between classical music and cognitive science. In particular, there are several parallels that exist between musical Romanticism and connectionist cognitive science. The most general similarity between the two is that both are reactions against the Cartesian view of the mind that dominated the Enlightenment.

Romantic composers wished to replace the calculated, rational form of music such as Bach's contrapuntal fugues (Gaines, 2005; Hofstadter, 1979) with a music that expressed intensity of feeling, which communicated the sublime. "It was a *retrogression* to the primitive relationship that man had had to music—to the mysterious, the exciting, the magical" (Einstein, 1947, p. 8). As a result, musical Romanticism championed purely instrumental music; music that was not paired with words. The instrumental music of the Romantics "became the choicest means of saying what could not be said, of expressing something deeper than the *word* had been able to express" (p. 32). In a famous 1813 passage, music critic E. T. A. Hoffman proclaimed instrumental music to be "the most romantic of all the arts—one might almost say, the only genuinely romantic one—for its sole subject is the infinite" (Strunk, 1950, p. 775).

Connectionist cognitive science too is a reaction against the rationalism and logicism of Cartesian philosophy. And one form of this reaction parallels Romantic music's move away from the word: many connectionists interpreted the ability of networks to accomplish classical tasks as evidence that cognitive science need not appeal to explicit rules or symbols (Bechtel & Abrahamsen, 1991; Horgan & Tienson, 1996; Ramsey, Stich, & Rumelhart, 1991; Rumelhart & McClelland, 1986a).

A second aspect of musical Romanticism's reaction against reason was its emphasis on the imaginary and the sublime. In general, the Romantic arts provided escape by longingly looking back at "unspoiled," preindustrial existences and by using settings that were wild and fanciful. Nature was a common inspiration. The untamed mountains and chasms of the Alps stood in opposition to the Enlightenment's view that the world was ordered and structured.

For example, in the novel *Frankenstein* (Shelley, 1985), after the death of Justine, Victor Frankenstein seeks solace in a mountain journey. The beauty of a valley through which he travelled "was augmented and rendered sublime by the mighty Alps, whose white and shining pyramids and domes towered above all, as belonging to another earth, the habitations of another race of beings" (p. 97). To be sublime was to reflect a greatness that could not be completely understood. "The immense mountains and precipices that overhung me on every side—the sound of the river raging among the rocks, and the dashing of the waterfalls around, spoke of a power mighty as Omnipotence" (p. 97).

Sublime Nature appeared frequently in musical Romanticism. Longyear's (1988, p. 12) examples include "the forest paintings in Weber's *Der Freischütz* or Wagner's; the landscapes and seascapes of Mendelssohn and Gade; the Alpine pictures in Schumann's or Tchaikovsky's *Manfred*" to name but a few.

Musical Romanticism also took great pains to convey the imaginary or the indescribable (Whittall, 1987). In some striking instances, Romantic composers followed the advice in John Keats' 1819 *Ode on a Grecian Urn*, "Heard melodies are sweet, but those unheard / Are sweeter." Consider Schumann's piano work *Humoreske* (Rosen, 1995). It uses three staves: one for the right hand, one for the left, and a third—containing the melody!—which is not to be played at all. Though inaudible, the melody "is embodied in the upper and lower parts as a kind of after resonance—out of phase, delicate, and shadowy" (p. 8). The effects of the melody emerge from playing the other parts.

In certain respects, connectionist cognitive science is sympathetic to musical Romanticism's emphasis on nature, the sublime, and the imaginary. Cartesian philosophy, and the classical cognitive science that was later inspired by it, view the mind as disembodied, being separate from the natural world. In seeking theories that are biologically plausible and neuronally inspired (McClelland & Rumelhart, 1986; Rumelhart & McClelland, 1986c), connectionists took a small step towards embodiment. Whereas Descartes completely separated the mind from the world, connectionists assume that brains cause minds (Searle, 1984).

Furthermore, connectionists recognize that the mental properties caused by brains may be very difficult to articulate using a rigid set of rules and symbols. One reason that artificial neural networks are used to study music is because they may capture regularities that cannot be rationally expressed (Bharucha, 1999; Rowe, 2001; Todd & Loy, 1991). These regularities emerge from the nonlinear interactions amongst network components (Dawson, 2004; Hillis, 1988). And the difficulty in explaining such interactions suggests that networks are sublime. Artificial neural networks seem to provide "the possibility of constructing intelligence without first understanding it" (Hillis, 1988, p. 176).

Musical Romanticism also celebrated something of a scale less grand than sublime Nature: the individual. Romantic composers broke away from the established system of musical patronage. They began to write music for its own (or for the composer's own) sake, instead of being written for commission (Einstein, 1947).

Beethoven's piano sonatas were so brilliant and difficult that they were often beyond the capabilities of amateur performers who had mastered Haydn and Mozart. His symphonies were intended to speak "to a humanity that the creative artist had raised to his own level" (p. 38). The subjectivity and individualism of musical Romanticism is one reason that there is no typical symphony, art-song, piano piece or composer from this era (Longyear, 1988).

Individualism was also reflected in the popularity of musical virtuosos, for whom the Romantic period was a golden age (Claudon, 1980). These included the violinists Paganini and Baillot, and the pianists Liszt, Chopin, and Schumann. They were famous not only for their musical prowess, but also for a commercialization of their character that exploited Romanticist ideals (Plantinga, 1984). Paganini and Liszt were "transformed by the Romantic imagination into a particular sort of hero: mysterious, sickly, and bearing the faint marks of dark associations with another world" (Plantinga, 1984, p. 185).

Individualism is also a fundamental characteristic of connectionism. It is not a characteristic of connectionist researchers themselves (but see below), but is instead a characteristic of the networks that they describe. When connectionist simulations are reported, the results are almost invariably provided for individual networks. This was demonstrated in Chapter 4; the interpretations of internal structure presented there are always of individual networks. This is because there are many sources of variation between networks as a result of the manner in which they are randomly initialized (Dawson, 2005). Thus it is unlikely that one network will be identical to another, even though both have learned the same task. Rather than exploring "typical" network properties, it is more expedient to investigate the interesting characteristics that can be found in one of the networks that were successfully trained.

There are famous individual networks that are analogous to musical virtuosos. These include the Jets-Sharks network used to illustrate the interactive activation with competition (IAC) architecture (McClelland & Rumelhart, 1988); a multilayered network that converted English verbs from present to past tense (Pinker & Prince, 1988; Rumelhart & McClelland, 1986a); and the NETTALK system that learned to read aloud (Sejnowski & Rosenberg, 1988).

Individualism revealed itself in another way in musical Romanticism. When Romantic composers wrote music for its own sake, they assumed that its audience would be found later (Einstein, 1947). Unfortunately, "few artists gained recognition without long, difficult struggles" (Riedel, 1969, p. 6). The isolation of the composer from the audience was an example of another Romantic invention: the composer was the misunderstood genius who idealistically pursued art for art's sake. "The Romantic musician . . . was proud of his isolation. In earlier centuries the idea of misunderstood genius was not only unknown; it was inconceivable" (Einstein, 1947, p. 16).

The isolated genius is a recurring character in modern histories of connectionism, one of which is presented as a fairy tale (Papert, 1988), providing an interesting illustration of the link between Romanticism and connectionism. According to the prevailing view of connectionist history (Anderson & Rosenfeld, 1998; Hecht-Nielsen, 1987; Medler, 1998; Olazaran, 1996), the isolation of the neural net researcher began with a crusade by Minsky and Papert, prior to the publication of *Perceptrons* (Minsky & Papert, 1969), against research funding for perceptron-like systems.

> Minsky and Papert's campaign achieved its purpose. The common wisdom that neural networks were a research dead-end became firmly established. Artificial intelligence researchers got all of the neural network research money and more. The world had been reordered. And neurocomputing had to go underground. (Hecht-Nielsen, 1987, p. 17)

Going underground, at least in North America, meant connectionist research was conducted sparingly by a handful of researchers, disguised by labels such as "adaptive pattern recognition" and "biological modelling" during the "quiet years" from 1967 until 1982 (Hecht-Nielsen, 1987). A handful of neural network researchers "struggled through the entire span of quiet years in obscurity." While it did not completely disappear, "neural-net activity decreased significantly and was displaced to areas outside AI (it was considered 'deviant' within AI)" (Olazaran, 1996, p. 642). Like the Romantic composers they resemble, these isolated connectionist researchers conducted science for science's sake, with little funding, waiting for an audience to catch up—which occurred with the 1980s rise of New Connectionism.

Even though Romanticism can be thought of as a musical revolution, it did not abandon the old forms completely. Instead, Romanticist composers adapted them, and explored them, for their own purposes. For example, consider the history of the symphony. In the early seventeenth century, the symphony was merely a short overture played before the raising of the curtains at an opera (Lee, 1916). Later, the more interesting of these compositions came to be performed to their own audiences outside the theatre. The modern symphony, which typically consists of four movements (each with an expected form and tempo), begins to be seen in the eighteenth-century compositions of Carl Philip Emmanuel Bach. Experiments with this structure were conducted in the later eighteenth century by Haydn and Mozart. When Beethoven wrote his symphonies in the early nineteenth century, the modern symphonic form was established—and likely perfected. "No less a person than Richard Wagner affirmed that the right of composing symphonies was abolished by Beethoven's Ninth'" (p. 172).

Beethoven is often taken to be the first Romantic composer because he also proved that the symphony had enormous expressive power. The Romantic composers who followed in his footsteps did not introduce dramatic changes in musical form;

rather they explored variations within this form in attempts to heighten its emotional expressiveness. "Strictly speaking, no doubt, musical Romanticism is more style than language" (Whittall, 1987, p. 17). Romantic composers developed radically new approaches to instrumentation, producing new tone colours (Ratner, 1992). The amount of sound was manipulated as an expressive tool; Romanticists increased "the compass, dynamic range, and timbral intensity of virtually all instruments" (Ratner, 1992, p. 9). New harmonic progressions were invented. But all of these expressive innovations involved relaxing, rather than replacing, classical conventions. "There can be little doubt that 'romantic' musical styles emanate from and comingle with 'classic' ones. There is no isolable time and place where one leaves off and the other begins" (Plantinga, 1984, p. 22).

Connectionist cognitive science has been portrayed as a revolution (Hanson & Olson, 1991) and as a paradigm shift (Schneider, 1987). However, it is important to remember that it, like musical Romanticism, also shares many of the characteristics of the classical school that it reacted against.

For instance, connectionists don't abandon the notion of information processing; they argue that the brain is just a different kind of information processor than is a digital computer (Churchland, Koch, & Sejnowski, 1990). Connectionists don't discard the need for representations; they instead offer different kinds, such as distributed representations (Hinton, McClelland, & Rumelhart, 1986). Connectionists don't dispose of symbolic accounts; they propose that they are approximations to subsymbolic regularities (Smolensky, 1988).

Furthermore, it was argued earlier in this book that connectionist cognitive science cannot be distinguished from classical cognitive science on many other dimensions, including the adoption of functionalism (Douglas & Martin, 1991) and the classical sandwich (Calvo & Gomila, 2008; Clark, 1997). When these two approaches are compared in the context of the multiple levels of investigation discussed in Chapter 2, there are many similarities between them:

> Indeed, the fact that the two can be compared in this way at all indicates a commitment to a common paradigm—an endorsement of the foundational assumption of cognitive science: cognition is information processing. (Dawson, 1998, p. 298)

Copland (1952, pp. 69–70) argued that the drama of European music was defined by two polar forces: "the pull of tradition as against the attraction of innovation." These competing forces certainly contributed to the contradictory variety found in musical Romanticism (Einstein, 1947); perhaps they too have shaped modern connectionist cognitive science. This issue can be explored by considering connectionist approaches to musical cognition and comparing them to the classical research on musical cognition that was described earlier in the current chapter.

6.4 The Connectionist Approach to Musical Cognition

Connectionist research on musical cognition is perhaps not as established as classical research, but it has nonetheless produced a substantial and growing literature (Bharucha, 1999; Fiske, 2004; Griffith & Todd, 1999; Todd & Loy, 1991). The purpose of this section is to provide a very brief orientation to this research. As the section develops, the relationship of connectionist musical cognition to certain aspects of musical Romanticism is illustrated.

By the late 1980s, New Connectionism had begun to influence research on musical cognition. The effects of this spreading influence have been documented in two collections of research papers (Griffith & Todd, 1999; Todd & Loy, 1991). Connectionist musical cognition has been studied with a wide variety of network architectures, and covers a broad range of topics, most notably classifying pitch and tonality, assigning rhythm and metre, classifying and completing melodic structure, and composing new musical pieces (Griffith & Todd, 1999).

Why use neural networks to study musical cognition? Bharucha (1999) provided five reasons. First, artificial neural networks can account for the learning of musical patterns via environmental exposure. Second, the type of learning that they describe is biologically plausible. Third, they provide a natural and biologically plausible account of contextual effects and pattern completion during perception. Fourth, they are particularly well suited to modelling similarity-based regularities that are important in theories of musical cognition. Fifth, they can discover regularities (e.g., in musical styles) that can elude more formal analyses.

To begin our survey of connectionist musical cognition, let us consider the artificial neural network classifications of pitch, tonality, and harmony (Griffith & Todd, 1999; Purwins et al., 2008). A wide variety of such tasks have been successfully explored: artificial neural networks have been trained to classify chords (Laden & Keefe, 1989; Yaremchuk & Dawson, 2005; Yaremchuk & Dawson, 2008), assign notes to tonal schema similar to the structures proposed by Krumhansl (1990) (Leman, 1991; Scarborough, Miller, & Jones, 1989), model the effects of expectation on pitch perception and other aspects of musical perception (Bharucha, 1987; Bharucha & Todd, 1989), add harmony to melodies (Shibata, 1991), determine the musical key of a melody (Griffith, 1995), and detect the chord patterns in a composition (Gjerdingen, 1992).

Artificial neural networks are well suited for this wide range of pitch-related tasks because of their ability to exploit contextual information, which in turn permits them to deal with noisy inputs. For example, networks are capable of pattern completion, which is replacing information that is missing from imperfect input patterns. In musical cognition, one example of pattern completion is virtual pitch (Terhardt, Stoll, & Seewann, 1982a, 1982b), the perception of pitches that are missing their fundamental frequency.

Consider a sine wave whose frequency is f. When we hear a musical sound, its pitch (i.e., its tonal height, or the note that we experience) is typically associated with this fundamental frequency (Helmholtz & Ellis, 1954; Seashore, 1967). The harmonics of this sine wave are other sine waves whose frequencies are integer multiples of f (i.e., $2f, 3f, 4f$ and so on). The timbre of the sound (whether we can identify a tone as coming from, for example, a piano versus a clarinet) is a function of the amplitudes of the various harmonics that are also audible (Seashore, 1967).

Interestingly, when a complex sound is filtered so that its fundamental frequency is removed, our perception of its pitch is not affected (Fletcher, 1924). It is as if the presence of the other harmonics provides enough information for the auditory system to fill in the missing fundamental, so that the correct pitch is heard—a phenomenon Schumann exploited in *Humoreske*. Co-operative interactions amongst neurons that detect the remaining harmonics are likely responsible for this effect (Cedolin & Delgutte, 2010; Smith et al., 1978; Zatorre, 2005).

Artificial neural networks can easily model such co-operative processing and complete the missing fundamental. For instance, one important connectionist system is called a Hopfield network (Hopfield, 1982, 1984). It is an autoassociative network that has only one set of processing units, which are all interconnected. When a pattern of activity is presented to this type of network, signals spread rapidly to all of the processors, producing dynamic interactions that cause the network's units to turn on or off over time. Eventually the network will stabilize in a least-energy state; dynamic changes in processor activities will come to a halt.

Hopfield networks can be used to model virtual pitch, because they complete the missing fundamental (Benuskova, 1994). In this network, each processor represents a sine wave of a particular frequency; if the processor is on, then this represents that the sine wave is present. If a subset of processors is activated to represent a stimulus that is a set of harmonics with a missing fundamental, then when the network stabilizes, the processor representing the missing fundamental will be also activated. Other kinds of self-organizing networks are also capable of completing the missing fundamental (Sano & Jenkins, 1989).

An artificial neural network's ability to deal with noisy inputs allows it to cope with other domains of musical cognition as well, such as assigning rhythm and metre (Desain & Honing, 1989; Griffith & Todd, 1999). Classical models of this type of processing hierarchically assign a structure of beats to different levels of a piece, employing rules that take advantage of the fact that musical rhythm and metre are associated with integer values (e.g., as defined by time signatures, or in the definition of note durations such as whole notes, quarter notes, and so on) (Lerdahl & Jackendoff, 1983; Temperley, 2001). However, in the actual performance of a piece, beats will be noisy or imperfect, such that perfect integer ratios of beats will not occur (Gasser, Eck, & Port, 1999). Connectionist models can correct for this problem, much as networks can restore absent information such as the missing fundamental.

For example, one network for assigning rhythm and metre uses a system of oscillating processors, units that fire at a set frequency (Large & Kolen, 1994). One can imagine having available a large number of such oscillators, each representing a different frequency. While an oscillator's frequency of activity is constant, its phase of activity can be shifted (e.g., to permit an oscillator to align itself with external beats of the same frequency). If the phases of these processors can also be affected by co-operative and competitive interactions between the processors themselves, then the phases of the various components of the system can become entrained. This permits the network to represent the metrical structure of a musical input, even if the actual input is noisy or imperfect. This notion can be elaborated in a self-organizing network that permits preferences for, or expectancies of, certain rhythmic patterns to determine the final representation that the network converges to (Gasser Eck, & Port, 1999).

The artificial neural network examples provided above illustrate another of Bharucha's (1999) advantages of such models: biological plausibility. Many neural network models are attempts to simulate some aspects of neural accounts of auditory and musical perception. For instance, place theory is the proposal that musical pitch is represented by places of activity along the basilar membrane in the cochlea (Helmholtz & Ellis, 1954; von Bekesy, 1928). The implications of place theory can be explored by using it to inspire spatial representations of musical inputs to connectionist networks (Sano & Jenkins, 1989).

The link between connectionist accounts and biological accounts of musical cognition is not accidental, because both reflect reactions against common criticisms. Classical cognitive scientist Steven Pinker is a noted critic of connectionist cognitive science (Pinker, 2002; Pinker & Prince, 1988). Pinker (1997) has also been a leading proponent of massive modularity, which ascribes neural modules to most cognitive faculties—except for music. Pinker excluded music because he could not see any adaptive value for its natural selection: "As far as biological cause and effect are concerned, music is useless. It shows no signs of design for attaining a goal such as long life, grandchildren, or accurate perception and prediction. of the world" (p. 528). The rise of modern research in the cognitive neuroscience of music (Cedolin & Delgutte, 2010; Peretz & Coltheart, 2003; Peretz & Zatorre, 2003; Purwins et al., 2008; Stewart et al., 2006; Warren, 2008) is a reaction against this classical position, and finds a natural ally in musical connectionism.

In the analogy laid out in the previous section, connectionism's appeal to the brain was presented as an example of its Romanticism. Connectionist research on musical cognition reveals other Romanticist parallels. Like musical Romanticism, connectionism is positioned to capture regularities that are difficult to express in language or by using formal rules (Loy, 1991).

For example, human subjects can accurately classify short musical selections into different genres or styles in a remarkably short period of time, within a quarter of a second (Gjerdingen & Perrott, 2008). But it is difficult to see how one could provide a classical account of this ability because of the difficulty in formally defining a genre or style for a classical model. "It is not likely that musical styles can be isolated successfully by simple heuristics and introspection, nor can they be readily modeled as a rule-solving problem" (Loy, 1991, p. 31).

However, many different artificial neural networks have been developed to classify music using categories that seem to defy precise, formal definitions. These include networks that can classify musical patterns as belonging to the early works of Mozart (Gjerdingen, 1990); classify selections as belonging to different genres of Western music (Mostafa & Billor, 2009); detect patterns of movement between notes in segments of music (Gjerdingen, 1994) in a fashion similar to a model of apparent motion perception (Grossberg & Rudd, 1989, 1992); evaluate the affective aesthetics of a melody (Coutinho & Cangelosi, 2009; Katz, 1995); and even predict the possibility that a particular song has "hit potential" (Monterola et al., 2009).

Categories such as genre or hit potential are obviously vague. However, even identifying a stimulus as being a particular song or melody may also be difficult to define formally. This is because a melody can be transposed into different keys, performed by different instruments or voices, or even embellished by adding improvisational flourishes.

Again, melody recognition can be accomplished by artificial neural networks that map, for instance, transposed versions of the same musical segment onto a single output representation (Benuskova, 1995; Bharucha & Todd, 1989; Page, 1994; Stevens & Latimer, 1992). Neural network melody recognition has implications for other aspects of musical cognition, such as the representational format for musical memories. For instance, self-organizing networks can represent the hierarchical structure of a musical piece in an abstract enough fashion so that only the "gist" is encoded, permitting the same memory to be linked to multiple auditory variations (Large, Palmer, & Pollack, 1995). Auditory processing organizes information into separate streams (Bregman, 1990); neural networks can accomplish this for musical inputs by processing relationships amongst pitches (Grossberg, 1999).

The insights into musical representation that are being provided by artificial neural networks have important implications beyond musical cognition. There is now wide availability of music and multimedia materials in digital format. How can such material be classified and searched? Artificial neural networks are proving to be useful in addressing this problem, as well as for providing adaptive systems for selecting music, or generating musical playlists, based on a user's mood or past preferences (Bugatti, Flammini, & Migliorati, 2002; Jun, Rho, & Hwang, 2010; Liu, Hsieh, & Tsai, 2010; Muñoz-Expósito et al., 2007).

Musical styles, or individual musical pieces, are difficult to precisely define, and therefore are problematic to incorporate into classical theories. "The fact that even mature theories of music are informal is strong evidence that the performer, the listener, and the composer do not operate principally as rule-based problem solvers" (Loy, 1991, p. 31). That artificial neural networks are capable of classifying music in terms of such vague categories indicates that "perhaps connectionism can show the way to techniques that do not have the liabilities of strictly formal systems" (p. 31). In other words, the flexibility and informality of connectionist systems allows them to cope with situations that may be beyond the capacity of classical models. Might not this advantage also apply to another aspect of musical cognition, composition?

Composition has in fact been one of the most successful applications of musical connectionism. A wide variety of composing networks have been developed. Networks have been developed to compose single-voiced melodies on the basis of learned musical structure (Mozer, 1991; Todd, 1989); to compose harmonized melodies or multiple-voice pieces (Adiloglu & Alpaslan, 2007; Bellgard & Tsang, 1994; Hoover & Stanley, 2009; Mozer, 1994); to learn jazz melodies and harmonies, and then to use this information to generate new melodies when presented with novel harmonies (Franklin, 2006); and to improvise by composing variations on learned melodies (Nagashima & Kawashima, 1997). The logic of network composition is that the relationship between successive notes in a melody, or between different notes played at the same time in a harmonized or multiple-voice piece, is not random, but is instead constrained by stylistic, melodic, and acoustic constraints (Kohonen et al., 1991; Lewis, 1991; Mozer, 1991, 1994). Networks are capable of learning such constraints and using them to predict, for example, what the next note should be in a new composition.

In keeping with musical Romanticism, however, composing networks are presumed to have internalized constraints that are difficult to formalize or to express in ordinary language. "Nonconnectionist algorithmic approaches in the computer arts have often met with the difficulty that 'laws' of art are characteristically fuzzy and ill-suited for algorithmic description" (Lewis, 1991, p. 212). Furthermore these "laws" are unlikely to be gleaned from analyzing the internal structure of a network, "since the hidden units typically compute some complicated, often uninterpretable function of their inputs" (Todd, 1989, p. 31). It is too early to label a composing network as an isolated genius, but it would appear that these networks are exploiting regularities that are in some sense sublime!

This particular parallel between musical Romanticism and connectionism, that both capture regularities that cannot be formalized, is apparent in another interesting characteristic of musical connectionism. The most popular algorithm for training artificial neural networks is the generalized delta rule (i.e., error backpropagation) (Chauvin & Rumelhart, 1995; Widrow & Lehr, 1990), and networks trained with this kind of supervised learning rule are the most likely to be found in the cognitive science literature. While self-organizing networks are present in this literature and have made important contributions to it (Amit, 1989; Carpenter & Grossberg, 1992; Grossberg, 1988; Kohonen, 1984, 2001), they are much less popular. However, this does not seem to be the case in musical connectionism.

For example, in the two collections that document advances in artificial neural network applications to musical cognition (Griffith & Todd, 1999; Todd & Loy, 1991), 23 papers describe new neural networks. Of these contributions, 9 involve supervised learning, while 14 describe unsupervised, self-organizing networks. This indicates a marked preference for unsupervised networks in this particular connectionist literature.

This preference is likely due to the view that supervised learning is not practical for musical cognition, either because many musical regularities can be acquired without feedback or supervision (Bharucha, 1991) or because for higher-level musical tasks the definition of the required feedback is impossible to formalize (Gjerdingen, 1989). "One wonders, for example, if anyone would be comfortable in claiming that one interpretation of a musical phrase is only 69 percent [as] true as another" (p. 67). This suggests that the musical Romanticism of connectionism is even reflected in its choice of network architectures.

6.5 The Embodied Nature of Modern Music[1]

European classical music is innovation constrained by tradition (Copland, 1952). By the end of the nineteenth century, composers had invented a market for instrumental music by refining established musical conventions (Rosen, 1988). "The European musician is forced into the position of acting as caretaker and preserver of other men's music, whether he likes it or no" (Copland, 1952, p. 69).

What are the general characteristics of European classical music? Consider the sonata-allegro form, which is based upon particular musical themes or melodies that are associated with a specific tonality. That is, they are written in a particular musical key. This tonality dictates harmonic structure; within a musical key, certain notes or chords will be consonant, while others will not be played because of their dissonance. The sonata-allegro form also dictates an expected order in which themes and musical keys are explored and a definite time signature to be used throughout.

---

[1] Much of the text in this section has been adapted from the second chapter of Dawson, Dupuis, and Wilson (2010).

The key feature from above is tonality, the use of particular musical keys to establish an expected harmonic structure. "Harmony is Western music's uniquely distinguishing element" (Pleasants, 1955, p. 97). It was a reaction against this distinguishing characteristic that led to what is known as modern music (Griffiths, 1994, 1995; Ross, 2007). This section further explores the analogy between classical music and cognitive science via parallels between modern music and embodied cognitive science.

In the early twentieth century, classical music found itself in a crisis of harmony (Pleasants, 1955). Composers began to abandon most of the characteristics of traditional European classical music in an attempt to create a new music that better reflected modern times. "'Is it not our duty,' [Debussy] asked, 'to find a symphonic means to express our time, one that evokes the progress, the daring and the victories of modern days? The century of the aeroplane deserves its music'" (Griffiths, 1994, p. 98).

Modern music is said to have begun with the *Prélude à L'après-midi d'un faune* composed by Claude Debussy between 1892 and 1894 (Griffiths, 1994). The *Prélude* breaks away from the harmonic relationships defined by strict tonality. It fails to logically develop themes. It employs fluctuating tempos and irregular rhythms. It depends critically on instrumentation for expression. Debussy "had little time for the thorough, continuous, symphonic manner of the Austro-German tradition, the 'logical' development of ideas which gives music the effect of a narrative" (p. 9).

> Debussy had opened the paths of modern music—the abandonment of traditional
> tonality, the development of new rhythmic complexity, the recognition of color
> as an essential, the creation of a quite new form for each work, the exploration of
> deeper mental processes. (Griffiths, 1994, p. 12)

In the twentieth century, composers experimented with new methods that further pursued these paths and exploited notions related to emergence, embodiment, and stigmergy.

To begin, let us consider how modern music addressed the crisis of harmony by composing deliberately atonal music. The possibility of atonality in music emerges from the definition of musical tonality. In Western music there are 12 possible notes available. If all of these notes are played in order from lowest to highest, with each successive note a semitone higher than the last, the result is a chromatic scale.

Different kinds of scales are created by invoking constraints that prevent some notes from being played, as addressed in the Chapter 4 discussion of jazz progressions. A major scale is produced when a particular set of 7 notes is played, and the remaining 5 notes are not played. Because a major scale does not include all of the notes in a chromatic scale, it has a distinctive sound—its tonality. A composition that had the tonal centre of A major only includes those notes that belong to the A-major scale.

This implies that what is required to produce music that is atonal is to include *all* of the notes from the chromatic scale. If all notes were included, then it would be impossible to associate this set of notes with a tonal centre. One method of ensuring atonality is the "twelve-tone technique," or dodecaphony, invented by Arnold Schoenberg.

When a dodecaphony is employed, a composer starts by listing all twelve possible notes in some desired order, called the tone row. The tone row is the basis for a melody: the composer begins to write the melody by using the first note in the tone row, for a desired duration, possibly with repetition. However, this note cannot be reused in the melody until the remaining notes have also been used in the order specified by the tone row. This ensures that the melody is atonal, because all of the notes that make up a chromatic scale have been included. Once all twelve notes have been used, the tone row is used to create the next section of the melody. At this time, it can be systematically manipulated to produce musical variation.

The first dodecaphonic composition was Schoenberg's 1923 *Suite for Piano, Op. 25*. Schoenberg and his students Alban Berg and Anton Webern composed extensively using the twelve-note technique. A later musical movement called serialism used similar systems to determine other parameters of a score, such as note durations and dynamics. It was explored by Olivier Messiaen and his followers, notably Pierre Boulez and Karlheinz Stockhausen (Griffiths, 1995).

Dodecaphony provided an alternative to the traditional forms of classical music. However, it still adhered to the Austro-German tradition's need for structure. Schoenberg invented dodecaphony because he needed a system to compose larger-scale atonal works; prior to its invention he was "troubled by the lack of system, the absence of harmonic bearings on which large forms might be directed. Serialism at last offered a new means of achieving order" (Griffiths, 1994, p. 81).

A new generation of American composers recognized that dodecaphony and serialism were still strongly tied to musical tradition: "To me, it was music of the past, passing itself off as music of the present" (Glass, 1987, p. 13). Critics accused serialist compositions of being mathematical or mechanical (Griffiths, 1994), and serialism did in fact make computer composition possible: in 1964 Gottfried Koenig created Project 1, which was a computer program that composed serial music (Koenig, 1999).

Serialism also shared the traditional approach's disdain for the audience. American composer Steve Reich (1974, p. 10) noted that "in serial music, the series itself is seldom audible," which appears to be a serial composer's intent (Griffiths, 1994). Bernstein (1976, p. 273) wrote that Schoenberg "produced a music that was extremely difficult for the listener to follow, in either form or content." This music's opacity, and its decidedly different or modern sound, frequently led to hostile receptions. One notable example is *The Agony of Modern Music*:

The vein for which three hundred years offered a seemingly inexhaustible yield of beautiful music has run out. What we know as modern music is the noise made by deluded speculators picking through the slag pile. (Pleasants, 1955, p. 3)

That serial music was derived from a new kind of formalism also fuelled its critics.

Faced with complex and lengthy analyses, baffling terminology and a total rejection of common paradigms of musical expression, many critics—not all conservative—found ample ammunition to back up their claims that serial music was a mere intellectual exercise which could not seriously be regarded as music at all. (Grant, 2001, p. 3)

Serialism revealed that European composers had difficulty breaking free of the old forms even when they recognized a need for new music (Griffiths, 1994). Schoenberg wrote, "I am at least as conservative as Edison and Ford have been. But I am, unfortunately, not quite as progressive as they were in their own fields" (Griffiths, 1995, p. 50).

American composers rejected the new atonal structures (Bernstein, 1976). Phillip Glass described his feelings about serialism so: "A wasteland, dominated by these maniacs, these creeps, who were trying to make everyone write this crazy creepy music" (Schwarz, 1996). When Glass attended concerts, the only "breaths of fresh air" that he experienced were when works from modern American composers such as John Cage were on the program (Glass, 1987). Leonard Bernstein (1976, p. 273) wrote that "free atonality was in itself a point of no return. It seemed to fulfill the conditions for musical progress. . . . But then: a dead end. Where did one go from here?" The new American music was more progressive than its European counterpart because its composers were far less shackled by musical traditions.

For instance, American composers were willing to relinquish the central control of the musical score, recognizing the improvisational elements of classical composition (Benson, 2003). Some were even willing to surrender the composer's control over the piece (Cage, 1961), recognizing that many musical effects depended upon the audience's perceptual processes (Potter, 2000; Schwarz, 1996). It was therefore not atonality itself but instead the American reaction to it that led to a classical music with clear links to embodied cognitive science.

Consider, for instance, the implications of relinquishing centralized control in modern music. John Cage was largely motivated by his desire to free musical compositions from the composer's will. He wrote that "when silence, generally speaking, is not in evidence, the will of the composer is. Inherent silence is equivalent to denial of the will" (Cage, 1961, p. 53). Cage's most famous example of relinquishing control is in his "silent piece," *4'33"*, first performed by pianist David Tudor in 1952 (Nyman, 1999). It consists of three parts; the entire score for each part reads "TACET," which instructs the performer to remain silent. Tudor signaled the start of each part by closing the keyboard lid, and opened the lid when the part was over.

4'33" places tremendous compositional responsibility upon its audience. Cage is quoted on this subject as saying:

> Most people think that when they hear a piece of music, they're not doing anything but something is being done to them. Now this is not true, and we must arrange our music, we must arrange our art, we must arrange everything, I believe, so that people realize that they themselves are doing it. (Nyman, 1999, p. 24)

This is contrary to the traditional disembodiment of classical music that treats audiences as being passive and unimportant.

Cage pioneered other innovations as he decentralized control in his compositions. From the early 1950s onwards, he made extended use of chance operations when he composed. Cage used dice rolls to determine the order of sounds in his 1951 piano piece *Music of Changes* (Ross, 2007). The stochastic nature of Cage's compositional practices did not produce music that sounded random. This is because Cage put tremendous effort into choosing interesting sound elements. "In the *Music of Changes* the effect of the chance operations on the structure (making very apparent its anachronistic character) was balanced by a control of the materials" (Cage, 1961, p. 26). Cage relaxed his influence on control—that is, upon which element to perform next—with the expectation that this, coupled with his careful choice of elements that could be chosen, would produce surprising and interesting musical results. Cage intended novel results to *emerge* from his compositions.

The combination of well-considered building blocks to produce emergent behaviours that surprise and inform is characteristic of embodied cognitive science (Braitenberg, 1984; Brooks, 1999; Dawson, 2004; Dawson, Dupuis, & Wilson, 2010; Pfeifer & Scheier, 1999; Webb & Consi, 2001).

> Advances in synthetic psychology come about by taking a set of components, by letting them interact, and by observing surprising emergent phenomena. However, the role of theory and prior knowledge in this endeavor is still fundamentally important, because it guides decisions about what components to select, and about the possible dynamics of their interaction. In the words of Cervantes, diligence is the mother of good luck. (Dawson, 2004, p. 22)

An emphasis on active audiences and emergent effects is also found in the works of other composers inspired by Cage (Schwarz, 1996). For instance, compositions that incorporated sounds recorded on magnetic tape were prominent in early minimalist music. Minimalist pioneer Terry Riley began working with tape technology in 1960 (Potter, 2000). He recorded a variety of sounds and made tape loops from them. A tape loop permitted a sound segment to be repeated over and over. He then mixed these tapes using a device called an echoplex that permitted the sounds "to be repeated in an ever-accumulating counterpoint against itself" (p. 98). Further complexities of sound were produced by either gradually or suddenly changing the speed of the tape to distort the tape loop's frequency. Riley's tape loop experiments led him to explore the effects of repetition, which was to become a centrally important feature of minimalist music.

Riley's work strongly influenced other minimalist composers. One of the most famous minimalist tape compositions is Steve Reich's 1965 *It's Gonna Rain*. Reich recorded a sermon of a famous street preacher, Brother Walter, who made frequent Sunday appearances in San Francisco's Union Square. From this recording, Reich made a tape loop of a segment of the sermon that contained the title phrase. Reich (2002) played two copies of this tape loop simultaneously on different tape machines, and made a profound discovery:

> In the process of trying to line up two identical tape loops in some particular relationship, I discovered that the most interesting music of all was made by simply lining the loops up in unison, and letting them slowly shift out of phase with each other. (Reich, 2002, p. 20)

He recorded the result of phase-shifting the loops, and composed his piece by phase-shifting a loop of this recording. Composer Brian Eno describes Reich's *It's Gonna Rain* thus:

> The piece is very, very interesting because it's tremendously simple. It's a piece of music that anybody could have made. But the results, sonically, are very complex. . . . What you become aware of is that you are getting a huge amount of material and experience from a very, very simple starting point. (Eno, 1996)

The complexities of *It's Gonna Rain* emerge from the dynamic combination of simple components, and thus are easily linked to the surrender of control that was begun by John Cage. However, they also depend to a large extent upon the perceptual processes of a listener when confronted with the continuous repetition of sound fragments. "The mind is mesmerized by repetition, put into such a state that small motifs can leap out of the music with a distinctness quite unrelated to their acoustic dominance" (Griffiths, 1994, p. 167). From a perceptual point of view, it is impossible to maintain a constant perception of a repeated sound segment. During the course of listening, the perceptual system will habituate to some aspects of it, and as a result—as if by chance—new regularities will emerge. "The listening experience itself can become aleatory in music[,] subject to 'aural illusions'" (p. 166).

Minimalism took advantage of the active role of the listener and exploited repetition to deliberately produce aural illusions. The ultimate effect of a minimalist composition is not a message created by the composer and delivered to a (passive) audience, but is instead a collaborative effort between musician and listener. Again, this mirrors the interactive view of world and agent that characterizes embodied cognitive science and stands opposed to the disembodied stance taken by both Austro-German music and classical cognitive science.

Minimalism became lastingly important when its composers discovered how their techniques, such as decentralized control, repetition, and phase shifting, could be communicated using a medium that was more traditional than tape loops. This was accomplished when Terry Riley realized that the traditional musical score could be reinvented to create minimalist music. Riley's 1964 composition *In C* is 53 bars of music written in the key of C major, indicating a return to tonal music. Each bar is extremely simple; the entire score fits onto a single page. Performers play each bar in sequence. However, they repeat a bar as many times as they like before moving on to the next. When they reach the final bar, they repeat it until all of the other performers have reached it. At that time, the performance is concluded.

Riley's *In C* can be thought of as a tape loop experiment realized as a musical score. Each performer is analogous to one of the tape loops, and the effect of the music arises from their interactions with one another. The difference, of course, is that each "tape loop" is not identical to the others, because each performer controls the number of times that they repeat each bar. Performers listen and react to *In C* as they perform it.

There are two compelling properties that underlie a performance of *In C*. First, each musician is an independent agent who is carrying out a simple act. At any given moment each musician is performing one of the bars of music. Second, what each musician does at the next moment is affected by the musical environment that the ensemble of musicians is creating. A musician's decision to move from one bar to the next depends upon what they are hearing. In other words, the musical environment being created is literally responsible for controlling the activities of the agents who are performing *In C*. This is a musical example of a concept that we discussed earlier as central to embodied cognitive science: stigmergy.

In stigmergy, the behaviours of agents are controlled by an environment in which they are situated, and which they also can affect. The performance of a piece like *In C* illustrates stigmergy in the sense that musicians decide what to play next on the basis of what they are hearing right now. Of course, what they decide to play will form part of the environment, and will help guide the playing decisions of other performers.

The stigmergic nature of minimalism contrasts with the classical ideal of a composer transcribing mental contents. One cannot predict what *In C* will sound like by examining its score. Only an actual performance will reveal what *In C*'s score represents. Reich (1974, p. 9) wrote: "Though I may have the pleasure of discovering musical processes and composing the musical material to run through them, once the process is set up and loaded it runs by itself."

Reich's idea of a musical process running by itself is reminiscent of synthetic psychology, which begins by defining a set of primitive abilities for an agent. Typically there are nonlinear interactions between these building blocks, and between the building blocks and the environment. As a result, complex and interesting behaviours emerge—results that far exceed behavioural predictions based on knowing the agent's makeup (Braitenberg, 1984). Human intelligence is arguably the emergent product of simple, interacting mental agents (Minsky, 1985). The minimalists have tacitly adopted this view and created a mode of composition that reflects it.

The continual evolution of modern technology has had a tremendous impact on music. Some of this technology has created situations in which musical stigmergy is front and centre. For example, consider a computer program called Swarm Music (Blackwell, 2003). In Swarm Music, there are one or more swarms of "particles." Each particle is a musical event: it exists in a musical space where the coordinates of the space define musical parameters such as pitch, duration, and loudness, and the particle's position defines a particular combination of these parameters. A swarm of particles is dynamic, and it is drawn to attractors that are placed in the space. The swarm can thus be converted into music. "The swarming behavior of these particles leads to melodies that are not structured according to familiar musical rules, but are nevertheless neither random nor unpleasant" (Blackwell & Young, 2004).

Swarm Music is made dynamic by coupling it with human performers in an improvised and stigmergic performance. The sounds created by the human performers are used to revise the positions of the attractors for the swarms, causing the music generated by the computer system to change in response to the other performers. The human musicians then change their performance in response to the computer.

Performers who have improvised with Swarm Music are affected by its stigmergic nature. Jazz singer Kathleen Willison,

> was surprised to find in the first improvisation that Swarm Music seemed to
> be imitating her: '(the swarm) hit the same note at the same time—the harmonies worked.' However, there was some tension; 'at times I would have liked it
> to slow down . . . it has a mind of its own . . . give it some space.' Her solution to
> the 'forward motion' of the swarms was to 'wait and allow the music to catch up'.
> (Blackwell, 2003, p. 47)

Another new technology in which musical stigmergy is evident is the reacTable (Jordà et al., 2007; Kaltenbrunner et al., 2007). The reacTable is an electronic synthesizer that permits several different performers to play it at the same time. The reacTable is a circular, translucent table upon which objects can be placed. Some objects generate waveforms, some perform algorithmic transformations of their inputs, and some control others that are nearby. Rotating an object, and using a fingertip to manipulate a visual interface that surrounds it, modulates a musical process (i.e., changes the frequency and amplitude of a sine wave). Visual signals displayed on the reacTable—and visible to all performers—indicate the properties of the musical event produced by each object as well as the flow of signals from one object to another.

The reacTable is an example of musical stigmergy because when multiple performers use it simultaneously, they are reacting to the existing musical events. These events are represented as physical locations of objects on the reacTable itself, the visual signals emanating from these objects, and the aural events that the reacTable is producing. By co-operatively moving, adding, or removing objects, the musicians collectively improvise a musical performance. The reacTable is an interface intended to provide a "combination of intimate and sensitive control, with a more macro-structural and higher level control which is intermittently shared, transferred and recovered between the performer(s) and the machine" (Jordà et al., 2007, p. 145). That is, the reacTable—along with the music it produces—provides control analogous to that provided by the nest-in-progress of an insect colony.

From the preceding discussion, we see that modern music shares many characteristics with the embodied reaction to classical cognitive science. With its decentralization of control, responsibility for the composition has "leaked" from the composer's mind. Its definition also requires contributions from both the performers and the audience, and not merely a score. This has implications for providing accounts of musical meaning, or of the goals of musical compositions. The classical notion of music communicating intended meanings to audiences is not easily applied to modern music.

Classical cognitive science's view of communication is rooted in cybernetics (Shannon, 1948; Wiener, 1948), because classical cognitive science arose from exploring key cybernetic ideas in a cognitivist context (Conrad, 1964b; Leibovic, 1969; Lindsay & Norman, 1972; MacKay, 1969; Selfridge, 1956; Singh, 1966). As a result, the cybernetic notion of communication—transfer of information from one location to another—is easily found in the classical approach.

The classical notion of communication is dominated by the conduit metaphor (Reddy, 1979). According to the conduit metaphor, language provides containers (e.g., sentences, words) that are packed with meanings and delivered to receivers, who unpack them to receive the intended message. Reddy provides a large number of examples of the conduit metaphor, including: "You still haven't *given me any idea* of what you mean"; "You have to *put each concept into words* very carefully"; and "The *sentence was filled with emotion.*"

The conduit metaphor also applies to the traditional view of classical music, which construes this music as a "hot medium" to which the listener contributes little (McLuhan, 1994): the composer places some intended meaning into a score, the orchestra brings the score to life exactly as instructed by the score, and the (passive) audience unpacks the delivered music to get the composer's message.

> We thus hear people say that music can only have meaning if it is seen to be a type of language, with elements akin to words, phrases and sentences, and with elements that refer beyond themselves to extramusical things, events, or ideas. (Johnson, 2007, p. 207)

In other words, the classical view of musical meaning is very similar to the view of meaning espoused by classical cognitive science: music is a symbolic, intentional medium.

The view of music as a symbolic medium that conveys intended meaning has generated a long history of resistance. The autonomist school of aesthetics (see Hanslick, 1957) argued against the symbolic theories of musical meaning, as well as against theories that music communicated emotion. Hanslick's (1957) position was that music was a medium whose elements were pure and nonrepresentational. Hanslick famously argued that "the essence of music is sound and motion" (p. 48). Modern positions that treat musical meaning in an embodied fashion are related to Hanslick's (Johnson, 2007; Leman, 2008).

Embodied alternatives to musical meaning become attractive because the conduit metaphor breaks down in modern music. If control is taken away from the score and the conductor, if the musicians become active contributors to the composition (Benson, 2003), if the audience is actively involved in completing the composition as well, and if music is actually a "cool medium," then what is the intended message of the piece?

Modern embodied theories of music answer this question by taking a position that follows naturally from Hanslick's (1957) musical aesthetics. They propose that the sound and motion of music literally have bodily effects that are meaningful. For instance, Johnson (2007) noted that,

> to hear music is just to be moved and to feel in the precise way that is defined by the patterns of musical motion. Those feelings are meaningful in the same way that any pattern of emotional flow is meaningful to us at a pre-reflective level of awareness. (Johnson, 2007, p. 239)

Similarly, Leman (2008, p. 17) suggested that "moving sonic forms do something with our bodies, and therefore have a signification through body action rather than through thinking." Some implications of this position are considered in the next section.

Minimalist composers themselves adopt a McLuhanesque view of the meaning of their compositions: the music doesn't deliver a message, but is itself the message. After being schooled in the techniques of serialism, which deliberately hid the underlying musical structures from the audience's perception, the minimalists desired to create a different kind of composition. When presented minimalist compositions, the audience would hear the musical processes upon which the pieces were built. Reich (2002, p. 34) said he was "interested in perceptible processes. I want to be able to hear the process happening throughout the sounding music."

Reich made processes perceptible by making them gradual. But this didn't make his compositions less musical.

> Even when all the cards are on the table and everyone hears what is gradually happening in a musical process, there are still enough mysteries to satisfy all. These mysteries are the impersonal, unintended, psychoacoustic by-products of the intended process. (Reich, 2002, p. 35)

Reich's recognition that the listener contributes to the composition—that classical music is a cool medium, not a hot one—is fundamental to minimalist music. Philip Glass (1987) was surprised to find that he had different experiences of different performances of Samuel Beckett's *Play*, for which Glass composed music. He realized that "Beckett's *Play* doesn't exist separately from its relationship to the viewer, who is included as part of the play's content" (p. 36). Audiences of Glass' *Einstein on the Beach* had similar experiences. "The point about *Einstein* was clearly not what it 'meant' but that it was *meaningful* as generally experienced by the people who saw it" (p. 33).

Modern music has many parallels to embodied cognitive science, and has many characteristics that distinguish it from other traditions of classical music. Alternative views of composition, the role of the audience, and the control of a performance are clearly analogous to embodied concepts such as emergence, embodiment, and stigmergy. They also lead to a very different notion of the purpose of music, in its transition from "hot" to "cool." Not surprisingly, the radical differences between classical and modern music are reflected in differences between classical and embodied cognitive science's study of musical cognition, as is discussed in the next section.

6.6 The Embodied Approach to Musical Cognition

A well-established modern view of classical music is that it has meaning, and that its purpose is to convey this meaning in a fashion that is consistent with Reddy's (1979) conduit metaphor.

> Composers and performers of all cultures, theorists of diverse schools and styles, aestheticians and critics of many different persuasions are all agreed that music has

> meaning and that this meaning is somehow communicated to both participants and listeners. (Meyer, 1956, p. 1)

Furthermore, there is a general consensus that the meaning that is communicated is affective, and not propositional, in nature. However, the means by which musical meaning is communicated is subject to a tremendous amount of debate (Meyer, 1956; Robinson, 1997).

One view of musical communication, consistent with classical cognitive science, is that music is a symbol system. For example, the semiotic view of music is that it is a system of signs that provides a narrative or a discourse (Agawu, 1991, 2009; Austerlitz, 1983; Lidov, 2005; Monelle, 2000; Pekkilä, Neumeyer, & Littlefield, 2006; Tarasti, 1995; Turino, 1999). From this perspective, musical signs are intentional: they are about the tensions or emotions they produce or release in listeners. This approach naturally leads to an exploration of the parallels between music and language (Austerlitz, 1983; Jackendoff, 2009; Lidov, 2005), as well as to the proposal of generative grammars of musical structure (Lerdahl, 2001; Lerdahl & Jackendoff, 1983; Sundberg & Lindblom, 1976). Potential parallels between language and music have led some researchers to describe brain areas for syntax and semantics that are responsible for processing both music and language (Koelsch et al., 2004; Patel, 2003).

A related view of musical communication, but one more consistent with connectionist than classical cognitive science, is that music communicates emotion but does so in a way that cannot be captured by set of formal rules or laws (Lewis, 1991; Loy, 1991; Minsky, 1981; Todd, 1989). Instead, musical meanings are presumed to be entwined in a complex set of interactions between past experiences and current stimulation, interactions that may be best captured by the types of learning exhibited by artificial neural networks. "Many musical problems that resist formal solutions may turn out to be tractable anyway, in future simulations that grow artificial musical semantic networks" (Minsky, 1981, p. 35).

Both views of musical meaning described above are consistent with the conduit metaphor, in that they agree that (1) music is intentional and content-bearing (although they disagree about formalizing this content) and (2) that the purpose of music is to communicate this content to audiences. A third approach to musical meaning, most consistent with embodied cognitive science, distinguishes itself from the other two by rejecting the conduit metaphor.

According to the embodied view (Clarke, 2005; Johnson, 2007; Leman, 2008), the purpose of music is not to acquire abstract or affective content, but to instead directly, interactively, and physically experience music. "People try to be involved with music because this involvement permits an experience of behavioral resonance with physical energy" (Leman, 2008, p. 4).

The emphasis on direct contact that characterizes the embodied view of music is a natural progression from the autonomist school of musical aesthetics that arose in the nineteenth century (Hanslick, 1957). Music critic Eduard Hanslick (1957) opposed the view that music was representative and that its purpose was to communicate content or affect. For Hanslick, a scientific aesthetics of music was made impossible by sentimental appeals to emotion: "The greatest obstacle to a scientific development of musical aesthetics has been the undue prominence given to the action of music on our feelings" (p. 89).

As noted previously, Hanslick (1957, p. 48) argued instead that "the essence of music is sound and motion." The modern embodied approach to music echoes and amplifies this perspective. Johnson (2007) agreed with Hanslick that music is not typically representative or intentional. Instead, Johnson argued that the dynamic nature of music—its motion, in Hanslick's sense—presents "the flow of human experience, feeling, and thinking in concrete, embodied forms" (p. 236). The motion of music is not communicative, it is causal. "To hear the music is just to be moved and to feel in the precise way that is defined by the patterns of the musical motion" (p. 239). The motion intrinsic to the structure of music is motion that we directly and bodily experience when it is presented to us. Johnson argues that this is why metaphors involving motion are so central to our conceptualization of music.

"Many people try to get into direct contact with music. Why do they do so? Why do people make great efforts to attend a concert? Why do they invest so much time in learning to play a musical instrument" (Leman, 2008, p. 3). If the meaning of music is the felt movement that it causes, then the need for direct experience of music is completely understandable. This is also reflected in an abandonment of the conduit metaphor. The embodied view of music does not accept the notion that music is a conduit for the transmission of propositional or affective contents. Indeed, it hypothesizes that the rational assessment of music might interfere with how it should best be experienced.

> Activities such as reasoning, interpretation, and evaluation may disturb the feeling of being directly involved because the mind gets involved in a representation of the state of the environment, which distracts the focus and, as a result, may break the 'magic spell' of being entrained. (Leman, 2008, p. 5)

Clearly embodied researchers have a very different view of music than do classical or connectionist researchers. This in turn leads to very different kinds of research on musical cognition than the examples that have been introduced earlier in this chapter.

To begin, let us consider the implication of the view that listeners should be directly involved with music (Leman, 2008). From this view, it follows that the full appreciation of music requires far more than the cognitive interpretation of auditory stimulation. "It is a matter of corporeal immersion in sound energy, which is a direct way of feeling musical reality. It is less concerned with cognitive reflection, evaluation, interpretation, and description" (Leman, 2008, p. 4). This suggests that cross-modal interactions may be critical determinants of musical experience.

Some research on musical cognition is beginning to explore this possibility. In one study (Vines et al., 2006) subjects were presented with performances by two clarinetists. Some subjects only heard, some subjects only saw, and some subjects both heard and saw the performances. Compared to the first two groups of subjects, those who both heard and saw the performances had very different experiences. The visual information altered the experience of tension at different points, and the movements of the performers provided additional information that affected the experienced phrasing as well as expectations about emotional content. "The auditory and visual channels mutually enhance one another to convey content, and . . . an emergent quality exists when a musician is both seen and heard" (p. 108).

In a more recent study, Vines et al. (2011) used a similar methodology, but they also manipulated the expressive style with which the stimulus (a solo clarinet piece composed by Stravinsky) was performed. Subjects were presented with the piece in restrained, standard, or exaggerated fashion. These manipulations of expressive style only affected the subjects who could see the performance. Again, interactions were evident when performances were both seen and heard. For instance, subjects in this condition had significantly higher ratings of "happiness" in comparison to other subjects.

> The visual component of musical performance makes a unique contribution to the communication of emotion from performer to audience. Seeing a musician can augment, complement, and interact with the sound to modify the overall experience of music. (Vines et al., 2011, p. 168)

Of course, the embodied approach to music makes much stronger claims than that there are interactions between hearing and seeing; it views cognition not as a medium for planning, but instead as a medium for acting. It is not surprising, then, to discover that embodied musical cognition has studied the relationships between music and actions, gestures, and motion in a variety of ways (Gritten & King, 2011).

One of the most prominent of these relationships involves the exploration of new kinds of musical instruments, called digital musical instruments. A digital musical instrument is a musical instrument that involves a computer and in which the generation of sound is separate from the control interface that chooses sound (Marshall et al. 2009). This distinction is important, because as Marshall et al. (2009) pointed out, there are many available sensors that can register a human agent's movements, actions, or gestures. These include force sensitive resistors, video cameras, accelerometers, potentiometers, and bend sensors, not to mention buttons and microphones.

The availability of digital sensors permits movements, actions, and gestures to be measured and used to control the sounds generated by a digital musical instrument. This requires that a mapping be defined from a measured action to a computer-generated sound (Verfaille, Wanderley, & Depalle, 2006). Of course, completely novel relationships between gesture and sound become possible within this framework (Sapir, 2002). This permits the invention of musical instruments that can be played by individuals with no training on an instrument, because they can interact with a digital musical instrument using everyday gestures and actions (Paradiso, 1999).

The development of digital musical instruments has resulted in the need to study a variety of topics quite different from those examined by classical and connectionist researchers. One important topic involves determining how to use measured actions to control sound production (Verfaille, Wanderley, & Depalle, 2006). However, an equally important topic concerns the nature of the gestures and actions themselves. In particular, researchers of digital musical instruments are concerned with exploring issues related to principles of good design (Dourish, 2001; Norman, 2002, 2004) in order to identify and evaluate possible interfaces between actions and instruments (Magnusson, 2010; O'Modhrain, 2011; Ungvary & Vertegaal, 2000; Wanderley & Orio, 2002). Another issue is to choose a set of actions that can be varied, so that a performer of a digital musical instrument can manipulate its expressiveness (Arfib, Couturier, & Kessous, 2005).

The development of digital musical instruments has also led to a reevaluation of the roles of composers, performers, and audience. In the acoustic paradigm (Bown, Eldridge, & McCormack, 2009), which adheres to the traditional view of classical music outlined earlier in this chapter, these three components have distinct and separable roles. Digital musical instruments result in the acoustic paradigm being disrupted. Bown, Eldridge, and McCormack (2009) argued that the software components should not be viewed as instruments, but instead as behavioural objects. A behavioural object is "an entity that can act as a medium for interaction between people through its dissemination and evolution, can develop interactively with individuals in processes of creative musical development, and can interact with other behavioral objects to produce musical output" (p. 193); it is behavioural in the sense that it can act and interact, but it is an object in the sense that it is a material thing that can be seen and touched.

In their role as behavioural objects, digital musical instruments blur the sharp distinctions between the roles defined by the acoustic paradigm (Bown, Eldridge, & McCormack, 2009). This is because their software components dramatically alter the interactions between composer, performer, and listener.

> Interaction does not involve the sharing simply of passive ideas or content, but of potentially active machines that can be employed for musical tasks. Whereas musical ideas may once have developed and circulated far more rapidly than the inanimate physical objects that define traditional musical instruments,

software objects can now evolve and move around at just as fast a pace. (Bown, Eldridge, & McCormack, 2009, p. 192)

The new interactions discussed by Bown, Eldridge, and McCormack (2009) suggested that digital musical instruments can affect musical thought. It has been argued that these new instruments actually scaffold musical cognition, and therefore they extend the musical mind (Magnusson, 2009). According to Magnusson, traditional acoustic instruments have been created in *bricoleur* fashion by exploring combinations of existing materials, and learning to play such an instrument involves exploring its affordances. "The physics of wood, strings and vibrating membranes were there to be *explored* and not invented" (p. 174). In contrast, the software of digital musical instruments permits many aspects of musical cognition to be extended into the instrument itself. Digital musical instruments,

> typically contain automation of musical patterns (whether blind or intelligent) that allow the performer to delegate musical actions to the instrument itself, such as playing arpeggios, generating rhythms, expressing spatial dimensions as scales (as opposed to pitches), and so on. (Magnusson, 2009, p. 168)

The embodied approach is not limited to the study of digital musical instruments. Actions are required to play traditional musical instruments, and such actions have been investigated. For instance, researchers have examined the fingering choices made by pianists as they sight read (Sloboda et al., 1998) and developed ergonomic models of piano fingering (Parncutt et al., 1997). Bowing and fingering movements for string instruments have also been the subject of numerous investigations (Baader, Kazennikov, & Wiesendanger, 2005; Kazennikov & Wiesendanger, 2009; Konczak, van der Velden, & Jaeger, 2009; Maestre et al., 2010; Rasamimanana & Bevilacqua, 2008; Turner-Stokes & Reid, 1999). This research has included the development of the MusicJacket, a worn device that analyzes the movement of a violin player and provides vibrotactile feedback to teach proper bowing (van der Linden et al., 2011). The relationship between alternative flute fingerings and their effect on produced tones have also been examined (Botros, Smith, & Wolfe, 2006; Verfaille, Depalle, & Wanderley, 2010).

The embodied approach is also actively exploring the possibility that gestural or other kinds of interactions can be used to retrieve digitized music (Casey et al., 2008; Leman, 2008). Personal music collections are becoming vast, and traditional methods of discovering music (i.e., record stores and radio stations) are being replaced by social networking sites and the World Wide Web. As a result, there is a growing need for these large digital collections of music to be searchable. However, the most common approach for cataloguing and searching these collections is to use textual metadata that provides an indirect description of the stored music, such as the name of the composer, the title of the song, or the genre of the music (Leman, 2008).

The embodied approach is interested in the possibility of using more direct aspects of music to guide such retrieval (Leman, 2008). Is it possible to access music on the basis of one's personal experience of music? Leman hypothesizes that human action can serve as the basis of a corporeal-based querying system for retrieving music. His idea is to use the body to convert a musical idea (e.g., a desire to retrieve a particular type of music) into musical physical energy that can be mapped onto the profiles of digitized music, permitting content-based retrieval. For instance, one could query a musical database by singing or playing a melody (De Mulder et al., 2006), by manipulating a spatial representation that maps the similarity of stored music (Cooper et al., 2006; Pampalk, Dixon, & Widmer, 2004), or even by making gestures (Ko & Byun, 2002).

Compared to the other two approaches described in this chapter, the embodied approach to musical cognition is fairly new, and it is not as established. "The hypothesis that musical communication is based on the encoding, transmission, and decoding of intended actions is, I believe, an attractive one. However, at this moment it is more a working hypothesis than an established fact" (Leman, 2008, p. 237). This "working hypothesis," though, has launched an interesting literature on the study of the relationship between music and action that is easily distinguished from the classical and connectionist research on musical cognition.

6.7 Cognitive Science and Classical Music

In the preceding sections of this chapter we have explored the analogy that cognitive science is like classical music. This analogy was developed by comparing the characteristics of three different types of classical music to the three different schools of cognitive science: Austro-German classical music to classical cognitive science, musical Romanticism to connectionist cognitive science, and modern music to embodied cognitive science.

We also briefly reviewed how each of the three different schools has studied topics in the cognition of music. One purpose of this review was to show that each school of cognitive science has already made important contributions to this research domain. Another purpose was to show that the topics in musical cognition studied by each school reflected different, tacit views of the nature of music. For instance, the emphasis on formalism in traditional classical music is reflected in classical cognitive science's attempt to create generative grammars of musical structure (Lerdahl & Jackendoff, 1983). Musical romanticism's affection for the sublime is reflected in connectionist cognitive science's use of unsupervised networks to capture regularities that cannot be formalized (Bharucha, 1999). Modern music's rejection of the classic distinctions between composer, performer, and audience is reflected in embodied cognitive science's exploration of how digital musical instruments can serve as behavioural objects to extend the musical mind (Bown, Eldridge, & McCormack, 2009; Magnusson, 2009).

The perspectives summarized above reflect a fragmentation of how cognitive science studies musical cognition. Different schools of cognitive science view music in dissimilar ways, and therefore they explore alternative topics using diverse methodologies. The purpose of this final section is to speculate on a different relationship between cognitive science and musical cognition, one in which the distinctions between the three different schools of thought become less important, and in which a hybrid approach to the cognition of music becomes possible.

One approach to drawing the different approaches to musical cognition together is to return to the analogy between cognitive science and classical music and to attempt to see whether the analogy itself provides room for co-operation between approaches. One of the themes of the analogy was that important differences between Austro-German music, Romantic music, and modern music existed, and that these differences paralleled those between the different schools of cognitive science. However, there are also similarities between these different types of music, and these similarities can be used to motivate commonalities between the various cognitive sciences of musical cognition. It was earlier noted that similarities existed between Austro-German classical music and musical Romanticism because the latter maintained some of the structures and traditions of the former. So let us turn instead to bridging a gap that seems much wider, the gap between Austro-German and modern music.

The differences between Austro-German classical music and modern music seem quite clear. The former is characterized by centralized control and formal structures; it is a hot medium (McLuhan, 1994) that creates marked distinctions between composer, performer, and a passive audience (Bown, Eldridge, & McCormack, 2009), and it applies the conduit metaphor (Reddy, 1979) to view the purpose of music as conveying content from composer to listener. In contrast, modern music seems to invert all of these properties. It abandons centralized control and formal structures; it is a cool medium that blurs the distinction between composer, performer, and an active audience; and it rejects the conduit metaphor and the intentional nature of music (Hanslick, 1957; Johnson, 2007).

Such dramatic differences between types of classical music suggest that it would not be surprising for very different theories to be required to explain such cognitive underpinnings. For instance, consider the task of explaining the process of musical composition. A classical theory might suffice for an account of composing Austro-German music, while a very different approach, such as embodied cognitive science, may be required to explain the composition of modern music.

One reason for considering the possibility of theoretical diversity is that in the cool medium of modern music, where control of the composition is far more decentralized, a modern piece seems more like an improvisation than a traditional composition. "A performance is essentially an *interpretation* of something that already exists, whereas improvisation presents us with something that only comes into being in the moment of its presentation" (Benson, 2003, p. 25). Jazz guitarist Derek Bailey (1992) noted that the ability of an audience to affect a composition is expected in improvisation: "Improvisation's responsiveness to its environment puts the performance in a position to be directly influenced by the audience" (p. 44). Such effects, and more generally improvisation itself, are presumed to be absent from the Austro-German musical tradition: "The larger part of classical composition is closed to improvisation and, as its antithesis, it is likely that it will always remain closed" (p. 59).

However, there is a problem with this kind of dismissal. One of the shocks delivered by modern music is that many of its characteristics also apply to traditional classical music.

For instance, Austro-German music has a long tradition of improvisation, particularly in church music (Bailey, 1992). A famous example of such improvisation occurred when Johann Sebastian Bach was summoned to the court of German Emperor Frederick the Great in 1747 (Gaines, 2005). The Emperor played a theme for Bach on the piano and asked Bach to create a three-part fugue from it. The theme was a trap, probably composed by Bach's son Carl Philipp Emanuel (employed by the Emperor), and was designed to resist the counterpoint techniques required to create a fugue. "Still, Bach managed, with almost unimaginable ingenuity, to do it, even alluding to the king's taste by setting off his intricate counterpoint with a few *gallant* flourishes" (Gaines, 2005, p. 9). This was pure improvisation, as Bach composed and performed the fugue on the spot.

Benson (2003) argued that much of traditional music is actually improvisational, though perhaps less evidently than in the example above. Austro-German music was composed within the context of particular musical and cultural traditions. This provided composers with a constraining set of elements to be incorporated into new pieces, while being transformed or extended at the same time.

> Composers are dependent on the 'languages' available to them and usually those languages are relatively well defined. What we call 'innovation' comes either from pushing the boundaries or from mixing elements of one language with another. (Benson, 2003, p. 43)

Benson argued that improvisation provides a better account of how traditional music is composed than do alternatives such as "creation" or "discovery," and then showed that improvisation also applies to the performance and the reception of pre-modern works.

The example of improvisation suggests that the differences between the different traditions of classical music are quantitative, not qualitative. That is, it is not the case that Austro-German music is (for example) formal while modern music is not; instead, it may be more appropriate to claim that the former is *more* formal (or more centrally controlled, or less improvised, or hotter) than the latter. The possibility of quantitative distinctions raises the possibility that different types of theories can be applied to the same kind of music, and it also suggests that one approach to musical cognition may benefit by paying attention to the concerns of another.

The likelihood that one approach to musical cognition can benefit by heeding the concerns of another is easily demonstrated. For instance, it was earlier argued that musical Romanticism was reflected in connectionism's assumption that artificial neural networks could capture regularities that cannot be formalized. One consequence of this assumption was shown to be a strong preference for the use of unsupervised networks.

However, unsupervised networks impose their own tacit restrictions upon what connectionist models can accomplish. One popular architecture used to study musical cognition is the Kohonen network (Kohonen, 1984, 2001), which assigns input patterns to winning (most-active) output units, and which in essence arranges these output units (by modifying weights) such that units that capture similar regularities are near one another in a two-dimensional map. One study that presented such a network with 115 different chords found that its output units arranged tonal centres in a pattern that reflected a noisy version of the circle of fifths (Leman, 1991).

A limitation of this kind of research is revealed by relating it to classical work on tonal organization (Krumhansl, 1990). As we saw earlier, Krumhansl found two circles of fifths (one for major keys, the other for minor keys) represented in a spiral representation wrapped around a toroidal surface. In order to capture this elegant representation, four dimensions were required (Krumhansl & Kessler, 1982). By restricting networks to representations of smaller dimensionality (such as a two-dimensional Kohonen feature map), one prevents them from detecting or representing higher-dimensional regularities. In this case, knowledge gleaned from classical research could be used to explore more sophisticated network architectures (e.g., higher-dimensional self-organized maps).

Of course, connectionist research can also be used to inform classical models, particularly if one abandons "gee whiz" connectionism and interprets the internal structure of musical networks (Dawson, 2009). When supervised networks are trained on tasks involving the recognition of musical chords (Yaremchuk & Dawson, 2005; Yaremchuk & Dawson, 2008), they organize notes into hierarchies that capture circles of major seconds and circles of major thirds, as we saw in the network analyses presented in Chapter 4. As noted previously, these so-called strange circles are rarely mentioned in accounts of music theory. However, once discovered, they are just as formal and as powerful as more traditional representations such as the circle of fifths. In other words, if one ignores the sublime nature of networks and seeks to interpret their internal structures, one can discover new kinds of formal representations that could easily become part of a classical theory.

Other, more direct integrations can be made between connectionist and classical approaches to music. For example, NetNeg is a hybrid artificial intelligence system for composing two voice counterpoint pieces (Goldman et al., 1999). It assumes that some aspects of musical knowledge are subsymbolic and difficult to formalize, while other aspects are symbolic and easily described in terms of formal rules. NetNeg incorporates both types of processes to guide composition. It includes a network component that learns to reproduce melodies experienced during a training phase and uses this knowledge to generate new melodies. It also includes two rule-based agents, each of which is responsible for composing one of the voices that make up the counterpoint and for enforcing the formal rules that govern this kind of composition.

There is a loose coupling between the connectionist and the rule-based agents in NetNeg (Goldman et al., 1999), so that both co-operate, and both place constraints, on the melodies that are composed. The network suggests the next note in the melody, for either voice, and passes this information on to a rule-based agent. This suggestion, combined with interactions between the two rule-based agents (e.g., to reach an agreement on the next note to meet some aesthetic rule, such as moving the melody in opposite directions), results in each rule-based agent choosing the next note. This selection is then passed back to the connectionist part of the system to generate the next melodic prediction as the process iterates.

Integration is also possible between connectionist and embodied approaches to music. For example, for a string instrument, each note in a composition can be played by pressing different strings in different locations, and each location can be pressed by a different finger (Sayegh, 1989). The choice of string, location, and fingering is usually not specified in the composition; a performer must explore a variety of possible fingerings for playing a particular piece. Sayegh has developed a connectionist system that places various constraints on fingering so the network can suggest the optimal fingering to use. A humorous—yet strangely plausible— account of linking connectionist networks with actions was provided in Garrison Cottrell's (1989) proposal of the "connectionist air guitar."

Links also exist between classical and embodied approaches to musical cognition, although these are more tenuous because such research is in its infancy. For example, while Leman (2008) concentrated on the direct nature of musical experience that characterizes the embodied approach, he recognized that indirect accounts—such as verbal descriptions of music—are both common and important. The most promising links are appearing in work on the cognitive neuroscience of music, which is beginning to explore the relationship between music perception and action.

Interactions between perception of music and action have already been established. For instance, when classical music is heard, the emotion associated with it can affect perceptions of whole-body movements directed towards objects (Van den Stock et al., 2009). The cognitive neuroscience of music has revealed a great deal of evidence for the interaction between auditory and motor neural systems (Zatorre, Chen, & Penhune, 2007).

Such evidence brings to mind the notion of simulation and the role of mirror neurons, topics that were raised in Chapter 5's discussion of embodied cognitive science. Is it possible that direct experience of musical performances engages the mirror system? Some researchers are considering this possibility (D'Ausilio, 2009; Lahav, Saltzman, & Schlaug, 2007). Lahav, Saltzman, and Schlaug (2007) trained non-musicians to play a piece of music. They then monitored their subjects' brain activity while they listened to this newly learned piece while not performing any movements. It was discovered that motor-related areas of the brain were activated during the listening. Less activity in these areas was noted if subjects heard the same notes that were learned, but presented in a different order (i.e., as a different melody).

The mirror system has also been shown to be involved in the observation and imitation of guitar chording (Buccino et al., 2004; Vogt et al., 2007); and musical expertise, at least for professional piano players, is reflected in more specific mirror neuron processing (Haslinger et al., 2005). It has even been suggested that the mirror system is responsible for listeners misattributing anger to John Coltrane's style of playing saxophone (Gridley & Hoff, 2006)!

A completely hybrid approach to musical cognition that includes aspects of all three schools of cognitive science is currently only a possibility. The closest realization of this possibility might be an evolutionary composing system (Todd & Werner, 1991). This system is an example of a genetic algorithm (Holland, 1992; Mitchell, 1996), which evolves a solution to a problem by evaluating the fitness of each member of a population, preserves the most fit, and then generates a new to-be-evaluated generation by combining attributes of the preserved individuals. Todd and Werner (1991) noted that such a system permits fitness to be evaluated by a number of potentially quite different critics; their model considers contributions of human, rule-based, and network critics.

Music is a complicated topic that has been considered at multiple levels of investigation, including computational or mathematical (Assayag et al., 2002; Benson, 2007; Harkleroad, 2006; Lerdahl & Jackendoff, 1983), algorithmic or behavioural (Bailey, 1992; Deutsch, 1999; Krumhansl, 1990; Seashore, 1967; Snyder, 2000), and implementational or biological (Jourdain, 1997; Levitin, 2006; Peretz & Zatorre, 2003). Music clearly is a domain that is perfectly suited to cognitive science. In this chapter, the analogy between classical music and cognitive science has been developed to highlight the very different contributions of classical, connectionist, and embodied cognitive science to the study of musical cognition. It raised the possibility of a more unified approach to musical cognition that combines elements of all three different schools of thought.

7

Marks of the Classical?

7.0 Chapter Overview

In the previous chapter, the characteristics of the three approaches to cognitive science were reviewed, highlighting important distinctions between the classical, connectionist, and embodied approaches. This was done by exploring the analogy between cognitive science and classical music. It was argued that each of the three approaches within cognitive science was analogous to one of three quite different traditions within classical music, and that these differences were apparent in how each approach studied music cognition. However, at the end of the chapter the possibility of hybrid theories of music cognition was raised.

The possibility of hybrid theories of music cognition raises the further possibility that the differences between the three approaches within cognitive science might not be as dramatic as could be imagined. The purpose of the current chapter is to explore this further possibility. It asks the question: are there marks of the classical? That is, is there a set of necessary and sufficient properties that distinguish classical theories from connectionist and embodied theories?

The literature suggests that there should be a large number of marks of the classical. It would be expected that classical theories appeal to centralized control, serial processing, local and internal representations, explicit rules, and a cognitive vocabulary that appeals to the contents of mental representations. It would also be

315

expected that both connectionist and embodied theories reject many, if not all, of these properties.

In the current chapter we examine each of these properties in turn and make the argument that they do not serve as marks of the classical. First, an examination of the properties of classical theories, as well as a reflection on the properties of the computing devices that inspired them, suggests that none of these properties are necessary classical components. Second, it would also appear that many of these properties are shared by other kinds of theories, and therefore do not serve to distinguish classical cognitive science from either the connectionist or the embodied approaches.

The chapter ends by considering the implications of this conclusion. I argue that the differences between the approaches within cognitive science reflect variances in emphasis, and not qualitative differences in kind, amongst the three kinds of theory. This sets the stage for the possibility of hybrid theories of the type examined in Chapter 8.

7.1 Symbols and Situations

As new problems are encountered in a scientific discipline, one approach to dealing with them is to explore alternative paradigms (Kuhn, 1970). One consequence of adopting this approach is to produce a clash of cultures, as the new paradigms compete against the old.

> The social structure of science is such that individual scientists will justify the
> claims for a new approach by emphasizing the flaws of the old, as well as the virtues
> and goodness of the new. Similarly, other scientists will justify the continuation of
> the traditional method by minimizing its current difficulties and by discounting the
> powers or even the novelty of the new. (Norman, 1993, p. 3)

In cognitive science, one example of this clash of cultures is illustrated in the rise of connectionism. Prior to the discovery of learning rules for multilayered networks, there was a growing dissatisfaction with the progress of the classical approach (Dreyfus, 1972). When trained multilayered networks appeared in the literature, there was an explosion of interest in connectionism, and its merits—and the potential for solving the problems of classical cognitive science—were described in widely cited publications (McClelland & Rumelhart, 1986, 1988; Rumelhart & McClelland, 1986c; Schneider, 1987; Smolensky, 1988). In response, defenders of classical cognitive science argued against the novelty and computational power of the new connectionist models (Fodor & McLaughlin, 1990; Fodor & Pylyshyn, 1988; Minsky & Papert, 1988; Pinker & Prince, 1988).

A similar clash of cultures, concerning the debate that arose as part of embodied cognitive science's reaction to the classical tradition, is explored in more detail in this section. One context for this clash is provided by the research of eminent AI researcher Terry Winograd. Winograd's PhD dissertation involved programming a computer to understand natural language, the SHRDLU system that operated in a restricted blocks world (Winograd, 1972a, 1972b). SHRDLU would begin with a representation of different shaped and coloured blocks arranged in a scene. A user would type in a natural language command to which the program would respond, either by answering a query about the scene or performing an action that changed the scene. For instance, if instructed "Pick up a big red block," SHRDLU would comprehend this instruction, execute it, and respond with "OK." If then told "Find a block that is taller than the one you are holding and put it in the box," then SHRDLU had to comprehend the words *one* and *it*; it would respond "By *it* I assume you mean the block which is taller than the one I am holding."

Winograd's (1972a) program was a prototypical classical system (Harnish, 2002). It parsed input strings into grammatical representations, and then it took advantage of the constraints of the specialized blocks world to map these grammatical structures onto a semantic interpretation of the scene. SHRDLU showed "that if the database was narrow enough the program could be made deep enough to display human-like interactions" (p. 121).

Winograd's later research on language continued within the classical tradition. He wrote what served as a bible to those interested in programming computers to understand language, *Language As a Cognitive Process, Volume 1: Syntax* (Winograd, 1983). This book introduced and reviewed theories of language and syntax, and described how those theories had been incorporated into working computer programs. As the title suggests, a second volume on semantics was planned by Winograd. However, this second volume never appeared.

Instead, Winograd's next groundbreaking book, *Understanding Computers and Cognition*, was one of the pioneering works in embodied cognitive science and launched a reaction against the classical approach (Winograd & Flores, 1987b). This book explained why Winograd did not continue with a text on the classical approach to semantics, because he had arrived at the opinion that classical accounts of language understanding would never be achieved. "Our position, in accord with the preceding chapters, is that computers cannot understand language" (p. 107).

The reason that Winograd and Flores (1987b) adopted this position was their view that computers are restricted to a rationalist notion of meaning that, in accordance with methodological solipsism (Fodor, 1980), must interpret terms independently of external situations or contexts. Winograd and Flores argued instead for an embodied, radically non-rational account of meaning: "Meaning always derives from an interpretation that is rooted in a situation" (Winograd & Flores, 1987b, p. 111). They took their philosophical inspiration from Heidegger instead of from Descartes.

Winograd and Flores' (1987b) book was impactful and divisive. For example, the journal *Artificial Intelligence* published a set of four widely divergent reviews of the book (Clancey, 1987; Stefik & Bobrow, 1987; Suchman, 1987; Vellino, 1987), prefaced by an introduction noting that "when new books appear to be controversial, we try to present multiple perspectives on them." Winograd and Flores (1987a) also published a response to the four reviews. In spite of its contentious reception, the book paved the way for research in situated cognition (Clancey, 1997), and it is one of the earliest examples of what is now well-established embodied cognitive science.

The rise of the embodied reaction is the first part of the clash of cultures in Norman's (1993) sociology of cognitive science. A second part is the response of classical cognitive science to the embodied movement, a response that typically involves questioning the adequacy and the novelty of the new paradigm. An excellent example of this aspect of the culture clash is provided in a series of papers published in the journal *Cognitive Science* in 1993.

This series began with a paper entitled "Situated action: A symbolic interpretation" (Vera & Simon, 1993), which provided a detailed classical response to theories of situated action (SA) or situated cognition, approaches that belong to embodied cognitive science. This response was motivated by Vera and Simon's (1993) observation that SA theories reject central assumptions of classical cognitive science: situated action research "denies that intelligent systems are correctly characterized as physical symbol systems, and especially denies that symbolic processing lies at the heart of intelligence" (pp. 7–8). Vera and Simon argued in favor of a much different conclusion: that situated action research is essentially classical in nature. "We find that there is no such antithesis: SA systems are symbolic systems, and some past and present symbolic systems are SA systems" (p. 8).

Vera and Simon (1993) began their argument by characterizing the important characteristics of the two positions that they aimed to integrate. Their view of classical cognitive science is best exemplified by the general properties of physical symbol systems (Newell, 1980) that were discussed in Chapter 3, with prototypical examples being early varieties of production systems (Anderson, 1983; Newell, 1973, 1990; Newell & Simon, 1972).

Vera and Simon (1993) noted three key characteristics of physical symbol systems: perceptual processes are used to establish the presence of various symbols or symbolic structures in memory; reasoning processes are used to manipulate internal symbol strings; and finally, the resulting symbol structures control motor actions on the external world. In other words, sense-think-act processing was explicitly articulated. "Sequences of actions can be executed with constant interchange among (a) receipt of information about the current state of the environment (perception), (b) internal processing of information (thinking), and (c) response to the environment (motor activity)" (p. 10).

Critical to Vera and Simon's (1993) attempt to cast situated action in a classical context was their notion of "symbol." First, symbols were taken to be some sort of pattern, so that pattern recognition processes could assert that some pattern is a token of a particular symbolic type (i.e., symbol recognition). Second, such patterns were defined as true symbols when,

> they can designate or denote. An information system can take a symbol token as input and use it to gain access to a referenced object in order to affect it or be affected by it in some way. Symbols may designate other symbols, but they may also designate patterns of sensory stimuli, and they may designate motor actions.
> (Vera & Simon, 1993, p. 9)

Vera and Simon (1993) noted that situated action or embodied theories are highly variable and therefore difficult to characterize. As a result, they provided a very general account of the core properties of such theories by focusing on a small number, including Winograd and Flores (1987b). Vera and Simon observed that situated action theories require accounts of behaviour to consider situations or contexts, particularly those involving an agent's environment. Agents must be able to adapt to ill-posed (i.e., difficult to formalize) situations, and do so via direct and continuously changing interactions with the environment.

Vera and Simon (1993) went on to emphasize six main claims that in their view characterized most of the situated action literature: (1) situated action requires no internal representations; (2) it operates directly with the environment (sense-act rather than sense-think-act); (3) it involves direct access to affordances; (4) it does not use productions; (5) it exploits a socially defined, not physically defined, environment; and (6) it makes no use of symbols. With this position, Vera and Simon were situated to critique the claim that the embodied approach is qualitatively different from classical cognitive science. They did so by either arguing against the import of some embodied arguments, or by in essence arguing for the formal equivalence of classical and SA theories. Both of these approaches are in accord with Norman's (1993) portrayal of a culture clash.

As an example of the first strategy, consider Vera and Simon's (1993) treatment of the notion of readiness-to-hand. This idea is related to Heidegger's (1962) concept of *Dasein*, or being-in-the-world, which is an agent's sense of being engaged with its world. Part of this engagement involves using "entities," which Heidegger called equipment, and which are experienced in terms of what cognitive scientists would describe as affordances or potential actions (Gibson, 1979). "Equipment is essentially 'something-in-order-to'" (Heidegger, 1962, p. 97).

Heidegger's (1962) position was that when agents experience the affordances of equipment, other properties—such as the physical nature of equipment—disappear. This is readiness-to-hand. "That with which our everyday dealings proximally dwell is not the tools themselves. On the contrary, that with which we concern ourselves primarily is the work" (p. 99). Another example of readiness-to-hand is the blind person's cane, which is not experienced as such when it is being used to navigate, but is instead experienced as an extension of the person themselves (Bateson, 1972, p. 465): "The stick is a pathway along which transforms of difference are being transmitted."

Heidegger's philosophy played a dominant role in the embodied theory proposed by Winograd and Flores (1987b). They took readiness-to-hand as evidence of direct engagement with the world; we only become aware of equipment itself when the structural coupling between world, equipment, and agent breaks down. Winograd and Flores took the goal of designing equipment, such as human-computer interfaces, to be creating artifacts that are invisible to us when they are used. "A successful word processing device lets a person operate on the words and paragraphs displayed on the screen, without being aware of formulating and giving commands" (Winograd & Flores, 1987b, p. 164). The invisibility of artifacts—the readiness-to-hand of equipment—is frequently characterized as being evidence of good design (Dourish, 2001; Norman, 1998, 2002, 2004).

Importantly, readiness-to-hand was also used by Winograd and Flores (1987b) as evidence for rejecting the need for classical representations, and to counter the claim that tool use is mediated by symbolic thinking or planning (Miller, Galanter, & Pribram, 1960). From the classical perspective, it might be expected that an agent is consciously aware of his or her plans; the absence of such awareness, or readiness-to-hand, must therefore indicate the absence of planning. Thus readiness-to-hand reflects direct, non-symbolic links between sensing and acting.

> If we focus on concernful activity instead of on detached contemplation, the status of this representation is called into question. In driving a nail with a hammer (as opposed to thinking about a hammer), I need not make use of any explicit representation of the hammer. (Winograd & Flores, 1987b, p. 33)

Vera and Simon (1993, p. 19) correctly noted, though, that our conscious awareness of entities is mute with respect to either the nature or the existence of representational formats: "Awareness has nothing to do with whether something is represented symbolically, or in some other way, or not at all." That is, consciousness of contents is not a defining feature of physical symbol systems. This position is a deft dismissal of using readiness-to-hand to support an anti-representational position.

After dealing with the implications of readiness-to-hand, Vera and Simon (1993) considered alternate formulations of the critiques raised by situated action researchers. Perhaps the prime concern of embodied cognitive science is that the classical approach emphasizes internal, symbolic processing to the near total exclusion of sensing and acting. We saw in Chapter 3 that production system pioneers admitted that their earlier efforts ignored sensing and acting (Newell, 1990). (We also saw an attempt to rectify this in more recent production system architectures [Meyer et al., 2001; Meyer & Kieras, 1997a, 1997b]).

Vera and Simon (1993) pointed out that the classical tradition has never disagreed with the claim that theories of cognition cannot succeed by merely providing accounts of internal processing. Action and environment are key elements of pioneering classical accounts (Miller, Galanter, & Pribram, 1960; Simon, 1969). Vera and Simon stress this by quoting the implications of Simon's (1969) own parable of the ant:

> The proper study of mankind has been said to be man. But . . . man—or at least the intellective component of man—may be relatively simple; . . . most of the complexity of his behavior may be drawn from his environment, from his search for good designs. (Simon, 1969, p. 83)

Modern critics of the embodied notion of the extended mind (Adams & Aizawa, 2008) continue to echo this response: "The orthodox view in cognitive science maintains that minds do interact with their bodies and their environments" (pp. 1–2).

Vera and Simon (1993) emphasized the interactive nature of classical models by briefly discussing various production systems designed to interact with the world. These included the Phoenix project, a system that simulates the fighting of forest fires in Yellowstone National Park (Cohen et al., 1989), as well as the Navlab system for navigating an autonomous robotic vehicle (Pomerleau, 1991; Thorpe, 1990). Vera and Simon also described a production system for solving the Towers of Hanoi problem, but it was highly scaffolded. That is, its memory for intermediate states of the problem was in the external towers and discs themselves; the production system had neither an internal representation of the problem nor a goal stack to plan its solution. Instead, it solved the problem perceptually, with its productions driven by the changing appearance of the problem over time.

The above examples were used to argue that at least some production systems are situated action models. Vera and Simon (1993) completed their argument by making the parallel argument that some notable situated action theories are symbolic because they are instances of production systems. One embodied theory that received this treatment was Rodney Brooks' behaviour-based robotics (Brooks, 1991, 1989, 1999, 2002), which was introduced in Chapter 5. To the extent that they agreed that Brooks' robots do not employ representations, Vera and Simon suggested that this limits their capabilities. "It is consequently unclear whether Brooks and his Creatures are on the right track towards fully autonomous systems that can function in a wider variety of environments" (Vera & Simon, 1993, p. 35).

However, Vera and Simon (1993) went on to suggest that even systems such as Brooks' robots could be cast in a symbolic mould. If a system has a state that is in some way indexed to a property or entity in the world, then that state should be properly called a symbol. As a result, a basic sense-act relationship that was part of the most simplistic subsumption architecture would be an example of a production for Vera and Simon.

Furthermore, Vera and Simon (1993) argued that even if a basic sense-act relationship is wired in, and therefore there is no need to view it as symbolized, it is symbolic nonetheless:

> On the condition end, the neural impulse aroused by the encoded incoming stimuli denotes the affordances that produced these stimuli, while the signals to efferent nerves denote the functions of the actions. There is every reason to regard these impulses and signals as symbols: A symbol can as readily consist of the activation of a neuron as it can of the creation of a tiny magnetic field. (Vera and Simon, 1993, p. 42)

Thus any situated action model can be described in a neutral, symbolic language—as a production system—including even the most reflexive, anti-representational instances of such models.

The gist of Vera and Simon's (1993) argument, then, was that there is no principled difference between classical and embodied theories, because embodied models that interact with the environment are in essence production systems. Not surprisingly, this position attracted a variety of criticisms.

For example, *Cognitive Science* published a number of articles in response to the original paper by Vera and Simon (Norman, 1993). One theme apparent in some of these papers was that Vera and Simon's definition of *symbol* was too vague to be useful (Agre, 1993; Clancey, 1993). Agre, for instance, accused Vera and Simon not of defending a well-articulated theory, but instead of exploiting an indistinct worldview. He argued that they "routinely claim vindication through some 'symbolic' gloss of whatever phenomenon is under discussion. The problem is that just about anything can seem 'symbolic' if you look at it right" (Agre, 1993, p. 62).

One example of such vagueness was Vera and Simon's (1993) definition of a symbol as a "designating pattern." What do they mean by *designate*? Designation has occurred if "an information system can take a symbol token as input and use it to gain access to a referenced object in order to affect it or to be affected by it in some way" (Vera & Simon, 1993, p. 9). In other words the mere establishment of a deictic or indexing relationship (Pylyshyn, 1994, 2000, 2001) between the world and some state of an agent is sufficient for Vera and Simon to deem that state "symbolic."

This very liberal definition of *symbolic* leads to some very glib characterizations of certain embodied positions. Consider Vera and Simon's (1993) treatment of affordances as defined in the ecological theory of perception (Gibson, 1979). In Gibson's theory, affordances—opportunities for action offered by entities in the world—are perceived directly; no intervening symbols or representations are presumed. "When I assert that perception of the environment is direct, I mean that it is not mediated by retinal pictures, neural pictures, or mental pictures" (p. 147). Vera and Simon (1993, p. 20) denied direct perception: "the thing that corresponds to an affordance is a symbol stored in central memory denoting the encoding in functional terms of a complex visual display, the latter produced, in turn, by the actual physical scene that is being viewed."

Vera and Simon (1993) adopted this representational interpretation of affordances because, by their definition, an affordance designates some worldly state of affairs and must therefore be symbolic. As a result, Vera and Simon redefined the sense-act links of direct perception as indirect sense-think-act processing. To them, affordances were symbols informed by senses, and actions were the consequence of the presence of motor representations. Similar accounts of affordances have been proposed in the more recent literature (Sahin et al., 2007).

While Vera and Simon's (1993) use of designation to provide a liberal definition of *symbol* permits a representational account of anti-representational theories, it does so at the expense of neglecting core assumptions of classical models. In particular, other leading classical cognitive scientists adopt a much more stringent definition of *symbol* that prevents, for instance, direct perception to be viewed as a classical theory. Pylyshyn has argued that cognitive scientists must adopt a cognitive vocabulary in their theories (Pylyshyn, 1984). Such a vocabulary captures regularities by appealing to the contents of representational states, as illustrated in adopting the intentional stance (Dennett, 1987) or in employing theory-theory (Gopnik & Meltzoff, 1997; Gopnik & Wellman, 1992).

Importantly, for Pylyshyn mere designation is not sufficient to define the content of symbols, and therefore is not sufficient to support a classical or cognitive theory. As discussed in detail in Chapter 8, Pylyshyn has developed a theory of vision that requires indexing or designation as a primitive operation (Pylyshyn, 2003c, 2007). However, this theory recognizes that designation occurs without representing the features of indexed entities, and therefore does not establish cognitive content. As a result, indexing is a critical component of Pylyshyn's theory—but it is also a component that he explicitly labels as being non-representational and non-cognitive.

Vera and Simon's (1993) vagueness in defining the symbolic has been a central concern in other critiques of their position. It has been claimed that Vera and Simon omit one crucial characteristic in their definition of *symbol system*: the capability of being a universal computing device (Wells, 1996). Wells (1996) noted in one example that devices such as Brooks' behaviour-based robots are not capable of universal computation, one of the defining properties of a physical symbol system (Newell & Simon, 1976). Wells argues that if a situated action model is not universal, then it cannot be a physical symbol system, and therefore cannot be an instance of the class of classical or symbolic theories.

The trajectory from Winograd's (1972a) early classical research to his pioneering articulation of the embodied approach (Winograd & Flores, 1987b) and the route from Winograd and Flores' book to Vera and Simon's (1993) classical account of situated action to the various responses that this account provoked raise a number of issues.

First, this sequence of publications nicely illustrates Norman's (1993) description of culture clashes in cognitive science. Dissatisfied with the perceived limits of the classical approach, Winograd and Flores highlighted its flaws and detailed the potential advances of the embodied approach. In reply, Vera and Simon (1993) discounted the differences between classical and embodied theories, and even pointed out how connectionist networks could be cast in the light of production systems.

Second, the various positions described above highlight a variety of perspectives concerning the relationships between different schools of thought in cognitive science. At one extreme, all of these different schools of thought are considered to be classical in nature, because all are symbolic and all fall under a production system umbrella (Vera & Simon, 1993). At the opposite extreme, there are incompatible differences between the three approaches, and supporters of one approach argue for its adoption and for the dismissal of the others (Chemero, 2009; Fodor & Pylyshyn, 1988; Smolensky, 1988; Winograd & Flores, 1987b).

In between these poles, one can find compromise positions in which hybrid models that call upon multiple schools of thought are endorsed. These include proposals in which different kinds of theories are invoked to solve different sorts of problems, possibly at different stages of processing (Clark, 1997; Pylyshyn, 2003c). These also include proposals in which different kinds of theories are invoked simultaneously to co-operatively achieve a full account of some phenomenon (McNeill, 2005).

Third, the debate between the extreme poles appears to hinge on core definitions used to distinguish one position from another. Is situated cognition classical? As we saw earlier, this depends on the definition of *symbolic*, which is a key classical idea, but it has not been as clearly defined as might be expected (Searle, 1992). It is this third point that is the focus of the remainder of this chapter. What are the key concepts that are presumed to distinguish classical cognitive science from its putative competitors? When one examines these concepts in detail, are they truly distinguished between positions? Or do they instead reveal potential compatibilities between the different approaches to cognitive science?

7.2 Marks of the Classical

In previous chapters, the elements of classical, of connectionist, and of embodied cognitive science have been presented. We have proceeded in a fashion that accentuated potential differences between these three schools of thought. However, now that the elements of all three approaches have been presented, we are in a position to explore how real and extensive these differences are. Is there one cognitive science, or many? One approach to answering this question is to consider whether the distinctions between the elements of the cognitive sciences are truly differences in *kind*.

The position of the current chapter is that there are strong relations amongst the three schools of thought in cognitive science; differences between these schools are more matters of degree than qualitative differences of kind. Let us set a context for this discussion by providing an argument similar in structure to the one framed by Adams and Aizawa (2008) against the notion of the extended mind.

One important critique of embodied cognitive science's proposal of the extended mind is based on an analysis of the mark of the cognitive (Adams & Aizawa, 2008). The mark of the cognitive is a set of necessary and sufficient features that distinguish cognitive phenomena from other phenomena. Adams and Aizawa's central argument against the extended mind is that it fails to provide the required features.

> If one thinks that cognitive processing is simply any sort of dynamical system process, then—so understood—cognitive processing is again likely to be found spanning the brain, body and environment. But, so understood, cognitive processing will also be found in the swinging of a pendulum of a grandfather clock or the oscillations of the atoms of a hydrogen molecule. Being a dynamical system is pretty clearly insufficient for cognition or even a cognitive system.
> (Adams & Aizawa, 2008, p. 23)

Connectionist and embodied approaches can easily be characterized as explicit reactions against the classical viewpoint. That is, they view certain characteristics of classical cognitive science as being incorrect, and they propose theories in which these characteristics have been removed. For instance, consider Rodney Brooks' reaction against classical AI and robotics:

> During my earlier years as a postdoc at MIT, and as a junior faculty member at Stanford, I had developed a heuristic in carrying out research. I would look at how everyone else was tackling a certain problem and find the core central thing that they all agreed on so much that they never even talked about it. I would negate the central implicit belief and see where it led. This often turned out to be quite useful.
> (Brooks, 2002, p. 37)

This reactive approach suggests a context for the current chapter: that there should be a mark of the classical, a set of necessary and sufficient features that distinguish the theories of classical cognitive science from the theories of either connectionist or of embodied cognitive science. Given the material presented in earlier chapters, a candidate set of such features can easily be produced: central control, serial processing, internal representations, explicit rules, the disembodied mind, and so on. Alternative approaches to cognitive science can be characterized as taking a subset of these features and inverting them in accordance with Brooks' heuristic.

In the sections that follow we examine candidate features that define the mark of the classical. It is shown that none of these features provide a necessary and sufficient distinction between classical and non-classical theories. For instance, central control is not a required property of a classical system, but was incorporated as an engineering convenience. Furthermore, central control is easily found in non-classical systems such as connectionist networks.

If there is no mark of the classical, then this indicates that there are not many cognitive sciences, but only one. Later chapters support this position by illustrating theories of cognitive science that incorporate elements of all three approaches.

7.3 Centralized versus Decentralized Control

Two of the key elements of a classical theory of cognitive science are a set of primitive symbols and a set of primitive processes for symbol manipulation. However, these two necessary components are not by themselves sufficient to completely define a working classical model. A third element is also required: a mechanism of control.

Control is required to determine "what to do next," to choose which primitive operation is to be applied at any given moment.

> Beyond the capability to execute the basic operations singly, a computing machine must be able to perform them according to the sequence—or rather, the logical pattern—in which they generate the solution of the mathematical problem that is the actual purpose of the calculation in hand. (von Neumann, 1958, p. 11)

The purpose of this section is to explore the notion of control from the perspective of the three schools of thought in cognitive science. This is done by considering cognitive control in the context of the history of the automatic control of computing devices. It is argued that while the different approaches in cognitive science may claim to have very different accounts of cognitive control, there are in fact no qualitative differences amongst these accounts.

One of the earliest examples of automatic control was Jacquard's punched card mechanism for, in essence, programming a loom to weave a particular pattern into silk fabric (Essinger, 2004), as discussed in Chapter 3. One punched card controlled the appearance of one thread row in the fabric. Holes punched in the card permitted rods to move, which raised specified threads to make them visible at this point in the fabric. The cards that defined a pattern were linked together as a belt that advanced one card at a time during weaving. A typical pattern to be woven was defined by around 2,000 to 4,000 different punched cards; very complex patterns required using many more cards. For instance, Jacquard's self-portrait in silk was defined by 24,000 different punched cards.

Jacquard patented his loom in 1804 (Essinger, 2004). By the end of the nineteenth century, punched cards inspired by his invention had a central place in the processing of information. However, their role was to represent this information, not to control how it was manipulated.

After Herman Hollerith graduated from Columbia School of Mines in 1879, he was employed to work on the 1880 United States Census, which was the first census to collect not only population data but also to be concerned with economic issues (Essinger, 2004). Hollerith's census experience revealed a marked need to automate the processing of the huge amount of information that had been collected.

> While engaged in work upon the tenth census, the writer's attention was called to the methods employed in the tabulation of population statistics and the enormous expense involved. These methods were at the time described as 'barbarous[;] some machine ought to be devised for the purpose of facilitating such tabulations'.
> (Hollerith, 1889, p. 239)

Hollerith's response was to represent census information using punched cards (Austrian, 1982; Comrie, 1933; Hollerith, 1889). A standard punched card, called a tabulating card, measured 18.7 cm by 8.3 cm, and its upper left hand corner was beveled to prevent the card from being incorrectly oriented. A blank tabulating card consisted of 80 vertical columns, with 12 different positions in each column through which a hole could be punched. The card itself acted as an electrical insulator and was passed through a wire brush and a brass roller. The brush and roller came in contact wherever a hole had been punched, completing an electrical circuit and permitting specific information to be read from a card and acted upon (Eckert, 1940).

Hollerith invented a set of different devices for manipulating tabulating cards. These included a card punch for entering data by punching holes in cards, a verifier for checking for data entry errors, a counting sorter for sorting cards into different groups according to the information punched in any column of interest, a tabulator or accounting machine for adding numbers punched into a set of cards, and a multiplier for taking two different numbers punched on a card, computing their product, and punching the product onto the same card. Hollerith's devices were employed during the 1890 census. They saved more than two years of work and $5 million dollars, and permitted complicated tables involving relationships between different variables to be easily created (Essinger, 2004).

In Hollerith's system, punched cards represented information, and the various specialized devices that he invented served as the primitive processes available for manipulating information. Control, however, was not mechanized—it was provided by a human operator of the various tabulating machines in a room. "The calculating process was done by passing decks of cards from one machine to the next, with each machine contributing something to the process" (Williams, 1997, p. 253). This approach was very powerful. In what has been described as the first book about computer programming, *Punched Card Methods in Scientific Computation* (Eckert, 1940), astronomer Wallace Eckert described how a set of Hollerith's machines—a punched card installation—could be employed for harmonic analysis, for solving differential equations, for computing planetary perturbations, and for performing many other complex calculations.

The human controller of a punched card installation was in a position analogous to a weaver in Lyon prior to the invention of Jacquard's loom. That is, both were human operators—or more precisely, human controllers—of machines responsible for producing complicated products. Jacquard revolutionized the silk industry by automating the control of looms. Modern computing devices arose from an analogous innovation, automating the control of Hollerith's tabulators (Ceruzzi, 1997, p. 8): The entire room comprising a punched card installation "including the people in it—and not the individual machines is what the electronic computer eventually replaced."

The first phase of the history of replacing punched card installations with automatically controlled computing devices involved the creation of calculating devices that employed mechanical, electromechanical, or relay technology (Williams, 1997). This phase began in the 1930s with the creation of the German calculators invented by Konrad Zuse (Zuse, 1993), the Bell relay computers developed by George Stibitz (Irvine, 2001; Stibitz & Loveday, 1967a, 1967b), and the Harvard machines designed by Howard Aiken (Aiken & Hopper, 1946).

The internal components of any one of these calculators performed operations analogous to those performed by the different Hollerith machines in a punched card installation. In addition, the actions of these internal components were automatically controlled. Completing the parallel with the Jacquard loom, this control was accomplished using punched tape or cards. The various Stibitz and Aiken machines read spools of punched paper tape; Zuse's machines were controlled by holes punched in discarded 35 mm movie film (Williams, 1997). The calculators developed during this era by IBM, a company that had been founded in part from Hollerith's Computer Tabulating Recording Company, were controlled by decks of punched cards (Williams, 1997).

In the 1940s, electromechanical or relay technology was replaced with much faster electronic components, leading to the next generation of computer devices. Vacuum tubes were key elements of both the Atanasoff-Berry computer (ABC), created by John Atanasoff and Clifford Berry (Burks & Burks, 1988; Mollenhoff, 1988; Smiley, 2010), and the ENIAC (Electronic Numerical Integrator and Computer) engineered by Presper Eckert and John Mauchly (Burks, 2002; Neukom, 2006).

The increase in speed of the internal components of electronic computers caused problems with paper tape or punched card control. The issue was that the electronic machines were 500 times faster than relay-based devices (Pelaez, 1999), which meant that traditional forms of control were far too slow.

This control problem was solved for Eckert and Mauchly's ENIAC by using a master controller that itself was an electronic device. It was a set of ten electronic switches that could each be set to six different values; each switch was associated with a counter that could be used to advance a switch to a new setting when a predefined value was reached (Williams, 1997). The switches would route incoming signals to particular components of ENIAC, where computations were performed; a change in a switch's state would send information to a different component of ENIAC. The control of this information flow was accomplished by using a plug board to physically wire the connections between switches and computer components. This permitted control to match the speed of computation, but at a cost:

> ENIAC was a fast but relatively inflexible machine. It was best suited for use in long and repetitious calculations. Once it was wired up for a particular program, it was in fact a special purpose machine. Adapting it to another purpose (a different problem) required manual intervention to reconfigure the electrical circuits. (Pelaez, 1999, p. 361)

Typically two full days of rewiring the plug board were required to convert ENIAC from one special purpose machine to another.

Thus the development of electronic computers led to a crisis of control. Punched tape provided flexible, easily changed, control. However, punched tape readers were too slow to take practical advantage of the speed of the new machines. Plug boards provided control that matched the speed of the new componentry, but was inflexible and time consuming to change. This crisis of control inspired another innovation, the stored program computer (Aspray, 1982; Ceruzzi, 1997; Pelaez, 1999).

The notion of the stored program computer was first laid out in 1945 by John von Neumann in a draft memo that described the properties of the EDVAC (Electronic Discrete Variable Automatic Computer), the computer that directly descended from the ENIAC (Godfrey & Hendry, 1993; von Neumann, 1993). One of the innovations of this design was the inclusion of a central controller. In essence, the instructions that ordinarily would be represented as a sequence on a punched tape would instead be represented *internally* in EDVAC's memory. The central controller had the task of fetching, interpreting, and executing an instruction from memory and then repeating this process after proceeding to the next instruction in the sequence.

There is no clear agreement about which particular device was the first stored program computer; several candidate machines were created in the same era. These include the EDVAC (created 1945–1950) (Reitwiesner, 1997; von Neumann, 1993; Williams, 1993), Princeton's IAS computer (created 1946–1951) (Burks, 2002; Cohen, 1999), and the Manchester machine (running in 1948) (Copeland, 2011; Lavington, 1980). Later work on the ENIAC also explored its use of stored programs (Neukom, 2006). Regardless of "firsts," all of these machines were functionally equivalent in the sense that they replaced external control—as by a punched tape—with internalizing tape instructions into memory.

The invention of the stored program computer led directly to computer science's version of the classical sandwich (Hurley, 2001). "Sensing" involves loading the computer's internal memory with both the program and the data to be processed. "Thinking" involves executing the program and performing the desired calculations upon the stored data. "Acting" involves providing the results of the calculations to the computer's operator, for instance by punching an output tape or a set of punched cards.

The classical sandwich is one of the defining characteristics of classical cognitive science (Hurley, 2001), and the proposal of a sense-act cycle to replace the sandwich's sense-think-act processing (Brooks, 1999, 2002; Clark, 1997, 2008; Pfeifer & Scheier, 1999) is one of the characteristic reactions of embodied cognitive science against the classical tradition (Shapiro, 2011). Classical cognitive science's adoption of the classical sandwich was a natural consequence of being inspired by computer science's approach to information processing, which, at the time that classical cognitive science was born, had culminated in the invention of the stored program computer.

However, we have seen from the history leading up to its invention that the stored program computer—and hence the classical sandwich—was not an in-principle requirement for information processing. It was instead the result of a practical need to match the speed of control with the speed of electronic components. In fact, the control mechanisms of a variety of information processing models that are central to classical cognitive science are in fact quite consistent with embodied cognitive science.

For example, the universal Turing machine is critically important to classical cognitive science, not only in its role of defining the core elements of symbol manipulation, but also in its function of defining the limits of computation (Dawson, 1998). However, in most respects a universal Turing machine is a device that highlights some of the key characteristics of the embodied approach.

For instance, the universal Turing machine is certainly not a stored program computer (Wells, 2002). If one were to actually build such a device—the original was only used as a theoretical model (Turing, 1936)—then the only internal memory that would be required would be for holding the machine table and the machine head's internal state. (That is, if any internal memory was required at all. Turing's notion of machine state was inspired by the different states of a typewriter's keys [Hodges, 1983], and thus a machine state may not be remembered or represented, but rather merely adopted. Similarly, the machine table would presumably be built from physical circuitry, and again would be neither represented nor remembered). The program executed by a universal Turing machine, and the data manipulations that resulted, were completely scaffolded. The machine's memory is literally an external notebook analogous to that used by Oscar in the famous argument for extending the mind (Clark & Chalmers, 1998). That is, the data and program for a universal Turing machine are both stored externally, on the machine's ticker tape.

Indeed, the interactions between a universal Turing machine's machine head and its ticker tape are decidedly of the sense-act, and not of the sense-think-act, variety. Every possible operation in the machine table performs an action (either writing something on the ticker tape or moving the tape one cell to the right or to the left) immediately after sensing the current symbol on the tape and the current state of the machine head. No other internal, intermediary processing (i.e., thinking) is required.

Similarly, external scaffolding was characteristic of later-generation relay computers developed at Bell labs, such as the Mark III. These machines employed more than one tape reader, permitting external tapes to be used to store tables of pre-computed values. This resulted in the CADET architecture ("Can't Add, Doesn't Even Try") that worked by looking up answers to addition and other problems instead of computing the result (Williams, 1997). This was possible because of a "hunting circuit" that permitted the computer to move to any desired location on a punched tape (Stibitz & Loveday, 1967b). ENIAC employed scaffolding as well, obtaining standard function values by reading them from cards (Williams, 1997).

From an engineering perspective, the difference between externally controlled and stored program computers was quantitative (e.g. speed of processing) and not qualitative (e.g. type of processing). In other words, to a computer engineer there may be no principled difference between a sense-act device such as a universal Turing machine and a sense-think-act computer such as the EDVAC. In the context of cognitive control, then, there may be no qualitative element that distinguishes the classical and embodied approaches.

Perhaps a different perspective on control may reveal sharp distinctions between classical and embodied cognitive science. For instance, a key element in the 1945 description of the EDVAC was the component called the central control unit (Godfrey & Hendry, 1993; von Neumann, 1993). It was argued by von Neumann that the most efficient way to control a stored program computer was to have a physical component of the device devoted to control (i.e., to the fetching, decoding, and executing of program steps). Von Neumann called this the "central control organ." Perhaps it is the notion that control is centralized to a particular location or organ of a classical device that serves as the division between classical and embodied models. For instance, behaviour-based roboticists often strive to decentralize control (Brooks, 1999). In Brooks' early six-legged walking robots like Attila, each leg of the robot was responsible for its own control, and no central control organ was included in the design (Brooks, 2002).

However, it appears that the need for a central control organ was tied again to pragmatic engineering rather than to a principled requirement for defining information processing. The adoption of a central controller reflected adherence to engineering's principle of modular design (Marr, 1976). According to this principle, "any large computation should be split up and implemented as a collection of small sub-parts that are as nearly independent of one another as the overall task allows" (p. 485). Failure to devise a functional component or process according to the principle of modular design typically means,

> that the process as a whole becomes extremely difficult to debug or to improve, whether by a human designer or in the course of natural evolution, because a small change to improve one part has to be accompanied by many simultaneous compensating changes elsewhere. (Marr, 1976, p. 485)

Digital computers were explicitly designed according to the principle of modular design, which von Neumann (1958) called "the principle of only one organ for each basic operation" (p. 13). Not only was this good engineering practice, but von Neumann also argued that this principle distinguished digital computers from their analog ancestors such as the differential analyzer (Bush, 1931).

The principle of modular design is also reflected in the architecture of the universal Turing machine. The central control organ of this device is its machine table (see Figure 3-8), which is separate and independent from the other elements of the device, such as the mechanisms for reading and writing the tape, the machine state, and so on. Recall that the machine table is a set of instructions; each instruction is associated with a specific input symbol and a particular machine state. When a Turing machine in physical state x reads symbol y from the tape, it proceeds to execute the instruction at coordinates (x, y) in its machine table.

Importantly, completely decentralized control results in a Turing machine when von Neumann's (1958) principle of only one organ for each basic operation is taken to the extreme. Rather than taking the entire machine table as a central control organ, one could plausibly design an uber-modular system in which each instruction was associated with its own organ. For example, one could replace the machine table with a production system in which each production was responsible for one of the machine table's entries. The conditions for each production would be a particular machine state and a particular input symbol, and the production's action would be the required manipulation of the ticker tape. In this case, the production system version of the Turing machine would behave identically to the original version. However, it would no longer have a centralized control organ.

In short, central control is not a necessary characteristic of classical information processing, and therefore does not distinguish between classical and embodied theories. Another way of making this point is to remember the Chapter 3 observation that production systems are prototypical examples of classical architectures (Anderson et al., 2004; Newell, 1973), but they, like many embodied models (Dawson, Dupuis, & Wilson, 2010; Holland & Melhuish, 1999; Susi & Ziemke, 2001; Theraulaz & Bonabeau, 1999), are controlled stigmergically. "Traditional production system control is internally stigmergic, because the contents of working memory determine which production will act at any given time" (Dawson, Dupuis, & Wilson, 2010, p. 76).

The discussion to this point has used the history of the automatic control of computers to argue that characteristics of control cannot be used to provide a principled distinction between classical and embodied cognitive science. Let us now examine connectionist cognitive science in the context of cognitive control.

Connectionists have argued that the nature of cognitive control provides a principled distinction between network models and models that belong to the classical tradition (Rumelhart & McClelland, 1986b). In particular, connectionist cognitive scientists claim that control in their networks is completely decentralized, and that this property is advantageous because it is biologically plausible: "There is one final aspect of our models which is vaguely derived from our understanding of brain functioning. This is the notion that there is *no central executive* overseeing the general flow of processing" (Rumelhart & McClelland, 1986b, p. 134).

However, the claim that connectionist networks are not under central control is easily refuted; Dawson and Schopflocher (1992a) considered a very simple connectionist system, the distributed memory or standard pattern associator described in Chapter 4 (see Figure 4-1). They noted that connectionist researchers typically describe such models as being autonomous, suggesting that the key operations of such a memory (namely learning and recall) are explicitly defined in its architecture, that is, in the connection weights and processors, as depicted in Figure 4-1.

However, Dawson and Schopflocher (1992a) proceeded to show that even in such a simple memory system, whether the network learns or recalls information depends upon instructions provided by an external controller: the programmer demonstrating the behaviour of the network. When instructed to learn, the components of the standard pattern associator behave one way. However, when instructed to recall, these same components behave in a very different fashion. The nature of the network's processing depends critically upon signals provided by a controller that is not part of the network architecture.

For example, during learning the output units in a standard pattern associator serve as a second bank of input units, but during recall they record the network's response to signals sent from the other input units. How the output units behave is determined by whether the network is involved in either a learning phase or a recall phase, which is signaled by the network's user, not by any of its architectural components. Similarly, during the learning phase connection weights are modified according to a learning rule, but the weights are not modified during the recall phase. How the weights behave is under the user's control. Indeed, the learning rule is defined outside the architecture of the network that is visible in Figure 4-1.

Dawson and Schopflocher (1992a) concluded that,

current PDP networks are not autonomous because their learning principles are not in fact directly realized in the network architecture. That is, networks governed by these principles require explicit signals from some external controller to determine when they will learn or when they will perform a learned task. (Dawson and Schopflocher 1992a, pp. 200–201)

This is not a principled limitation, for Dawson and Schopflocher presented a much more elaborate architecture that permits a standard pattern associator to learn and recall autonomously, that is, without the need for a user's intervention. However, this architecture is not typical; standard pattern associators like the one in Figure 4-1 demand executive control.

The need for such control is not limited to simple distributed memories. The same is true for a variety of popular and more powerful multilayered network architectures, including multilayered perceptrons and self-organizing networks (Roy, 2008). "There is clearly a *central executive that oversees* the operation of the back-propagation algorithm" (p. 1436). Roy (2008) proceeded to argue that such control is itself required by brain-like systems, and therefore biologically plausible networks demand not only an explicit account of data transformation, but also a biological theory of executive control.

In summary, connectionist networks generally require the same kind of control that is a typical component of a classical model. Furthermore, it was argued earlier that there does not appear to be any principled distinction between this kind of control and the type that is presumed in an embodied account of cognition. Control is a key characteristic of a cognitive theory, and different schools of thought in cognitive science are united in appealing to the same type of control mechanisms. In short, central control is not a mark of the classical.

7.4 Serial versus Parallel Processing

Classical cognitive science was inspired by the characteristics of digital computers; few would deny that the classical approach exploits the digital computer metaphor (Pylyshyn, 1979a). Computers are existence proofs that physical machines are capable of manipulating, with infinite flexibility, semantically interpretable expressions (Haugeland, 1985; Newell, 1980; Newell & Simon, 1976). Computers illustrate how logicism can be grounded in physical mechanisms.

The connectionist and the embodied reactions to classical cognitive science typically hold that the digital computer metaphor is not appropriate for theories of cognition. It has been argued that the operations of traditional electronic computers are qualitatively different from those of human cognition, and as a result the classical models they inspire are doomed to fail, as are attempts to produce artificial intelligence in such machines (Churchland & Sejnowski, 1992; Dreyfus, 1972, 1992; Searle, 1980).

In concert with rejecting the digital computer metaphor, connectionist and embodied cognitive scientists turn to qualitatively different notions in an attempt to distinguish their approaches from the classical theories that preceded them. However, their attempt to define the mark of the classical, and to show how this mark does not apply to their theories, is not always successful.

For example, it was argued in the previous section that when scholars abandoned the notion of centralized control, they were in fact reacting against a concept that was not a necessary condition of classical theory, but was instead an engineering convenience. Furthermore, mechanisms of control in connectionist and embodied theories were shown not to be radically different from those of classical models. The current section provides another such example.

One of the defining characteristics of classical theory is serial processing, the notion that only one operation can be executed at a time. Opponents of classical cognitive science have argued that this means classical models are simply too slow to be executed by the sluggish hardware that makes up the brain (Feldman & Ballard, 1982). They suggest that what is instead required is parallel processing, in which many operations are carried out simultaneously. Below it is argued that characterizing digital computers or classical theories as being serial in nature is not completely accurate. Furthermore, characterizing alternative schools of thought in cognitive science as champions of parallel processing is also problematic. In short, the difference between serial and parallel processing may not provide a clear distinction between different approaches to cognitive science.

It cannot be denied that serial processing has played an important role in the history of modern computing devices. Turing's (1936) original account of computation was purely serial: a Turing machine processed only a single symbol at a time, and did so by only executing a single operation at a time. However, the purpose of Turing's proposal was to provide an uncontroversial notion of "definite method"; serial processing made Turing's notion of computation easy to understand, but was not a necessary characteristic.

A decade later, the pioneering stored program computer EDVAC was also a serial device in two different ways (Ceruzzi, 1997; von Neumann, 1993). First, it only executed one command at a time. Second, even though it used 44 bits to represent a number as a "word," it processed these words serially, operating on them one bit at a time. Again, though, this design was motivated by a desire for simplicity—in this case, simplicity of engineering. "The device should be as simple as possible, that is, contain as few elements as possible. This can be achieved by never performing two operations simultaneously, if this would cause a significant increase in the number of elements required" (von Neumann, 1993, p. 8).

Furthermore, the serial nature of EDVAC was also dictated by engineering constraints on the early stored program machines. The existence of such devices depended upon the invention of new kinds of memory components (Williams, 1997). EDVAC used a delay line memory system, which worked by delaying a series of pulses (which represented a binary number) for a few milliseconds, and then by feeding these pulses back into the delay line so that they persisted in memory. Crucially, delay line memories only permitted stored information to be accessed in serial, one bit at a time.

EDVAC's simple, serial design reflected an explicit decision against parallel processing that von Neumann (1993) called telescoping processes.

> It is also worth emphasizing that up to now all thinking about high speed digital computing devices has tended in the opposite direction: Towards acceleration by telescoping processes at the price of multiplying the number of elements required. It would therefore seem to be more instructive to try to think out as completely as possible the opposite viewpoint. (von Neumann, 1993, p. 8)

EDVAC's opposite viewpoint was only practical because of the high speed of its vacuum tube components.

Serial processing was an attractive design decision because it simplified the architecture of EDVAC. However, it was not a *necessary* design decision. The telescoping of processes was a common design decision in older computing devices that used slower components. Von Neumann was well aware that many of EDVAC's ancestors employed various degrees of parallel processing.

> In all existing devices where the element is not a vacuum tube the reaction time of the element is sufficiently long to make a certain telescoping of the steps involved in addition, subtraction, and still more in multiplication and division, desirable. (von Neumann, 1993, p. 6)

For example, the Zuse computers performed arithmetic operations in parallel, with one component manipulating the exponent and another manipulating the mantissa of a represented number (Zuse, 1993). Aiken's Mark II computer at Harvard also had multiple arithmetic units that could be activated in parallel, though this was not common practice because coordination of its parallel operations were difficult to control (Williams, 1997). ENIAC used 20 accumulators as mathematical operators, and these could be run simultaneously; it was a parallel machine (Neukom, 2006).

In spite of von Neumann's (1993) championing of serial processing, advances in computer memory permitted him to adopt a partially parallel architecture in the machine he later developed at Princeton (Burks, Goldstine, & Von Neumann, 1989). Cathode ray tube memories (Williams & Kilburn, 1949) allowed all of the bits of a word in memory to be accessed in parallel, though operations on this retrieved information were still conducted in serial.

> To get a word from the memory in this scheme requires, then, one switching mechanism to which all 40 tubes are connected in parallel. Such a switching scheme seems to us to be simpler than the technique needed in the serial system and is, of course, 40 times faster. We accordingly adopt the parallel procedure and thus are led to consider a so-called *parallel machine*, as contrasted with the serial principles being considered for the EDVAC. (Burks, Goldstine & von Neumann, 1989, p. 44)

Interestingly, the extreme serial design in EDVAC resurfaced in the pocket calculators of the 1970s, permitting them to be simple and small (Ceruzzi, 1997).

The brief historical review provided above indicates that while some of the early computing devices were serial processors, many others relied upon a certain degree of parallel processing. The same is true of some prototypical architectures proposed by classical cognitive science. For example, production systems (Newell, 1973, 1990; Newell & Simon, 1972) are serial in the sense that only one production manipulates working memory at a time. However, all of the productions in such a system scan the working memory in parallel when determining whether the condition that launches their action is present.

An alternative approach to making the case that the serial processing is not a mark of the classical is to note that serial processing also appears in non-classical architectures. The serial versus parallel distinction is typically argued to be one of the key differences between connectionist and classical theories. For instance, parallel processing is required to explain how the brain is capable of performing complex calculations in spite of the slowness of neurons in comparison to electronic components (Feldman & Ballard, 1982; McClelland, Rumelhart, & Hinton, 1986; von Neumann, 1958). In comparing brains to digital computers, von Neumann (1958, p. 50) noted that "the natural componentry favors automata with more, but slower, organs, while the artificial one favors the reverse arrangement of fewer, but faster organs."

It is certainly the case that connectionist architectures have a high degree of parallelism. For instance, all of the processing units in the same layer of a multilayered perceptron are presumed to operate simultaneously. Nevertheless, even prototypical parallel distributed processing models reveal the presence of serial processing.

One reason that the distributed memory or the standard pattern associator requires external, central control (Dawson & Schopflocher, 1992a) is because this kind of model is not capable of simultaneous learning and recalling. This is because one of its banks of processors is used as a set of input units during learning, but is used completely differently, as output units, during recall. External control is used to determine how these units are employed and therefore determines whether the machine is learning or recalling. External control also imposes seriality in the sense that during learning input, patterns are presented in sequence, and during recall, presented cues are again presented one at a time. Dawson and Schopflocher (1992a) demonstrated how true parallel processing could be accomplished in such a network, but only after substantially elaborating the primitive components of the connectionist architecture.

A degree of serial processing is also present in multilayered networks. First, while all processors in one layer can be described as operating in parallel, the flow of information from one layer to the next is serial. Second, the operations of an individual processor are intrinsically serial. A signal cannot be output until internal activation has been computed, and internal activation cannot be computed until the net input has been determined.

Parallel processing is not generally proposed as a characteristic that distinguishes embodied from classical models. However, some researchers have noted the advantages of decentralized computation in behaviour-based robots (Brooks, 1999).

Again, though, embodied theories seem to exploit a mixture of parallel and serial processing. Consider the early insect-like walking robots of Rodney Brooks (1989, 1999, 2002). Each leg in the six-legged robot Genghis is a parallel processor, in the sense that each leg operates autonomously. However, the operations of each leg can be described as a finite state automaton (see the appendix on Genghis in Brooks, 2002), which is an intrinsically serial device.

The stigmergic control of the swarm intelligence that emerges from a collection of robots or social insects (Beni, 2005; Bonabeau & Meyer, 2001; Hinchey, Sterritt, & Rouff, 2007; Sharkey, 2006; Tarasewich & McMullen, 2002) also appears to be a mixture of parallel and serial operations. A collective operates in parallel in the sense that each member of the collective is an autonomous agent. However, the behaviour of each agent is often best characterized in serial: first the agent does one thing, and then it does another, and so on. For instance, in a swarm capable of creating a nest by blind bulldozing (Parker, Zhang, & Kube, 2003), agents operate in parallel. However, each agent moves in serial from one state (e.g., plowing, colliding, finishing) to another.

In summary, serial processing has been stressed more in classical models, while parallel processing has received more emphasis in connectionist and embodied approaches. However, serial processing cannot be said to be a mark of the classical.

First, serial processing in classical information processing systems was adopted as an engineering convenience, and many digital computers included a certain degree of parallel processing. Second, with careful examination serial processing can also be found mixed in with the parallel processing of connectionist networks or of collective intelligences.

7.5 Local versus Distributed Representations

Classical and connectionist cognitive scientists agree that theories of cognition must appeal to internal representations (Fodor & Pylyshyn, 1988). However, they appear to have strong disagreements about the nature of such representations. In particular, connectionist cognitive scientists propose that their networks exploit distributed representations, which provide many advantages over the local representations that they argue characterize the classical approach (Bowers, 2009). That is, distributed representations are often taken to be a mark of the connectionist, and local representations are taken to be a mark of the classical.

There is general, intuitive agreement about the differences between distributed and local representations. In a connectionist distributed representation, "knowledge is coded as a pattern of activation across many processing units, with each unit contributing to multiple, different representations. As a consequence, there is no one unit devoted to coding a given word, object, or person" (Bowers, 2009, p. 220). In contrast, in a classical local representation, "individual words, objects, simple concepts, and the like are coded distinctly, with their own dedicated representation" (p. 22).

However, when the definition of *distributed representation* is examined more carefully (van Gelder, 1991), two facts become clear. First, this term is used by different connectionists in different ways. Second, some of the uses of this term do not appear to differentiate connectionist from classical representations.

Van Gelder (1991) noted, for instance, that one common sense of *distributed representation* is that it is extended: a distributed representation uses many units to represent each item, while local representations do not. "To claim that a node is distributed is presumably to claim that its states of activation correspond to patterns of neural activity—to aggregates of neural 'units'—rather than to activations of single neurons" (Fodor & Pylyshyn, 1988, p. 19). It is this sense of an extended or distributed representation that produces connectionist advantages such as damage resistance, because the loss of one of the many processors used to represent a concept will not produce catastrophic loss of represented information.

However, the use of *extended* to define *distributed* does not segregate connectionist representations from their classical counterparts. For example, the mental image is an important example of a classical representation (Kosslyn, 1980; Kosslyn, Thompson, & Ganis, 2006; Paivio, 1971, 1986). It would be odd to think of a mental image as being distributed, particularly in the context of the connectionist use of this term. However, proponents of mental imagery would argue that they are extended, functionally in terms of being extended over space, and physically in terms of being extended over aggregates of neurons in topographically organized areas of the cortex (Kosslyn, 1994; Kosslyn, Ganis, & Thompson, 2003;

Kosslyn et al., 1995). "There is good evidence that the brain depicts representations literally, using space on the cortex to represent space in the world" (Kosslyn, Thompson, & Ganis, 2006, p. 15).

Another notion of distributed representation considered by van Gelder (1991) was the coarse code (Feldman & Ballard, 1982; Hinton, McClelland, & Rumelhart, 1986). Again, a coarse code is typically presented as distinguishing connectionist networks from classical models. A coarse code is extended in the sense that multiple processors are required to do the representing. These processors have two properties. First, their receptive fields are wide—that is, they are very broadly tuned, so that a variety of circumstances will lead to activation in a processor. Second, the receptive fields of different processors overlap. In this kind of representation, a high degree of accuracy is possible by pooling the responses of a number of broadly tuned (i.e., coarse) processors (Dawson, Boechler, & Orsten, 2005; Dawson, Boechler, & Valsangkar-Smyth, 2000).

While coarse coding is an important kind of representation in the connectionist literature, once again it is possible to find examples of coarse coding in classical models as well. For example, one way that coarse coding of spatial location is presented by connectionists (Hinton, McClelland, & Rumelhart, 1986) can easily be recast in terms of Venn diagrams. That is, each non-empty set represents the coarse location of a target in a broad spatial area; the intersection of overlapping non-empty sets provides more accurate target localization.

However, classical models of syllogistic reasoning can be cast in similar fashions that include Euler circles and Venn diagrams (Johnson-Laird, 1983). Indeed, Johnson-Laird's (1983) more modern notion of mental models can themselves be viewed as an extension of these approaches: syllogistic statements are represented as a tableau of different instances; the syllogism is solved by combining (i.e., intersecting) tableaus for different statements and examining the relevant instances that result. In other words, mental models can be considered to represent a classical example of coarse coding, suggesting that this concept does not necessarily distinguish connectionist from classical theories.

After his more detailed analysis of the concept, van Gelder (1991) argued that a stronger notion of *distributed* is required, and that this can be accomplished by invoking the concept of superposition. Two different concepts are superposed if the same resources are used to provide their representations. "Thus in connectionist networks we can have different items stored as patterns of activity over the same set of units, or multiple different associations encoded in one set of weights" (p. 43).

Van Gelder (1991) pointed out that one issue with superposition is that it must be defined in degrees. For instance, it may be the case that not all resources are used simultaneously to represent all contents. Furthermore, operationalizing the notion of superposition depends upon how resources are defined and measured.

Finally, different degrees of superposition may be reflected in the number of different contents that a given resource can represent. For example, it is well known that one kind of artificial neural network, the Hopfield network (Hopfield, 1982), is of limited capacity, where if the network is comprised of N processors, it will be only to be able to represent in the order of $0.18N$ distinct memories (Abu-Mostafa & St. Jacques, 1985; McEliece, et al., 1987).

Nonetheless, van Gelder (1991) expressed confidence that the notion of superposition provides an appropriate characteristic for defining a distributed representation. "It is strong enough that very many kinds of representations do not count as superposed, yet it manages to subsume virtually all paradigm cases of distribution, whether these are drawn from the brain, connectionism, psychology, or optics" (p. 54).

Even if van Gelder's (1991) definition is correct, it is still the case that the concept of superposition does not universally distinguish connectionist representations from classical ones. One example of this is when concepts are represented as collections of features or microfeatures. For instance, in an influential PDP model called an interactive activation and competition network (McClelland & Rumelhart, 1988), most of the processing units represent the presence of a variety of features. Higher-order concepts are defined as sets of such features. This is an instance of superposition, because the same feature can be involved in the representation of multiple networks. However, the identical type of representation—that is, superposition of featural elements—is also true of many prototypical classical representations, including semantic networks (Collins & Quillian, 1969, 1970a, 1970b) and feature set representations (Rips, Shoben, & Smith, 1973; Tversky, 1977; Tversky & Gati, 1982).

The discussion up to this point has considered a handful of different notions of distributed representation, and has argued that these different definitions do not appear to uniquely separate connectionist and classical concepts of representation. To wrap up this discussion, let us take a different approach, and consider why in some senses connectionist researchers may still need to appeal to local representations.

One problem of considerable interest within cognitive neuroscience is the issue of assigning specific behavioural functions to specific brain regions; that is, the localization of function. To aid in this endeavour, cognitive neuroscientists find it useful to distinguish between two qualitatively different types of behavioural deficits. A single dissociation consists of a patient performing one task extremely poorly while performing a second task at a normal level, or at least very much better than the first. In contrast, a double dissociation occurs when one patient performs the first task significantly poorer than the second, and another patient (with a different brain injury) performs the second task significantly poorer than the first (Shallice, 1988).

Cognitive neuroscientists have argued that double dissociations reflect damages to localized functions (Caramazza, 1986; Shallice, 1988). The view that dissociation data reveals internal structures that are local in nature has been named the locality assumption (Farah, 1994).

However, Farah (1994) hypothesized that the locality assumption may be unwarranted for two reasons. First, its validity depends upon the additional assumption that the brain is organized into a set of functionally distinct modules (Fodor, 1983). Farah argued that the modularity of the brain is an unresolved empirical issue. Second, Farah noted that it is possible for nonlocal or distributed architectures, such as parallel distributed processing (PDP) networks, to produce single or double dissociations when lesioned. As the interactive nature of PDP networks is "directly incompatible with the locality assumption" (p. 46), the locality assumption may not be an indispensable tool for cognitive neuroscientists.

Farah (1994) reviewed three areas in which neuropsychological dissociations had been used previously to make inferences about the underlying local structure. For each she provided an alternative architecture—a PDP network. Each of these networks, when locally damaged, produced (local) behavioural deficits analogous to the neuropsychological dissociations of interest. These results led Farah to conclude that one cannot infer that a specific behavioural deficit is associated with the loss of a local function, because the prevailing view is that PDP networks are, by definition, distributed and therefore nonlocal in structure.

However, one study challenged Farah's (1994) argument both logically and empirically (Medler, Dawson, & Kingstone, 2005). Medler, Dawson, and Kingstone (2005) noted that Farah's whole argument was based on the assumption that connectionist networks exhibit universally distributed internal structure. However, this assumption needs to be empirically supported; Medler and colleagues argued that this could only be done by interpreting the internal structure of a network and by relating behavioural deficits to interpretations of ablated components. They noted that it was perfectly possible for PDP networks to adopt internal representations that were more local in nature, and that single and double dissociations in lesioned networks may be the result of damaging local representations.

Medler, Dawson, and Kingstone (2005) supported their position by training a network on a logic problem and interpreting the internal structure of the network, acquiring evidence about how local or how nonlocal the function of each hidden unit was. They then created different versions of the network by lesioning one of its 16 hidden units, assessing behavioural deficits in each lesioned network. They found that the more local a hidden unit was the more profound and specific was the behavioural deficit that resulted when the unit was lesioned. "For a double dissociation to occur within a computational model, the model must have some form of functional localization" (p. 149).

We saw earlier that one of the key goals of connectionist cognitive science was to develop models that were biologically plausible. Clearly one aspect of this is to produce networks that are capable of reflecting appropriate deficits in behaviour when damaged, such as single or double dissociations. Medler, Dawson, and Kingstone (2005) have shown that the ability to do so, even in PDP networks, requires local representations. This provides another line of evidence against the claim that distributed representations can be used to distinguish connectionist from classical models. In other words, local representations do not appear to be a mark of the classical.

7.6 Internal Representations

One of the key properties of classical cognitive science is its emphasis on sense-think-act processing. Classical cognitive scientists view the purpose of cognition as planning action on the basis of input information. This planning typically involves the creation and manipulation of internal models of the external world. Is the classical sandwich (Hurley, 2001) a mark of the classical?

Sense-think-act processing does not distinguish classical models from connectionist networks. The distributed representations within most modern networks mediate all relationships between input units (sensing) and output units (responding). This results in what has been described as the connectionist sandwich (Calvo & Gomila, 2008). Sense-think-act processing is a mark of both the classical and the connectionist.

While sense-think-act processing does not distinguish classical cognitive science from connectionism, it may very well differentiate it from embodied cognitive science. Embodied cognitive scientists have argued in favor of sense-act processing that abandons using internal models of the world (Pfeifer & Scheier, 1999). The purpose of cognition might not be to plan, but instead to control action on the world (Clark, 1997). Behaviour-based robots arose as an anti-representational reaction to classical research in artificial intelligence (Brooks, 1991). The direct link between perception and action—a link often described as circumventing internal representation—that characterized the ecological approach to perception (Gibson, 1979; Turvey et al., 1981) has been a cornerstone of embodied theory (Chemero, 2009; Chemero & Turvey, 2007; Neisser, 1976; Noë, 2004; Winograd & Flores, 1987a).

The distinction between sense-think-act processing and sense-act processing is a putative differentiator between classical and embodied approaches. However, it is neither a necessary nor sufficient one. This is because in both classical and embodied approaches, mixtures of both types of processing can readily be found.

For example, it was earlier shown that the stored program computer—a digital computer explicitly designed to manipulate internal representations—emerged from technical convenience, and did not arise because classical information processing demanded internal representations. Prototypical classical machines, such as the

Turing machine, can easily be described as pure sense-act processors (Wells, 1996). Also, earlier electromechanical computers often used external memories to scaffold processing because of the slow speed of their componentry.

Furthermore, prototypical classical architectures in cognitive science appeal to processes that are central to the embodied approach. For example, modern production systems have been extended to include sensing and acting, and have used these extensions to model (or impose) constraints on behaviour, such as our inability to use one hand to do two tasks at the same time (Kieras & Meyer, 1997; Meyer et al., 2001; Meyer & Kieras, 1997a, 1997b, 1999; Meyer et al., 1995). A production system for solving the Towers of Hanoi problem also has been formulated that uses the external towers and discs as the *external* representation of the problem (Vera & Simon, 1993). Some have argued that the classical emphasis on internal thinking, at the expense of external sense-acting, simply reflects the historical development of the classical approach and does not reflect its intrinsic nature (Newell, 1990).

Approaching this issue from the opposite direction, many embodied cognitive scientists are open to the possibility that the representational stance of classical cognitive science may be required to provide accounts of some cognitive phenomena. For instance, Winograd and Flores (1987a) made strong arguments for embodied accounts of cognition. They provided detailed arguments of how classical views of cognition are dependent upon the disembodied view of the mind that has descended from Descartes. They noted that "detached contemplation can be illuminating, but it also obscures the phenomena themselves by isolating and categorizing them" (pp. 32–33). However, in making this kind of observation, they admitted the existence of a kind of reasoning called detached contemplation. Their approach offers an alternative to representational theories, but does not necessarily completely abandon the possibility of internal representations.

Similarly, classical cognitive scientists who appeal exclusively to internal representations and embodied cognitive scientists who completely deny internal representations might be staking out extreme and radical positions to highlight the differences between their approaches (Norman, 1993). Some embodied cognitive scientists have argued against this radical polarization of cognitive science, such as Clark (1997):

> Such radicalism, I believe, is both unwarranted and somewhat counterproductive. It invites competition where progress demands cooperation. In most cases, at least, the emerging emphasis on the roles of body and world can be seen as complementary to the search for computational and representational understandings.
> (Clark, 1997, p. 149)

Clark (1997) adopted this position because he realized that representations may be critical to cognition, provided that appeals to representation do not exclude appeals to other critical, embodied elements: "We should not be too quick to reject the more traditional explanatory apparatuses of computation and representation. Minds may be essentially embodied and embedded and *still* depend crucially on brains which compute and represent" (p. 143).

The reason that an embodied cognitive scientist such as Clark may be reluctant to eliminate representations completely is because one can easily consider situations in which internal representations perform an essential function. Clark (1997) suggested that some problems might be representation hungry, in the sense that the very nature of these problems requires their solutions to employ internal representations. A problem might be representation hungry because it involves features that are not reliably present in the environment, as in reasoning about absent states, or in counterfactual reasoning. A problem might also be representation hungry if it involves reasoning about classes of objects that are extremely abstract, because there is a wide variety of different physical realizations of class instances (for instance, reasoning about "computers"!).

The existence of representation-hungry problems leaves Clark (1997) open to representational theories in cognitive science, but these theories must be placed in the context of body and world. Clark didn't want to throw either the representational or embodied babies out with the bathwater (Hayes, Ford, & Agnew, 1994). Instead, he viewed a co-operative system in which internal representations can be used when needed, but the body and the world can also be used to reduce internal cognitive demands by exploiting external scaffolds. "We will not discover the *right* computational and representational stories unless we give due weight to the role of body and local environment—a role that includes both problem definition and, on occasion, problem solution" (Clark, 1997, p. 154).

It would seem, then, that internal representations are not a mark of the classical, and some cognitive scientists are open to the possibility of hybrid accounts of cognition. That is, classical researchers are extending their representational theories by paying more attention to actions on the world, while embodied researchers are open to preserving at least some internal representations in their theories. An example hybrid theory that appeals to representations, networks, and actions (Pylyshyn, 2003c, 2007) is presented in detail in Chapter 8.

7.7 Explicit Rules versus Implicit Knowledge

Connectionists have argued that one mark of the classical is its reliance on explicit rules (McClelland, Rumelhart, & Hinton, 1986). For example, it has been claimed that all classical work on knowledge acquisition "shares the assumption that the goal of learning is to formulate explicit rules (proposition, productions, etc.) which capture powerful generalizations in a succinct way" (p. 32).

Explicit rules may serve as a mark of the classical because it has also been argued that they are *not* characteristic of other approaches in cognitive science, particularly connectionism. Many researchers assume that PDP networks acquire implicit knowledge. For instance, consider this claim about a network that learns to convert verbs from present to past tense:

> The model learns to behave in accordance with the rule, not by explicitly noting that most words take -*ed* in the past tense in English and storing this rule away explicitly, but simply by building up a set of connections in a pattern associator through a long series of simple learning experiences. (McClelland, Rumelhart, & Hinton, 1986, p. 40)

One problem that immediately arises in using explicit rules as a mark of the classical is that the notions of explicit rules and implicit knowledge are only vaguely defined or understood (Kirsh, 1992). For instance, Kirsh (1992) notes that the distinction between explicit rules and implicit knowledge is often proposed to be similar to the distinction between local and distributed representations. However, this definition poses problems for using explicit rules as a mark of the cognitive. This is because, as we have already seen in an earlier section of this chapter, the distinction between local and distributed representations does not serve well to separate classical cognitive science from other approaches.

Furthermore, defining explicit rules in terms of locality does not eliminate connectionism's need for them (Hadley, 1993). Hadley (1993) argued that there is solid evidence of the human ability to instantaneously learn and apply rules.

> *Some* rule-like behavior cannot be the product of 'neurally-wired' rules whose structure is embedded in particular networks, for the simple reason that humans can often apply rules (with considerable accuracy) as soon as they are told the rules. (Hadley, 1993, p. 185)

Hadley proceeded to argue that connectionist architectures need to exhibit such (explicit) rule learning. "The foregoing conclusions present the connectionist with a formidable scientific challenge, which is, to show how general purpose rule following mechanisms may be implemented in a connectionist architecture" (p. 199).

Why is it that, on more careful consideration, it seems that explicit rules are not a mark of the cognitive? It is likely that the assumption that PDP networks acquire implicit knowledge is an example of what has been called gee whiz connectionism (Dawson, 2009). That is, connectionists *assume* that the internal structure of their networks is neither local nor rule-like, and they *rarely* test this assumption by conducting detailed interpretations of network representations. When such interpretations are conducted, they can reveal some striking surprises. For instance, the internal structures of networks have revealed classical rules of logic (Berkeley et al., 1995) and classical production rules (Dawson et al., 2000).

The discussion in the preceding paragraphs raises the possibility that connectionist networks can acquire explicit rules. A complementary point can also be made to question explicit rules as a mark of the classical: classical models may not themselves require explicit rules. For instance, classical cognitive scientists view an explicit rule as an encoded representation that is part of the algorithmic level. Furthermore, the reason that it is explicitly represented is that it is not part of the architecture (Fodor & Pylyshyn, 1988). In short, classical theories posit a combination of explicit (algorithmic, or stored program) and implicit (architectural) determinants of cognition. As a result, classical debates about the cognitive architecture can be construed as debates about the implicitness or explicitness of knowledge:

> Not only is there no reason why Classical models are required to be rule-explicit but—as a matter of fact—arguments over which, if any, rules are explicitly mentally represented have raged for decades *within* the Classicist camp.
> (Fodor & Pylyshyn, p. 60)

To this point, the current section has tacitly employed the context that the distinction between explicit rules and implicit knowledge parallels the distinction between local and distributed representations. However, other contexts are also plausible. For example, classical models may be characterized as employing explicit rules in the sense that they employ a structure/process distinction. That is, classical systems characteristically separate their symbol-holding memories from the rules that modify stored contents.

For instance, the Turing machine explicitly distinguishes its ticker tape memory structure from the rules that are executed by its machine head (Turing, 1936). Similarly, production systems (Anderson, 1983; Newell, 1973) separate their symbolic structures stored in working memory from the set of productions that scan and manipulate expressions. The von Neumann (1958, 1993) architecture by definition separates its memory organ from the other organs that act on stored contents, such as its logical or arithmetical units.

To further establish this alternative context, some researchers have claimed that PDP networks or other connectionist architectures do *not* exhibit the structure/process distinction. For instance, a network can be considered to be an active data structure that not only stores information, but at the same time manipulates it (Hillis, 1985). From this perspective, the network is both structure and process.

However, it is still the case that the structure/process distinction fails to provide a mark of the classical. The reason for this was detailed in this chapter's earlier discussion of control processes. That is, almost all PDP networks are controlled by external processes—in particular, learning rules (Dawson & Schopflocher, 1992a; Roy, 2008). This external control takes the form of rules that are as explicit as any to be found in a classical model.

To bring this discussion to a close, I argue that a third context is possible for distinguishing explicit rules from implicit knowledge. This context is the difference between digital and analog processes. Classical rules may be explicit in the sense that they are digital: consistent with the neural all-or-none law (Levitan & Kaczmarek, 1991; McCulloch & Pitts, 1943), as the rule either executes or does not. In contrast, the continuous values of the activation functions used in connectionist networks permit knowledge to be applied to varying degrees. From this perspective, networks are analog, and are not digital.

Again, however, this context also does not successfully provide a mark of the classical. First, one consequence of Church's thesis and the universal machine is that digital and analogical devices are functionally equivalent, in the sense that one kind of computer can simulate the other (Rubel, 1989). Second, connectionist models themselves can be interpreted as being either digital or analog in nature, depending upon task demands. For instance, when a network is trained to either respond or not, as in pattern classification (Lippmann, 1989) or in the simulation of animal learning (Dawson, 2008), output unit activation is treated as being digital. However, when one is interested in solving a problem in which continuous values are required, as in function approximation (Hornik, Stinchcombe, & White, 1989; Kremer, 1995; Medler & Dawson, 1994) or in probability matching (Dawson et al., 2009), the same output unit activation function is treated as being analog in nature.

In conclusion, though the notion of explicit rules has been proposed to distinguish classical models from other kinds of architectures, a more careful consideration suggests that this approach is flawed. Our analysis suggests, however, that the use of explicit rules does not appear to be a reliable mark of the classical. Regardless of how the notion of explicit rules is defined, it appears that classical architectures do not use such rules exclusively, and it also appears that such rules need to be part of connectionist models of cognition.

7.8 The Cognitive Vocabulary

The goal of cognitive science is to explain cognitive phenomena. One approach to such explanation is to generate a set of laws or principles that capture the regularities that are exhibited by members that belong to a particular class. Once it is determined that some new system belongs to a class, then it is expected that the principles known to govern that class will also apply to the new system. In this sense, the laws governing a class capture generalizations (Pylyshyn, 1984).

The problem that faced cognitive science in its infancy was that the classes of interest, and the laws that captured generalizations about their members, depended upon which level of analysis was adopted (Marr, 1982). For instance, at a physical level of investigation, electromechanical and digital computers do not belong to the same class. However, at a more abstract level of investigation (e.g., at the architectural level described in Chapter 2), these two very different types of physical devices belong to the same class, because their components are functionally equivalent: "Many of the electronic circuits which performed the basic arithmetic operations [in ENIAC] were simply electronic analogs of the same units used in mechanical calculators and the commercial accounting machines of the day" (Williams, 1997, p. 272).

The realization that cognitive systems must be examined from multiple levels of analysis motivated Marr's (1982) tri-level hypothesis. According to this hypothesis, cognitive systems must be explained at three different levels of analysis: physical, algorithmic, and computational.

> It is not enough to be able to predict locally the responses of single cells, nor is it enough to be able to predict locally the results of psychophysical experiments. Nor is it enough to be able to write computer programs that perform approximately in the desired way. One has to do all these things at once and also be very aware of the additional level of explanation that I have called the level of computational theory. (Marr, 1982, pp. 329–330)

The tri-level hypothesis provides a foundation for cognitive science and accounts for its interdisciplinary nature (Dawson, 1998). This is because each level of analysis uses a qualitatively different vocabulary to ask questions about cognitive systems and uses very different methods to provide the answers to these questions. That is, each level of analysis appeals to the different languages and techniques of distinct scientific disciplines. The need to explain cognitive systems at different levels of analysis forces cognitive scientists to be interdisciplinary.

Marr's (1982) tri-level hypothesis can also be used to compare the different approaches to cognitive science. Is the tri-level hypothesis equally applicable to the three different schools of thought? Provided that the three levels are interpreted at a moderately coarse level, it would appear that this question could be answered affirmatively.

At Marr's (1982) implementational level, cognitive scientists ask how information processes are physically realized. For a cognitive science of biological agents, answers to implementational-level questions are phrased in a vocabulary that describes biological mechanisms. It would appear that all three approaches to cognitive science are materialist and as a result are interested in conducting implementational-level analyses. Differences between the three schools of thought at this level might only be reflected in the scope of biological mechanisms that are of interest. In particular, classical and connectionist cognitive scientists will emphasize neural mechanisms, while embodied cognitive scientists are likely to be interested not only in the brain but also in other parts of the body that interact with the external world.

At Marr's (1982) algorithmic level, cognitive scientists are interested in specifying the procedures that are used to solve particular information processing problems. At this level, there are substantial technical differences amongst the three schools of thought. For example, classical and connectionist cognitive scientists would appeal to very different kinds of representations in their algorithmic accounts (Broadbent, 1985; Rumelhart & McClelland, 1985). Similarly, an algo-rithmic account of internal planning would be quite different from an embodied account of controlled action, or of scaffolded, cognition. In spite of such technical differences, though, it would be difficult to claim that one approach to cognitive sci-ence provides procedural accounts, while another does not. All three approaches to cognitive science are motivated to investigate at the algorithmic level.

At Marr's (1982) computational level, cognitive scientists wish to determine the nature of the information processing problems being solved by agents. Answering these questions usually requires developing proofs in some formal language. Again, all three approaches to cognitive science are well versed in posing computational-level questions. The differences between them are reflected in the formal lan-guage used to explore answers to these questions. Classical cognitive science often appeals to some form of propositional logic (Chomsky, 1959a; McCawley, 1981; Wexler & Culicover, 1980), the behaviour of connectionist networks lends itself to being described in terms of statistical mechanics (Amit, 1989; Grossberg, 1988; Smolensky, 1988; Smolensky & Legendre, 2006), and embodied cognitive scientists have a preference for dynamical systems theory (Clark, 1997; Port & van Gelder, 1995b; Shapiro, 2011).

Marr's (1982) tri-level hypothesis is only one example of exploring cognition at multiple levels. Precursors of Marr's approach can be found in core writings that appeared fairly early in cognitive science's modern history. For instance, philosopher Jerry Fodor (1968b) noted that one cannot establish any kind of equivalence between the behaviour of an organism and the behaviour of a simulation without first specify-ing a level of description that places the comparison in a particular context.

Marr (1982) himself noted that an even stronger parallel exists between the tri-level hypothesis and Chomsky's (1965) approach to language. To begin with, Chomsky's notion of an innate and universal grammar, as well as his idea of a "language organ" or a "faculty of language," reflect a materialist view of language. Chomsky clearly expects that language can be investigated at the implementational level. The language faculty is due "to millions of years of evolution or to principles of neural organization that may be even more deeply grounded in physical law" (p. 59). Similarly, "the study of innate mechanisms leads us to universal grammar, but also, of course, to investigation of the biologically determined principles that underlie language use" (Chomsky, 1980, p. 206).

Marr's (1982) algorithmic level is mirrored by Chomsky's (1965) concept of linguistic performance. Linguistic performance is algorithmic in the sense that a performance theory should account for "the actual use of language in concrete situ-ations" (Chomsky, 1965, p. 4). The psychology of language can be construed as being primarily concerned with providing theories of performance (Chomsky, 1980). That is, psychology's "concern is the processes of production, interpretation, and the like, which make use of the knowledge attained, and the processes by which transition takes place from the initial to the final state, that is, language acquisition" (pp. 201–202). An account of the processes that underlie performance requires an investigation at the algorithmic level.

Finally, Marr (1982) noted that Chomsky's notion of linguistic competence parallels the computational level of analysis. A theory of linguistic competence specifies an ideal speaker-listener's knowledge of language (Chomsky, 1965). A grammar is a theory of competence; it provides an account of the nature of language that "is unaffected by such grammatically irrelevant conditions as memory limitations, distractions, shifts of attention and interest, and errors (random or characteristic) in applying . . . knowledge of the language in actual performance" (p. 3). As a computational-level theory, a grammar accounts for what in principle *could* be said or understood; in contrast, a performance theory accounts for language behaviours that actually occurred (Fodor, 1968b). Marr (1982) argued that influential criticisms of Chomsky's theory (Winograd, 1972a) mistakenly viewed transformational grammar as an algorithmic, and not a computational, account. "Chomsky's theory of transformational grammar is a true computational theory . . . concerned solely with specifying what the syntactic decomposition of an English sentence should be, and not at all with how that decomposition should be achieved" (Marr, 1982, p. 28).

The notion of the cognitive vocabulary arises by taking a *different* approach to linking Marr's (1982) theory of vision to Chomsky's (1965) theory of language. In addition to proposing the tri-level hypothesis, Marr detailed a sequence of different types of representations of visual information. In the early stages of visual processing, information was represented in the primal sketch, which provided a spatial representation of visual primitives such as boundaries between surfaces. Operations on the primal sketch produced the 2½-D sketch, which represents the properties, including depth, of all visible surfaces. Finally, operations on the 2½-D sketch produce the 3-D model, which represents the three-dimensional properties of objects (including surfaces not directly visible) in a fashion that is independent of view.

Chomsky's (1965) approach to language also posits different kinds of representations (Jackendoff, 1987). These include representations of phonological structure, representations of syntax, and representations of semantic or conceptual structures. Jackendoff argued that Marr's (1982) theory of vision could be directly linked to Chomsky's theory of language by a mapping between 3-D models and conceptual structures. This link permits the output of visual processing to play a critical role in fixing the semantic content of linguistic representations (Jackendoff, 1983, 1990).

One key element of Jackendoff's (1987) proposal is the distinction that he imposed between syntax and semantics. This type of separation is characteristic of classical cognitive science, which strives to separate the formal properties of symbols from their content-bearing properties (Haugeland, 1985).

For instance, classical theorists define symbols as physical patterns that bear meaning because they denote or designate circumstances in the real world (Vera & Simon, 1993). The *physical pattern* part of this definition permits symbols to be manipulated in terms of their shape or form: all that is required is that the physical nature of a pattern be sufficient to identify it as a token of some symbolic type. The *designation* aspect of this definition concerns the meaning or semantic content of the symbol and is completely separate from its formal or syntactic nature.

> To put it dramatically, interpreted formal tokens lead two lives: SYNTACTICAL LIVES, in which they are meaningless markers, moved according to the rules of some self-contained game; and SEMANTIC LIVES, in which they have meanings and symbolic relations to the outside world. (Haugeland, 1985, p. 100)

In other words, when cognitive systems are viewed representationally (e.g., as in Jackendoff, 1987), they can be described at different levels, but these levels are not identical to those of Marr's (1982) tri-level hypothesis. Representationally, one level is physical, involving the physical properties of symbols. A second level is formal, concerning the logical properties of symbols. A third level is semantic, regarding the meanings designated by symbols. Again, each of these levels involves using a particular vocabulary to capture its particular regularities.

This second sense of levels of description leads to a position that some researchers have used to distinguish classical cognitive science from other approaches. In particular, it is first proposed that a cognitive vocabulary is used to capture regularities at the semantic level of description. It is then argued that the cognitive vocabulary is a mark of the classical, because it is a vocabulary that is used by classical cognitive scientists, but which is not employed by their connectionist or embodied counterparts.

The cognitive vocabulary is used to capture regularities at the cognitive level that cannot be captured at the physical or symbolic levels (Pylyshyn, 1984). "But what sort of regularities can these be? The answer has already been given: precisely the regularities that tie goals, beliefs, and actions together in a rational manner" (p. 132). In other words, the cognitive vocabulary captures regularities by describing meaningful (i.e., rational) relations between the contents of mental representations. It is the vocabulary used when one adopts the intentional stance (Dennett, 1987) to predict future behaviour or when one explains an agent at the knowledge level (Newell, 1982, 1993).

> To treat a system at the knowledge level is to treat it as having some knowledge and some goals, and believing it will do whatever is within its power to attain its goals, in so far as its knowledge indicates. (Newell, 1982, p. 98)

The power of the cognitive vocabulary is that it uses meaningful relations between mental contents to explain intelligent behaviour (Fodor & Pylyshyn, 1988). For instance, meaningful, complex tokens are possible because the semantics of such expressions are defined in terms of the contents of their constituent symbols as well as the structural relationships that hold between these constituents. The cognitive vocabulary's exploitation of constituent structure leads to the systematicity of classical theories: if one can process some expressions, then it is guaranteed that other expressions can also be processed because of the nature of constituent structures. This in turn permits classical theories to be productive, capable of generating an infinite variety of expressions from finite resources.

Some classical theorists have argued that other approaches in cognitive science do not posit the structural relations between mental contents that are captured by the cognitive vocabulary (Fodor & Pylyshyn, 1988). For instance, Fodor and Pylyshyn (1988) claimed that even though connectionist theories are representational, they are not cognitive because they exploit a very limited kind of relationship between represented contents.

> Classical theories disagree with Connectionist theories about what primitive relations hold among these content-bearing entities. Connectionist theories acknowledge only causal connectedness as a principled relation among nodes; when you know how activation and inhibition flow among them, you know everything there is to know about how the nodes in a network are related. (Fodor and Pylyshyn, 1988, p. 12)

As a result, Fodor and Pylyshyn argued, connectionist models are not componential, nor systematic, nor even productive. In fact, because they do not use a cognitive vocabulary (in the full classical sense), connectionism is not cognitive.

Related arguments can be made against positions that have played a central role in embodied cognitive science, such as the ecological approach to perception advocated by Gibson (1979). Fodor and Pylyshyn (1981) have argued against the notion of direct perception, which attempts to construe perception as involving the direct pick-up of information about the layout of a scene; that is, acquiring this information without the use of inferences from cognitive contents: "The fundamental difficulty for Gibson is that 'about' (as in 'information *about* the layout in the light') is a semantic relation, and Gibson has no account *at all* of what it is to recognize a semantic relation" (p. 168). Fodor and Pylyshyn argued that Gibson's only notion of information involves the correlation between states of affairs, and that this notion is insufficient because it is not as powerful as the classical notion of structural relations among cognitive contents. "The semantic notion of information that Gibson needs depends, so far as anyone knows, on precisely the mental representation construct that he deplores" (p. 168).

It is clear from the discussion above that Pylyshyn used the cognitive vocabulary to distinguish classical models from connectionist and embodied theories. This does not mean that he believed that non-classical approaches have no contributions to make. For instance, in Chapter 8 we consider in detail his theory of seeing and visualizing (Pylyshyn, 2003c, 2007); it is argued that this is a hybrid theory, because it incorporates elements from all three schools of thought in cognitive science.

However, one of the key elements of Pylyshyn's theory is that vision is quite distinct from cognition; he has made an extended argument for this position. When he appealed to connectionist networks or embodied access to the world, he did so in his account of visual, and not cognitive, processes. His view has been that such processes can only be involved in vision, because they do not appeal to the cognitive vocabulary and therefore cannot be viewed as cognitive processes. In short, the cognitive vocabulary is viewed by Pylyshyn as a mark of the classical.

Is the cognitive vocabulary a mark of the classical? It could be—provided that the semantic level of explanation captures regularities that cannot be expressed at either the physical or symbolic levels. Pylyshyn (1984) argued that this is indeed the case, and that the three different levels are independent:

> The reason we need to postulate representational content for functional states is to explain the existence of certain distinctions, constraints, and regularities in the behavior of at least human cognitive systems, which, in turn, appear to be expressible only in terms of the semantic content of the functional states of these systems. Chief among the constraints is some principle of rationality. (Pylyshyn, 1984, p. 38)

However, it is not at all clear that in the practice of classical cognitive science—particularly the development of computer simulation models—the cognitive level is distinct from the symbolic level. Instead, classical researchers adhere to what is known as the formalist's motto (Haugeland, 1985). That is, the semantic regularities of a classical model emerge from the truth-preserving, but syntactic, regularities at the symbolic level.

> If the formal (syntactical) rules specify the relevant texts and if the (semantic) interpretation must make sense of all those texts, then simply playing by the rules is itself a surefire way to make sense. Obey the formal rules of arithmetic, for instance, and your answers are sure to be true. (Haugeland, 1985, p. 106)

If this relation holds between syntax and semantics, then the cognitive vocabulary is not capturing regularities that cannot be captured at the symbolic level.

The formalist's motto is a consequence of the physical symbol system hypothesis (Newell, 1980; Newell & Simon, 1976) that permitted classical cognitive science to replace Cartesian dualism with materialism. Fodor and Pylyshyn (1988, p. 13) adopt the physical symbol system hypothesis, and tacitly accept the formalist's motto: "Because Classical mental *representations* have combinatorial structure, it is possible for Classical mental *operations* to apply to them by reference to their form." Note that in this quote, operations are concerned with formal and not semantic properties; semantics is preserved provided that there is a special relationship between constraints on symbol manipulations and constraints on symbolic content.

To summarize this section: The interdisciplinary nature of cognitive science arises because cognitive systems require explanations at multiple levels. Two multiple level approaches are commonly found in the cognitive science literature. The first is Marr's (1982) tri-level hypothesis, which requires cognitive systems to be explained at the implementational, algorithmic, and computational levels. It is argued above that all three schools of thought in cognitive science adhere to the tri-level hypothesis. Though at each level there are technical differences to be found between classical, connectionist, and embodied cognitive science, all three approaches seem consistent with Marr's approach. The tri-level hypothesis cannot be used to distinguish one cognitive science from another.

The second is a tri-level approach that emerges from the physical symbol system hypothesis. It argues that information processing requires explanation at three independent levels: the physical, the symbolic, and the semantic (Dennett, 1987; Newell, 1982; Pylyshyn, 1984). The physical and symbolic levels in this approach bear a fairly strong relationship to Marr's (1982) implementational and algorithmic levels. The semantic level, though, differs from Marr's computational level in calling for a cognitive vocabulary that captures regularities by appealing to the contents of mental representations. This cognitive vocabulary has been proposed as a mark of the classical that distinguishes classical theories from those proposed by connectionist and embodied researchers. However, it has been suggested that this view may not hold, because the formalist's motto makes the proposal of an independent cognitive vocabulary difficult to defend.

7.9 From Classical Marks to Hybrid Theories

Vera and Simon's (1993) analysis of situated action theories defines one extreme pole of a continuum for relating different approaches in cognitive science. At this end of the continuum, all theories in cognitive science—including situated action theories and connectionist theories—are classical or symbolic in nature. "It follows that there is no need, contrary to what followers of SA seem sometimes to claim, for cognitive psychology to adopt a whole new language and research agenda, breaking completely from traditional (symbolic) cognitive theories" (p. 46).

The position defined by Vera and Simon's (1993) analysis unites classical, connectionist, and cognitive science under a classical banner. However, it does so because key terms, such as *symbolic*, are defined so vaguely that their value becomes questionable. Critics of their perspective have argued that anything can be viewed as symbolic given Vera and Simon's liberal definition of what symbols are (Agre, 1993; Clancey, 1993).

The opposite pole of the continuum for relating different approaches in cognitive science is defined by theories that propose sharp differences between different schools of thought, and which argue in favor of adopting one while abandoning others (Chemero, 2009; Fodor & Pylyshyn, 1988; Smolensky, 1988; Winograd & Flores, 1987b).

One problem with this end of the continuum, an issue that is the central theme of the current chapter, is that it is very difficult to define marks of the classical, features that uniquely distinguish classical cognitive science from competing approaches. Our examination of the modern computing devices that inspired classical cognitive science revealed that many of these machines lacked some of the properties that are often considered marks of the classical. That is, it is not clear that properties such as central control, serial processing, local and internal representations, explicit rules, and the cognitive vocabulary are characteristics that distinguish classical theories from other kinds of models.

The failure to find clear marks of the classical may suggest that a more profitable perspective rests somewhere along the middle of the continuum for relating different approaches to cognitive science, for a couple of reasons. For one, the extent to which a particular theory is classical (or connectionist, or embodied) may be a matter of degrees. That is, any theory in cognitive science may adopt features such as local vs. distributed representations, internal vs. external memories, serial vs. parallel processes, and so on, to varying degrees. Second, differences between approaches may be important in the middle of the continuum, but may not be so extreme or distinctive that alternative perspectives cannot be co-operatively coordinated to account for cognitive phenomena.

To say this differently, rather than seeking marks of the classical, perhaps we should find arcs that provide links between different theoretical perspectives. One phenomenon might not nicely lend itself to an explanation from one school of thought, but be more easily accounted for by applying more than one school of thought at the same time. This is because the differing emphases of the simultaneously applied models may be able to capture different kinds of regularities. Cognitive science might be unified to the extent that it permits different theoretical approaches to be combined in hybrid models.

A hybrid model is one in which two or more approaches are applied simultaneously to provide a complete account of a whole phenomenon. The approaches might be unable to each capture the entirety of the phenomenon, but—in a fashion analogous to coarse coding—provide a complete theory when the different aspects that they capture are combined. One example of such a theory is provided in David McNeill's (2005) *Gesture And Thought*.

McNeill (2005) noted that the focus of modern linguistic traditions on competence instead of performance (Chomsky, 1965) emphasizes the study of static linguistic structures. That is, such traditions treat language as a thing, not as a process. In contrast to this approach, other researchers have emphasized the dynamic nature of language (Vygotsky, 1986), treating it as a process, not as a thing. One example of a dynamic aspect of language of particular interest to McNeill (2005) is gesture, which in McNeill's view is a form of imagery. Gestures that accompany language are dynamic because they are extended through time with identifiable beginnings, middles, and ends. McNeill's proposal was that a complete account of language requires the simultaneous consideration of its static and dynamic elements.

McNeill (2005) argued that the static and dynamic elements of language are linked by a dialectic. A dialectic involves some form of opposition or conflict that is resolved through change; it is this necessary change that makes dialectic dynamic. The dialectic of language results because speech and gesture provide very different formats for encoding meaning. For instance,

> in speech, ideas are separated and arranged sequentially; in gesture, they are instantaneous in the sense that the meaning of the gesture is not parceled out over time (even though the gesture may take time to occur, its full meaning is immediately present). (McNeill, 2005, p. 93)

As well, speech involves analytic meaning (i.e., based on parts), pre-specified pairings between form and meaning, and the use of forms defined by conventions. In contrast, gestures involve global meaning, imagery, and idiosyncratic forms that are created on the fly.

McNeill (2005) noted that the dialectic of language arises because there is a great deal of evidence suggesting that speech and gesture are synchronous. That is, gestures do not occur during pauses in speech to fill in meanings that are difficult to utter; both occur at the same time. As a result, two very different kinds of meaning are presented simultaneously. "Speech puts different semiotic modes together at the same moment of the speaker's cognitive experience. This is the key to the dialectic" (p. 94).

According to McNeill (2005), the initial co-occurrence of speech and gesture produces a growth point, which is an unstable condition defined by the dialectic. This growth point is unpacked in an attempt to resolve the conflict between dynamic and static aspects of meaning. This unpacking is a move from the unstable to the stable. This is accomplished by creating a static, grammatical structure. "Change seeks repose. A grammatically complete sentence (or its approximation) is a state of repose par excellence, a natural stopping point, intrinsically static and reachable from instability" (p. 95). Importantly, the particular grammatical structure that is arrived at when stability is achieved depends upon what dynamic or gestural information was present during speech.

McNeill's (2005) theory is intriguing because it exploits two different kinds of theories simultaneously: a classical theory of linguistic competence and an embodied theory of gestured meaning. Both the static/classical and dynamic/embodied parts of McNeill's theory are involved with conveying meaning. They occur at the same time and are therefore co-expressive, but they are not redundant: "gesture and speech express the same underlying idea unit but express it in their own ways—their own aspects of it, and when they express overlapping aspects they do so in distinctive ways" (p. 33). By exploiting two very different approaches in cognitive science, McNeill is clearly providing a hybrid model.

One hybrid model different in nature from McNeill's (2005) is one in which multiple theoretical approaches are applied in succession. For example, theories of perception often involve different stages of processing (e.g., visual detection, visual cognition, object recognition [Treisman, 1988]). Perhaps one stage of such processing is best described by one kind of theory (e.g., a connectionist theory of visual detection) while a later stage is best described by a different kind of theory (e.g., a symbolic model of object recognition). One such theory of seeing and visualizing favoured by Pylyshyn (2003c, 2007) is discussed in detail as an example of a hybrid cognitive science in Chapter 8.

8

Seeing and Visualizing

8.0 Chapter Overview

Zenon Pylyshyn is one of the leading figures in the study of the foundations of cognitive science. His own training was highly interdisciplinary; he earned degrees in engineering-physics, control systems, and experimental psychology. In 1994, he joined Rutgers University as Board of Governors Professor of Cognitive Science and Director of the Rutgers Center for Cognitive Science. Prior to his arrival at Rutgers he was Professor of Psychology, Professor of Computer Science, Director of the University of Western Ontario Center for Cognitive Science, and an honorary professor in the departments of Philosophy and Electrical Engineering at Western. I myself had the privilege of having Pylyshyn as my PhD supervisor when I was a graduate student at Western.

Pylyshyn is one of the key proponents of classical cognitive science (Dedrick & Trick, 2009). One of the most important contributions to classical cognitive science has been his analysis of its foundations, presented in his classic work *Computation and Cognition* (Pylyshyn, 1984). Pylyshyn's (1984) book serves as a manifesto for classical cognitive science, in which cognition is computation: the manipulation of formal symbols. It stands as one of the pioneering appeals for using the multiple levels of investigation within cognitive science. It provides an extremely cogent argument for the need to use a cognitive vocabulary to capture explanatory generalizations in the study of cognition. In it, Pylyshyn also argued for establishing the strong equivalence of a cognitive theory by determining the characteristics of the cognitive architecture.

As a champion of classical cognitive science, it should not be surprising that Pylyshyn has published key criticisms of other approaches to cognitive science. Fodor and Pylyshyn's (1988) *Cognition* article "Connectionism and cognitive architecture" is one of the most cited critiques of connectionist cognitive science that has ever appeared. Fodor and Pylyshyn (1981) have also provided one of the major critiques of direct perception (Gibson, 1979). This places Pylyshyn securely in the camp against embodied cognitive science; direct perception in its modern form of active perception (Noë, 2004) has played a major role in defining the embodied approach. Given the strong anti-classical, anti-representational perspective of radical embodied cognitive science (Chemero, 2009), it is far from surprising to be able to cite Pylyshyn's work in opposition to it.

In addition to pioneering classical cognitive science, Pylyshyn has been a crucial contributor to the literature on mental imagery and visual cognition. He is well known as a proponent of the propositional account of mental imagery, and he has published key articles critiquing its opponent, the depictive view (Pylyshyn, 1973, 1979b, 1981a, 2003b). His 1973 article "What the mind's eye tells the mind's brain: A critique of mental imagery" is a science citation classic that is responsible for launching the imagery debate in cognitive science. In concert with his analysis of mental imagery, Pylyshyn has developed a theory of visual cognition that may serve as an account of how cognition connects to the world (Pylyshyn, 1989, 1999, 2000, 2001, 2003c, 2007; Pylyshyn & Storm, 1988). The most extensive treatments of this theory can be found in his 2003 book *Seeing and Visualizing*—which inspired the title of the current chapter—and in his 2007 book *Things and Places*.

The purpose of the current chapter is to provide a brief introduction to Pylyshyn's theory of visual cognition, in part because this theory provides a wonderful example of the interdisciplinary scope of modern cognitive science. A second, more crucial reason is that, as argued in this chapter, this theory contains fundamental aspects of all three approaches—in spite of Pylyshyn's position as a proponent of classical cognitive science and as a critic of both connectionist and embodied cognitive science. Thus Pylyshyn's account of visual cognition provides an example of the type of hybrid theory that was alluded to in the previous two chapters: a theory that requires classical, connectionist, and embodied elements.

8.1 The Transparency of Visual Processing

Some researchers are concerned that many perceptual theorists tacitly assume a snapshot conception of experience (Noë, 2002) or a video camera theory of vision (Frisby, 1980). Such tacit assumptions are rooted in our phenomenal experience of an enormously high-quality visual world that seems to be delivered to us effortlessly. "You open your eyes and—*presto!*—you enjoy a richly detailed picture-like experience of the world, one that represents the world in sharp focus, uniform detail and high resolution from the centre out to the periphery" (Noë, 2002, p. 2).

Indeed, our visual experience suggests that perception puts us in direct contact with reality. Perception is transparent; when we attempt to attend to perceptual processing, we miss the processing itself and instead experience the world around us (Gendler & Hawthorne, 2006). Rather than experiencing the world as picture-like (Noë, 2002), it is as if we simply experience the world (Chalmers, 2006; Merleau-Ponty, 1962). Merleau-Ponty (1962, p. 77) noted that "our perception ends in objects, and the object[,] once constituted, appears as the reason for all the experiences of it which we have had or could have." Chalmers (2006) asserts that,

> in the Garden of Eden, we had unmediated contact with the world. We were directly acquainted with objects in the world and with their properties. Objects were presented to us without causal mediation, and properties were revealed to us in their true intrinsic glory. (Chalmers, 2006, p. 49)

To say that visual processing is transparent is to say that we are only aware of the contents that visual processes deliver. This was a central assumption to the so-called New Look theory of perception. For instance, Bruner (1957, p. 124) presumed that "all perceptual experience is necessarily the end product of a categorization process." Ecological perception (Gibson, 1979), a theory that stands in strong opposition in almost every respect to the New Look, also agrees that perceptual processes are transparent. "What one becomes aware of by holding still, closing one eye, and observing a frozen scene are not visual sensations but only *the surfaces of the world that are viewed now from here*" (p. 286, italics original).

That visual processing is transparent is not a position endorsed by all. For instance, eighteenth-century philosopher George Berkeley and nineteenth-century art critic John Ruskin both argued that it was possible to recover the "innocence of the eye" (Gombrich, 1960). According to this view, it is assumed that at birth humans have no concepts, and therefore cannot experience the world in terms of objects or categories; "what we really see is only a medley of colored patches such as Turner paints" (p. 296). Seeing the world of objects requires learning about the required categories. It was assumed that an artist could return to the "innocent eye": "the painter must clear his mind of all he knows about the object he sees, wipe the slate clean, and make nature write her own story" (p. 297).

Most modern theories of visual perception take the middle ground between the New Look and the innocent eye by proposing that our experience of visual categories is supported by, or composed of, sensed information (Mach, 1959). Mach (1959) proclaimed that, thus, perceptions, presentations, volitions, and emotions, in short the whole inner and outer world, are put together, in combinations of varying evanescence and permanence, out of a small number of homogeneous elements. Usually, these elements are called sensations. (Mach, 1959, p. 22)

From this perspective, a key issue facing any theory of seeing or visualizing is determining where sensation ends and where perception begins.

Unfortunately, the demarcation between sensation and perception is not easily determined by introspection. Subjective experience can easily lead us to the intentional fallacy in which a property of the content of a mental representation is mistakenly attributed to the representation itself (Pylyshyn, 2003c). We see in the next section that the transparency of visual processing hides from our awareness a controversial set of processes that must cope with tremendously complex information processing problems.

8.2 The Poverty of the Stimulus

Some researchers have noted a striking tension between experience and science (Varela, Thompson, & Rosch, 1991). On the one hand, our everyday experience provides a compelling and anchoring sense of self-consciousness. On the other hand, cognitive science assumes a fundamental self-fragmentation, because much of thought is putatively mediated by mechanisms that are modular, independent, and completely incapable of becoming part of conscious experience. "Thus cognitivism challenges our conviction that consciousness and the mind either amount to the same thing or [that] there is an essential or necessary connection between them" (p. 49).

The tension between experience and science is abundantly evident in vision research. It is certainly true that the scientific study of visual perception relies heavily on the analysis of visual experience (Pylyshyn, 2003c). However, researchers are convinced that this analysis must be performed with caution and be supplemented by additional methodologies. This is because visual experience is not complete, in the sense that it does not provide direct access to or experience of visual processing. Pylyshyn (2003b) wrote,

what we do [experience] is misleading because it is always the *world* as it appears to us that we see, not the real work that is being done by the mind in going from the proximal stimuli, generally optical patterns on the retina, to the familiar experience of seeing (or imagining) the world. (Pylyshyn, 2003b, p. xii)

Vision researchers have long been aware that the machinery of vision is not a part of our visual experience. Helmholtz noted that "it might seem that nothing could be easier than to be conscious of one's own sensations; and yet experience shows that for the discovery of subjective sensations some special talent is needed" (Helmholtz & Southall, 1962b, p. 6). Cognitive psychologist Roger Shepard observed that,

> we do not first experience a two-dimensional image and then consciously calcu-
> late or infer the three-dimensional scene that is most likely, given that image. The
> first thing we experience is the three-dimensional world—as our visual system has
> already inferred it for us on the basis of the two-dimensional input. (Shepard, 1990,
> p. 168)

In the nineteenth century, Hermann von Helmholtz argued that our visual experience results from the work of unconscious mechanisms. "The psychic activities that lead us to infer that there in front of us at a certain place there is a certain object of a certain character, are generally not conscious activities, but unconscious ones" (Helmholtz & Southall, 1962b, p. 4). However, the extent and nature of this unconscious processing was only revealed when researchers attempted to program computers to see. It was then discovered that visual processes face a difficult problem that also spurred advances in modern linguistic theory: the poverty of the stimulus.

Generative linguistics distinguished between those theories of language that were descriptively adequate and those that were explanatorily adequate (Chomsky, 1965). A descriptively adequate theory of language provided a grammar that was capable of describing the structure of any possible grammatical sentence in a language and incapable of describing the structure of any sentence that did not belong to this language. A more powerful explanatorily adequate theory was descriptively adequate but also provided an account of how that grammar was learned. "To the extent that a linguistic theory succeeds in selecting a descriptively adequate grammar on the basis of primary linguistic data, we can say that it meets the condition of *explanatory adequacy*" (p. 25).

Why did Chomsky use the ability to account for language learning as a defining characteristic of explanatory adequacy? It was because Chomsky realized that language learning faced the poverty of the stimulus. The poverty-of-the-stimulus argument is the claim that primary linguistic data—that is, the linguistic utterances heard by a child—do not contain enough information to uniquely specify the grammar used to produce them.

> It seems that a child must have the ability to 'invent' a generative grammar that
> defines well-formedness and assigns interpretations to sentences even though the
> primary linguistic data that he uses as a basis for this act of theory construction
> may, from the point of view of the theory he constructs, be deficient in various
> respects. (Chomsky, 1965, p. 201)

The poverty of the stimulus is responsible for formal proofs that text learning of a language is not possible if the language is defined by a complex grammar (Gold, 1967; Pinker, 1979; Wexler & Culicover, 1980).

Language acquisition can be described as solving the projection problem: determining the mapping from primary linguistic data to the acquired grammar (Baker, 1979; Peters, 1972). When language learning is so construed, the poverty of the stimulus becomes a problem of underdetermination. That is, the projection from data to grammar is not unique, but is instead one-to-many: one set of primary linguistic data is consistent with many potential grammars.

For sighted individuals, our visual experience makes us take visual perception for granted. We have the sense that we simply look at the world and see it. Indeed, the phenomenology of vision led artificial intelligence pioneers to expect that building vision into computers would be a straightforward problem. For instance, Marvin Minsky assigned one student, as a summer project, the task of programming a computer to see (Horgan, 1993). However, failures to develop computer vision made it apparent that the human visual system was effortlessly solving, in real time, enormously complicated information processing problems. Like language learning, vision is dramatically underdetermined. That is, if one views vision as the projection from primary visual data (the proximal stimulus on the retina) to the internal interpretation or representation of the distal scene, this projection is one-to-many. A single proximal stimulus is consistent with an infinite number of different interpretations (Gregory, 1970; Marr, 1982; Pylyshyn, 2003c; Rock, 1983; Shepard, 1990).

One reason that vision is underdetermined is because the distal world is arranged in three dimensions of space, but the primary source of visual information we have about it comes from patterns of light projected onto an essentially two dimensional surface, the retina. "According to a fundamental theorem of topology, the relations between objects in a space of three dimensions cannot all be preserved in a two-dimensional projection" (Shepard, 1990, pp. 173–175).

This source of underdetermination is illustrated in Figure 8-1, which illustrates a view from the top of an eye observing a point in the distal world as it moves from position X_1 to position Y_1 over a given interval of time.

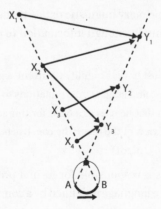

Figure 8-1. Underdetermination of projected movement.

The primary visual data caused by this movement is the motion, from point A to point B, of a point projected onto the back of the retina. The projection from the world to the back of the eye is uniquely defined by the laws of optics and of projective geometry.

However, the projection in the other direction, from the retina to the distal world, is not unique. If one attempts to use the retinal information alone to identify the distal conditions that caused it, then infinitely many possibilities are available. *Any* of the different paths of motion in the world (occurring over the same duration) that are illustrated in Figure 8-1 are consistent with the proximal information projected onto the eye. Indeed, movement from *any* position along the dashed line through the X-labelled points to *any* position along the other dashed line is a potential cause of the proximal stimulus.

One reason for the poverty of the visual stimulus, as illustrated in Figure 8-1, is that information is necessarily lost when an image from a three-dimensional space is projected onto a two-dimensional surface.

> We are so familiar with seeing, that it takes a leap of imagination to realize that there are problems to be solved. But consider it. We are given tiny distorted upside-down images in the eyes, and we see separate solid objects in surrounding space. From the patterns of stimulation on the retinas we perceive the world of objects, and this is nothing short of a miracle. (Gregory, 1978, p. 9)

A second reason for the poverty of the visual stimulus arises because the neural circuitry that mediates visual perception is subject to the limited order constraint (Minsky & Papert, 1969). There is no single receptor that takes in the entire visual stimulus in a glance. Instead, each receptor processes only a small part of the primary visual data. This produces deficiencies in visual information. For example, consider the aperture problem that arises in motion perception (Hildreth, 1983), illustrated in Figure 8-2.

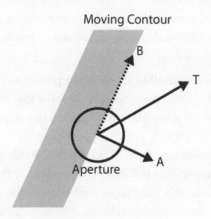

Figure 8-2. The aperture problem in motion perception.

In this situation, a motion detector's task is to detect the movement of a contour, shown in grey. However, the motion detector is of limited order: its window on the moving contour is the circular aperture in the figure, an aperture that is much smaller than the contour it observes.

Because of its small aperture, the motion detector in Figure 8-2 can only be sensitive to the component of the contour's motion that is perpendicular to the edge of the contour, vector A. It is completely blind to any motion parallel to the contour, the dashed vector B. This is because movement in this direction will not change the appearance of anything within the aperture. As a result, the motion detector is unable to detect the true movement of the contour, vector T.

The limited order constraint leads to a further source of visual underdetermination. If visual detectors are of limited order, then our interpretation of the proximal stimulus must be the result of combining many different (and deficient) local measurements together. However, many different global interpretations exist that are consistent with a single set of such measurements. The local measurements by themselves cannot uniquely determine the global perception that we experience.

Consider the aperture problem of Figure 8-2 again. Imagine one, or many, local motion detectors that deliver vector A at many points along that contour. How many true motions of the contour could produce this situation? In principle, one can create an infinite number of different possible vector Ts by choosing any desired length of vector B—to which any of the detectors are completely blind—and adding it to the motion that is actually detected, i.e., vector A.

Pylyshyn (2003b, 2007) provided many arguments against the theory that vision constructs a representation of the world, which is depictive in nature. However, the theory that Pylyshyn opposed is deeply entrenched in accounts of visual processing.

> For years the common view has been that a large-scope inner image is built up by
> superimposing information from individual glances at the appropriate coordinates
> of the master image: as the eye moves over a scene, the information on the retina
> is transmitted to the perceptual system, which then projects it onto an inner screen
> in the appropriate location, thus painting the larger scene for the mind side to
> observe. (Pylyshyn, 2003b, pp. 16–17)

Proponents of this view face another source of the poverty of the visual stimulus. It is analogous to the limited order constraint, in the sense that it arises because vision proceeds by accessing small amounts of information in a sequence of fragmentary glimpses.

Although we experience our visual world as a rich, stable panorama that is present in its entirety, this experience is illusory (Dennett, 1991; Pylyshyn, 2003c, 2007). Evidence suggests that we only experience fragments of the distal world a glance at a time. For instance, we are prone to change blindness, where we fail to notice a substantial visual change even though it occurs in plain sight (O'Regan et al., 2000).

A related phenomenon is inattentional blindness, in which visual information that should be obvious is not noticed because attention is not directed to it (even though the gaze is!). In one famous experiment (Simons & Chabris, 1999), subjects watched a video of a basketball game and were instructed to count the number of times that the teams changed possession of the ball. In the midst of the game a person dressed in a gorilla suit walked out onto the court and danced a jig. Amazingly, most subjects failed to notice this highly visible event because they were paying attention to the ball.

If the visual system collects fragments of visual information a glance at a time, then our visual experience further suggests that these different fragments are "stitched together" to create a stable panorama. In order for this to occur, the fragments have to be inserted in the correct place, presumably by identifying components of the fragment (in terms of visible properties) in such a way that it can be asserted that "object x in one location in a glimpse collected at time $t + 1$ is the same thing as object y in a different location in a glimpse collected at an earlier time t." This involves computing correspondence, or tracking the identities of objects over time or space, a problem central to the study of binocular vision (Marr, Palm, & Poggio, 1978; Marr & Poggio, 1979) and motion perception (Dawson, 1991; Dawson & Pylyshyn, 1988; Ullman, 1978, 1979).

However, the computing of correspondence is a classic problem of underdetermination. If there are N different elements in two different views of a scene, then there are at least $N!$ ways to match the identities of elements across the views. This problem cannot be solved by image matching—basing the matches on the appearance or description of elements in the different views—because the dynamic nature of the world, coupled with the loss of information about it when it is projected onto the eyes, means that there are usually radical changes to an object's proximal stimulus over even brief periods of time.

> How do we know which description uniquely applies to a particular individual and, what's more important, how do we know which description will be unique at some time in the future when we will need to find the representation of that particular token again in order to add some newly noticed information to it? (Pylyshyn, 2007, p. 12)

To summarize, visual perception is intrinsically underdetermined because of the poverty of the visual stimulus. If the goal of vision is to construct representations of the distal world, then proximal stimuli do not themselves contain enough information to accomplish this goal. In principle, an infinite number of distal scenes could be the cause of a single proximal stimulus. "And yet we do not perceive a range of possible alternative worlds when we look out at a scene. We invariably see a single unique layout. Somehow the visual system manages to select one of the myriad logical possibilities" (Pylyshyn, 2003b, p. 94). Furthermore, the interpretation selected by the visual system seems—from our success in interacting with the world—to almost always be correct. "What is remarkable is that we err so seldom" (Shepard, 1990, p. 175).

How does the visual system compensate for the poverty of the stimulus as well as generate unique and accurate solutions to problems of underdetermination? In the following sections we consider two very different answers to this question, both of which are central to Pylyshyn's theory of visual cognition. The first of these, which can be traced back to Helmholtz (Helmholtz & Southall, 1962b) and which became entrenched with the popularity of the New Look in the 1950s (Bruner, 1957, 1992), is that visual perception is full-fledged cognitive processing. "Given the slenderest clues to the nature of surrounding objects we identify them and act not so much according to what is directly sensed, *but to what is believed*" (Gregory, 1970, p. 11).

8.3 Enrichment via Unconscious Inference

Hermann von Helmholtz was not aware of problems of visual underdetermination of the form illustrated in Figures 8-1 and 8-2. However, he was aware that visual sensors could be seriously misled. One example that he considered at length (Helmholtz & Southall, 1962a, 1962b) was the mechanical stimulation of the eye (e.g., slight pressure on the eyeball made by a blunt point), which produced a sensation of light (a pressure-image or phosphene) even though a light stimulus was not present. From this he proposed a general rule for determining the "ideas of vision":

> Such objects are always imagined as being present in the field of vision as
> would have to be there in order to produce the same impression on the nerv-
> ous mechanism, the eyes being used under ordinary normal conditions.
> (Helmholtz & Southall, 1962b, p. 2)

Helmholtz's studies of such phenomena forced him to explain the processes by which such a rule could be realized. He first noted that the visual system does not have direct access to the distal world, but instead that primary visual data was retinal activity. He concluded that *inference* must be involved to transform retinal activity into visual experience. "It is obvious that we can never emerge from the world of our sensations to the apperception of an external world, except by inferring from the changing sensation that external objects are the causes of this change" (Helmholtz & Southall, 1962b, p. 33). This theory allowed Helmholtz to explain visual illusions as the result of mistaken reasoning rather than as the product of malfunctions in the visual apparatus: "It is rather simply an illusion in the judgment of the material presented to the senses, resulting in a false idea of it" (p. 4).

Helmholtz argued that the accuracy of visual inferences is due to an agent's constant exploration and experimentation with the world, determining how actions in the world such as changing viewpoints alter visual experience.

Spontaneously and by our own power, we vary some of the conditions under which the object has been perceived. We know that the changes thus produced in the way that objects look depend solely on the movements we have executed. Thus we obtain a different series of apperceptions of the same object, by which we can be convinced with experimental certainty that they are simply apperceptions and that it is the common cause of them all. (Helmholtz & Southall, 1962b, p. 31)

Helmholtz argued that the only difference between visual inference and logical reasoning was that the former was unconscious while the latter was not, describing "the psychic acts of ordinary perception as *unconscious conclusions*" (Helmholtz & Southall, 1962b, p. 4). Consciousness aside, seeing and reasoning were processes of the same kind: "There can be no doubt as to the similarity between the results of such unconscious conclusions and those of conscious conclusions" (p. 4).

A century after Helmholtz, researchers were well aware of the problem of underdetermination with respect to vision. Their view of this problem was that it was based in the fact that certain information is missing from the proximal stimulus, and that additional processing is required to supply the missing information. With the rise of cognitivism in the 1950s, researchers proposed a top-down, or theory-driven, account of perception in which general knowledge of the world was used to disambiguate the proximal stimulus (Bruner, 1957, 1992; Bruner, Postman, & Rodrigues, 1951; Gregory, 1970, 1978; Rock, 1983). This approach directly descended from Helmholtz's discussion of unconscious conclusions because it equated visual perception with cognition.

> One of the principal characteristics of perceiving [categorization] is a characteristic of cognition generally. There is no reason to assume that the laws governing inferences of this kind are discontinuous as one moves from perceptual to more conceptual activities. (Bruner, 1957, p. 124)

The cognitive account of perception that Jerome Bruner originated in the 1950s came to be known as the New Look. According to the New Look, higher-order cognitive processes could permit beliefs, expectations, and general knowledge of the world to provide additional information for disambiguation of the underdetermining proximal stimulus. "We not only believe what we see: to some extent we see what we believe" (Gregory, 1970, p. 15). Hundreds of studies provided experimental evidence that perceptual experience was determined in large part by a perceiver's beliefs or expectations. (For one review of this literature see Pylyshyn, 2003b.) Given the central role of cognitivism since the inception of the New Look, it is not surprising that this type of theory has dominated the modern literature.

> The belief that perception is thoroughly contaminated by such cognitive factors as expectations, judgments, beliefs, and so on, became the received wisdom in much of psychology, with virtually all contemporary elementary texts in human information processing and vision taking that point of view for granted. (Pylyshyn, 2003b, p. 56)

To illustrate the New Look, consider a situation in which I see a small, black and white, irregularly shaped, moving object. This visual information is not sufficient to uniquely specify what in the world I am observing. To deal with this problem, I use general reasoning processes to disambiguate the situation. Imagine that I am inside my home. I know that I own a black and white cat, I believe that the cat is indoors, and I expect that I will see this cat in the house. Thus I experience this visual stimulus as "seeing my cat Phoebe." In a different context, different expectations exist. For instance, if I am outside the house on the street, then the same proximal stimulus will be disambiguated with different expectations; "I see my neighbour's black and white dog Shadow." If I am down walking in the forest by the creek, then I may use different beliefs to "see a skunk."

> It would seem that a higher agency of the mind, call it the executive agency, has available to it the proximal input, which it can scan, and it then behaves in a manner very like a thinking organism in selecting this or that aspect of the stimulus as representing the outer object or event in the world. (Rock, 1983, p. 39)

The New Look in perception is a prototypical example of classical cognitive science. If visual perception is another type of cognitive processing, then it is governed by the same laws as are reasoning and problem solving. In short, a crucial consequence of the New Look is that visual perception is *rational*, in the sense that vision's success is measured in terms of the truth value of the representations it produces.

For instance, Richard Gregory (1970, p. 29, italics added) remarked that "it is surely remarkable that out of the infinity of possibilities the perceptual brain generally hits on just about the *best* one." Gregory (1978, p. 13, italics added) also equated visual perception to problem solving, describing it as "a dynamic searching for the *best interpretation* of the available data." The cognitive nature of perceptual processing allows,

> past experience and anticipation of the future to play a large part in augmenting sensory information, so that we do not perceive the world merely from the sensory information available at any given time, but rather we use this information to test hypotheses of what lies before us. Perception becomes a matter of suggesting and testing hypotheses. (Gregory, 1978, p. 221)

In all of these examples, perception is described as a process that delivers representational contents that are most (semantically) consistent with visual sensations and other intentional contents, such as beliefs and desires.

The problem with the New Look is this rational view of perception. Because of its emphasis on top-down influences, the New Look lacks an account of links between the world and vision that are causal and independent of beliefs. If all of our perceptual experience was belief dependent, then we would never see anything that we did not expect to see. This would not contribute to our survival, which often depends upon noticing and reacting to surprising circumstances in the environment.

Pylyshyn's (2003b, 2007) hybrid theory of visual cognition rests upon the assumption that there exists a cognitively impenetrable visual architecture that is separate from general cognition. This architecture is data-driven in nature, governed by causal influences from the visual world and insulated from beliefs and expectations. Such systems can solve problems of underdetermination without requiring assumptions of rationality, as discussed in the next section.

8.4 Natural Constraints

Some researchers would argue that perception is a form of cognition, because it uses inferential reasoning or problem solving processing to go beyond the information given. However, this kind of account is not the only viable approach for dealing with the poverty of the visual stimulus. Rock (1983, p. 3) wrote: "A phenomenon may appear to be intelligent, but the mechanism underlying it may have no common ground with the mechanisms underlying reasoning, logical thought, or problem solving." The natural computation approach to vision (Ballard, 1997; Marr, 1982; Richards, 1988) illustrates the wisdom of Rock's quote, because it attempts to solve problems of underdetermination by using bottom-up devices that apply built-in constraints to filter out incorrect interpretations of an ambiguous proximal stimulus.

The central idea underlying natural computation is constraint propagation. Imagine a set of locations to which labels can be assigned, where each label is a possible property that is present at a location. Underdetermination exists when more than one label is possible at various locations. However, constraints can be applied to remove these ambiguities. Imagine that if some label x is assigned to one location then this prevents some other label y from being assigned to a neighbouring location. Say that there is good evidence to assign label x to the first location. Once this is done, a constraint can propagate outwards from this location to its neighbours, removing label y as a possibility for them and therefore reducing ambiguity.

Constraint propagation is part of the science underlying the popular Sudoku puzzles (Delahaye, 2006). A Sudoku puzzle is a 9×9 grid of cells, as illustrated in Figure 8-3. The grid is further divided into a 3×3 array of smaller 3×3 grids called cages. In Figure 8-3, the cages are outlined by the thicker lines. When the puzzle is solved, each cell will contain a digit from the range 1 to 9, subject to three constraints. First, a digit can occur only once in each row of 9 cells across the grid. Second, a digit can only occur once in each column of 9 cells along the grid. Third, a digit can only occur once in each cage in the grid. The puzzle begins with certain numbers already assigned to their cells, as illustrated in Figure 8-3. The task is to fill in the remaining digits in such a way that none of the three constraining rules are violated.

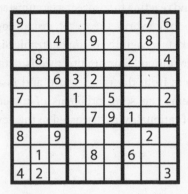

Figure 8-3. An example Sudoku puzzle.

A Sudoku puzzle can be considered as a problem to be solved by relaxation labelling. In relaxation labelling, sets of possible labels are available at different locations. For instance, at the start of the puzzle given in Figure 8-3 the possible labels at every blank cell are *1, 2, 3, 4, 5, 6, 7, 8*, and *9*. There is only one possible label (given in the figure) that has already been assigned to each of the remaining cells. The task of relaxation labelling is to iteratively eliminate extra labels at the ambiguous locations, so that at the end of processing only one label remains.

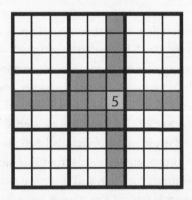

Figure 8-4. The "there can be only one" constraint propagating from the cell labelled *5*

In the context of relaxation labelling, Sudoku puzzles can be solved by propagating different constraints through the grid; this causes potential labels to be removed from ambiguous cells. One key constraint, called "there can be only one," emerges from the primary definition of a Sudoku puzzle. In the example problem given in Figure 8-3, the digit 5 has been assigned at the start to a particular location, which is also shown in Figure 8-4. According to the rules of Sudoku, this means that this digit cannot appear anywhere else in the column, row, or cage that contains this location. The affected locations are shaded dark grey in Figure 8-4. One can propagate the "there can be only one" constraint through these locations, removing the digit 5 as a possible label for any of them.

This constraint can be propagated iteratively through the puzzle. During one iteration, any cell with a unique label can be used to eliminate that label from all of the other cells that it controls (e.g., as in Figure 8-4). When this constraint is applied in this way, the result may be that some new cells have unique labels. In this case the constraint can be applied again, from these newly unique cells, to further disambiguate the Sudoku puzzle.

The "there can be only one" constraint is important, but it is not powerful enough on its own to solve any but the easiest Sudoku problems. This means that other constraints must be employed as well. Another constraint is called "last available label," and is illustrated in Figure 8-5.

Figure 8-5A illustrates one of the cages of the Figure 8-3 Sudoku problem partway through being solved (i.e., after some iterations of "there can be only one"). The cells containing a single number have been uniquely labelled. The other cells still have more than one possible label, shown as multiple digits within the cell. Note the one cell at the bottom shaded in grey. It has the possible labels *1, 3, 5,* and *9*. However, this cell has the "last available label" of *9*—the label *9* is not available in any other cell in the cage. Because a *9* is required to be in this cage, this means that this label must be assigned here and the cell's other three possible labels can be removed. Note that when this is done, the "last available label" constraint applies to a second cell (shown in grey in Figure 8-5B), meaning that it can be uniquely assigned the label *1* by applying this constraint a second time.

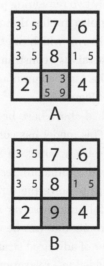

Figure 8-5. The "last available label" constraint.

After two applications of the "last available label" constraint, the cage illustrated in Figure 8-5A becomes the cage shown at the top of Figure 8-6. Note that this cage has only two ambiguous cells, each with the possible labels *3* and *5*. These two cells define what Sudoku solvers call a naked pair, which can be used to define a third rule called the "naked pair constraint."

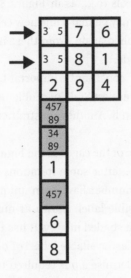

Figure 8-6. The "naked pair constraint."

In the naked pair pointed out by the two arrows in Figure 8-6, it is impossible for one cell to receive the label *3* and for the other cell not to receive the label *5*. This is because these two cells have only two remaining possible labels, and both sets of labels are identical. However, this also implies that the labels *3* and *5* cannot exist elsewhere in the part of the puzzle over which the two cells containing the naked pair have control. Thus one can use this as a constraint to remove the possible labels *3* and *5* from the other cells in the same column as the naked pair, i.e., the cells shaded in grey in the lower part of Figure 8-6.

The three constraints described above have been implemented as a working model in an Excel spreadsheet. This model has confirmed that by applying only these three constraints one can solve a variety of Sudoku problems of easy and medium difficulty, and can make substantial progress on difficult problems. (These three constraints are not sufficient to solve the difficult Figure 8-3 problem.) In order to develop a more successful Sudoku solver in this framework, one would have to identify additional constraints that can be used. A search of the Internet for "Sudoku tips" reveals a number of advanced strategies that can be described as constraints, and which could be added to a relaxation labelling model.

For our purposes, though, the above Sudoku example illustrates how constraints can be propagated to solve problems of underdetermination. Furthermore, it shows that such solutions can be fairly mechanical in nature, not requiring higher-order reasoning or problem solving. For instance, the "there can be only one" constraint could be instantiated as a simple set of interconnected switches: turning the 5 on in Figure 8-4 would send a signal that would turn the 5 off at all of the other grey-shaded locations.

The natural computation approach to vision assumes that problems of visual underdetermination are also solved by non-cognitive processes that use constraint propagation. However, the constraints of interest to such researchers are not formal rules of a game. Instead, they adopt naïve realism, and they assume that the external world is structured and that some aspects of this structure must be true of nearly every visual scene. Because the visual system has evolved to function in this structured environment, it has internalized those properties that permit it to solve problems of underdetermination. "The perceptual system has internalized the most pervasive and enduring regularities of the world" (Shepard, 1990, p. 181).

The regularities of interest to researchers who endorse natural computation are called natural constraints. A natural constraint is a property that is almost invariably true of any location in a visual scene. For instance, many visual properties of three-dimensional scenes, such as depth, colour, texture, and motion, vary smoothly. This means that two locations in the three-dimensional scene that are very close together are likely to have very similar values for any of these properties, while this will not be the case for locations that are further apart. Smoothness can therefore be used to constrain interpretations of a proximal stimulus: an interpretation whose properties vary smoothly is much more likely to be true of the world than interpretations in which property smoothness is not maintained.

Natural constraints are used to solve visual problems of underdetermination by imposing additional restrictions on scene interpretations. In addition to being consistent with the proximal stimulus, the interpretation of visual input must also be consistent with the natural constraints. With appropriate natural constraints, only a single interpretation will meet both of these criteria (for many examples, see Marr, 1982). A major research goal for those who endorse the natural computation approach to vision is identifying natural constraints that filter out correct interpretations from all the other (incorrect) possibilities.

For example, consider the motion correspondence problem (Ullman, 1979), which is central to Pylyshyn's (2003b, 2007) hybrid theory of visual cognition. In the motion correspondence problem, a set of elements is seen at one time, and another set of elements is seen at a later time. In order for the visual system to associate a sense of movement to these elements, their identities must be tracked over time. The assertion that some element x, seen at time t, is the "same thing" as some other element y, seen at time $t + 1$, is called a motion correspondence match. However, the assignment of motion correspondence matches is underdetermined. This is illustrated in Figure 8-7 as a simple apparent motion stimulus in which two squares (dashed outlines) are presented at one time, and then later presented in different locations (solid outlines). For this display there are two logical sets of motion correspondence matches that can be assigned, shown in B and C of the figure. Both sets of matches are consistent with the display, but they represent radically different interpretations of the identities of the elements over time. Human observers of this display will invariably experience it as Figure 8-7B, and never as Figure 8-7C. Why is this interpretation preferred over the other one, which seems just as logically plausible?

The natural computation approach answers this question by claiming that the interpretation illustrated in Figure 8-7B is consistent with additional natural constraints, while the interpretation in Figure 8-7C is not. A number of different natural constraints on the motion correspondence problem have been identified and then incorporated into computer simulations of motion perception (Dawson, 1987, 1991; Dawson, Nevin-Meadows, & Wright, 1994; Dawson & Pylyshyn, 1988; Dawson & Wright, 1989, 1994; Ullman, 1979).

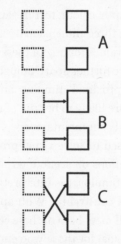

Figure 8-7. The motion correspondence problem.

One such constraint is called the nearest neighbour principle. The visual system prefers to assign correspondence matches that represent short element displacements (Burt & Sperling, 1981; Ullman, 1979). For example, the two motion correspondence matches in Figure 8-7B are shorter than the two in Figure 8-7C; they are therefore more consistent with the nearest neighbour principle.

The nearest neighbour principle is a natural constraint because it arises from the geometry of the typical viewing conditions for motion (Ullman, 1979, pp. 114–118). When movement in a three-dimensional world is projected onto a two-dimensional surface (e.g., the retina), slower movements occur with much higher probability on the retina than do faster movements. A preference for slower movement is equivalent to exploiting the nearest neighbour principle, because a short correspondence match represents slow motion, while a long correspondence match represents fast motion.

Another powerful constraint on the motion correspondence problem is called the relative velocity principle (Dawson, 1987, 1991). To the extent that visual elements arise from physical features on solid surfaces, the movement of neighbouring elements should be similar. According to the relative velocity principle, motion correspondence matches should be assigned in such a way that objects located near one another will be assigned correspondence matches consistent with movements of similar direction and speed. This is true of the two matches illustrated in Figure 8-7B, which are of identical length and direction, but not of the two matches illustrated in Figure 8-7C, which are of identical length but represent motion in different directions.

Like the nearest neighbour constraint, the relative velocity principle is a natural constraint. It is a variant of the property that motion varies smoothly across a scene (Hildreth, 1983; Horn & Schunk, 1981). That is, as objects in the real world move, locations near to one another should move in similar ways. Furthermore, Hildreth (1983) has proven that solid objects moving arbitrarily in three-dimensional space project unique smooth patterns of retinal movement. The relative velocity principle exploits this general property of projected motion.

Other natural constraints on motion correspondence have also been proposed. The element integrity principle is a constraint in which motion correspondence matches are assigned in such a way that elements only rarely split into two or fuse together into one (Ullman, 1979). It is a natural constraint in the sense that the physical coherence of surfaces implies that the splits or fusions are unlikely. The polarity matching principle is a constraint in which motion correspondence matches are assigned between elements of identical contrast (e.g., between two elements that are both light against a dark background, or between two elements that are both dark against a light background) (Dawson, Nevin-Meadows, & Wright, 1994). It is a natural constraint because movement of an object in the world might change its shape and colour, but is unlikely to alter the object's contrast relative to its background.

The natural computation approach to vision is an alternative to a classical approach called unconscious inference, because natural constraints can be exploited by systems that are not cognitive, that do not perform inferences on the basis of cognitive contents. In particular, it is very common to see natural computation models expressed in a very anti-classical form, namely, artificial neural networks (Marr, 1982). Indeed, artificial neural networks provide an ideal medium for propagating constraints to solve problems of underdetermination.

The motion correspondence problem provides one example of an artificial neural network approach to solving problems of underdetermination (Dawson, 1991; Dawson, Nevin-Meadows, & Wright, 1994). Dawson (1991) created an artificial neural network that incorporated the nearest neighbour, relative velocity, element integrity, and polarity matching principles. These principles were realized as patterns of excitatory and inhibitory connections between processors, with each processor representing a possible motion correspondence match. For instance, the connection between two matches that represented movements similar in distance and direction would have an excitatory component that reflected the relative velocity principle. Two matches that represented movements of different distances and directions would have an inhibitory component that reflected the same principle. The network would start with all processors turned on to similar values (indicating that each match was initially equally likely), and then the network would iteratively send signals amongst the processors. The network would quickly converge to a state in which some processors remained on (representing the preferred correspondence matches) while all of the others were turned off. This model was shown to be capable of modelling a wide variety of phenomena in the extensive literature on the perception of apparent movement.

The natural computation approach is defined by another characteristic that distinguishes it from classical cognitive science. Natural constraints are not psychological properties, but are instead properties of the world, or properties of how the world projects itself onto the eyes. "The visual constraints that have been discovered so far are based almost entirely on principles that derive from laws of optics and projective geometry" (Pylyshyn, 2003b, p. 120). Agents exploit natural constraints—or more precisely, they internalize these constraints in special processors that constitute what Pylyshyn calls early vision—because they are generally true of the world and therefore work.

To classical theories that appeal to unconscious inference, natural constraints are merely "heuristic bags of tricks" that happen to work (Anstis, 1980; Ramachandran & Anstis, 1986); there is no attempt to ground these tricks in the structure of the world. In contrast, natural computation theories are embodied, because they appeal to structure in the external world and to how that structure impinges on perceptual agents. As naturalist Harold Horwood (1987, p. 35) writes, "If you look attentively at a fish you can see that the water has shaped it. The fish is not merely in the water: the qualities of the water itself have called the fish into being."

8.5 Vision, Cognition, and Visual Cognition

It was argued earlier that the classical approach to underdetermination, unconscious inference, suffered from the fact that it did not include any causal links between the world and internal representations. The natural computation approach does not suffer from this problem, because its theories treat vision as a data-driven or bottom-up process. That is, visual information from the world comes into contact with visual modules—special purpose machines—that automatically apply natural constraints and deliver uniquely determined representations. How complex are the representations that can be delivered by data-driven processing? To what extent could a pure bottom-up theory of perception succeed?

On the one hand, the bottom-up theories are capable of delivering a variety of rich representations of the visual world (Marr, 1982). These include the primal sketch, which represents the proximal stimulus as an array of visual primitives, such as oriented bars, edges, and terminators (Marr, 1976). Another is the 2½-D sketch, which makes explicit the properties of visible surfaces in viewer-centred coordinates, including their depth, colour, texture, and orientation (Marr & Nishihara, 1978). The information made explicit in the 2½-D sketch is available because data-driven processes can solve a number of problems of underdetermination, often called "shape from" problems, by using natural constraints to determine three-dimensional shapes and distances of visible elements. These include structure from motion (Hildreth, 1983; Horn & Schunk, 1981; Ullman, 1979; Vidal & Hartley, 2008), shape from shading (Horn & Brooks, 1989), depth from binocular disparity (Marr, Palm, & Poggio, 1978; Marr & Poggio, 1979), and shape from texture (Lobay & Forsyth, 2006; Witkin, 1981).

> It would not be a great exaggeration to say that early vision—part of visual processing that is prior to access to general knowledge—computes just about everything that might be called a 'visual appearance' of the world *except* the identities and names of the objects. (Pylyshyn, 2003b, p. 51)

On the other hand, despite impressive attempts (Biederman, 1987), it is generally acknowledged that the processes proposed by natural computationalists cannot deliver representations rich enough to make full contact with semantic knowledge of the world. This is because object recognition—assigning visual information to semantic categories—requires identifying object parts and determining spatial relationships amongst these parts (Hoffman & Singh, 1997; Singh & Hoffman, 1997). However, this in turn requires directing attention to specific entities in visual representations (i.e., individuating the critical parts) and using serial processes to determine spatial relations amongst the individuated entities (Pylyshyn, 1999, 200 1, 2003c, 2007; Ullman, 1984). The data-driven, parallel computations that characterize natural computation theories of vision are poor candidates for computing relationships between individuated objects or their parts. As a result, what early vision "does not do is *identify* the things we are looking at, in the sense of relating them to things we have seen before, the contents of our memory. And it does not make judgments about how things really are" (Pylyshyn, 2003b, p. 51).

Thus it appears that a pure, bottom-up natural computation theory of vision will not suffice. Similarly, it was argued earlier that a pure, top-down cognitive theory of vision is also insufficient. A complete theory of vision requires co-operative interactions between both data-driven and top-down processes. As philosopher Jerry Fodor (1985, p. 2) has noted, "perception is smart like cognition in that it is typically inferential, it is nevertheless dumb like reflexes in that it is typically encapsulated." This leads to what Pylyshyn calls the independence hypothesis: the proposal that some visual processing must be independent of cognition. However, because we are consciously aware of visual information, a corollary of the independence hypothesis is that there must be some interface between visual processing that is not cognitive and visual processing that is.

This interface is called visual cognition (Enns, 2004; Humphreys & Bruce, 1989; Jacob & Jeannerod, 2003; Ullman, 2000), because it involves visual attention (Wright, 1998). Theories in visual cognition about both object identification (Treisman, 1988; Ullman, 2000) and the interpretation of motion (Wright & Dawson, 1994) typically describe three stages of processing: the precognitive delivery of visual information, the attentional analysis of this visual information, and the linking of the results of these analyses to general knowledge of the world.

One example theory in visual cognition is called feature integration theory (Treisman, 1986, 1988; Treisman & Gelade, 1980). Feature integration theory arose from two basic experimental findings. The first concerned search latency functions, which represent the time required to detect the presence or absence of a target as a function of the total number of display elements in a visual search task. Pioneering work on visual search discovered the so-called "pop-out effect": for some targets, the search latency function is essentially flat. This indicated that the time to find a target is independent of the number of distractor elements in the display. This result was found for targets defined by a unique visual feature (e.g., colour, contrast, orientation, movement), which seemed to pop out of a display, automatically drawing attention to the target (Treisman & Gelade, 1980). In contrast, the time to detect a target defined by a unique *combination* of features generally increases with the number of distractor items, producing search latency functions with positive slopes.

The second experimental finding that led to feature integration theory was the discovery of illusory conjunctions (Treisman & Schmidt, 1982). Illusory conjunctions occur when features are mistakenly combined. For instance, subjects might be presented a red triangle and a green circle in a visual display but experience an illusory conjunction: a green triangle and a red circle.

Feature integration theory arose to explain different kinds of search latency functions and illusory conjunctions. It assumes that vision begins with a first, non-cognitive stage of feature detection in which separate maps for a small number of basic features, such as colour, orientation, size, or movement, record the presence and location of detected properties. If a target is uniquely defined in terms of possessing one of these features, then it will be the only source of activity in that feature map and will therefore pop out, explaining some of the visual search results.

A second stage of processing belongs properly to visual cognition. In this stage, a spotlight of attention is volitionally directed to a particular spot on a master map of locations. This attentional spotlight enables the visual system to integrate features by bringing into register different feature maps at the location of interest. Different features present at that location can be conjoined together in a temporary object representation called an object file (Kahneman, Treisman, & Gibbs, 1992; Treisman, Kahneman, & Burkell, 1983). Thus in feature integration theory, searching for objects defined by unique *combinations* of features requires a serial scan of the attentional spotlight from location to location, explaining the nature of search latency functions for such objects. This stage of processing also explains illusory conjunctions, which usually occur when the attentional processing is divided, impairing the ability of correctly combining features into object files.

A third stage of processing belongs to higher-order cognition. It involves using information about detected objects (i.e., features united in object files) as links to general knowledge of the world.

> Conscious perception depends on temporary object representations in which the different features are collected from the dimensional modules and inter-related, then matched to stored descriptions in a long-term visual memory to allow recognition. (Treisman, 1988, p. 204)

Another proposal that relies on the notion of visual cognition concerns visual routines (Ullman, 1984). Ullman (1984) noted that the perception of spatial relations is central to visual processing. However, many spatial relations cannot be directly delivered by the parallel, data-driven processes postulated by natural computationalists, because these relations are not defined over entire scenes, but are instead defined over particular entities in scenes (i.e., objects or their parts). Furthermore, many of these relations must be computed using serial processing of the sort that is not proposed to be part of the networks that propagate natural constraints.

For example, consider determining whether some point x is inside a contour y. Ullman (1984) pointed out that there is little known about how the relation *inside* (x, y) is actually computed, and argued that it most likely requires serial processing in which activation begins at x, spreading outward. It can be concluded that x is inside y if the spreading activation is contained by y. Furthermore, before *inside* (x, y) can be computed, the two entities, x and y, have to be individuated and selected—*inside* makes no sense to compute without their specification. "What the visual system needs is a way to refer to individual elements *qua* token individuals" (Pylyshyn, 2003b, p. 207).

With such considerations in mind, Ullman (1984) developed a theory of visual routines that shares many of the general features of feature integration theory. In an initial stage of processing, data-driven processes deliver early representations of the visual scene. In the second stage, visual cognition executes visual routines at specified locations in the representations delivered by the first stage of processing. Visual routines are built from a set of elemental operations and used to establish spatial relations and shape properties. Candidate elemental operations include indexing a salient item, spreading activation over a region, and tracing boundaries. A visual routine is thus a program, assembled out of elemental operations, which is activated when needed to compute a necessary spatial property. Visual routines are part of visual cognition because attention is used to select a necessary routine (and possibly create a new one), and to direct the routine to a specific location of interest. However, once the routine is activated, it can deliver its spatial judgment without requiring additional higher-order resources.

In the third stage, the spatial relations computed by visual cognition are linked, as in feature integration theory, to higher-order cognitive processes. Thus Ullman (1984) sees visual routines as providing an interface between the representations created by data-driven visual modules and the content-based, top-down processing of cognition. Such an interface permits data-driven and theory-driven processes to be combined, overcoming the limitations that such processes would face on their own.

> Visual routines operate in the middle ground that, unlike the bottom-up creation of the base representations, is a part of the top-down processing and yet is independent of object-specific knowledge. Their study therefore has the advantage of going beyond the base representations while avoiding many of the additional complications associated with higher level components of the system. (Ullman, 1984, p. 119)

The example theories of visual cognition presented above are hybrid theories in the sense that they include both bottom-up and top-down processes, and they invoke attentional mechanisms as a link between the two. In the next section we see that Pylyshyn's (2003b, 2007) theory of visual indexing is similar in spirit to these theories and thus exhibits their hybrid characteristics. However, Pylyshyn's theory of visual cognition is hybrid in another important sense: it makes contact with classical, connectionist, and embodied cognitive science.

Pylyshyn's theory of visual cognition is classical because one of the main problems that it attempts to solve is how to identify or re-identify individuated entities. Classical processing is invoked as a result, because "individuating and re-identifying in general require the heavy machinery of concepts and descriptions"

(Pylyshyn, 2007, p. 32). Part of Pylyshyn's theory of visual cognition is also connectionist, because he appeals to non-classical mechanisms to deliver visual representations (i.e., natural computation), as well as to connectionist networks (in particular, to winner-take-all mechanisms; see Feldman & Ballard, 1982) to track entities after they have been individuated with attentional tags (Pylyshyn, 2001, 2003c). Finally, parts of Pylyshyn's theory of visual cognition draw on embodied cognitive science. For instance, the reason that tracking element identities—solving the correspondence problem—is critical is because Pylyshyn assumes a particular embodiment of the visual apparatus, a limited-order retina that cannot take in all information in a glance. Similarly, Pylyshyn uses the notion of cognitive scaffolding to account for the spatial properties of mental images.

8.6 Indexing Objects in the World

Pylyshyn's theory of visual cognition began in the late 1970s with his interest in explaining how diagrams were used in reasoning (Pylyshyn, 2007). Pylyshyn and his colleagues attempted to investigate this issue by building a computer simulation that would build and inspect diagrams as part of deriving proofs in plane geometry.

From the beginning, the plans for this computer simulation made contact with two of the key characteristics of embodied cognitive science. First, the diagrams created and used by the computer simulation were intended to be external to it and to scaffold the program's geometric reasoning.

> Since we wanted the system to be as psychologically realistic as possible we did not want all aspects of the diagram to be 'in its head' but, as in real geometry problem-solving, remain on the diagram it was drawing and examining. (Pylyshyn, 2007, p. 10)

Second, the visual system of the computer was also assumed to be psychologically realistic in terms of its embodiment. In particular, the visual system was presumed to be a moving fovea that was of limited order: it could only examine the diagram in parts, rather than all at once.

> We also did not want to assume that all properties of the entire diagram were available at once, but rather that they had to be *noticed* over time as the diagram was being drawn and examined. If the diagram were being inspected by moving the eyes, then the properties should be within the scope of the moving fovea. (Pylyshyn, 2007, p. 10)

These two intersections with embodied cognitive science—a scaffolding visual world and a limited order embodiment—immediately raised a fundamental information processing problem. As different lines or vertices were added to a diagram, or as these components were scanned by the visual system, their different identities had to be maintained or tracked over time. In order to function as intended, the program had to be able to assert, for example, that "this line observed here" is the same as "that line observed there" when the diagram is being scanned. In short, in considering how to create this particular system, Pylyshyn recognized that it required two core abilities: to be able to individuate visual entities, and to be able to track or maintain the identities of visual entities over time.

To maintain the identities of individuated elements over time is to solve the correspondence problem. How does one keep track of the identities of different entities perceived in different glances? According to Pylyshyn (2003b, 2007), the classical answer to this question must appeal to the contents of representations. To assert that some entity seen in a later glance was the same as one observed earlier, the descriptions of the current and earlier entities must be compared. If the descriptions matched, then the entities should be deemed to be the same. This is called the image matching solution to the correspondence problem, which also dictates how entities must be individuated: they must be uniquely described, when observed, as a set of properties that can be represented as a mental description, and which can be compared to other descriptions.

Pylyshyn rejects the classical image matching solution to the correspondence problem for several reasons. First, multiple objects can be tracked as they move to different locations, even if they are identical in appearance (Pylyshyn & Storm, 1988). In fact, multiple objects can be tracked as their properties change, even when their location is constant and shared (Blaser, Pylyshyn, & Holcombe, 2000). These results pose problems for image matching, because it is difficult to individuate and track identical objects by using their descriptions!

Second, the poverty of the stimulus in a dynamic world poses severe challenges to image matching. As objects move in the world or as we (or our eyes) change position, a distal object's projection as a proximal stimulus will change properties, even though the object remains the same. "If objects can change their properties, we don't know under what description the object was last stored" (Pylyshyn, 2003b, p. 205).

A third reason to reject image matching comes from the study of apparent motion, which requires the correspondence problem to be solved before the illusion of movement between locations can be added (Dawson, 1991; Wright & Dawson, 1994). Studies of apparent motion have shown that motion correspondence is mostly insensitive to manipulations of figural properties, such as shape, colour, or spatial frequency (Baro & Levinson, 1988; Cavanagh, Arguin, & von Grunau, 1989; Dawson, 1989; Goodman, 1978; Kolers, 1972; Kolers & Green, 1984; Kolers & Pomerantz, 1971; Kolers & von Grunau, 1976; Krumhansl, 1984; Navon, 1976; Victor & Conte, 1990). This insensitivity to form led Nelson Goodman (1978, p. 78) to conclude that "plainly the visual system is persistent, inventive, and sometimes rather perverse in building a world according to its own lights." One reason for this perverseness may be that the neural circuits for processing motion are largely independent of those for processing form (Botez, 1975; Livingstone & Hubel, 1988; Maunsell & Newsome, 1987; Ungerleider & Mishkin, 1982).

A fourth reason to reject image matching is that it is a purely cognitive approach to individuating and tracking entities. "Philosophers typically assume that in order to individuate something we must conceptualize its relevant properties. In other words, we must first represent (or cognize or conceptualize) the *relevant conditions of individuation*" (Pylyshyn, 2007, p. 31). Pylyshyn rejected this approach because it suffers from the same core problem as the New Look: it lacks causal links to the world.

Pylyshyn's initial exploration of how diagrams aided reasoning led to his realization that the individuation and tracking of visual entities are central to an account of how vision links us to the world. For the reasons just presented, he rejected a purely classical approach—mental descriptions of entities—for providing these fundamental abilities. He proposed instead a theory that parallels the structure of the examples of visual cognition described earlier. That is, Pylyshyn's (2003b, 2007) theory of visual cognition includes a non-cognitive component (early vision), which delivers representations that can be accessed by visual attention (visual cognition), which in turn deliver representations that can be linked to general knowledge of the world (cognition).

On the one hand, the early vision component of Pylyshyn's (2003b, 2007) theory of visual cognition is compatible with natural computation accounts of perception (Ballard, 1997; Marr, 1982). For Pylyshyn, the role of early vision is to provide causal links between the world and the perceiving agent without invoking cognition or inference:

> Only a highly constrained set of properties can be selected by early vision, or can
> be directly 'picked up.' Roughly, these are what I have elsewhere referred to as
> 'transducable' properties. These are the properties whose detection does not require
> accessing memory and drawing inferences. (Pylyshyn, 2003b, p. 163)

The use of natural constraints to deliver representations such as the primal sketch and the 2½-D sketch is consistent with Pylyshyn's view.

On the other hand, Pylyshyn (2003b, 2007) added innovations to traditional natural computation theories that have enormous implications for explanations of seeing and visualizing. First, Pylyshyn argued that one of the primitive processes of early vision is individuation—the picking out of an entity as being distinct from others. Second, he used evidence from feature integration theory and cognitive neuroscience to claim that individuation picks out objects, but not on the basis of their locations. That is, preattentive processes can detect elements or entities via primitive features but simultaneously not deliver the location of the features, as is the case in pop-out. Third, Pylyshyn argued that an individuated entity—a visual object—is preattentively tagged by an index, called a FINST ("for finger instantiation"), which can only be used to access an individuated object (e.g., to retrieve its properties when needed). Furthermore, only a limited number (four) of FINSTs are available. Fourth, once assigned to an object, a FINST remains attached to it even as the object changes its location or other properties. Thus a primitive component of early vision is the solution of the correspondence problem, where the role of this solution is to maintain the link between FINSTs and dynamic, individuated objects.

The revolutionary aspect of FINSTs is that they are presumed to individuate and track visual objects without delivering a description of them and without fixing their location. Pylyshyn (2007) argued that this is the visual equivalent of the use of indexicals or demonstratives in language: "Think of demonstratives in natural language—typically words like *this* or *that*. Such words allow us to refer to things without specifying what they are or what properties they have" (p. 18). FINSTs are visual indices that operate in exactly this way. They are analogous to placing a finger on an object in the world, and, while not looking, keeping the finger in contact with it as the object moved or changed— thus the term *finger instantiation*. As long as the finger is in place, the object can be referenced ("this thing that I am pointing to now"), even though the finger does not deliver any visual properties.

There is a growing literature that provides empirical support for Pylyshyn's FINST hypothesis. Many of these experiments involve the multiple object tracking paradigm (Flombaum, Scholl, & Pylyshyn, 2008; Franconeri et al., 2008; Pylyshyn, 2006; Pylyshyn & Annan, 2006; Pylyshyn et al., 2008; Pylyshyn & Storm, 1988; Scholl, Pylyshyn, & Feldman, 2001; Sears & Pylyshyn, 2000). In the original version of this paradigm (Pylyshyn & Storm, 1988), subjects were shown a static display made up of a number of objects of identical appearance. A subset of these objects blinked for a short period of time, indicating that they were to-be-tracked targets. Then the blinking stopped, and all objects in the display began to move independently and randomly for a period of about ten seconds. Subjects had the task of tracking the targets, with attention only; a monitor ended trials in which eye movements were detected. At the end of a trial, one object blinked and subjects had to indicate whether or not it was a target.

The results of this study (see Pylyshyn & Storm, 1988) indicated that subjects could simultaneously track up to four independently moving targets with high accuracy. Multiple object tracking results are explained by arguing that FINSTs are allocated to the flashing targets prior to movement, and objects are tracked by the primitive mechanism that maintains the link from visual object to FINST. This link permits subjects to judge targethood at the end of a trial.

The multiple object tracking paradigm has been used to explore some of the basic properties of the FINST mechanism. Analyses indicate that this process is parallel, because up to four objects can be tracked, and tracking results cannot be explained by a model that shifts a spotlight of attention serially from target to target

(Pylyshyn & Storm, 1988). However, the fact that no more than four targets can be tracked also shows that this processing has limited capacity. FINSTs are assigned to objects, and not locations; objects can be tracked through a location-less feature space (Blase, Pylyshyn, & Holcombe, 2000). Using features to make the objects distinguishable from one another does not aid tracking, and object properties can actually change during tracking without subjects being aware of the changes (Bahrami, 2003; Pylyshyn, 2007). Thus FINSTs individuate and track visual objects but do not deliver descriptions of the properties of the objects that they index.

Another source of empirical support for the FINST hypothesis comes from studies of subitizing (Trick & Pylyshyn, 1993, 1994). Subitizing is a phenomenon in which the number of items in a set of objects (the cardinality of the set) can be effortlessly and rapidly detected if the set has four or fewer items (Jensen, Reese, & Reese, 1950; Kaufman et al., 1949). Larger sets cannot be subitized; a much slower process is required to serially count the elements of larger sets. Subitizing necessarily requires that the items to be counted are individuated from one another. Trick and Pylyshyn (1993, 1994) hypothesized that subitizing could be accomplished by the FINST mechanism; elements are preattentively individuated by being indexed, and counting simply requires accessing the number of indices that have been allocated.

Trick and Pylyshyn (1993, 1994) tested this hypothesis by examining subitizing in conditions in which visual indexing was not possible. For instance, if the objects in a set are defined by conjunctions of features, then they cannot be preattentively FINSTed. Importantly, they also cannot be subitized. In general, subitizing does not occur when the elements of a set that are being counted are defined by properties that require serial, attentive processing in order to be detected (e.g., sets of concentric contours that have to be traced in order to be individuated; or sets of elements defined by being on the same contour, which also require tracing to be identified).

At the core of Pylyshyn's (2003b, 2007) theory of visual cognition is the claim that visual objects can be preattentively individuated and indexed. Empirical support for this account of early vision comes from studies of multiple object tracking and of subitizing. The need for such early visual processing comes from the goal of providing causal links between the world and classical representations, and from embodying vision in such a way that information can only be gleaned a glimpse at a time. Thus Pylyshyn's theory of visual cognition, as described to this point, has characteristics of both classical and embodied cognitive science. How does the theory make contact with connectionist cognitive science? The answer to this question comes from examining Pylyshyn's (2003b, 2007) proposals concerning preattentive mechanisms for individuating visual objects and tracking them. The mechanisms that Pylyshyn proposed are artificial neural networks.

For instance, Pylyshyn (2000, 2003b) noted that a particular type of artificial neural network, called a winner-take-all network (Feldman & Ballard, 1982), is ideally suited for preattentive individuation. Many versions of such a network have been proposed to explain how attention can be automatically drawn to an object or to a distinctive feature (Fukushima, 1986; Gerrissen, 1991; Grossberg, 1980; Koch & Ullman, 1985; LaBerge Carter, & Brown, 1992; Sandon, 1992). In a winner-take-all network, an array of processing units is assigned to different objects or to feature locations. For instance, these processors could be distributed across the preattentive feature maps in feature integration theory (Treisman, 1988; Treisman & Gelade, 1980). Typically, a processor will have an excitatory connection to itself and will have inhibitory connections to its neighbouring processors. This pattern of connectivity results in the processor that receives the most distinctive input becoming activated and at the same time turning off its neighbours.

That such mechanisms might be involved in individuation is supported by results that show that the time course of visual search can be altered by visual manipulations that affect the inhibitory processing of such networks (Dawson & Thibodeau, 1998). Pylyshyn endorses a modified winner-take-all network as a mechanism for individuation; the modification permits an object indexed by the network to be interrogated in order to retrieve its properties (Pylyshyn, 2000).

Another intersection between Pylyshyn's (2003b, 2007) theory of visual cognition and connectionist cognitive science comes from his proposals about preattentive tracking. How can such tracking be accomplished without the use of image matching? Again, Pylyshyn noted that artificial neural networks, such as those that have been proposed for solving the motion correspondence problem (Dawson, 1991; Dawson, Nevin-Meadows, & Wright, 1994; Dawson & Pylyshyn, 1988; Dawson & Wright, 1994), would serve as tracking mechanisms. This is because such models belong to the natural computation approach and have shown how tracking can proceed preattentively via the exploitation of natural constraints that are implemented as patterns of connectivity amongst processing units.

Furthermore, Dawson (1991) has argued that many of the regularities that govern solutions to the motion correspondence problem are consistent with the hypothesis that solving this problem is equivalent to tracking assigned visual tags. For example, consider some observations concerning the location of motion correspondence processing and attentional tracking processes in the brain. Dawson argued that motion correspondence processing is most likely performed by neurons located in Area 7 of the parietal cortex, on the basis of motion signals transmitted from earlier areas, such as the motion-sensitive area MT. Area 7 of the parietal cortex is also a good candidate for the locus of tracking of individuated entities.

First, many researchers have observed cells that appear to mediate object tracking in Area 7, such as visual fixation neurons and visual tracking neurons. Such cells are not evident earlier in the visual pathway (Goldberg & Bruce, 1985; Hyvarinen & Poranen, 1974; Lynch et al., 1977; Motter & Mountcastle, 1981; Robinson, Goldberg,

& Stanton, 1978; Sakata et al., 1985).

Second, cells in this area are also governed by extraretinal (i.e., attentional) influences—they respond to attended targets, but not to unattended targets, even when both are equally visible (Robinson, Goldberg, & Stanton, 1978). This is required of mechanisms that can pick out and track targets from identically shaped distractors, as in a multiple object tracking task.

Third, Area 7 cells that appear to be involved in tracking appear to be able to do so across sensory modalities. For instance, hand projection neurons respond to targets to which hand movements are to be directed and do not respond when either the reach or the target are present alone (Robinson Goldberg, & Stanton, 1978). Similarly, there exist many Area Y cells that respond during manual reaching, tracking, or manipulation, and which also have a preferred direction of reaching (Hyvarinen & Poranen, 1974). Such cross-modal coordination of tracking is critical, because as we see in the next section, Pylyshyn's (2003b, 2007) theory of visual cognition assumes that indices can be applied, and tracked, in different sensory modalities, permitting seeing agents to point at objects that have been visually individuated.

The key innovation and contribution of Pylyshyn's (2003b, 2007) theory of visual cognition is the proposal of preattentive individuation and tracking. This proposal can be seamlessly interfaced with related proposals concerning visual cognition. For instance, once objects have been tagged by FINSTs, they can be operated on by visual routines (Ullman, 1984, 2000). Pylyshyn (2003b) pointed out that in order to execute, visual routines require such individuation:

> The visual system must have some mechanism for picking out and referring to particular elements in a display in order to decide whether two or more such elements form a pattern, such as being *collinear*, or being *inside*, *on*, or *part of* another element, so on. Pylyshyn (2003b, pp. 206–207)

In other words, visual cognition can direct attentional resources to FINSTed entities.

Pylyshyn's (2003b, 2007) theory of visual cognition also makes contact with classical cognition. He noted that once objects have been tagged, the visual system can examine their spatial properties by applying visual routines or using focal attention to retrieve visual features. The point of such activities by visual cognition would be to update descriptions of objects stored as object files (Kahneman, Treisman, & Gibbs, 1992). The object file descriptions can then be used to make contact with the semantic categories of classical cognition. Thus the theory of visual indexing provides a causal grounding of visual concepts:

> Indexes may serve as the basis for real individuation of physical objects. While it is clear that you cannot individuate objects in the full-blooded sense without a conceptual apparatus, it is also clear that you cannot individuate them with *only*

a conceptual apparatus. Sooner or later concepts must be grounded in a primitive causal connection between thoughts and things. (Pylyshyn, 2001, p. 154)

It is the need for such grounding that has led Pylyshyn to propose a theory of visual cognition that includes characteristics of classical, connectionist, and embodied cognitive science.

8.7 Situation, Vision, and Action

Why is Pylyshyn's (2003b, 2007) proposal of preattentive visual indices important? It has been noted that one of the key problems facing classical cognitive science is that it needs some mechanism for referring to the world that is preconceptual, and that the impact of Pylyshyn's theory of visual cognition is that it provides an account of exactly such a mechanism (Fodor, 2009). How this is accomplished is sketched out in Figure 8-8, which provides a schematic of the various stages in Pylyshyn's theory of visual cognition.

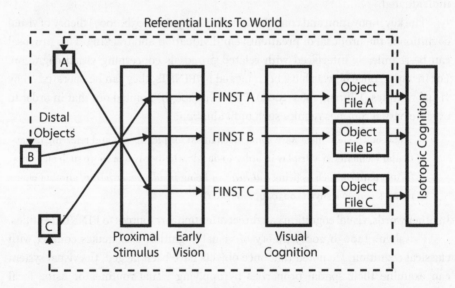

Figure 8-8. Pylyshyn's theory of preattentive visual indexing provides referential links from object files to distal objects in the world.

The initial stages of the theory posit causal links from distal objects arrayed in space in a three-dimensional world and mental representations that are produced from these links. The laws of optics and projective geometry begin by creating a proximal stimulus—a pattern of stimulation on the retina—that is uniquely determined, but because of the problem of underdetermination cannot be uniquely inverted. The problem of underdetermination is initially dealt with by a variety of visual modules that compose early vision, and which use natural constraints to deliver unique and useful representations of the world (e.g., the primal sketch and the 2½-D sketch). Pylyshyn's theory of visual cognition elaborates Marr's (1982) natural computation view of vision. In addition to using Marr's representations, Pylyshyn claims that early vision can individuate visual objects by assigning them one of a limited number of tags (FINSTs). Furthermore, preattentive processes permit these tags to remain attached, even if the properties of the tagged objects change. This result of early vision is illustrated in Figure 8-8 as the sequences of solid arrows that link each visual object to its own internal FINST.

Once objects have been individuated by the assignment of visual indices, the operations of visual cognition can be applied (Treisman, 1986, 1988; Ullman, 1984, 2000). Attention can be directed to individuated elements, permitting visual properties to be detected or spatial relations amongst individuated objects to be computed. The result is that visual cognition can be used to create a description of an individuated object in its object file (Kahneman, Treisman, & Gibbs, 1992). As shown in Figure 8-8, visual cognition has created an internal object file for each of the three distal objects involved in the diagram.

Once object files have been created, general knowledge of the world—isotropic cognitive processes (Fodor, 1983) can be exploited. Object files can be used to access classical representations of the world, permitting semantic categories to be applied to the visual scene.

However, object files permit another important function in Pylyshyn's theory of visual cognition because of the preattentive nature of the processes that created them: a referential link from an object file to a distal object in the world. This is possible because the object files are associated with FINSTs, and the FINSTs themselves were the end product of a causal, non-cognitive chain of events:

> An index corresponds to two sorts of links or relations: on the one hand, it corresponds to a *causal chain* that goes from visual objects to certain tokens in the representation of the scene being built (perhaps an *object file*), and on the other hand, it is also a *referential relationship* that enables the visual system to refer to those particular [visual objects]. The second of these functions is possible because the first one exists and has the right properties. (Pylyshyn, 2003b, p. 269)

The referential links back to the distal world are illustrated as the dashed lines in Figure 8-8.

The availability of the referential links provides Pylyshyn's theory of visual cognition (2003b, 2007) with distinct advantages over a purely classical model. Recall that a top-down model operates by creating and maintaining internal descriptions of distal objects. It was earlier noted that one problem with this approach is that the projected information from an object is constantly changing, in spite of the fact that the object's identity is constant. This poses challenges for solving the correspondence problem by matching descriptions. However, this also leads a classical model directly into what is known as the frame problem (Ford & Pylyshyn, 1996; Pylyshyn, 1987). The frame problem faces any system that has to update classical descriptions of a changing world. This is because as a property changes, a classical system must engage in a series of deductions to determine the implications of the change. The number of possible deductions is astronomical, resulting in the computational intractability of a purely descriptive system.

The referential links provide a solution to the frame problem. This is because the tracking of a FINSTed object and the perseverance of the object file for that object occur without the need of constantly updating the object's description. The link between the FINST and the world is established via the causal link from the world through the proximal stimulus to the operation of early vision. The existence of the referential link permits the contents of the object file to be refreshed or updated—not constantly, but only when needed. "One of the purposes of a tag was to allow the visual system to revisit the tagged object to encode some new property" (Pylyshyn, 2003b, p. 208).

The notion of revisiting an indexed object in order to update the contents of an object file when needed, combined with the assumption that visual processing is embodied in such a way to be of limited order, link Pylyshyn's (2003b, 2007) theory of visual cognition to a different theory that is central to embodied cognitive science, enactive perception (Noë, 2004). Enactive perception realizes that the detailed phenomenal experience of vision is an illusion because only a small amount of visual information is ever available to us (Noë, 2002). Enactive perception instead views perception as a sensorimotor skill that can access information in the world when it is needed. Rather than building detailed internal models of the world, enactive perception views the world as its own representation (Noë, 2009); we don't encode an internal model of the world, we inspect the outer world when required or desired. This account of enactive perception mirrors the role of referential links to the distal world in Pylyshyn's theory of visual cognition.

Of course, enactive perception assumes much more than information in the world is accessed, and not encoded. It also assumes that the goal of perception is to guide bodily actions upon the world. "Perceiving is a way of acting. Perception is not something that happens to us, or in us. It is something we do" (Noë, 2004, p. 1). This view of perception arises because enactive perception is largely inspired by Gibson's (1966, 1979) ecological approach to perception. Actions on the world were central to Gibson. He proposed that perceiving agents "picked up" the affordances of objects in the world, where an affordance is a possible action that an agent could perform on or with an object.

Actions on the world (ANCHORs) provide a further link between Pylyshyn's (2003b, 2007) theory of visual cognition and enactive perception, and consequently with embodied cognitive science. Pylyshyn's theory also accounts for such actions, because FINSTs are presumed to exist in different sensory modalities. In particular, ANCHORs are analogous to FINSTs and serve as indices to places in motor-command space, or in proprioceptive space (Pylyshyn, 1989). The role of ANCHORs is to serve as indices to which motor movements can be directed. For instance, in the 1989 version of his theory, Pylyshyn hypothesized that ANCHORs could be used to direct the gaze (by moving the fovea to the ANCHOR) or to direct a pointer.

The need for multimodal indexing is obvious because we can easily point at what we are looking at. Conversely, if we are not looking at something, it cannot be indexed, and therefore cannot be pointed to as accurately. For instance, when subjects view an array of target objects in a room, close their eyes, and then imagine viewing the objects from a novel vantage point (a rotation from their original position), their accuracy in pointing to the targets decreases (Rieser, 1989). Similarly, there are substantial differences between reaches towards visible objects and reaches towards objects that are no longer visible but are only present through imagery or memory (Goodale, Jakobson, & Keillor, 1994). Likewise, when subjects reach towards an object while avoiding obstacles, visual feedback is exploited to optimize performance; when visual feedback is not available, the reaching behaviour changes dramatically (Chapman & Goodale, 2010).

In Pylyshyn's (2003b, 2007) theory of visual cognition, coordination between vision and action occurs via interactions between visual and motor indices, which generate mappings between the spaces of the different kinds of indices. Requiring transformations between spatial systems makes the location of indexing and tracking mechanisms in parietal cortex perfectly sensible. This is because there is a great deal of evidence suggesting that parietal cortex instantiates a variety of spatial mappings, and that one of its key roles is to compute transformations between different spatial representations (Andersen et al., 1997; Colby & Goldberg, 1999; Merriam, Genovese, & Colby, 2003; Merriam & Colby, 2005). One such transformation could produce coordination between visual FINSTs and motor ANCHORs.

One reason that Pylyshyn's (2003b, 2007) theory of visual cognition is also concerned with visually guided action is his awareness of Goodale's work on visuomotor modules (Goodale, 1988, 1990, 1995; Goodale & Humphrey, 1998; Goodale et al., 1991), work that was introduced earlier in relation to embodied cognitive science. The evidence supporting Goodale's notion of visuomotor modules clearly indicates that some of the visual information used to control actions is not available to isotropic cognitive processes, because it can affect actions without requiring or producing conscious awareness. It seems very natural, then, to include motor indices (i.e., ANCHORs) in a theory in which such tags are assigned and maintained preattentively.

The discussion in this section would seem to place Pylyshyn's (2003b, 2007) theory of visual cognition squarely in the camp of embodied cognitive science. Referential links between object files and distal objects permit visual information to be accessible without requiring the constant updating of descriptive representations. The postulation of indices that can guide actions and movements and the ability to coordinate these indices with visual tags place a strong emphasis on action in Pylyshyn's approach.

However, Pylyshyn's theory of visual cognition has many properties that make it impossible to pigeonhole as an embodied position. In particular, a key difference between Pylyshyn's theory and enactive perception is that Pylyshyn does not believe that the sole goal of vision is to guide action. Vision is also concerned with descriptions and concepts—the classical cognition of represented categories:

> Preparing for action is not the only purpose of vision. Vision is, above all, a way to find out about the world, and there may be many reasons why an intelligent organism may wish to know about the world, apart from wanting to act upon it. (Pylyshyn, 2003b, p. 133)

8.8 Scaffolding the Mental Image

In Chapter 3 we introduced the imagery debate, which concerns two different accounts of the architectural properties of mental images. One account, known as the depictive theory (Kosslyn, 1980, 1994; Kosslyn, Thompson, & Ganis, 2006), argues that we experience the visual properties of mental images because the format of these images is quasi-pictorial, and that they literally depict visual information.

The other account, propositional theory, proposes that images are not depictive, but instead describe visual properties using a logical or propositional representation (Pylyshyn, 1973, 1979b, 1981a, 2003b). It argues that the privileged properties of mental images proposed by Kosslyn and his colleagues are actually the result of the intentional fallacy: the spatial properties that Kosslyn assigns to the format of images should more properly be assigned to their contents.

The primary support for the depictive theory has come from relative complexity evidence collected from experiments on image scanning (Kosslyn, 1980) and mental rotation (Shepard & Cooper, 1982). This evidence generally shows a linear relationship between the time required to complete a task and a spatial property of an image transformation. For instance, as the distance between two locations on an image increases, so too does the time required to scan attention from one location to the other. Similarly, as the amount of rotation that must be applied to an image increases, so too does the time required to judge that the image is the same or different from another. Proponents of propositional theory have criticized these results by demonstrating that they are cognitively penetrable (Pylyshyn, 2003c): a change in tacit information eliminates the linear relationship between time and image transformation, which would not be possible if the depictive properties of mental images were primitive.

If a process such as image scanning is cognitively penetrable, then this means that subjects have the choice not to take the time to scan attention across the image. But this raises a further question: "Why should people persist on using this method when scanning entirely in their imagination where the laws of physics and the principles of spatial scanning do not apply (since there is no real space)?" (Pylyshyn, 2003b, p. 309). Pylyshyn's theory of visual cognition provides a possible answer to this question that is intriguing, because it appeals to a key proposal of the embodied approach: cognitive scaffolding.

Pylyshyn's scaffolding approach to mental imagery was inspired by a general research paradigm that investigated whether visual processing and mental imagery shared mechanisms. In such studies, subjects superimpose a mental image over other information that is presented visually, in order to see whether the different sources of information can interact, for instance by producing a visual illusion (Bernbaum & Chung, 1981; Finke & Schmidt, 1977; Goryo, Robinson, & Wilson, 1984; Ohkuma, 1986). This inspired what Pylyshyn (2007) called the index projection hypothesis. This hypothesis brings Pylyshyn's theory of visual cognition into contact with embodied cognitive science, because it invokes cognitive scaffolding via the visual world.

According to the index projection hypothesis, mental images are scaffolded by visual indices that are assigned to real world (i.e., to visually present) entities. For instance, consider Pylyshyn's (2003b) application of the index projection hypothesis to the mental map paradigm used to study image scanning:

> If, for example, you imagine the map used to study mental scanning superimposed
> over one of the walls in the room you are in, you can use the visual features of
> the wall to anchor various objects in the imagined map. In this case, the increase
> in time it takes to access information from loci that are further apart is easily
> explained since the 'images,' or, more neutrally, 'thoughts' of these objects *are actu-*
> *ally* located further apart. (Pylyshyn, 2003b, p. 376, p. 374)

In other words, the spatial properties revealed in mental scanning studies are not due to mental images per se, but instead arise from "the real spatial nature of the sensory world onto which they are 'projected'" (p. 374).

If the index projection hypothesis is valid, then how does it account for mental scanning results when no external world is visible? Pylyshyn argued that in such conditions, the linear relationship between distance on an image and the time to scan it may not exist. For instance, evidence indicates that when no external information is visible, smooth attentional scanning may not be possible (Pylyshyn & Cohen, 1999).

As well, the exploration of mental images is accompanied by eye movements similar to those that occur when a real scene is explored (Brandt & Stark, 1997). Pylyshyn (2007) pointed out that this result is exactly what would be predicted by the index projection hypothesis, because the eye movements would be directed to real world entities that have been assigned visual indices.

The cognitive scaffolding of mental images may not merely concern their manipulation, but might also be involved when images are created. There is a long history of the use of mental images in the art of memory (Yates, 1966). One important technique is the ancient method of loci, in which mental imagery is used to remember a sequence of ideas (e.g., ideas to be presented in a speech).

The memory portion of the *Rhetorica ad Herrenium*, an anonymous text that originated in Rome circa 86 BC and reached Europe by the Middle Ages, teaches the method of loci as follows. A well-known building is used as a "wax tablet" onto which memories are to be "written." As one mentally moves, in order, through the rooms of the building, one places an image representing some idea or content in each locus—that is, in each imagined room. During recall, one mentally walks through the building again, and "sees" the image stored in each room. "The result will be that, reminded by the images, we can repeat orally what we have committed to the *loci*, proceeding in either direction from any *locus* we please" (Yates, 1966, p. 7).

In order for the method of loci to be effective, a great deal of effort must be used to initially create the loci to be used to store memories (Yates, 1966). Ancient rules of memory taught students the most effective way to do this. According to the *Rhetorica ad Herrenium*, each fifth locus should be given a distinguishing mark. A locus should not be too similar to the others, in order to avoid confusion via resemblance. Each locus should be of moderate size and should not be brightly lit, and the intervals between loci should also be moderate (about thirty feet). Yates (1966, p. 8) was struck by "the astonishing visual precision which [the classical rules of memory] imply. In a classically trained memory the space between the *loci* can be measured, the lighting of the *loci* is allowed for."

How was such a detailed set of memory loci to be remembered? The student of memory was taught to use what we would now call cognitive scaffolding. They should lay down a set of loci by going to an *actual* building, and by *literally* moving through it from locus to locus, carefully committing each place to memory as they worked (Yates, 1966). Students were advised to visit secluded buildings in order to avoid having their memorization distracted by passing crowds. *The Phoenix*, a memory manual published by Peter of Ravenna in 1491, recommended visiting unfrequented churches for this reason. These classical rules for the art of memory "summon up a vision of a forgotten social habit. Who is that man moving slowly in the lonely building, stopping at intervals with an intent face? He is a rhetoric student forming a set of memory *loci*" (Yates, 1966, p. 8).

According to the index projection hypothesis, "by anchoring a small number of imagined objects to real objects in the world, the imaginal world inherits much of the geometry of the real world" (Pylyshyn, 2003b, p. 378). The classical art of memory, the method of loci, invokes a similar notion of scaffolding, attempting not only to inherit the real world's geometry, but to also inherit its permanence.

8.9 The Bounds of Cognition

The purpose of this chapter was to introduce Pylyshyn's (2003b, 2007) theory of visual cognition. This theory is of interest because different aspects of it make contact with classical, connectionist, or embodied cognitive science.

The classical nature of Pylyshyn's theory is found in his insistence that part of the purpose of vision is to make contact with perceptual categories that can be involved in general cognitive processing (e.g., inference and problem solving). The connectionist nature of Pylyshyn's theory is found in his invocation of artificial neural networks as the mechanisms for assigning and tracking indices as part of early vision. The embodied nature of Pylyshyn's theory is found in referential links between object files and distal objects, the use of indices to coordinate vision and action, and the use of indices and of referential links to exploit the external world as a scaffold for seeing and visualizing.

However, the hybrid nature of Pylyshyn's theory of visual cognition presents us with a different kind of puzzle. How is this to be reconciled with Pylyshyn's position as a champion of classical cognitive science and as a critic of connectionist (Fodor & Pylyshyn, 1988) and embodied (Fodor & Pylyshyn, 1981) traditions? The answer to this question is that when Pylyshyn writes of *cognition*, this term has a very technical meaning that places it firmly in the realm of classical cognitive science, and which—by this definition—separates it from both connectionist and embodied cognitive science.

Recall that Pylyshyn's (2003b, 2007) theory of visual cognition was motivated in part by dealing with some of the problems facing purely cognitive theories of perception such as the New Look. His solution was to separate early vision from cognition and to endorse perceptual mechanisms that solve problems of underdetermination without requiring inferential processing.

> I propose a distinction between vision and cognition in order to try to carve nature at her joints, that is, to locate components of the mind/brain that have some principled boundaries or some principled constraints in their interactions with the rest of the mind. (Pylyshyn, 2003b, p. 39)

The key to the particular "carving" of the system in his theory is that early vision, which includes preattentive mechanisms for individuating and tracking objects, does not do so by using concepts, categories, descriptions, or inferences. Time and again in his accounts of seeing and visualizing, Pylyshyn describes early vision as being "preconceptual" or "non-conceptual."

This is important because of Pylyshyn's (1984) characterization of the levels of analysis of cognitive science. Some of the levels of analysis that he invoked—in particular, the implementational and algorithmic levels—are identical to those levels as discussed in Chapter 2 in this volume. However, Pylyshyn's version of the computational level of analysis is more restrictive than the version that was also discussed in that earlier chapter.

For Pylyshyn (1984), a computational-level analysis requires a cognitive vocabulary. A cognitive vocabulary captures generalizations by appealing to the contents of representations, and it also appeals to lawful principles governing these contents (e.g., rules of inference, the principle of rationality). "The cognitive vocabulary is roughly similar to the one used by what is undoubtedly the most successful predictive scheme available for human behavior—folk psychology" (p. 2).

When Pylyshyn (2003b, 2007) separates early vision from cognition, he is proposing that the cognitive vocabulary cannot be productively used to explain early vision, because early vision is not cognitive, it is preconceptual. Thus it is no accident that when his theory of visual cognition intersects connectionist and embodied cognitive science, it does so with components that are part of Pylyshyn's account of early vision. Connectionism and embodiment are appropriate in this component of Pylyshyn's theory because his criticism of these approaches is that they are not cognitive, because they do not or cannot use a cognitive vocabulary!

Towards a Cognitive Dialectic

9.0 Chapter Overview

In the philosophy of G. W. F. Hegel, ideas developed by following a dialectical progression. They began as theses that attempted to explain some truth; deficiencies in theses permitted alternative ideas to be formulated. These alternatives, or antitheses, represented the next stage of the progression. A final stage, synthesis, approached truth by creating an emergent combination of elements from theses and antitheses. It has been argued that cognitive science provides an example of a dialectical progression. The current chapter begins by casting classical cognitive science as the thesis and considering both connectionist cognitive science and embodied cognitive science as viable antitheses. This argument is supported by reviewing some of the key differences amongst these three approaches. What remains is considering whether synthesis of these various approaches is possible.

Some of the arguments from previous chapters, including the possibility of hybrid accounts of cognition, are used to support the claim that synthesis in cognitive science is possible, though it has not yet been achieved. It is further argued that one reason synthesis has been impeded is because modern cognitivism, which exemplifies the classical approach, arose as a violent reaction against behaviourist psychology. Some of the core elements of cognitive antitheses, such as exploiting associations between ideas as well as invoking environmental control, were also foundations of the behaviourist school of thought. It is suggested that this has worked against synthesis, because exploring such ideas has the ideological impact of abandoning the cognitive revolution.

In this chapter I then proceed to consider two approaches for making the completion of a cognitive dialectic more likely. One approach is to consider the successes of the natural computation approach to vision, which developed influential theories that reflect contributions of all three approaches to cognitive science. It was able to do so because it had no ideological preference of one approach over the others. The second approach is for classical cognitive science to supplement its analytical methodologies with forward engineering. It is argued that such a synthetic methodology is likely to discover the limits of a "pure" paradigm, producing a tension that may only be resolved by exploring the ideas espoused by other positions within cognitive science.

9.1 Towards a Cognitive Dialectic

A dialectic involves conflict which generates tension and is driven by this tension to a state of conflict resolution (McNeill, 2005). According to philosopher G. W. F. Hegel (1931), ideas evolve through three phases: thesis, antithesis, and synthesis. Different approaches to the study of cognition can be cast as illustrating a dialectic (Sternberg, 1999).

> Dialectical progression depends upon having a critical tradition that allows current beliefs (theses) to be challenged by alternative, contrasting, and sometimes even radically divergent views (antitheses), which may then lead to the origination of new ideas based on the old (syntheses). (Sternberg, 1999, p. 52)

The first two aspects of a dialectic, thesis and antithesis, are easily found throughout the history of cognitive science. Chapters 3, 4, and 5 present in turn the elements of classical, connectionist, and embodied cognitive science. I have assigned both connectionist and embodied approaches with the role of antitheses to the classical thesis that defined the earliest version of cognitive science. One consequence of antitheses arising against existing theses is that putative inadequacies of the older tradition are highlighted, and the differences between the new and the old approaches are emphasized (Norman, 1993). Unsurprisingly, it is easy to find differences between the various cognitive sciences and to support the position that cognitive science is fracturing in the same way that psychology did in the early twentieth century. The challenge to completing the dialectic is exploring a synthesis of the different cognitive sciences.

One kind of tool that is becoming popular for depicting and organizing large amounts of information, particularly for various Internet sites, is the tag cloud or word cloud (Dubinko et al., 2007). A word cloud is created from a body of text; it summarizes that text visually by using size, colour, and font. Typically, the more frequently a term appears in a text, the larger is its depiction in a word cloud. The goal of a word cloud is to summarize a document in a glance. As a way to illustrate contrasts between classical, connectionist, and embodied cognitive sciences, I compare word clouds created for each of chapters 3, 4, and 5. Figure 9-1 presents the word cloud generated for Chapter 3 on classical cognitive science. Note that it highlights words that are prototypically classical, such as *physical, symbol, system, language, grammar, information, expression*, as well as key names like *Turing* and *Newell*.

Figure 9-1. Word cloud generated from the text of Chapter 3 on classical cognitive science.

An alternative word cloud emerges from Chapter 4 on connectionist cognitive science, as shown in Figure 9-2. This word cloud picks out key connectionist elements such as *network, input, hidden, output, units, connections, activity, learning, weights*, and *neural*; names found within the cloud are *McCulloch, Berkeley, Rescorla-Wagner*, and *Rumelhart*. Interestingly, the words *connectionist* and *classical* are equally important in this cloud, probably reflecting the fact that connectionist properties are typically introduced by contrasting them with (problematic) classical characteristics. The word clouds in Figures 9-1 and 9-2 differ strikingly from one another.

A third word cloud that is very different from the previous two is provided in Figure 9-3, which was compiled from Chapter 5 on embodied cognitive science. The words that it highlights include *behaviour, world, environment, control, agent, robot, body, nature, extended*, and *mind*; names captured include *Grey Walter, Clark*, and *Ashby*. Once again, *embodied* and *classical* are both important terms in the chapter, reflecting that the embodied approach is an antithesis to the classical thesis, and is often presented in direct contrast to classical cognitive science.

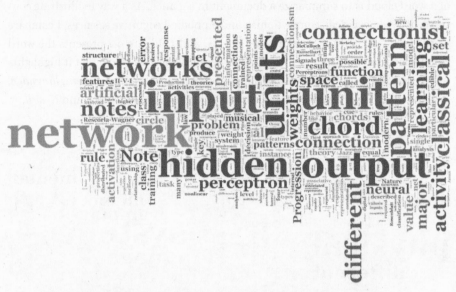

Figure 9-2. Word cloud generated from the text of Chapter 4 on connectionist cognitive science.

Figure 9-3. Word cloud generated from the text of Chapter 4 on embodied cognitive science.

Another way to illustrate the differences between the different approaches to cognitive science is to consider a set of possible dimensions or features and to characterize each approach to cognitive science in terms of each dimension. Table 9-1 presents one example of this manoeuvre. The dimensions used in this table—*core ideas, preferred formalism, tacit assumption*, and so on—were selected because I viewed them as being important, but the list of these features could be extended.

	Classical Cognitive Science	Connectionist Cognitive Science	Embodied Cognitive Science
Core Ideas	Mind as a physical symbol system Mind as digital computer Mind as planner Mind as creator and manipulator of models of the world Mind as sense-think-act processing	Mind as information processor, but not as a digital computer Mind as a parallel computer Mind as pattern recognizer Mind as a statistical engine Mind as biologically plausible mechanism	Mind as controller of action Mind emerging from situation and embodiment, or being-in-the-world Mind as extending beyond skull into world Mind as sense-act processing
Preferred Formalism	Symbolic logic	Nonlinear optimization	Dynamical systems theory
Tacit Assumption	Nativism, naïve realism	Empiricism	Embodied interaction
Type of Processing	Symbol manipulation	Pattern recognition	Acting on the world
Prototypical Architecture	Production system (Newell, 1973)	Multilayer perceptron (Rumelhart et al., 1986b)	Behaviour-based robot (Brooks, 1989)
Prototypical Domain	Language Problem solving	Discrimination learning Perceptual categorization	Locomotion Social interaction

	Classical Cognitive Science	Connectionist Cognitive Science	Embodied Cognitive Science
Philosophical Roots	Hobbes	Aristotle	Vico
	Descartes	Locke	Dewey
	Leibniz	Hume	Heidegger
	Craik	James	Merleau-Ponty
Some Key Modern Theorists	Chomsky	J.A. Anderson	Brooks
	Dennett	Hinton	Clark
	Fodor	Kohonen	Noë
	Pylyshyn	McClelland	Wilson
Some Pioneering Works	*Plans And The Structure Of Behavior* (Miller et al., 1960)	*Principles Of Neurodynamics* (Rosenblatt, 1962)	*Cognition And Reality* (Neisser, 1976)
	Aspects Of The Theory Of Syntax (Chomsky, 1965)	*Parallel Models Of Associative Memory* (Hinton & Anderson, 1981)	*The Ecological Approach To Visual Perception* (Gibson, 1979)
	Human Problem Solving (Newell & Simon, 1972)	*Parallel Distributed Processing* (McClelland & Rumelhart, 1986; Rumelhart & McClelland, 1986c)	*Understanding Computers And Cognition* (Winograd & Flores, 1987b)

Table 9-1. Contrasts between the three schools of thought in cognitive science.

An examination of Table 9-1 once again reveals marked differences between the three approaches as described in this volume. Other features could be added to this table, but I suspect that they too would reveal striking differences between the three views of cognitive science, and would be less likely to reveal striking similarities.

The illustrations so far—with the word clouds and with the table—definitely point towards the existence of theses and antitheses. An obvious tension exists within cognitive science. How might a synthesis be achieved to alleviate this tension? One approach to achieving synthesis in the cognitive dialectic may involve considering why the differences highlighted in Table 9-1 have arisen.

One context for considering Table 9-1 is the Indian fable of the six blind men and the elephant, the subject of a famous nineteenth-century poem by John Godfrey Saxe (Saxe, 1868). Each blind man feels a different part of the elephant, and comes away with a very different sense of the animal. The one who touched the tusk likens an elephant to a spear, the one who felt the knee compares the animal to a tree, the one who grabbed the tail likens it to a rope, and so on. After each has explored their part of the elephant, they reconvene to discuss its nature, and find that each has a dramatically different concept of the animal. The result is a heated, and ultimately unresolved, dispute: "And so these men of Indostan / Disputed loud and long, / Each in his own opinion / Exceeding stiff and strong, / Though each was partly in the right, / And all were in the wrong!" (p. 260).

To apply the moral of this story to the differences highlighted in Table 9-1, it is possible that the different approaches to cognitive science reflect differences that arise because each pays attention to different aspects of cognition, and none directs its attention to the complete picture. This view is consistent with one characterization of cognitive science that appeared at the cusp of the connectionist revolution (Norman, 1980).

Norman (1980) characterized a mature classical cognitive science that had decomposed human cognition into numerous information processing subsystems that defined what Norman called the pure cognitive system. The core of the pure cognitive system was a physical symbol system.

Norman's (1980) concern, though, was that the classical study of the pure cognitive system was doomed to fail because it, like one of the blind men, was paying attention to only one component of human cognition. Norman, prior to the rise of either connectionist or embodied cognitive science, felt that more attention had to be paid to the biological mechanisms and the surrounding environments of cognitive agents.

> The human is a physical symbol system, yes, with a component of pure cognition describable by mechanisms. . . . But the human is more: the human is an animate organism, with a biological basis and an evolutionary and cultural history. Moreover, the human is a social animal, interacting with others, with the environment, and with itself. The core disciplines of cognitive science have tended to ignore these aspects of behavior. (Norman, 1980, pp. 2–4)

Norman (1980) called for cognitive scientists to study a variety of issues that would extend their focus beyond the study of purely classical cognition. This included returning to a key idea of cybernetics, feedback between agents and their environments. "The concept has been lost from most of cognitive studies, in part because of the lack of study of output and of performance" (p. 6). For Norman, cognitive science had to consider "different aspects of the entire system, including the parts that are both internal and external to the cognizer" (p. 9).

Norman's (1980) position points out one perspective for unifying the diversity illustrated in Table 9-1: recognize that each school of cognitive science is, like each blind man in the fable, investigating an incomplete aspect of cognition and take advantage of this by combining these different perspectives. "I believe in the value of multiple philosophies, multiple viewpoints, multiple approaches to common issues. I believe a virtue of Cognitive Science is that it brings together heretofore disparate disciplines to work on common themes" (pp. 12–14).

One illustration of the virtue of exploring multiple viewpoints in the study of single topics is Norman's own work on design (Norman, 1998, 2002, 2004). Another illustration is the hybrid theory of seeing and visualizing (Pylyshyn, 2003c, 2007) described in Chapter 8, which draws on all three approaches to cognitive science in an attempt to arrive at a more complete account of a broad and diverse topic. The key to such successful examples is the acknowledgment that there is much to be gained from a co-operative view of different approaches; there is no need to view each approach to cognitive science as being mutually exclusive competitors.

9.2 Psychology, Revolution, and Environment

Norman (1980) called for cognitive science to extend its domain beyond the investigation of pure cognition, suggesting, for example, a return to some of the topics that were central to cybernetics, such as feedback between agents and environments. This was not the first time that such a suggestion had been made.

Twenty years earlier, in *Plans and the Structure of Behavior* (Miller et al., 1960), cognitive psychologist George Miller, mathematical psychologist Eugene Galanter, and neuropsychologist Karl Pribram argued for cognitivism to revisit the contributions of cybernetics. The reason for this was that Miller, Galanter, and Pribram, like Norman, were worried that if cognitivism focused exclusively on mental representations, then it would be incomplete. Such a perspective "left an organism more in the role of a spectator than of a participant in the drama of living. Unless you can use your Image to do something, you are like a man who collects maps but never makes a trip" (p. 2).

A related perspective was the theme of another key work that preceded Norman (1980), Ulric Neisser's (1976) *Cognition and Reality*. Neisser, an eminent pioneer of cognitivism (Neisser, 1967), argued that the relevance of cognitive psychology required it to be concerned with factors that lay beyond mental representations. "Perception and cognition are usually not just operations in the head, but transactions with the world. These transactions do not merely *in*form the perceiver, they also *trans*form him" (Neisser, 1976, p. 11). Rather than being inspired by cybernetics, Neisser was interested in reformulating cognitivism in the context of Gibson's (1966, 1979) theory of ecological perception. "Because perception and action take place in continuous dependence on the environment, they cannot be understood without an understanding of that environment itself" (Neisser, 1976, p. 183).

It would appear, then, that there is an extended history of important cognitivists calling for cognitive science to extend itself beyond the study of what Norman (1980) called the pure cognitive system. It is equally clear that this message has not had the desired impact. For instance, had the main theme of Miller, Galanter, and Pribram (1960) been widely accepted, then there would have been no need for similar proposals to appear decades later, as with Neisser (1976) and Norman (1980).

Why has cognitive science stubbornly held firm to the classical approach, emphasizing the study of pure cognition? One possible answer to this question is that the development of cognitivism in one of cognitive science's key contributors, psychology, occurred in a combative context that revealed thesis and antithesis but was not conducive to synthesis. This answer is considered in more detail below.

It is often claimed that cognitive science is chiefly concerned with the human cognitive capacities (Gardner, 1984; von Eckardt, 1995). Ironically, the one discipline that would be expected to have the most to say about human mental phenomena—experimental psychology—was one of the last to accept cognitivism. This was because around the time cognitive science emerged, experimental psychology was dominated by behaviourism.

Behaviourists argued that a scientific psychology must restrict itself to the study of observable behaviour and avoid invoking theoretical constructs that could not be directly observed, such as mental representation.

> So long as behaviorism held sway—that is, during the 1920s, 1930s, and 1940s—
> questions about the nature of human language, planning, problem solving, imagi-
> nation and the like could only be approached stealthily and with difficulty, if they
> were tolerated at all. (Gardner, 1984, p. 12)

Other disciplines were quicker to endorse cognitivism and to draw upon the insights of diverse fields of study because they were not restricted by the behaviourist yoke. For instance, mathematician Norbert Wiener (1948) created the field of cybernetics after realizing that problems involving communication, feedback, and information were general enough to span many disciplines. He held "the conviction that the most fruitful areas for the growth of the sciences were those which had been neglected as a no-man's land between the various fields" (p. 8).

Wiener realized that progress in cybernetics required interaction between researchers trained in different disciplines. He was a key organizer of the first joint meeting concerning cybernetics, held at Princeton in 1944, which included engineers, physiologists, and mathematicians. This in turn led to the Macy conferences on cybernetics that occurred regularly from 1946 through 1953 (Conway &

Siegelman, 2005). The Macy conferences broadened the range of participants who attended the 1944 Princeton meeting to include psychologists, sociologists, and anthropologists.

The success of the Macy meetings prepared the way for a variety of similar interdisciplinary conferences that in turn set the stage for cognitive science. One of these was a 1956 conference organized by MIT's Special Interest Group in Information Theory. This conference included presentations by Newell and Simon on their logic machine, and by Chomsky on generative grammar (Miller, 2003). Thus conference participant George Miller, trained in the behaviourist tradition, would have heard computer scientists and linguists freely using representational terms to great effect.

The success of cognitivism in other disciplines, communicated to psychologists who participated in these interdisciplinary conferences, led to a reaction against behaviourism in psychology. "No longer were psychologists restricted in their explanatory accounts to events that could either be imposed on a subject or observed in one's behavior; psychologists were now willing to consider the representation of information in the mind" (Gardner, 1984, p. 95).

George Miller (2003) has provided a personal account of this transition. His first book, *Language and Communication* (Miller, 1951), deliberately employed a behaviourist framework, a framework that he would completely abandon within a few years because of the influence of the cognitivist work of others. "In 1951, I apparently still hoped to gain scientific respectability by swearing allegiance to behaviorism. Five years later, inspired by such colleagues as Noam Chomsky and Jerry Bruner, I had stopped pretending to be a behaviorist" (Miller, 2003, p. 141).

However, because cognitivism arose as a reaction against behaviourism in North American experimental psychology, cognitive psychology developed by taking an antagonistic approach to almost all of the central behaviourist positions (Bruner, 1990; Sperry, 1993). "We were not out to 'reform' behaviorism, but to replace it" said Bruner (1990, p. 3). In psychology, the cognitive revolution,

> was not one of finding new positives to support the important role of cognition, many of which were already long evident. Rather, the story is one of discovering an alternative logic by which to refute the seemingly incontestable reasoning that heretofore required science to ostracize mind and consciousness. (Sperry, 1993, p. 881)

Consider but one example that illustrates the tone within psychology during the cognitive revolution. Skinner's (1957) account of language, *Verbal Behavior*, elicited a review by Noam Chomsky (1959b) that serves as one of the pioneering articles in cognitivism and is typically viewed as the turning point against psychological behaviourism (MacCorquodale, 1970; Schlinger, 2008). Some researchers, though, have objected to the tone of Chomsky's review: "It is ungenerous to a fault; condescending, unforgiving, obtuse, and ill-humored" (MacCorquodale, 1970, p. 84).

On the other side of the antagonism, behaviourists have never accepted the impact of Chomsky's review or the outcome of the cognitive revolution. Schlinger (2008, p. 335) argued that fifty years after its publication, *Verbal Behavior* (and behaviourism) was still vital because it worked: "It seems absurd to suggest that a book review could cause a paradigmatic revolution or wreak all the havoc that Chomsky's review is said to have caused to *Verbal Behavior* or to behavioral psychology."

The tone of the debate about *Verbal Behavior* is indicative of the tension and conflict that characterized cognitivism's revolt against behaviourist psychology. As noted earlier, cognitivists such as Bruner viewed their goal as replacing, and not revising, behaviourist tenets: "It was not a revolution against behaviorism with the aim of transforming behaviorism into a better way of pursuing psychology by adding a little mentalism to it. Edward Tolman had done that, to little avail" (Bruner, 1990, p. 2).

One behaviourist position that was strongly reacted against by cognitivism "was the belief in the supremacy and the determining power of the environment" (Gardner, 1984, p. 11). Cognitive psychologists turned almost completely away from environmental determinism. Instead, humans were viewed as active information processors (Lindsay & Norman, 1972; Reynolds & Flagg, 1977). For instance, the New Look in perception was an argument that environmental stimulation could be overridden by the contents of beliefs, desires, and expectations (Bruner, 1957). In cognitivism, mind triumphed over environmental matter.

Cognitive psychology's radical rejection of the role of the environment was a departure from the earlier cybernetic tradition, which placed a strong emphasis on the utility of feedback between an agent and its world. Cyberneticists had argued that,

> for effective action on the outer world, it is not only essential that we possess good effectors, but that the performance of these effectors be properly monitored back to the central nervous system, and that the readings of these monitors be properly combined with the other information coming in from the sense organs to produce a properly proportioned output to the effectors. (Wiener, 1948, p. 114)

Some cognitivists still agreed with the view that the environment was an important contributor to the complexity of behaviour, as shown by Simon's parable of the ant (Simon, 1969; Vera & Simon, 1993). Miller, Galanter, and Pribram (1960) acknowledged that humans and other organisms employed internal representations of the world. However, they were also "disturbed by a theoretical vacuum between cognition and action" (p. 11). They attempted to fill this vacuum by exploring the relevance of key cybernetic ideas, particularly the notion of environmental feedback, to cognitive psychology.

However, it is clear that Miller, Galanter, and Pribram's (1960) message about the environment had little substantive impact. Why else would Norman (1980) be conveying the same message twenty years later? It is less clear why this was the case. One possibility is that as cognitivism took root in experimental psychology, and as cognitive psychology in turn influenced empirical research within cognitive science, interest in the environment was a minority position. Cognitive psychology was clearly in a leading position to inform cognitive science about its prototypical domain (i.e., adult human cognition; see von Eckardt, 1995). Perhaps this informing included passing along antagonist views against core behaviourist ideas.

Of course, cognitive psychology's antagonism towards behaviourism and the behaviourist view of the environment is not the only reason for cognitive science's rise as a classical science. Another reason is that cognitive science was not so much inspired by cybernetics, but was instead inspired by computer science and the implications of the digital computer. Furthermore, the digital computer that inspired cognitive science—the von Neumann architecture, or the stored-program computer (von Neumann, 1993)—was a device that was primarily concerned with the manipulation of internal representations.

Finally, the early successes in developing classical models of a variety of high-level cognitive phenomena such as problem solving (Newell et al., 1958; Newell & Simon, 1961, 1972), and of robots that used internal models to plan before executing actions on the world (Nilsson, 1984), were successes achieved without worrying much about the relationship between world and agent. Sense-think-act processing, particularly the sort that heavily emphasized thinking or planning, was promising new horizons for the understanding of human cognition. Alternative approaches, rooted in older traditions of cybernetics or behaviourism, seemed to have been completely replaced.

One consequence of this situation was that cognitive science came to be *defined* in a manner that explicitly excluded non-classical perspectives. For example, consider von Eckardt's (1995) attempt to characterize cognitive science. Von Eckardt argued that this can be done by identifying a set of domain-specifying assumptions, basic research questions, substantive assumptions, and methodological assumptions. Importantly, the specific members of these sets that von Eckardt identified reflect a prototypical classical cognitive science and seem to exclude both connectionist and embodied varieties.

Consider just one feature of von Eckardt's (1995) project. She began by specifying the identification assumption for cognitive science—its assumed domain of study. According to von Eckardt, the best statement of this assumption is to say that cognitive science's domain is human cognitive capacities. Furthermore, her discussion of this assumption—and of possible alternatives to it—rejects non-classical variants of cognitive science.

For instance, Simon's (1969) early consideration of the sciences of the artificial cast intelligence as being the ability to adapt behaviour to changing demands of the environment. Von Eckardt (1995) considered this idea as being a plausible alternative to her preferred identification assumption. However, her analysis of Simon's proposal can be dismissed because it is too broad: "for there are cases of adaptive behavior (in Simon's sense) mediated by fairly low-level biological mechanisms that are not in the least bit cognitive and, hence, do not belong within the domain of cognitive science" (p. 62). This view would appear to reject connectionism as being cognitive science, in the sense that it works upward from low-level biological mechanisms (Dawson, 2004) and that connectionism rejects the classical use of an explanatory cognitive vocabulary (Fodor & Pylyshyn, 1988; Smolensky, 1988).

Similarly, von Eckardt (1995) also rejected an alternative to her definition of cognitive science's identification assumption, which would include in cognitive science the study of core embodied issues, such as cognitive scaffolding.

> Human beings represent and use their 'knowledge' in many ways, only some of which involve the human mind. What we know is represented in books, pictures, computer databases, and so forth. Clearly, cognitive science does not study the representation and the use of knowledge in all these forms. (von Eckardt, 1995, p. 67)

If cognitive science does not study external representations, then by von Eckardt's definition the embodied approach does not belong to cognitive science.

The performance of classical simulations of human cognitive processes led researchers to propose in the late 1950s that within a decade most psychological theories would be expressed as computer programs (Simon & Newell, 1958). The classical approach's failure to deliver on such promises led to pessimism (Dreyfus, 1992), which resulted in critical assessments of the classical assumptions that inspired alternative approaches (Rumelhart & McClelland, 1986c; Winograd & Flores, 1987b). The preoccupation of classical cognitivism with the manipulation of internal models of the world may have prevented it from solving problems that depend on other factors, such as a cybernetic view of the environment.

> As my colleague George Miller put it some years later, 'We nailed our new credo to the door, and waited to see what would happen. All went very well, so well, in fact, that in the end we may have been the victims of our success.' (Bruner, 1990, pp. 2–3)

How has classical cognitivism been a victim of its success? Perhaps its success caused it to be unreceptive to completing the cognitive dialectic. With the rise of the connectionist and embodied alternatives, cognitive science seems to have been in the midst of conflict between thesis and antithesis, with no attempt at synthesis. Fortunately there are pockets of research within cognitive science that can illustrate a path towards synthesis, a path which requires realizing that each of the schools of thought we have considered here has its own limits, and that none of these schools of thought should be excluded from cognitive science by definition. One example domain in which synthesis is courted is computational vision.

9.3 Lessons from Natural Computation

To sighted human perceivers, visual perception seems easy: we simply look and see. Perhaps this is why pioneers of computer vision took seeing for granted. One student of Marvin Minsky was assigned—as a summer project—the task of programming vision into a computer (Horgan, 1993). Only when such early projects were attempted, and had failed, did researchers realize that the visual system was effortlessly solving astronomically difficult information processing problems.

Visual perception is particularly difficult when one defines its goal as the construction of internal models of the world (Horn, 1986; Marr, 1976, 1982; Ullman, 1979). Such representations, called distal stimuli, must make explicit the three-dimensional structure of the world. However, the information from which the distal stimulus is constructed—the proximal stimulus—is not rich enough to uniquely specify 3-D structure. As discussed in Chapter 8, the poverty of proximal stimuli underdetermines visual representations of the world. A single proximal stimulus is consistent with, in principle, an infinitely large number of different world models. The underdetermination of vision makes computer vision such a challenge to artificial intelligence researchers because information has to be added to the proximal stimulus to choose the correct distal stimulus from the many that are possible.

The cognitive revolution in psychology led to one approach for dealing with this problem: the New Look in perception proposed that seeing is a form of problem solving (Bruner, 1957, 1992; Gregory, 1970, 1978; Rock, 1983). General knowledge of the world, as well as beliefs, expectations, and desires, were assumed to contribute to our visual experience of the world, providing information that was missing from proximal stimuli.

The New Look also influenced computer simulations of visual perception. Knowledge was loaded into computer programs to be used to guide the analysis of visual information. For instance, knowledge of the visual appearance of the components of particular objects, such as an air compressor, could be used to guide the segmentation of a raw image of such a device into meaningful parts (Tenenbaum & Barrow, 1977). That is, the computer program could see an air compressor by exploiting its pre-existing knowledge of what it looked like. This general approach—using pre-existing knowledge to guide visual perception—was widespread in the computer science literature of this era (Barrow & Tenenbaum, 1975). Barrow and Tenenbaum's (1975) review of the state of the art at that time concluded that image segmentation was a low-level interpretation that was guided by knowledge, and they argued that the more knowledge the better.

Barrow and Tenenbaum's (1975) review described a New Look within computer vision:

> Higher levels of perception could involve partitioning the picture into 'meaningful' regions, based on models of particular objects, classes of objects, likely events in the world, likely configurations, and even on nonvisual events. Vision might be viewed as a vast, multi-level optimization problem, involving a search for the best interpretation simultaneously over all levels of knowledge. (Barrow & Tenenbaum, 1975, p. 2)

However, around the same time a very different data-driven alternative to computer vision emerged (Waltz, 1975).

Waltz's (1975) computer vision system was designed to assign labels to regions and line segments in a scene produced by drawing lines and shadows. "These labels describe the edge geometry, the connection or lack of connection between adjacent regions, the orientation of each region in three dimensions, and the nature of the illumination for each region" (p. 21). The goal of the program was to assign one and only one label to each part of a scene that could be labelled, except in cases where a human observer would find ambiguity.

Waltz (1975) found that extensive, general knowledge of the world was not required to assign labels. Instead, all that was required was a propagation of local constraints between neighbouring labels. That is, if two to-be-labelled segments were connected by a line, then the segments had to be assigned consistent labels. Two ends of a line segment could not be labelled in such a way that one end of the line would be given one interpretation and the other end a different interpretation that was incompatible with the first. Waltz found that this approach was very powerful and could be easily applied to novel scenes, because it did not depend on specialized, scene-specific knowledge. Instead, all that was required was a method to determine what labels were possible for any scene location, followed by a method for comparisons between possible labels, in order to choose unique and compatible labels for neighbouring locations.

The use of constraints to filter out incompatible labels is called relaxation labelling (Rosenfeld, Hummel, & Zucker, 1976); as constraints propagate through neighbouring locations in a representation, the representation moves into a stable, lower-energy state by removing unnecessary labels. The discussion of solving Sudoku problems in Chapter 7 illustrates an application of relaxation labelling. Relaxation labelling proved to be a viable data-driven approach to dealing with visual underdetermination.

Relaxation labelling was the leading edge of a broad perspective for understanding vision. This was the natural computation approach to vision (Hildreth, 1983; Marr, 1976, 1982; Marr & Hildreth, 1980; Marr & Nishihara, 1978; Marr, Palm, & Poggio, 1978; Marr & Poggio, 1979; Marr & Ullman, 1981; Richards, 1988; Ullman, 1979). Researchers who endorse the natural computation approach to vision use naïve realism to solve problems of underdetermination. They hypothesize that the visual world is intrinsically structured, and that some of this structure is true of any visual scene. They assume that a visual system that has evolved in such a structured world is able to take advantage of these visual properties to solve problems of underdetermination.

The properties of interest to natural computation researchers are called natural constraints. A natural constraint is a property of the visual world that is almost always true of any location in any scene. For example, a great many visual properties of three-dimensional scenes (depth, texture, colour, shading, motion) vary smoothly. This means that two locations very near one another in a scene are very likely to have very similar values for any of these properties. Locations that are further apart will not be as likely to have similar values for these properties.

Natural constraints can be used to solve visual problems of underdetermination by imposing restrictions on scene interpretations. Natural constraints are properties that must be true of an interpretation of a visual scene. They can therefore be used to filter out interpretations consistent with the proximal stimulus but not consistent with the natural constraint. For example, an interpretation of a scene that violated the smoothness constraint, because its visual properties did not vary smoothly in the sense described earlier, could be automatically rejected and never experienced.

The natural computation approach triumphed because it was able to identify a number of different natural constraints for solving a variety of visual problems of underdetermination (for many examples, see Marr, 1982). As in the scene labelling approach described above, the use of natural constraints did not require scene-specific knowledge. Natural computation researchers did not appeal to problem solving or inference, in contrast to the knowledge-based models of an earlier generation (Barrow & Tenenbaum, 1975; Tenenbaum & Barrow, 1977). This was because natural constraints could be exploited using data-driven algorithms, such as neural networks. For instance, one can exploit natural constraints for scene labelling by using processing units to represent potential labels and by defining natural constraints between labels using the connection weights between processors (Dawson, 1991). The dynamics of the signals sent through this network will turn on the units for labels consistent with the constraints and turn off all of the other units.

In the context of the current discussion of the cognitive sciences, the natural computation approach to vision offers an interesting perspective on how a useful synthesis of divergent perspectives is possible. This is because the natural computation approach appeals to elements of classical, connectionist, and embodied cognitive science.

Initially, the natural computation approach has strong classical characteristics. It views visual perception as a prototypical representational phenomenon, endorsing sense-think-act processing.

> The study of vision must therefore include not only the study of how to extract from images the various aspects of the world that are useful to us, but also an inquiry into the nature of the internal representations by which we capture this information and thus make it available as a basis for decisions about our thoughts and actions. (Marr, 1982, p. 3)

Marr's theory of early vision proposed a series of different kinds of representations of visual information, beginning with the raw primal sketch and ending with the 2½-D sketch that represented the three-dimensional locations of all visible points and surfaces.

However representational it is, though, the natural computation approach is certainly not limited to the study of what Norman (1980) called the pure cognitive system. For instance, unlike New Look theories of human perception, natural computation theories paid serious attention to the structure of the world. Indeed, natural constraints are not psychological properties, but are instead properties of the world. They are not identified by performing perceptual experiments, but are instead discovered by careful mathematical analyses of physical structures and their optical projections onto images. "The major task of Natural Computation is a formal analysis and demonstration of how unique and correct interpretations can be inferred from sensory data by exploiting lawful properties of the natural world" (Richards, 1988, p. 3). The naïve realism of the natural computation approach forced it to pay careful attention to the structure of the world.

In this sense, the natural computation approach resembles a cornerstone of embodied cognitive science, Gibson's (1966, 1979) ecological theory of perception. Marr (1982) himself saw parallels between his natural computation approach and Gibson's theory, but felt that natural computation addressed some flaws in ecological theory. Marr's criticism was that Gibson rejected the need for representation, because Gibson underestimated the complexity of detecting invariants: "Visual information processing is actually very complicated, and Gibson was not the only thinker who was misled by the apparent simplicity of the act of seeing" (p. 30). In Marr's view, detecting visual invariants required exploiting natural constraints to build representations from which invariants could be detected and used. For instance, detecting the invariants available in a key Gibsonian concept, the optic flow field, requires applying smoothness constraints to local representations of detected motion (Hildreth, 1983; Marr, 1982).

Strong parallels also exist between the natural computation approach and connectionist cognitive science, because natural computation researchers were highly motivated to develop computer simulations that were biologically plausible. That is, the ultimate goal of a natural computation theory was to provide computational, algorithmic, and implementational accounts of a visual process. The requirement that a visual algorithm be biologically implementable results in a preference for parallel, co-operative algorithms that permit local constraints to be propagated through a network. As a result, most natural computation theories can be translated into connectionist networks.

How is it possible for the natural computation approach to endorse elements of each school of thought in cognitive science? In general, this synthesis of ideas is the result of a very pragmatic view of visual processing. Natural computation researchers recognize that "pure" theories of vision will be incomplete. For instance, Marr (1982) argued that vision must be representational in nature. However, he also noted that these representations are impossible to understand without paying serious attention to the structure of the external world.

Similarly, Marr's (1982) book, *Vision*, is a testament to the extent of visual interpretation that can be achieved by data-driven processing. However, data-driven processes cannot deliver a complete visual interpretation. At some point—when, for instance, the 2½-D sketch is linked to a semantic category—higher-order cognitive processing must be invoked. This openness to different kinds of processing is why a natural computation researcher such as Shimon Ullman can provide groundbreaking work on an early vision task such as computing motion correspondence matches (1979) and also be a pioneer in the study of higher-order processes of visual cognition (1984, 2000).

The search for biologically plausible algorithms is another example of the pragmatism of the natural computation approach. Classical theories of cognition have been criticized as being developed in a biological vacuum (Clark, 1989). In contrast, natural computation theories have no concern about eliminating low-level biological accounts from their theories. Instead, the neuroscience of vision is used to inform natural computation algorithms, and computational accounts of visual processing are used to provide alternative interpretations of the functions of visual neurons. For instance, it was only because of his computational analysis of the requirements of edge detection that Marr (1982) was able to propose that the centre-surround cells of the lateral geniculate nucleus were convolving images with difference-of-Gaussian filters.

The pragmatic openness of natural computation researchers to elements of the different approaches to cognitive science seems to markedly contrast with the apparent competition that seems to characterize modern cognitive science (Norman, 1993). One account of this competition might be to view it as a conflict between scientific paradigms (Kuhn, 1970). From this perspective, some antagonism between perspectives is necessary, because newer paradigms are attempting to show how they are capable of replacing the old and of solving problems beyond the grasp of the established framework. If one believes that they are engaged in such an endeavour, then a fervent and explicit rejection of including any of the old paradigm within the new is to be expected.

According to Kuhn (1970), a new paradigm will not emerge unless a crisis has arisen in the old approach. Some may argue that this is exactly the case for classical cognitive science, whose crises have been identified by its critics (Dreyfus, 1972, 1992), and which have led to the new connectionist and embodied paradigms. However, it is more likely that it is premature for paradigms of cognitive science to be battling one another, because cognitive science may very well be pre-paradigmatic, in search of a unifying body of belief that has not yet been achieved.

The position outlined in Chapter 7, that it is difficult to identify a set of core tenets that distinguish classical cognitive science from the connectionist and the embodied approaches, supports this view. Such a view is also supported by the existence of approaches that draw on the different "paradigms" of cognitive science, such as the theory of seeing and visualizing (Pylyshyn, 2003c, 2007) discussed in Chapter 8, and the natural computation theory of vision. If cognitive science were not pre-paradigmatic, then it should be easy to distinguish its different paradigms, and theories that draw from different paradigms should be impossible.

If cognitive science is pre-paradigmatic, then it is in the process of identifying its core research questions, and it is still deciding upon the technical requirements that must be true of its theories. My suspicion is that a mature cognitive science will develop that draws on core elements of all three approaches that have been studied. Cognitive science is still in a position to heed the call of a broadened cognitivism (Miller, Galanter, & Pribram, 1960; Norman, 1980). In order to do so, rather than viewing its current approaches as competing paradigms, it would be better served by adopting the pragmatic approach of natural computation and exploiting the advantages offered by all three approaches to cognitive phenomena.

9.4 A Cognitive Synthesis

Modern experimental psychology arose around 1860 (Fechner, 1966), and more than a century and a half later is viewed by many as still being an immature, pre-paradigmatic discipline (Buss, 1978; Leahey, 1992). The diversity of its schools of thought and the breadth of topics that it studies are a testament to experimental psychology's youth as a science. "In the early stages of the development of any science different men confronting the same range of phenomena, but not usually all the same particular phenomena, describe and interpret them in different ways" (Kuhn, 1970, p. 17).

Cognitive science was born in 1956 (Miller, 2003). Because it is about a century younger than experimental psychology, it would not be surprising to discover that cognitive science is also pre-paradigmatic. This might explain the variety of opinions about the nature of cognition, introduced earlier as the competing elements of classical, connectionist, and embodied cognitive science. "The pre-paradigm period, in particular, is regularly marked by frequent and deep debates over legitimate methods, problems, and standards of solution, though these serve rather to define schools than produce agreement" (Kuhn, 1970, pp. 47–48).

The current state of cognitive science defines an as yet incomplete dialectic. Competition amongst classical, connectionist, and embodied cognitive science reflects existing tensions between thesis and antithesis. What is missing is a state of synthesis in which cognitive science integrates key ideas from its competing schools of thought. This integration is necessary, because it is unlikely that, for instance, a classical characterization of the pure cognitive system will provide a complete explanation of cognition (Miller, Galanter, & Pribram, 1960; Neisser, 1976; Norman, 1980).

In the latter chapters of the current book, several lines of evidence are presented to suggest that synthesis within cognitive science is possible. First, it is extremely difficult to find marks of the classical, that is, characteristics that uniquely distinguish classical cognitive science from either the connectionist or embodied approaches. For instance, classical cognitive science was inspired by the digital computer, but a variety of digital computers incorporated processes consistent with connectionism (such as parallel processing) and with embodied cognitive science (such as external representations).

A second line of evidence is that there is a high degree of methodological similarity between the three approaches. In particular, each school of cognitive science can be characterized as exploring four different levels of investigation: computational, algorithmic, architectural, and implementational. We see in Chapter 6 that the different approaches have disagreements about the technical details within each level. Nevertheless, all four levels are investigated by all three approaches within cognitive science. Furthermore, when different approaches are compared at each level, strong similarities can be identified. This is why, for instance, that it has been claimed that the distinction between classical and connectionist cognitive science is blurred (Dawson, 1998).

A third line of evidence accounts for the methodological similarity amongst the different approaches: cognitive scientists from different schools of thought share many core assumptions. Though they may disagree about its technical details, all cognitive scientists view cognition as a form of information processing. For instance, each of the three schools of thought appeals to the notion of representation, while at the same time debating its nature. Are representations symbols, distributed patterns, or external artifacts? All cognitive scientists have rejected Cartesian dualism and are seeking materialist explanations of cognition.

More generally, all three approaches in cognitive science agree that cognition involves interactions between the world and states of agents. This is why a pioneer of classical cognitive science can make the following embodied claim: "A man, viewed as a behaving system, is quite simple. The apparent complexity of his behavior over time is largely a reflection of the complexity of the environment in which he finds himself" (Simon, 1969, p. 25). However, it is again fair to say that the contributions of world, body, and mind receive different degrees of emphasis within the three approaches to cognitive science. We saw earlier that production system pioneers admitted that they emphasized internal planning and neglected perception and action (Anderson et al., 2004; Newell, 1990). Only recently have they turned to including sensing and acting in their models (Kieras & Meyer, 1997; Meyer et al., 2001; Meyer & Kieras, 1997a, 1997b, 1999; Meyer et al., 1995). Even so, they are still very reluctant to include sense-act processing—links between sensing and acting that are not mediated by internal representations—to their sense-think-act production systems (Dawson, Dupuis, & Wilson, 2010).

A fourth line of evidence is the existence of hybrid theories, such as natural computation (Marr, 1982) or Pylyshyn's (2003) account of visual cognition. These theories explicitly draw upon concepts from each approach to cognitive science. Hybrid theories are only possible when there is at least tacit recognition that each school of thought within cognitive science has important, co-operative contributions to make. Furthermore, the existence of such theories completely depends upon the need for such co-operation: no one school of thought provides a sufficient explanation of cognition, but each is a necessary component of such an explanation.

It is one thing to note the possibility of a synthesis in cognitive science. It is quite another to point the way to bringing such a synthesis into being. One required component, discussed earlier in this chapter, is being open to the possible contributions of the different schools of thought, an openness demonstrated by the pragmatic and interdisciplinary natural computation theory of perception.

A second component, which is the topic of this final section of the book, is being open to a methodological perspective that pervaded early cognitive science and its immediate ancestors, but which has become less favored in more recent times. Synthesis in cognitive science may require a return, at least in part, to the practice of synthetic psychology.

Present-day cognitive science for the most part employs analytic, and not synthetic, methodological practices. That is, most cognitive scientists are in the business of carrying out reverse engineering (Dennett, 1998). They start with a complete, pre-existing cognitive agent. They then observe its behaviour, not to mention how the behaviour is affected by various experimental manipulations. The results of these observations are frequently used to create theories in the form of computer simulations (Newell & Simon, 1961). For instance, Newell and Simon (1972) collected data in the form of verbal protocols, and then used these protocols to derive working production systems. In other words, when analytic methodologies are used, the collection of data precedes the creation of a model.

The analytic nature of most cognitive science is reflected in its primary methodology, functional analysis, a prototypical example of reverse engineering (Cummins, 1975, 1983). Functional analysis dictates a top-down decomposition from the broad and abstract (i.e., computational specification of functions) to the narrower and more concrete (i.e., architecture and implementation).

Even the natural computation approach in vision endorsed a top-down analytic approach, moving from computational to implementational analyses instead of in the opposite direction. This was because higher-level analyses were used to guide interpretations of the lower levels.

> In order to understand why the receptive fields are as they are—why they are circularly symmetrical and why their excitatory and inhibitory regions have characteristic shapes and distributions—we have to know a little of the theory of differential operators, band-pass channels, and the mathematics of the uncertainty principle. (Marr, 1982, p. 28)

An alternative approach is synthetic, not analytic; it is bottom-up instead of top-down; and it applies forward engineering instead of reverse engineering. This approach has been called synthetic psychology (Braitenberg, 1984). In synthetic psychology, one takes a set of primitive building blocks of interest and creates a working system from them. The behaviour of this system is observed in order to determine what surprising phenomena might emerge from simple components, particularly when they are embedded in an interesting or complex environment. As a result, in synthetic psychology, models precede data, because they are the source of data.

The forward engineering that characterizes synthetic psychology proceeds as a bottom-up construction (and later exploration) of a cognitive model. Braitenberg (1984) argued that this approach would produce simpler theories than those produced by analytic methodologies, because analytic models fail to recognize the influence of the environment, falling prey to what is known as the frame of reference problem (Pfeifer & Scheier, 1999). Also, analytic techniques have only indirect access to internal components, in contrast to the complete knowledge of such structures that is possessed by a synthetic designer.

It is pleasurable and easy to create little machines that do certain tricks. It is also quite easy to observe the full repertoire of behavior of these machines—even if it goes beyond what we had originally planned, as it often does. But it is much more difficult to start from the outside and try to guess internal structure just from the observation of the data. (Braitenberg, 1984, p. 20)

Although Braitenberg proposed forward engineering as a novel methodology in 1984, it had been widely practised by cyberneticists beginning in the late 1940s. For instance, the original autonomous robots, Grey Walter's (1950a, 1950b, 1951, 1963) Tortoises, were created to observe whether complex behaviour would be supported by a small set of cybernetic principles. Ashby's (1956, 1960) Homeostat was created to study feedback relationships between simple machines; after it was constructed, Ashby observed that this device demonstrated interesting and complicated adaptive relationships to a variety of environments. This kind of forward engineering is currently prevalent in one modern field that has inspired embodied cognitive science, behaviour-based robotics (Brooks, 1999; Pfeifer & Scheier, 1999; Sharkey, 2006).

Forward engineering is not limited to the creation of autonomous robots. It has been argued that the synthetic approach characterizes a good deal of connectionism (Dawson, 2004). The thrust of this argument is that the building blocks being used are the components of a particular connectionist architecture. These are put together into a working system whose behaviour can then be explored. In the connectionist case, the synthesis of a working network involves using a training environment to modify a network by applying a general learning rule.

Classical cognitive science is arguably the most commonly practised form of cognitive science, and it is also far less likely to adopt synthetic methodologies. However, this does not mean that classical cognitive scientists have not usefully employed forward engineering. One prominent example is in the use of production systems to study human problem solving (Newell & Simon, 1972). Clearly the analysis of verbal protocols provided a set of potential productions to include in a model. However, this was followed by a highly synthetic phase of model development.

This synthetic phase proceeded as follows: Newell and Simon (1972) used verbal protocols to rank the various productions available in terms of their overall usage. They then began by creating a production system model that was composed of only a single production, the one most used. The performance of this simple system was then compared to the human protocol. The next step was to create a new production system by adding the next most used production to the original model, and examining the behaviour of the new two-production system. This process would continue, usually revealing better performance of the model (i.e., a better fit to human data) as the model was elaborated by adding each new production.

Forward engineering, in all of the examples alluded to above, provides a systematic exploration of what an architecture can produce "for free." That is, it is not used to create a model that fits a particular set of data. Instead, it is used to show how much surprising and complex behaviour can be generated from a simple set of components—particularly when that architecture is embedded in an interesting environment. It is used to explore the limits of a system—how many unexpected complexities appear in its behaviour? What behaviours are still beyond the system's capability? While reverse engineering encourages the derivation of a model constrained by data, forward engineering is concerned with a much more liberating process of model design. "Only about 1 in 20 [students] 'gets it'—that is, the idea of thinking about psychological problems by inventing mechanisms for them and then trying to see what they can and cannot do" (Minsky, 1995, personal communication).

The liberating aspect of forward engineering is illustrated in the development of the LEGO robot AntiSLAM (Dawson, Dupuis, & Wilson, 2010). Originally, this robot was created as a sonar-based version of one of Braitenberg's (1984) simple thought experiments, Vehicle 2. Vehicle 2 used two light sensors to control the speeds of two separate motors and generated photophobic or photophilic behaviour depending upon its wiring. We replaced the light sensors with two sonar sensors, which itself was a departure from convention, because the standard view was that the two sensors would interfere with one another (Boogaarts, 2007). However, we found that the robot generated nimble behaviours and effortlessly navigated around many different kinds of obstacles at top speed. A slight tweak of the robot's architecture caused it to follow along a wall on its right. We then realized that if the environment for the robot became a reorientation arena, then it would generate rotational error. The forward engineering of this very simple robot resulted in our discovery that it generated navigational regularities "for free."

The appeal of forward engineering, though, is not just the discovery of unexpected behaviour. It is also appealing because it leads to the discovery of an architecture's limits. Not only do you explore what a system can do, but you discover its failures as well. It has been argued that in the analytic tradition, failures often lead to abandoning a model (Dawson, 2004), because failures amount to an inability to fit a desired set of data. In the synthetic approach, which is not driven by data fitting, failures lead to tinkering with the architecture, usually by adding new capabilities to it (Brooks, 1999, 2002). The synthetic design of cognitive models is a prototypical instance of *bricolage* (Dawson, Dupuis, & Wilson, 2010; Turkle, 1995).

For instance, while the early version of AntiSLAM (Dawson, Dupuis, & Wilson, 2010) produced rotational error, it could not process competing geometric and local cues, because it had no capability of detecting local cues. After realizing that the robot was capable of reorientation, this issue was solved by adding a light sensor to the existing architecture, so that a corner's brightness could serve as a rudimentary feature. The robot is still inadequate, though, because it does not learn. We are currently exploring how this problem might be solved by adding a modifiable connection-ist network to map relations between sensors and motors. Note that this approach requires moving beyond a pure embodied account and taking advantage of connectionist concepts.

In my opinion, it is the limitations inevitably encountered by forward engineers that will provide incentive for a cognitive synthesis. Consider the strong anti-representational positions of radical embodied cognitive scientists (Chemero, 2009; Noë, 2004). It is certainly astonishing to see how much interesting behaviour can be generated by systems with limited internal representations. But how much of cognition can be explained in a data-driven, antirepresentational manner before researchers have to appeal to representations? For instance, is a radical embodied cognitive science of natural language possible? If embodied cognitive scientists take their theories to their limits, and are then open—as are natural computation researchers—to classical or connectionist concepts, then an interesting and productive cognitive synthesis is inevitable. That some embodied researchers (Clark, 1997) have long been open to a synthesis between embodied and classical ideas is an encouraging sign.

Similarly, radical connectionist researchers have argued that a great deal of cognition can be accomplished without the need for explicit symbols and explicit rules (Rumelhart & McClelland, 1986a; Smolensky, 1988). Classical researchers have acknowledged the incredible range of phenomena that have yielded to the fairly simple PDP architecture (Fodor & Pylyshyn, 1988). But, again, how much can connectionists explain from a pure PDP perspective, and what phenomena will elude their grasp, demanding that classical ideas be reintroduced? Might it be possible to treat networks as dynamic symbols, and then manipulate them with external rules that are different from the learning rules that are usually applied? Once again, recent ideas seem open to co-operative use of connectionist and classical ideas (Smolensky & Legendre, 2006).

The synthetic approach provides a route that takes a cognitive scientist to the limits of their theoretical perspective. This in turn will produce a theoretical tension that will likely only be resolved when core elements of alternative perspectives are seriously considered. Note that such a resolution will require a theorist to be open to admitting different kinds of ideas. Rather than trying to show that their architecture can do everything cognitive, researchers need to find what their architectures cannot do, and then expand their theories by including elements of alternative, possibly radically different, views of cognition.

This is not to say that the synthetic approach is the only methodology to be used. Synthetic methods have their own limitations, and a complete cognitive science requires interplay between synthesis and analysis (Dawson, 2004). In particular, cognitive science ultimately is in the business of explaining the cognition of biological agents. To do so, its models—including those developed via forward engineering—must be validated. Validating a theory requires the traditional practices of the analytic approach, seeking equivalences between computations, algorithms, and architectures. It is hard to imagine such validation not proceeding by adopting analytic methods that provide relative complexity, error, and intermediate state evidence. It is also hard to imagine that a complete exploration of a putative cognitive architecture will not exploit analytic evidence from the neurosciences.

Indeed, it may be that the inability to use analytic evidence to validate a "pure" model from one school of thought may be the primary motivation to consider alternative perspectives, fueling a true synthesis within cognitive science. According to Kuhn (1970), paradigms are born by discovering anomalies. The analytic techniques of cognitive science are well equipped to discover such problems. What is then required for synthesis is a willingness amongst cognitive scientists to admit that competing views of cognition might be able to be co-operatively applied in order to resolve anomalies.

References

Abu-Mostafa, Y. S. (1990). Learning from hints in neural networks. *Journal of Complexity, 6*, 192–198.

Abu-Mostafa, Y. S., & St. Jacques, J. M. (1985). Information capacity of the Hopfield model. *IEEE Transactions on Information Theory, 31*(4), 461–464.

Ackley, D. H., Hinton, G. E., & Sejnowski, T. J. (1985). A learning algorithm for Boltzman machines. *Cognitive Science, 9*, 147–169.

Adams, F., & Aizawa, K. (2008). *The bounds of cognition.* Malden, MA: Blackwell Publishers.

Adiloglu, K., & Alpaslan, F. N. (2007). A machine learning approach to two-voice counterpoint composition. *Knowledge-Based Systems, 20*(3), 300–309.

Agawu, V. K. (1991). *Playing with signs: A semiotric interpretation of classic music.* Princeton, NJ: Princeton University Press.

Agawu, V. K. (2009). *Music as discourse: Semiotic adventures in romantic music.* New York, NY: Oxford University Press.

Agre, P. E. (1993). The symbolic worldview: Reply to Vera and Simon. *Cognitive Science, 17*(1), 61–69.

Agre, P. E. (1997). *Computation and human experience.* New York, NY: Cambridge University Press.

Agulló, M., Carlson, D., Clague, K., Ferrari, G., Ferrari, M., & Yabuki, H. (2003). *LEGO Mindstorms masterpieces: Building and programming advanced robots.* Rockland, MA: Syngress Publishing.

Aiken, H. H., & Hopper, G. (1946). The automatic sequence controlled calculator. *Electrical Engineering, 65*, 384–391, 449–354, 522–528.

Aldenderfer, M. S., & Blashfield, R. K. (1984). *Cluster analysis* (Vol. 07-044). Beverly Hills, CA: Sage Publications.

Alerstam, T. (2006). Conflicting evidence about long-distance animal navigation. *Science, 313*(5788), 791–794.

Alexander, R. M. (2005). Walking made simple. *Science, 308*(5718), 58–59.

Allan, L. G. (1980). A note on measurement of contingency between two binary variables in judgment tasks. *Bulletin of the Psychonomic Society, 15*(3), 147–149.

Alossa, N., & Castelli, L. (2009). Amusia and musical functioning. *European Neurology, 61*(5), 269–277.

Amari, S. (1967). A theory of adaptive pattern classifiers. *IEEE Transactions on Electronic Computers, Ec16*(3), 299–307.

Amit, D. J. (1989). *Modeling brain function: The world of attractor neural networks.* Cambridge, MA: Cambridge University Press.

Andersen, R. A., Snyder, L. H., Bradley, D. C., & Xing, J. (1997). Multimodal representation of space in the posterior parietal cortex and its use in planning movements. *Annual Review of Neuroscience, 20*, 303–330.

Anderson, J. A. (1972). A simple neural network generating an interactive memory. *Mathematical Biosciences, 14*, 197–220.

Anderson, J. A. (1995). *An introduction to neural networks*. Cambridge, MA: MIT Press.

Anderson, J. A., & Rosenfeld, E. (1998). *Talking nets: An oral history of neural networks*. Cambridge, MA: MIT Press.

Anderson, J. A., Silverstein, J. W., Ritz, S. A., & Jones, R. S. (1977). Distinctive features, categorical perception and probability learning: Some applications of a neural model. *Psychological Review, 84*, 413–451.

Anderson, J. R. (1978). Arguments concerning representations for mental imagery. *Psychological Review, 85*, 249–277.

Anderson, J. R. (1983). *The architecture of cognition*. Cambridge, MA: Harvard University Press.

Anderson, J. R. (1985). *Cognitive psychology and its implications* (2nd ed.). New York, NY: W. H. Freeman.

Anderson, J. R., Bothell, D., Byrne, M. D., Douglass, S., Lebiere, C., & Qin, Y. L. (2004). An integrated theory of the mind. *Psychological Review, 111*(4), 1036–1060.

Anderson, J. R., & Bower, G. H. (1973). *Human associative memory*. Hillsdale, NJ: Lawrence Erlbaum Associates.

Anderson, J. R., & Matessa, M. (1997). A production system theory of serial memory. *Psychological Review, 104*(4), 728–748.

Anellis, I. (2004). The genesis of the truth-table device. *Russell: The Journal of Bertrand Russell Studies, 24*, 55–70.

Anstis, S. M. (1980). The perception of apparent movement. *Philosophical Transactions of the Royal Society of London, 290B*, 153–168.

Arfib, D., Couturier, J. M., & Kessous, L. (2005). Expressiveness and digital musical instrument design. *Journal of New Music Research, 34*(1), 125–136.

Arkin, R. C. (1998). *Behavior-based robotics*. Cambridge, MA: MIT Press.

Ashby, W. R. (1956). *An introduction to cybernetics*. London, UK: Chapman & Hall.

Ashby, W. R. (1960). *Design for a brain* (2nd ed.). New York, NY: John Wiley & Sons.

Asimov, I. (2004). *I, robot* (Bantam hardcover ed.). New York, NY: Bantam Books.

Aspray, W. F. (1982). Pioneer day, NCC 82: History of the stored-program concept. *Annals of the History of Computing, 4*(4), 358–361.

Assayag, G., Feichtinger, H. G., Rodrigues, J.-F., & European Mathematical Society. (2002). *Mathematics and music: a Diderot Mathematical Forum*. Berlin, Germany; New York, NY: Springer.

Aune, B. (1970). *Rationalism, empiricism, and pragmatism: An introduction*. New York, NY: Random House.

Austerlitz, R. (1983). Meaning in music: Is music like language and if so, how? *American Journal of Semiotics, 2*(3), 1–11.

Austrian, G. (1982). *Herman Hollerith: Forgotten giant of information processing*. New York, NY: Columbia University Press.

Ayotte, J., Peretz, I., & Hyde, K. (2002). Congenital amusia: A group study of adults afflicted with a music-specific disorder. *Brain, 125*, 238–251.

Baddeley, A. D. (1986). *Working memory*. Oxford, UK: Oxford University Press.

Baddeley, A. D. (1990). *Human memory: Theory and practice.* Needham Heights, MA: Allyn & Bacon.

Baddeley, A. D. (2003). Working memory: Looking back and looking forward. *Nature Reviews Neuroscience, 4*(10), 829–839.

Baader, A. P., Kazennikov, O., & Wiesendanger, M. (2005). Coordination of bowing and fingering in violin playing. *Cognitive Brain Research, 23*(2–3), 436–443.

Bahrami, B. (2003). Object property encoding and change blindness in multiple object tracking. *Visual Cognition, 10*(8), 949–963.

Bailey, D. (1992). *Improvisation: Its nature and practice in music.* New York, NY: Da Capo Press.

Bain, A. (1855). *The senses and the intellect.* London, UK: John W. Parker & Son.

Baker, C. L. (1979). Syntactic theory and the projection problem. *Linguistic Inquiry, 10*(4), 533–581.

Baker, K. (1982). *Chords and progressions for jazz and popular keyboard.* London, UK; New York, NY: Amsco Publications.

Balch, T., & Parker, L. E. (2002). *Robot teams.* Natick, MA: A. K. Peters.

Ballard, D. H. (1986). Cortical structures and parallel processing: Structure and function. *The Behavioral and Brain Sciences, 9*, 67–120.

Ballard, D. H. (1997). *An introduction to natural computation.* Cambridge, MA: MIT Press.

Bankes, S. C., & Margoliash, D. (1993). Parametric modeling of the temporal dynamics of neuronal responses using connectionist architectures. *Journal of Neurophysiology, 69*(3), 980–991.

Barkow, J. H., Cosmides, L., & Tooby, J. (1992). *The adapted mind: Evolutionary psychology and the generation of culture.* New York, NY: Oxford University Press.

Barlow, H. B. (1972). Single units and sensation: A neuron doctrine for perceptual psychology? *Perception, 1*(371–394).

Barlow, H. B. (1995). The neuron doctrine in perception. In M. S. Gazzaniga (Ed.), *The cognitive neurosciences* (pp. 415–435). Cambridge, MA: MIT Press.

Barlow, H. B., Fitzhugh, R., & Kuffler, S. (1957). Changes in organization of the receptive fields of the cat's retina during dark adaptation. *Journal of Physiology, 137*, 327–337.

Baro, J. A., & Levinson, E. (1988). Apparent motion can be perceived between patterns with dissimilar spatial frequencies. *Vision Research, 28*, 1311–1313.

Barrett, H. C., & Kurzban, R. (2006). Modularity in cognition: Framing the debate. *Psychological Review, 113*(3), 628–647.

Barrow, H. G., & Tenenbaum, J. M. (1975). Representation and use of knowledge in vision. *SIGART, 52*, 2–8.

Bateson, G. (1972). *Steps to an ecology of mind.* New York, NY: Ballantine Books.

Baumgartner, P., & Payr, S. (1995). *Speaking minds: Interviews with twenty eminent cognitive scientists.* Princeton, NJ: Princeton University Press.

Bechtel, W. (1985). Contemporary connectionism: Are the new parallel distributed processing models cognitive or associationist? *Behaviorism, 13*, 53–61.

Bechtel, W. (1994). Natural deduction in connectionist systems. *Synthese, 101*, 433–463.

Bechtel, W., & Abrahamsen, A. (1991). *Connectionism and the mind: Parallel processing, dynamics, and evolution in networks.* Cambridge, MA: Blackwell.

Bechtel, W., & Abrahamsen, A. (2002). *Connectionism and the mind: Parallel processing, dynamics, and evolution in networks* (2nd ed.). Malden, MA: Blackwell Publishers.

Bechtel, W., Graham, G., & Balota, D. A. (1998). *A companion to cognitive science*. Malden, MA: Blackwell Publishers.

Beer, R. D. (2003). The dynamics of active categorical perception in an evolved model agent. *Adaptive Behavior, 11*(4), 209–243.

Beer, R. D. (2010). Fitness space structure of a neuromechanical system. *Adaptive Behavior, 18*(2), 93–115.

Behrend, E. R., & Bitterman, M. E. (1961). Probability-matching in the fish. *American Journal of Psychology, 74*(4), 542–551.

Bellgard, M. I., & Tsang, C. P. (1994). Harmonizing music the Boltzmann way. *Connection Science, 6*, 281–297.

Bellmore, M., & Nemhauser, G. L. (1968). The traveling salesman problem: A survey. *Operations Research, 16*(3), 538–558.

Beni, G. (2005). From swarm intelligence to swarm robotics. *Swarm Robotics, 3342*, 1–9.

Beni, G., & Wang, J. (1991, April 9–11). Theoretical problems for the realization of distributed robotic systems. Paper presented at the IEEE International Conference on Robotics and Automation, Sacramento, CA.

Bennett, L. J. (1990). Modularity of mind revisited. *British Journal for the Philosophy of Science, 41*(3), 429–436.

Benson, B. (2003). *The improvisation of musical dialogue: A phenomenology of music*. Cambridge, UK; New York, NY: Cambridge University Press.

Benson, D. J. (2007). *Music: A mathematical offering*. Cambridge, UK; New York, NY: Cambridge University Press.

Bentin, S., & Golland, Y. (2002). Meaningful processing of meaningless stimuli: The influence of perceptual experience on early visual processing of faces. *Cognition, 86*(1), B1–B14.

Benuskova, L. (1994). Modeling the effect of the missing fundamental with an attractor neural network. *Network: Computation in Neural Systems, 5*(3), 333–349.

Benuskova, L. (1995). Modeling transpositional invariancy of melody recognition with an attractor neural network. *Network: Computation in Neural Systems, 6*(3), 313–331.

Bergmann, M., Moor, J., & Nelson, J. (1990). *The logic book*. New York, NY: McGraw Hill.

Berkeley, G. (1710). *A treatise concerning the principles of human knowledge*. Dublin, Ireland: Printed by A. Rhames for J. Pepyat.

Berkeley, I. S. N., Dawson, M. R. W., Medler, D. A., Schopflocher, D. P., & Hornsby, L. (1995). Density plots of hidden value unit activations reveal interpretable bands. *Connection Science, 7*, 167–186.

Berkeley, I. S. N., & Gunay, C. (2004). Conducting banding analysis with trained networks of sigmoid units. *Connection Science, 16*(2), 119–128.

Berlinski, D. (2000). *The advent of the algorithm*. San Diego, CA: Harcourt, Inc.

Bermúdez, J. L. (2010). *Cognitive science: An introduction to the science of the mind*. Cambridge, UK; New York, NY: Cambridge University Press.

Bernbaum, K., & Chung, C. S. (1981). Müller-Lyer illusion induced by imagination. *Journal of Mental Imagery, 5*, 125–128.

Bernstein, L. (1976). *The unanswered question: Six talks at Harvard*. Cambridge, MA: Harvard University Press.

Best, J. B. (1995). *Cognitive psychology*. St. Paul, MN: West Publishing.

Bever, T. G., Fodor, J. A., & Garrett, M. (1968). A formal limitation of associationism. In T. R. Dixon & D. L. Horton (Eds.), *Verbal behavior and general behavior theory* (pp. 582–585). Englewood Cliffs, NJ: Prentice Hall.

Bharucha, J. J. (1984). Anchoring effects in music: The resolution of dissonance. *Cognitive Psychology, 16*(4), 485–518.

Bharucha, J. J. (1987). Music cognition and perceptual facilitation: A connectionist framework. *Music Perception, 5*(1), 1–30.

Bharucha, J. J. (1991). Pitch, harmony, and neural nets: A psychological perspective. In P. M. Todd & D. G. Loy (Eds.), *Music and connectionism* (pp. 84–99). Cambridge, MA: MIT Press.

Bharucha, J. J. (1999). Neural nets, temporal composites, and tonality. In D. Deutsch (Ed.), *The psychology of music* (2nd ed., pp. 413–440). San Diego, CA: Academic Press.

Bharucha, J. J., & Todd, P. M. (1989). Modeling the perception of tonal structure with neural nets. *Computer Music Journal, 13*(4), 44–53.

Bickle, J. (1996). New wave psychophysical reductionism and the methodological caveats. *Philosophy and Phenomenological Research, LVI*, 57–78.

Biederman, I. (1987). Recognition by components: A theory of human image understanding. *Psychological Review, 94*, 115–147.

Blaauw, G. A., & Brooks, F. P. (1997). *Computer architecture: Concepts and evolution*. Reading, MA: Addison-Wesley.

Blackwell, T. (2003). Swarm music: Improvised music with multiswarms. Paper presented at the AISB Symposium on Artificial Intelligence and Creativity in Arts and Science, Aberystwyth, Wales.

Blackwell, T., & Young, M. (2004). Self-organised music. *Organised Sound, 9*, 123–136.

Bladin, P. F. (2006). W. Grey Walter, pioneer in the electroencephalogram, robotics, cybernetics, artificial intelligence. *Journal of Clinical Neuroscience, 13*(2), 170–177.

Blakemore, S. J., Winston, J., & Frith, U. (2004). Social cognitive neuroscience: Where are we heading? *Trends in Cognitive Sciences, 8*(5), 216–222.

Blaser, E., Pylyshyn, Z. W., & Holcombe, A. O. (2000). Tracking an object through feature space. *Nature, 408*(6809), 196–199.

Block, N. (1981). *Imagery*. Cambridge, MA: MIT Press.

Boden, M. A. (2006). *Mind as machine: A history of cognitive science*. New York, NY: Clarendon Press.

Bonabeau, E., & Meyer, C. (2001). Swarm intelligence: A whole new way to think about business. *Harvard Business Review, 79*(5), 106–114.

Bonabeau, E., Theraulaz, G., Deneubourg, J. L., Franks, N. R., Rafelsberger, O., Joly, J. L., et al. (1998). A model for the emergence of pillars, walls and royal chambers in termite nests. *Philosophical Transactions of the Royal Society of London Series B-Biological Sciences, 353*(1375), 1561–1576.

Bonds, M. E. (2006). *Music as thought: Listening to the symphony in the age of Beethoven*. Princeton, NJ: Princeton University Press.

Boogaarts, M. (2007). *The LEGO Mindstorms NXT idea book: Design, invent, and build*. San Francisco, CA: No Starch Press.

Boole, G. (2003). *The laws of thought.* Amherst, NY: Prometheus Books. (Original work published 1854)

Borges, J. L. (1962). *Labyrinths: Selected stories and other writings* (1st ed.). New York, NY: New Directions.

Boring, E. G. (1950). *A history of experimental psychology.* New York, NY: Appleton-Century-Crofts.

Botez, M. I. (1975). Two visual systems in clinical neurology: Readaptive role of the primitive system in visual agnosis patients. *European Neurology, 13,* 101–122.

Botros, A., Smith, J., & Wolfe, J. (2006). The virtual flute: An advanced fingering guide generated via machine intelligence. *Journal of New Music Research, 35*(3), 183–196.

Bower, G. H. (1993). The fragmentation of psychology. *American Psychologist, 48*(8), 905–907.

Bowers, J. S. (2009). On the biological plausibility of grandmother cells: Implications for neural network theories in psychology and neuroscience. *Psychological Review, 116*(1), 220–251.

Bown, O., Eldridge, A., & McCormack, J. (2009). Understanding interaction in contemporary digital music: From instruments to behavioural objects. *Organised Sound, 14*(2), 188–196.

Braitenberg, V. (1984). *Vehicles: Explorations in synthetic psychology.* Cambridge, MA: MIT Press.

Brandt, S. A., & Stark, L. W. (1997). Spontaneous eye movements during visual imagery reflect the content of the visual scene. *Journal of Cognitive Neuroscience, 9*(1), 27–38.

Braun, H. (1991). On solving traveling salesman problems by genetic algorithms. *Lecture Notes in Computer Science, 496,* 129–133.

Breazeal, C. (2002). *Designing sociable robots.* Cambridge, MA: MIT Press.

Breazeal, C. (2003). Toward sociable robots. *Robotics and Autonomous Systems, 42*(3–4), 167–175.

Breazeal, C. (2004). Social interactions in HRI: The robot view. *IEEE Transactions on Systems Man and Cybernetics Part C-Applications and Reviews, 34*(2), 181–186.

Breazeal, C., Gray, J., & Berlin, M. (2009). An embodied cognition approach to mindreading skills for socially intelligent robots. *International Journal of Robotics Research, 28*(5), 656–680.

Bregman, A. S. (1990). *Auditory scene analysis: The perceptual organization of sound.* Cambridge, MA: MIT Press.

Brentano, F. C. (1995). *Psychology from an empirical standpoint* (D. B. Terrell, A. C. Rancurello, & L.L. McAlister, Trans.). London, UK; New York, NY: Routledge. (Original work published 1874)

Broadbent, D. (1985). A question of levels: Comment on McClelland and Rumelhart. *Journal of Experimental Psychology: General, 114,* 189–192.

Brooks, F. P. (1962). Architectural philosophy. In W. Buchholz (Ed.), *Planning a computer system: Project stretch* (pp. 5–16). New York, NY: McGraw-Hill.

Brooks, R. A. (1989). A robot that walks; Emergent behaviours from a carefully evolved network. *Neural Computation, 1,* 253–262.

Brooks, R. A. (1991). Intelligence without representation. *Artificial Intelligence, 47,* 139–159.

Brooks, R. A. (1999). *Cambrian intelligence: The early history of the new AI.* Cambridge, MA: MIT Press.

Brooks, R. A. (2002). *Flesh and machines: How robots will change us.* New York, NY: Pantheon Books.

Brooks, R. A., Breazeal, C., Marjanovic, M., Scassellati, S., & Williamson, M. (1999). The Cog project: Building a humanoid robot. In C. Nehaniv (Ed.), *Computation for metaphors, analogy, and agents* (pp. 52–87). Berlin, Germany: Springer-Verlag.

Brooks, R. A., & Flynn, A. M. (1989). Fast, cheap and out of control: A robot invasion of the solar system. *Journal of The British Interplanetary Society, 42*, 478–485.

Brown, A. A., Spetch, M. L., & Hurd, P. L. (2007). Growing in circles: Rearing environment alters spatial navigation in fish. *Psychological Science, 18*, 569–573.

Brown, T. H. (1990). Hebbian synapses: Biophysical mechanisms and algorithms. *Annual Review of Neuroscience, 13*, 475–511.

Bruner, J. S. (1957). On perceptual readiness. *Psychological Review, 64*, 123–152.

Bruner, J. S. (1973). *Beyond the information given.* New York, NY: W.W. Norton & Company.

Bruner, J. S. (1990). *Acts of meaning.* Cambridge, MA: Harvard University Press.

Bruner, J. S. (1992). Another look at New Look 1. *American Psychologist, 47*(6), 780–783.

Bruner, J. S., Postman, L., & Rodrigues, J. (1951). Expectation and the perception of color. *American Journal of Psychology, 64*(2), 216–227.

Buccino, G., Binkofski, F., Fink, G. R., Fadiga, L., Fogassi, L., Gallese, V., et al. (2001). Action observation activates premotor and parietal areas in a somatotopic manner: An fMRI study. *European Journal of Neuroscience, 13*(2), 400–404.

Buccino, G., Vogt, S., Ritzl, A., Fink, G. R., Zilles, K., Freund, H. J., et al. (2004). Neural circuits underlying imitation learning of hand actions: An event-related fMRI study. *Neuron, 42*(2), 323–334.

Buck, G. H., & Hunka, S. M. (1999). W. Stanley Jevons, Allan Marquand, and the origins of digital computing. *IEEE Annals of the History of Computing, 21*(4), 21–27.

Bugatti, A., Flammini, A., & Migliorati, P. (2002). Audio classification in speech and music: A comparison between a statistical and a neural approach. *Eurasip Journal on Applied Signal Processing, 2002*(4), 372–378.

Burgess, N., Recce, M., & O'Keefe, J. (1995). Spatial models of the hippocampus. In M. A. Arbib (Ed.), *The handbook of brain theory and neural networks* (pp. 468–472). Cambridge, MA: MIT Press.

Burks, A. R., & Burks, A. W. (1988). *The first electronic computer: The Atanasoff story.* Ann Arbor, MI: University of Michigan Press.

Burks, A. W. (1975). Logic, biology and automata: Some historical reflections. *International Journal of Man-Machine Studies, 7*(3), 297–312.

Burks, A. W. (2002). The invention of the universal electronic computer: How the electronic computer revolution began. *Future Generation Computer Systems, 18*(7), 871–892.

Burks, A. W., Goldstine, H. H., & Von Neumann, J. (1989). Preliminary discussion of the logical design of an electronic computing instrument. In Z. Pylyshyn & L. Bannon (Eds.), *Perspectives on the computer revolution* (pp. 39–48). Norwood, NJ: Ablex. (Original work published 1946)

Burnod, Y. (1990). *An adaptive neural network: The cerebral cortex.* Englewood Cliffs, NJ: Prentice Hall.

Burt, P., & Sperling, G. (1981). Time, distance and feature trade-offs in visual apparent motion. *Psychological Review, 88*, 137–151.

Bush, V. (1931). The differential analyzer: A new machine for solving differential equations. *Journal of the Franklin Institute, 212*, 447–488.

Buss, A. R. (1978). The structure of psychological revolutions. *Journal of the History of the Behavioral Sciences, 14*(1), 57–64.

Byrne, David. (1980). *Seen and not seen*. On *Remain in light*. [CD]. New York City: Sire Records.

Cabeza, R., & Kingstone, A. (2006). *Handbook of functional neuroimaging of cognition* (2nd ed.). Cambridge, MA: MIT Press.

Cabeza, R., & Nyberg, L. (2000). Imaging cognition II: An empirical review of 275 PET and fMRI studies. *Journal of Cognitive Neuroscience, 12*(1), 1–47.

Cage, J. (1961). *Silence: Lectures and writings* (1st ed.). Middletown, CN: Wesleyan University Press.

Calderbank, R., & Sloane, N. J. A. (2001). Obituary: Claude Shannon (1916–2001). *Nature, 410*(6830), 768.

Calvin, W. H., & Ojemann, G. A. (1994). *Conversations with Neil's brain*. Reading, MA: Addison-Wesley.

Calvo, P., & Gomila, A. (2008). Directions for an embodied cognitive science: Toward an integrated approach. In P. Calvo & A. Gomila (Eds.), *Handbook of cognitive science: An embodied approach* (pp. 1–25). Oxford, UK: Elsevier.

Caramazza, A. (1986). On drawing inferences about the structure of normal cognitive systems from the analysis of patterns of impaired performance: The case for single-patient studies. *Brain and Cognition, 5*, 41–66.

Carpenter, G. A. (1989). Neural network models for pattern recognition and associative memory. *Neural Networks, 2*, 243–257.

Carpenter, G. A., & Grossberg, S. (1992). *Neural networks for vision and image processing*. Cambridge, MA: MIT Press.

Carruthers, P. (2006). *The architecture of the mind: Massive modularity and the flexibility of thought*. Oxford, UK: Oxford University Press.

Casey, M. A., Veltkamp, R., Goto, M., Leman, M., Rhodes, C., & Slaney, M. (2008). Content-based music information retrieval: Current directions and future challenges. *Proceedings of the IEEE, 96*(4), 668–696.

Caudill, M. (1992). *In our own image : Building an artificial person*. New York, NY: Oxford University Press.

Caudill, M., & Butler, B. (1992a). *Understanding neural networks* (Vol. 1). Cambridge, MA; MIT Press.

Caudill, M., & Butler, B. (1992b). *Understanding neural networks* (Vol. 2). Cambridge, MA; MIT Press.

Cavanagh, P., Arguin, M., & von Grunau, M. (1989). Interattribute apparent motion. *Vision Research, 29*, 1197–1204.

Cedolin, L., & Delgutte, B. (2010). Spatiotemporal representation of the pitch of harmonic complex tones in the auditory nerve. *Journal of Neuroscience, 30*(38), 12712–12724.

Ceruzzi, P. (1997). Crossing the divide: Architectural issues and the emergence of the stored program computer, 1935–1955. *IEEE Annals of the History of Computing, 19*(1), 5–12.

Chalmers, D. (2006). Perception and the fall from Eden. In T. Gendler & J. Hawthorne (Eds.), *Perceptual experience* (pp. 49–125). Oxford, UK: Oxford University Press.

Chapman, C. S., & Goodale, M. A. (2010). Seeing all the obstacles in your way: The effect of visual feedback and visual feedback schedule on obstacle avoidance while reaching. *Experimental Brain Research, 202*(2), 363–375.

Chapman, G. B., & Robbins, S. J. (1990). Cue interaction in human contingency judgment. *Memory & Cognition, 1 i8*(5), 537–545.

Chase, V. M., Hertwig, R., & Gigerenzer, G. (1998). Visions of rationality. *Trends in Cognitive Sciences*, *2*(6), 206–214.

Chater, N., & Oaksford, M. (1999). Ten years of the rational analysis of cognition. *Trends in Cognitive Sciences*, *3*(2), 57–65.

Chauvin, Y., & Rumelhart, D. E. (1995). *Backpropagation: Theory, architectures, and applications*. Hillsdale, NJ: Lawrence Erlbaum Associates.

Chemero, A. (2009). *Radical embodied cognitive science*. Cambridge, MA: MIT Press.

Chemero, A., & Turvey, M. T. (2007). Gibsonian affordances for roboticists. *Adaptive Behavior*, *15*(4), 473–480.

Cheng, K. (1986). A purely geometric module in the rat's spatial representation. *Cognition*, *23*, 149–178.

Cheng, K. (2005). Reflections on geometry and navigation. *Connection Science*, *17*(1–2), 5–21.

Cheng, K. (2008). Whither geometry? Troubles of the geometric module. *Trends in Cognitive Sciences*, *12*(9), 355–361.

Cheng, K., & Newcombe, N. S. (2005). Is there a geometric module for spatial orientation? Squaring theory and evidence. *Psychonomic Bulletin & Review*, *12*(1), 1–23.

Cheng, P. W. (1997). From covariation to causation: A causal power theory. *Psychological Review*, *104*(2), 367–405.

Cheng, P. W., & Holyoak, K. J. (1995). Complex adaptive systems as intuitive statisticians: Causality, contingency, and prediction. In H. L. Roitblat & J.-A. Meyer (Eds.), *Comparative approaches to cognitive science* (pp. 271–302). Cambridge, MA: MIT Press.

Cheng, P. W., & Novick, L. R. (1990). A probabilistic contrast model of causal induction. *Journal of Personality and Social Psychology*, *58*(4), 545–567.

Cheng, P. W., & Novick, L. R. (1992). Covariation in natural causal induction. *Psychological Review*, *99*(2), 365–382.

Chomsky, N. (1957). *Syntactic structures* (2nd ed.). Berlin, Germany; New York, NY: Mouton de Gruyter.

Chomsky, N. (1959a). On certain formal properties of grammars. *Information and Control*, *2*, 137–167.

Chomsky, N. (1959b). A review of B. F. Skinner's *Verbal Behavior*. *Language*, *35*, 26–58.

Chomsky, N. (1965). *Aspects of the theory of syntax*. Cambridge, MA: MIT Press.

Chomsky, N. (1966). *Cartesian linguistics: A chapter in the history of rationalist thought* (1st ed.). New York, NY: Harper & Row.

Chomsky, N. (1980). *Rules and representations*. New York, NY: Columbia University Press.

Christensen, S. M., & Turner, D. R. (1993). *Folk psychology and the philosophy of mind*. Hillsdale, NJ: Lawrence Erlbaum Associates.

Churchland, P. M. (1985). Reduction, qualia, and the direct introspection of brain states. *The Journal of Philosophy*, *LXXXII*, 8–28.

Churchland, P. M. (1988). *Matter and consciousness* (Rev. ed.). Cambridge, MA: MIT Press.

Churchland, P. M., & Churchland, P. S. (1990). Could a machine think? *Scientific American*, *262*, 32–37.

Churchland, P. S. (1986). *Neurophilosophy*. Cambridge, MA: MIT Press.

Churchland, P. S., Koch, C., & Sejnowski, T. J. (1990). What is computational neuroscience? In E. L. Schwartz (Ed.), *Computational neuroscience* (pp. 46–55). Cambridge, MA: MIT Press.

Churchland, P. S., & Sejnowski, T. J. (1989). Neural representation and neural computation. In L. Nadel, L. A. Cooper, P. Culicover & R. M. Harnish (Eds.), *Neural connections, mental computation* (pp. 15-48). Cambridge, MA: MIT Press.

Churchland, P. S., & Sejnowski, T. J. (1992). *The computational brain*. Cambridge, MA: MIT Press.

Clancey, W. J. (1987). T. Winograd, F. Flores, Understanding computers and cognition: A new foundation for design. *Artificial Intelligence, 31*(2), 232-250.

Clancey, W. J. (1993). Situated action: A neuropsychological interpretation response. *Cognitive Science, 17*(1), 87-116.

Clancey, W. J. (1997). *Situated cognition*. Cambridge, UK: Cambridge University Press.

Clark, A. (1989). *Microcognition*. Cambridge, MA: MIT Press.

Clark, A. (1993). *Associative engines*. Cambridge, MA: MIT Press.

Clark, A. (1997). *Being there: Putting brain, body, and world together again*. Cambridge, MA: MIT Press.

Clark, A. (1999). An embodied cognitive science? *Trends in Cognitive Sciences, 3*(9), 345-351.

Clark, A. (2003). *Natural-born cyborgs*. Oxford, UK; New York, NY: Oxford University Press.

Clark, A. (2008). *Supersizing the mind: Embodiment, action, and cognitive extension*. Oxford, UK; New York, NY: Oxford University Press.

Clark, A., & Chalmers, D. (1998). The extended mind (Active externalism). *Analysis, 58*(1), 7-19.

Clark, H. H. (1996). *Using language*. Cambridge, UK; New York, NY: Cambridge University Press.

Clarke, E. F. (2005). *Ways of listening: An ecological approach to the perception of musical meaning*. Oxford, UK; New York, NY: Oxford University Press.

Claudon, F. (1980). *The concise encyclopedia of romanticism*. Secaucus, NJ: Chartwell Books.

Cognitive Science Society. (2013) *Welcome to the Cognitive Science Society website*. Retrieved from http://www. http://cognitivesciencesociety.org/index.html

Cohen, I. B. (1999). *Howard Aiken: Portrait of a computer pioneer*. Cambridge, MA: MIT Press.

Cohen, P. R., Greenberg, M. L., Hart, D. M., & Howe, A. E. (1989). Trial by fire: Understanding the design requirements for agents in complex environments. *AI Magazine, 10*(3), 32-48.

Colby, C. L., & Goldberg, M. E. (1999). Space and attention in parietal cortex. *Annual Review of Neuroscience, 22*, 319-349.

Colby, K. M., Hilf, F. D., Weber, S., & Kraemer, H. C. (1972). Turing-like indistinguishability tests for validation of a computer simulation of paranoid processes. *Artificial Intelligence, 3*(2), 199-221.

Cole, J. (1998). *About face*. Cambridge, MA: MIT Press.

Collier, C. P., Wong, E. W., Belohradsky, M., Raymo, F. M., Stoddart, J. F., Kuekes, P. J., et al. (1999). Electronically configurable molecular-based logic gates. *Science, 285*(5426), 391-394.

Collins, A. M., & Quillian, M. R. (1969). Retrieval time from semantic memory. *Journal of Verbal Learning and Verbal Behavior, 8*, 240-247.

Collins, A. M., & Quillian, M. R. (1970a). Does category size affect categorization time? *Journal of Verbal Learning and Verbal Behavior, 9*(4), 432-438.

Collins, A. M., & Quillian, M. R. (1970b). Facilitating retrieval from semantic memory: Effect of repeating part of an inference. *Acta Psychologica, 33*, 304-314.

Collins, S. H., Ruina, A., Tedrake, R., & Wisse, M. (2005). Efficient bipedal robots based on passive-dynamic walkers. *Science, 307*(5712), 1082-1085.

Comrie, L. J. (1933). *The Hollerith and Powers tabulating machines*. London, UK: Printed for private circulation.

Conrad, R. (1964a). Acoustic confusions in immediate memory. *British Journal of Psychology, 55*(1), 75–84.

Conrad, R. (1964b). Information, acoustic confusion, and memory span. *British Journal of Psychology, 55*, 429–432.

Conway, F., & Siegelman, J. (2005). *Dark hero of the information age: In search of Norbert Wiener, the father of cybernetics*. New York, NY: Basic Books.

Cook, V. J., & Newson, M. (1996). *Chomsky's universal grammar: An introduction* (2nd ed.). Oxford, UK: Wiley-Blackwell.

Cooper, L. A., & Shepard, R. N. (1973a). Chronometric studies of the rotation of mental images. In W. G. Chase (Ed.), *Visual information processing* (pp. 75–176). New York, NY: Academic Press.

Cooper, L. A., & Shepard, R. N. (1973b). The time required to prepare for a rotated stimulus. *Memory & Cognition, 1*(3), 246–250.

Cooper, M., Foote, J., Pampalk, E., & Tzanetakis, G. (2006). Visualization in audio-based music information retrieval. *Computer Music Journal, 30*(2), 42–62.

Cooper, R. (1977). Obituary, W. Grey Walter. *Nature, 268*, 383–384.

Copeland, B. J. (2011). The Manchester computer: A revised history, Part 1: The memory. *IEEE Annals of the History of Computing, 33*(1), 4–21.

Copland, A. (1939). *What to listen for in music*. New York, NY; London, UK: Whittlesey House, McGraw-Hill Book Company.

Copland, A. (1952). *Music and imagination*. Cambridge, MA: Harvard University Press.

Cotter, N. E. (1990). The Stone-Weierstrass theorem and its application to neural networks. *IEEE transactions on Neural Networks, 1*, 290–295.

Cottrell, G. W. (1989). The connectionist air guitar: A dream come true. *Connection Science, 1*, 413.

Coutinho, E., & Cangelosi, A. (2009). The use of spatio-temporal connectionist models in psychological studies of musical emotions. *Music Perception, 27*(1), 1–15.

Couzin, I. D., Krause, J., Franks, N. R., & Levin, S. A. (2005). Effective leadership and decision-making in animal groups on the move. *Nature, 433*(7025), 513–516.

Craik, K. J. M. (1943). *The nature of explanation*. Cambridge, UK: Cambridge University Press.

Crick, F., & Asanuma, C. (1986). Certain aspects of the anatomy and physiology of the cerebral cortex. In J. McClelland & D. E. Rumelhart (Eds.), *Parallel distributed processing* (Vol. 2, pp. 333–371). Cambridge, MA: MIT Press.

Cummins, R. (1975). Functional analysis. *Journal of Philosophy, 72*, 741–760.

Cummins, R. (1983). *The nature of psychological explanation*. Cambridge, MA: MIT Press.

Cummins, R. (1989). *Meaning and mental representation*. Cambridge, MA: MIT Press.

Cybenko, G. (1989). Approximation by superpositions of a sigmoidal function. *Mathematics of Control, Signals, and Systems, 2*, 303–314.

Danks, D. (2003). Equilibria of the Rescorla-Wagner model. *Journal of Mathematical Psychology, 47*(2), 109–121.

Dasgupta, S. (1989). *Computer architecture: A modern synthesis*. New York, NY: Wiley.

D'Ausilio, A. (2009). Mirror-like mechanisms and music. *The Scientific World Journal, 9*, 1415–1422.

Davies, M., & Stone, T. (1995a). *Folk psychology: The theory of mind debate*. Oxford, UK: Wiley-Blackwell.

Davies, M., & Stone, T. (1995b). *Mental simulation: Evaluations and applications*. Oxford, UK; Cambridge, MA: Wiley-Blackwell.

Davis, P. J., & Hersh, R. (1981). *The mathematical experience*. Boston: Birkhäuser.

Davison, M., & McCarthy, D. (1988). *The matching law: A research review*. Hillsdale, NJ: Lawrence Erlbaum Associates.

Dawson, M. R. W. (1987). Moving contexts do affect the perceived direction of apparent motion in motion competition displays. *Vision Research, 27*, 799–809.

Dawson, M. R. W. (1989). Apparent motion and element connectedness. *Spatial Vision, 4*, 241–251.

Dawson, M. R. W. (1991). The how and why of what went where in apparent motion: Modeling solutions to the motion correspondence process. *Psychological Review, 98*, 569–603.

Dawson, M. R. W. (1998). *Understanding cognitive science*. Oxford, UK: Wiley-Blackwell.

Dawson, M. R. W. (2004). *Minds and machines: Connectionism and psychological modeling*. Malden, MA: Blackwell Publishers.

Dawson, M. R. W. (2005). *Connectionism: A hands-on approach* (1st ed.). Oxford, UK ; Malden, MA: Blackwell Publishers.

Dawson, M. R. W. (2008). Connectionism and classical conditioning. *Comparative Cognition and Behavior Reviews, 3*(Monograph), 1–115.

Dawson, M. R. W. (2009). Computation, cognition—and connectionism. In D. Dedrick & L. Trick (Eds.), *Cognition, computation, and Pylyshyn* (pp. 175–199). Cambridge, MA: MIT Press.

Dawson, M. R. W., Berkeley, I. S. N., Medler, D. A., & Schopflocher, D. P. (1994). Density plots of hidden value unit activations reveal interpretable bands and microbands. In B. MacDonald, R. Holte, and C. Ling (Eds.), *Proceedings of the Machine Learning Workshop at AI/GI/VI 1994* (pp.iii-1–iii-9). Calgary, AB: University of Calgary Press.

Dawson, M. R. W., & Boechler, P. M. (2007). Representing an intrinsically nonmetric space of compass directions in an artificial neural network. *International Journal of Cognitive Informatics and Natural Intelligence, 1*, 53–65.

Dawson, M. R. W., Boechler, P. M., & Orsten, J. (2005). An artificial neural network that uses coarse allocentric coding of direction to represent distances between locations in a metric space. *Spatial Cognition and Computation, 5*, 29–67.

Dawson, M. R. W., Boechler, P. M., & Valsangkar-Smyth, M. (2000). Representing space in a PDP network: Coarse allocentric coding can mediate metric and nonmetric spatial judgements. *Spatial Cognition and Computation, 2*, 181–218.

Dawson, M. R. W., & Di Lollo, V. (1990). Effects of adapting luminance and stimulus contrast on the temporal and spatial limits of short-range motion. *Vision Research, 30*, 415–429.

Dawson, M. R. W., Dupuis, B., Spetch, M. L., & Kelly, D. M. (2009). Simple artificial networks that match probability and exploit and explore when confronting a multiarmed bandit. *IEEE Transactions on Neural Networks, 20*(8), 1368–1371.

Dawson, M. R. W., Dupuis, B., & Wilson, M. (2010). *From bricks to brains: The embodied cognitive science of LEGO robots*. Edmonton, AB: Athabasca University Press.

Dawson, M. R. W., Kelly, D. M., Spetch, M. L., & Dupuis, B. (2010). Using perceptrons to explore the reorientation task. *Cognition, 114*(2), 207–226.

Dawson, M. R. W., Kremer, S., & Gannon, T. (1994). Identifying the trigger features for hidden units in a PDP model of the early visual pathway. In R. Elio (Ed.), *Tenth Canadian conference on artificial intelligence* (pp. 115–119). San Francisco, CA: Morgan Kaufmann.

Dawson, M. R. W., Medler, D. A., & Berkeley, I. S. N. (1997). PDP networks can provide models that are not mere implementations of classical theories. *Philosophical Psychology, 10*, 25–40.

Dawson, M. R. W., Medler, D. A., McCaughan, D. B., Willson, L., & Carbonaro, M. (2000). Using extra output learning to insert a symbolic theory into a connectionist network. *Minds and Machines: Journal for Artificial Intelligence, Philosophy and Cognitive Science, 10*, 171–201.

Dawson, M. R. W., Nevin-Meadows, N., & Wright, R. D. (1994). Polarity matching in the Ternus configuration. *Vision Research, 34*, 3347–3359.

Dawson, M. R. W., & Piercey, C. D. (2001). On the subsymbolic nature of a PDP architecture that uses a nonmonotonic activation function. *Minds and Machines: Journal for Artificial Intelligence, Philosophy and Cognitive Science, 11*, 197–218.

Dawson, M. R. W., & Pylyshyn, Z. W. (1988). Natural constraints on apparent motion. In Z. W. Pylyshyn (Ed.), *Computational processes in human vision: An interdisciplinary perspective* (pp. 99–120). Norwood, NJ: Ablex.

Dawson, M. R. W., & Schopflocher, D. P. (1992a). Autonomous processing in PDP networks. *Philosophical Psychology, 5*, 199–219.

Dawson, M. R. W., & Schopflocher, D. P. (1992b). Modifying the generalized delta rule to train networks of nonmonotonic processors for pattern classification. *Connection Science, 4*, 19–31.

Dawson, M. R. W., & Shamanski, K. S. (1994). Connectionism, confusion and cognitive science. *Journal of Intelligent Systems, 4*, 215–262.

Dawson, M. R. W., & Spetch, M. L. (2005). Traditional perceptrons do not produce the overexpectation effect. *Neural Information Processing: Letters and Reviews, 7*(1), 11–17.

Dawson, M. R. W., & Thibodeau, M. H. (1998). The effect of adapting luminance on the latency of visual search. *Acta Psychologica, 99*, 115–139.

Dawson, M. R. W., & Wright, R. D. (1989). The consistency of element transformations affects the visibility but not the direction of illusory motion. *Spatial Vision, 4*, 17–29.

Dawson, M. R. W., & Wright, R. D. (1994). Simultaneity in the Ternus configuration: Psychophysical data and a computer model. *Vision Research, 34*, 397–407.

Debus, A. G. (1978). *Man and nature in the Renaissance.* Cambridge, UK; New York, NY: Cambridge University Press.

Dedekind, R. (1901). *Essays on the theory of numbers I: Continuity and irrational numbers: II. The nature and meaning of numbers* (pp. 8–115). Chicago, IL: Open Court. (Original work published 1888)

Dedrick, D., & Trick, L. (2009). *Computation, cognition, and Pylyshyn.* Cambridge, MA: MIT Press.

Delahaye, J. P. (2006). The science behind Sudoku. *Scientific American, 294*(6), 80–87.

de Latil, P. (1956). *Thinking by machine: A study of cybernetics.* London, UK: Sidgwick and Jackson.

De Mulder, T., Martens, J. P., Pauws, S., Vignoli, F., Lesaffre, M., Leman, M., et al. (2006). Factors affecting music retrieval in query-by-melody. *IEEE Transactions on Multimedia, 8*(4), 728–739.

Dennett, D. C. (1978). *Brainstorms*. Cambridge, MA: MIT Press.

Dennett, D. C. (1987). *The intentional stance*. Cambridge, MA: MIT Press.

Dennett, D. C. (1991). *Consciousness explained*. Boston, MA: Little, Brown.

Dennett, D. C. (1998). Cognitive science as reverse engineering: Several meanings of "top–down" and "bottom-up." In D. Dennett (Ed.), *Brainchildren: Essays on designing minds* (pp. 249–260). Cambridge, MA: MIT Press.

Dennett, D. C. (2005). *Sweet dreams: Philosophical obstacles to a science of consciousness*. Cambridge, MA: MIT Press.

Derrington, A. M., & Lennie, P. (1982). The influence of temporal frequency and adaptation level on receptive field organization of retinal ganglion cells in the cat. *Journal of Physiology, 333*, 343–366.

Desain, P., & Honing, H. (1989). The quantization of musical time: A connectionist approach. *Computer Music Journal, 13*(3), 56–66.

Descartes, R. (1960). *Discourse on method and meditations*. (L.J. LaFleur, Trans.). Indianapolis, IN: Bobbs-Merrill. (Original work published 1637)

Descartes, R. (1996). *Meditations on first philosophy* (Rev. ed.). (J. Cottingham, Trans.). New York, NY: Cambridge University Press. (Original work published 1641)

Descartes, R. (2006). *A discourse on the method of correctly conducting one's reason and seeking truth in the sciences*. (I. Maclean, Trans.). Oxford, UK; New York, NY: Oxford University Press. (Original work published 1637)

Deutsch, D. (1999). *The psychology of music* (2nd ed.). San Diego, CA: Academic Press.

de Villiers, P. (1977). Choice in concurrent schedules and a quantitative formulation of the law of effect. In W. K. Honig & J. E. R. Staddon (Eds.), *Handbook of operant behavior* (pp. 233–287). Englewood Cliffs, NJ: Prentice Hall.

de Villiers, P., & Herrnstein, R. J. (1976). Toward a law of response strength. *Psychological Bulletin, 83*(6), 1131–1153.

Devlin, K. (1996). Good-bye Descartes? *Mathematics Magazine, 69*, 344–349.

Dewdney, C. (1998). *Last flesh: Life in the transhuman era* (1st ed.). Toronto, ON: HarperCollins.

Dewey, J. (1929). *Experience and nature* (2nd ed.). Chicago, IL: Open Court Publishing Company.

De Wilde, P. (1997). Neural network models, (2nd ed.). London, UK: Springer.

Dick, P. K. (1968). *Do androids dream of electric sheep?* (1st ed.). Garden City, NY: Doubleday.

Di Pellegrino, G., Fadiga, L., Fogassi, L., Gallese, V., & Rizzolatti, G. (1992). Understanding motor events: A neurophysiological study. *Experimental Brain Research, 91*(1), 176–180.

Dorigo, M., & Gambardella, L. M. (1997). Ant colonies for the travelling salesman problem. *Biosystems, 43*(2), 73–81.

Douglas, R. J., & Martin, K. A. C. (1991). Opening the grey box. *Trends in Neuroscience, 14*, 286–293.

Dourish, P. (2001). *Where the action is: The foundations of embodied interaction*. Cambridge, MA: MIT Press.

Downing, H. A., & Jeanne, R. L. (1986). Intraspecific and interspecific variation in nest architecture in the paper wasp *Polistes* (Hymenoptera, Vespidae). *Insectes Sociaux, 33*(4), 422–443.

Downing, H. A., & Jeanne, R. L. (1988). Nest construction by the paper wasp, *Polistes*: A test of stigmergy theory. *Animal Behaviour, 36*, 1729–1739.

Dreyfus, H. L. (1972). *What computers can't do: A critique of artificial reason* (1st ed.). New York, NY: Harper & Row.

Dreyfus, H. L. (1992). *What computers still can't do.* Cambridge, MA: MIT Press.

Dreyfus, H. L., & Dreyfus, S. E. (1988). Making a mind versus modeling the brain: Artificial intelligence back at the branchpoint. In S. Graubard (Ed.), *The artificial intelligence debate.* Cambridge, MA: MIT Press.

Driver-Linn, E. (2003). Where is psychology going? Structural fault lines, revealed by psychologists' use of Kuhn. *American Psychologist, 58*(4), 269–278.

Drob, S. L. (2003). Fragmentation in contemporary psychology: A dialectical solution. *Journal of Humanistic Psychology, 43*(4), 102–123.

Dubinko, M., Kumar, R., Magnani, J., Novak, J., Raghavan, P., & Tomkins, A. (2007). Visualizing tags over time. *Acm Transactions on the Web, 1*(2).

Duch, W., & Jankowski, N. (1999). Survey of neural transfer functions. *Neural Computing Surveys, 2*, 163–212.

Dutton, J. M., & Starbuck, W. H. (1971). *Computer simulation of human behavior.* New York, NY: John Wiley & Sons.

Dyer, J. R. G., Ioannou, C. C., Morrell, L. J., Croft, D. P., Couzin, I. D., Waters, D. A., et al. (2008). Consensus decision making in human crowds. *Animal Behaviour, 75*, 461–470.

Dyer, J. R. G., Johansson, A., Helbing, D., Couzin, I. D., & Krause, J. (2009). Leadership, consensus decision making and collective behaviour in humans. *Philosophical Transactions of the Royal Society B-Biological Sciences, 364*(1518), 781–789.

Eckert, W. J. (1940). *Punched card methods in scientific computation.* New York, NY: The Thomas J. Watson Astronomical Computing Bureau, Columbia University.

Edsinger-Gonzales, A., & Weber, J. (2004, November). Domo: A force sensing humanoid robot for manipulation research. Paper presented at the 4th IEEE/RAS International Conference on Humanoid Robots, Santa Monica, CA.

Eich, J. M. (1982). A composite holographic associative recall model. *Psychological Review, 89*, 627–661.

Einstein, A. (1947). *Music in the Romantic Era.* New York, NY: W. W. Norton & Company.

Ellis, H. D., & Florence, M. (1990). Bodamer (1947) paper on prosopagnosia. *Cognitive Neuropsychology, 7*(2), 81–105.

Elman, J. L., Bates, E. A., Johnson, M. H., Karmiloff-Smith, A., Parisi, D., & Plunkett, K. (1996). *Rethinking innateness.* Cambridge, MA: MIT Press.

Endicott, R. P. (1998). Collapse of the new wave. *The Journal of Philosophy, XCV*, 53–72.

Enns, J. T. (2004). *The thinking eye, the seeing brain: Explorations in visual cognition* (1st ed.). New York, NY: W.W. Norton.

Eno, Brian. (1996). *Evolving metaphors, in my opinion, is what artists do.* Retrieved from http://www.inmotionmagazine.com/eno1.html

Enquist, M., & Ghirlanda, S. (2005). *Neural networks and animal behavior.* Princeton, NJ: Princeton University Press.

Ericsson, K. A., & Simon, H. A. (1984). *Protocol analysis: Verbal reports as data*. Cambridge, MA: MIT Press.

Essinger, J. (2004). *Jacquard's web: How a hand loom led to the birth of the Information Age*. Oxford, UK; New York, NY: Oxford University Press.

Estes, W. K. (1975). Some targets for mathematical psychology. *Journal of Mathematical Psychology, 12*, 263–282.

Estes, W. K., & Straughan, J. H. (1954). Analysis of a verbal conditioning situation in terms of statistical learning theory. *Journal of Experimental Psychology, 47*(4), 225–234.

Etcoff, N. L., & Magee, J. J. (1992). Categorical perception of facial expressions. *Cognition, 44*(3), 227–240.

Evans, H. E. (1966). Behavior patterns of solitary wasps. *Annual Review of Entomology, 11*, 123–154.

Evans, H. E., & West-Eberhard, M. J. (1970). *The wasps*. Ann Arbor, MI: University of Michigan Press.

Evans, J. S. T. (2003). In two minds: Dual-process accounts of reasoning. *Trends in Cognitive Sciences, 7*(10), 454–459.

Everitt, B. (1980). *Cluster analysis*. New York, NY: Halsted.

Ewald, W. B. (1996). *From Kant to Hilbert: A source book on the foundations of mathematics*. Oxford, UK: Oxford University Press.

Farah, M. J. (1994). Neuropsychological evidence with an interactive brain: A critique of the "locality" assumption. *Behavioral and Brain Sciences, 17*, 43–104.

Faria, J. J., Dyer, J. R. G., Tosh, C. R., & Krause, J. (2010). Leadership and social information use in human crowds. *Animal Behaviour, 79*(4), 895–901.

Fechner, G. T. (1966). *Elements of psychophysics*. (H. E. Adler, Trans.). New York, NY: Holt. (Original work published 1860)

Feigenbaum, E. A., & Feldman, J. (1995). *Computers and thought*. Cambridge, MA: MIT Press.

Feigenbaum, E. A., & McCorduck, P. (1983). *The fifth generation*. Reading, MA: Addison-Wesley.

Feldman, J. A., & Ballard, D. H. (1982). Connectionist models and their properties. *Cognitive Science, 6*, 205–254.

Ferguson, K. (2008). *The music of Pythagoras* (1st U.S. ed.). New York, NY: Walker.

Ferrari, P. F., Gallese, V., Rizzolatti, G., & Fogassi, L. (2003). Mirror neurons responding to the observation of ingestive and communicative mouth actions in the monkey ventral premotor cortex. *European Journal of Neuroscience, 17*(8), 1703–1714.

Feyerabend, P. (1975). *Against method: Outline of an anarchistic theory of knowledge*. Atlantic Highlands, NJ: Humanities Press.

Finke, R. A., & Schmidt, M. J. (1977). Orientation-specific color aftereffects following imagination. *Journal of Experimental Psychology: Human Perception and Performance, 3*(4), 599–606.

Fischer, M. E., Couvillon, P. A., & Bitterman, M. E. (1993). Choice in honeybees as a function of the probability of reward. *Animal Learning & Behavior, 21*(3), 187–195.

Fiske, H. E. (2004). *Connectionist models of musical thinking*. Lewiston, NY: E. Mellen Press.

Fitch, W. T., Hauser, M. D., & Chomsky, N. (2005). The evolution of the language faculty: Clarifications and implications. *Cognition, 97*(2), 179–210.

Fletcher, G. J. O. (1995). *The scientific credibility of folk psychology*. Hillsdale, NJ: Lawrence Erlbaum Associates.

Fletcher, H. (1924). The physical criterion for determining the pitch of a musical tone. *Physical Review*, 23(3), 427–437.

Flombaum, J. I., Scholl, B. J., & Pylyshyn, Z. W. (2008). Attentional resources in visual tracking through occlusion: The high-beams effect. *Cognition*, 107(3), 904–931.

Fodor, J. A. (1968a). Appeal to tacit knowledge in psychological explanation. *Journal of Philosophy*, 65(20), 627–640.

Fodor, J. A. (1968b). *Psychological explanation: An introduction to the philosophy of psychology*. New York, NY: Random House.

Fodor, J. A. (1975). *The language of thought*. Cambridge, MA: Harvard University Press.

Fodor, J. A. (1980). Methodological solipsism considered as a research strategy in cognitive psychology. *Behavioral and Brain Sciences*, 3(1), 63–73.

Fodor, J. A. (1983). *The modularity of mind*. Cambridge, MA: MIT Press.

Fodor, J. A. (1985). Précis of the modularity of mind. *Behavioral & Brain Sciences*, 8, 1–42.

Fodor, J. A. (2000). *The mind doesn't work that way: The scope and limits of computational psychology*. Cambridge, MA: MIT Press.

Fodor, J. A. (2009). What's so good about Pylyshyn? In D. Dedrick & L. Trick (Eds.), *Computation, cognition, and Pylyshyn* (pp. ix–xvii). Cambridge, MA: MIT Press.

Fodor, J. A., & McLaughlin, B. P. (1990). Connectionism and the problem of systematicity: Why Smolensky's solution doesn't work. *Cognition*, 35, 183–204.

Fodor, J. A., & Pylyshyn, Z. W. (1981). How direct is visual perception? Some reflections on Gibson's ecological approach. *Cognition*, 9, 139–196.

Fodor, J. A., & Pylyshyn, Z. W. (1988). Connectionism and cognitive architecture. *Cognition*, 28, 3–71.

Fogel, D. B. (1988). An evolutionary approach to the traveling salesman problem. *Biological Cybernetics*, 60(2), 139–144.

Fong, T., Nourbakhsh, I., & Dautenhahn, K. (2003). A survey of socially interactive robots. *Robotics and Autonomous Systems*, 42(3–4), 143–166.

Ford, K. M., & Pylyshyn, Z. W. (1996). *The robot's dilemma revisited: The frame problem in artificial intelligence*. Norwood, NJ: Ablex Pub.

Francès, R. (1988). *The perception of music*. (W. J. Dowling, Trans.). Hillsdale, NJ: Lawrence Erlbaum Associates.

Franconeri, S. L., Lin, J. Y., Pylyshyn, Z. W., Fisher, B., & Enns, J. T. (2008). Evidence against a speed limit in multiple-object tracking. *Psychonomic Bulletin & Review*, 15(4), 802–808.

Franklin, J. A. (2006). Jazz melody generation using recurrent networks and reinforcement learning. *International Journal on Artificial Intelligence Tools*, 15(4), 623–650.

French, P. A., & Wettstein, H. K. (2007). *Philosophy and the empirical*. Malden, MA: Blackwell Publishers

French, R. M. (2000). The Turing test: The first 50 years. *Trends in Cognitive Sciences*, 4(3), 115–122.

Freud, S. (1976). The uncanny. (J. Strachey, Trans.). *New Literary History*, 7(3), 619–645. (Original work published 1919)

Friedmann, M. L. (1990). *Ear training for twentieth-century music*. New Haven, CT: Yale University Press.

Frisby, J. P. (1980). *Seeing: Illusion, brain, and mind*. Oxford: Oxford University Press.

Fukushima, K. (1986). A neural network model for selective attention in visual pattern recognition. *Biological Cybernetics, 55*, 5–15.

Funahashi, K. (1989). On the approximate realization of continuous mappings by neural networks. *Neural Networks, 2*, 183–192.

Gaines, J. R. (2005). *Evening in the Palace of Reason: Bach meets Frederick the Great in the Age of Enlightenment*. London, UK; New York, NY: Fourth Estate.

Gallagher, S. (2005). *How the body shapes the mind*. Oxford, UK; New York, NY: Clarendon Press.

Gallant, S. I. (1993). *Neural network learning and expert systems*. Cambridge, MA: MIT Press.

Gallese, V., Fadiga, L., Fogassi, L., & Rizzolatti, G. (1996). Action recognition in the premotor cortex. *Brain, 119*, 593–609.

Gallese, V., & Goldman, A. (1998). Mirror neurons and the simulation theory of mind-reading. *Trends in Cognitive Sciences, 2*(12), 493–501.

Gallese, V., Keysers, C., & Rizzolatti, G. (2004). A unifying view of the basis of social cognition. *Trends in Cognitive Sciences, 8*(9), 396–403.

Gallistel, C. R. (1990). *The organization of learning*. Cambridge, MA: MIT Press.

Gardner, H. (1984). *The mind's new science*. New York, NY: Basic Books.

Garfield, J. L. (1987). *Modularity in knowledge representation and natural-language understanding*. Cambridge, MA: MIT Press.

Gasser, M., Eck, D., & Port, R. (1999). Meter as mechanism: A neural network model that learns metrical patterns. *Connection Science, 11*(2), 187–216.

Gazzaniga, M. S. (2000). *Cognitive neuroscience: A reader*. Malden, MA: Blackwell Publishers.

Gendler, T., & Hawthorne, J. (2006). *Perceptual experience*. Oxford, UK: Oxford University Press.

Gerkey, B. P., & Mataric, M. J. (2002). Sold!: Auction methods for multirobot coordination. *IEEE Transactions on Robotics and Automation, 18*(5), 758–768.

Gerkey, B. P., & Mataric, M. J. (2004). A formal analysis and taxonomy of task allocation in multi-robot systems. *International Journal of Robotics Research, 23*(9), 939–954.

Gerrissen, J. F. (1991). On the network-based emulation of human visual search. *Neural Networks, 4*, 543–564.

Gerstner, W., & Kistler, W. M. (2002). Mathematical formulations of Hebbian learning. *Biological Cybernetics, 87*(5–6), 404–415.

Gibbs, R. W. (2006). *Embodiment and cognitive science*. Cambridge, UK: Cambridge University Press.

Gibson, J. J. (1966). *The senses considered as perceptual systems*. Boston, MA: Houghton Mifflin.

Gibson, J. J. (1979). *The ecological approach to visual perception*. Boston, MA: Houghton Mifflin.

Gilbert, D. (2002). Are psychology's tribes ready to form a nation? *Trends in Cognitive Sciences, 6*(1), 3–3.

Girosi, F., & Poggio, T. (1990). Networks and the best approximation property. *Biological Cybernetics, 63*, 169–176.

Gjerdingen, R. O. (1989). Using connectionist models to explore complex musical patterns. *Computer Music Journal, 13*(3), 67–75.

Gjerdingen, R. O. (1990). Categorization of musical patterns by self-organizing neuron-like networks. *Music Perception, 7*(4), 339–369.

Gjerdingen, R. O. (1992). Learning syntactically significant temporal patterns of chords: A masking field embedded in an ART-3 architecture. *Neural Networks, 5*(4), 551–564.

Gjerdingen, R. O. (1994). Apparent motion in music. *Music Perception, 11*(4), 335–370.

Gjerdingen, R. O., & Perrott, D. (2008). Scanning the dial: The rapid recognition of music genres. *Journal of New Music Research, 37*(2), 93–100.

Glanzer, M. (1972). Storage mechanisms in free recall. In G. H. Bower (Ed.), *The psychology of learning and motivation: Advances in research and theory.* New York, NY: Academic Press.

Glanzer, M., & Cunitz, A. R. (1966). Two storage mechanisms in free recall. *Journal of Verbal Learning and Verbal Behavior, 5*(4), 351–360.

Glass, P. (1987). *Music by Philip Glass* (1st ed.). New York, NY: Harper & Row.

Gleitman, L. R., & Liberman, M. (1995). *An invitation to cognitive science: Language* (Vol. 1, 2nd ed.). Cambridge, MA: MIT Press.

Gluck, M. A., & Bower, G. H. (1988). From conditioning to category learning: An adaptive network model. *Journal of Experimental Psychology-General, 117*(3), 227–247.

Gluck, M. A., & Myers, C. (2001). *Gateway to memory: An introduction to neural network modeling of the hippocampus and learning.* Cambridge, MA: MIT Press.

Godfrey, M. D., & Hendry, D. F. (1993). The computer as von Neumann planned it. *IEEE Annals of the History of Computing, 15*(1), 11–21.

Goertzen, J. R. (2008). On the possibility of unification: The reality and nature of the crisis in psychology. *Theory & Psychology, 18*(6), 829–852.

Gold, E. M. (1967). Language identification in the limit. *Information and Control, 10*, 447–474.

Goldberg, M. E., & Bruce, C. J. (1985). Cerebral cortical activity associated with the orientation of visual attention in the rhesus monkey. *Vision Research, 25*, 471–481.

Goldman, A. I. (2006). *Simulating minds: The philosophy, psychology, and neuroscience of mindreading.* Oxford, UK; New York, NY: Oxford University Press.

Goldman, C. V., Gang, D., Rosenschein, J. S., & Lehmann, D. (1999). NetNeg: A connectionist-agent integrated system for representing musical knowledge. *Annals of Mathematics and Artificial Intelligence, 25*(1–2), 69–90.

Goldstine, H. H. (1993). *The computer: From Pascal to von Neumann.* Princeton, NJ: Princeton University Press.

Gombrich, E. H. (1960). *Art and illusion: A study in the psychology of pictorial representation.* New York, NY: Pantheon Books.

Goodale, M. A. (1988). Modularity in visuomotor control: From input to output. In Z. W. Pylyshyn (Ed.), *Computational processes in human vision: An interdisciplinary perspective* (pp. 262–285). Norwood, NJ: Ablex.

Goodale, M. A. (1990). *Vision and action: The control of grasping.* Norwood, NJ: Ablex.

Goodale, M. A. (1995). The cortical organization of visual perception and visuomotor control. In S. M. Kosslyn & D. N. Osherson (Eds.), *An invitation to cognitive science: Visual cognition* (Vol. 2, pp. 167–213). Cambridge, MA: MIT Press.

Goodale, M. A., & Humphrey, G. K. (1998). The objects of action and perception. *Cognition, 67*, 181–207.

Goodale, M. A., Jakobson, L. S., & Keillor, J. M. (1994). Differences in the visual control of pantomimed and natural grasping movements. *Neuropsychologia, 32*(10), 1159–1178.

Goodale, M. A., Milner, A. D., Jakobson, L. S., & Carey, D. P. (1991). A neurological dissociation between perceiving objects and grasping them. *Nature, 349*(6305), 154–156.

Goodman, N. (1978). *Ways of worldmaking*. Indianapolis, IN: Hacket Publishing.

Gopnik, A., & Meltzoff, A. N. (1997). *Words, thoughts, and theories*. Cambridge, MA: MIT Press.

Gopnik, A., Meltzoff, A. N., & Kuhl, P. K. (1999). *The scientist in the crib: Minds, brains, and how children learn* (1st ed.). New York, NY: William Morrow & Co.

Gopnik, A., & Wellman, H. (1992). Why the child's theory of mind really is a theory. *Mind & Language, 7*, 145–171.

Gordon, R. M. (1986). Folk psychology as simulation. *Mind & Language, 1*(158–171).

Gordon, R. M. (1992). The simulation theory: Objections and misconceptions. *Mind & Language, 7*, 11–34.

Gordon, R. M. (1995). Sympathy, simulation, and the impartial spectator. *Ethics, 105*(4), 727–742.

Gordon, R. M. (1999). Simulation theory vs. theory-theory. In R. A. Wilson & F. C. Keil (Eds.), *The MIT encyclopedia of the cognitive sciences* (pp. 765–766). Cambridge, MA: MIT Press.

Gordon, R. M. (2005a). Intentional agents like myself. In S. Hurley & N. Chater (Eds.), *Perspectives on imitation: From mirror neurons to memes* (Vol. 1, pp. 95–106). Cambridge MA: MIT Press.

Gordon, R. M. (2005b). Simulation and systematic errors in prediction. *Trends in Cognitive Sciences, 9*(8), 361–362.

Gordon, R. M. (2007). Ascent routines for propositional attitudes. *Synthese, 159*(2), 151–165.

Gordon, R. M. (2008). Beyond mindreading. *Philosophical Explorations, 11*(3), 219–222.

Goryo, K., Robinson, J. O., & Wilson, J. A. (1984). Selective looking and the Muller-Lyer illusion: The effect of changes in the focus of attention on the Muller-Lyer illusion. *Perception, 13*(6), 647–654.

Goss, S., Aron, S., Deneubourg, J. L., & Pasteels, J. M. (1989). Self-organized shortcuts in the Argentine ant. *Naturwissenschaften, 76*(12), 579–581.

Graf, V., Bullock, D. H., & Bitterman, M. E. (1964). Further experiments on probability-matching in the pigeon. *Journal of the Experimental Analysis of Behavior, 7*(2), 151–157.

Grant, M. J. (2001). *Serial music, serial aesthetics: Compositional theory in post-war Europe*. New York, NY: Cambridge University Press.

Grasse, P. P. (1959). La reconstruction du nid et les coordinations interindividuelles chez *Bellicositermes natalensis* et *Cubitermes* sp. la théorie de la stigmergie: Essai d'interprétation du comportement des termites constructeurs. *Insectes Sociaux, 6*(1), 41–80.

Green, D. W. (1996). *Cognitive science: An introduction*. Oxford, UK: Wiley-Blackwell.

Green, E. A. H., & Malko, N. A. (1975). *The conductor and his score*. Englewood Cliffs, NJ: Prentice Hall.

Greeno, J. G., & Moore, J. L. (1993). Situativity and symbols: Response to Vera and Simon. *Cognitive Science, 17*, 49–59.

Greenwood, J. D. (1991). *The future of folk psychology: Intentionality and cognitive science.* Cambridge, UK; New York, NY: Cambridge University Press.

Greenwood, J. D. (1999). Simulation, theory-theory and cognitive penetration: No "instance of the fingerpost." *Mind & Language, 14*(1), 32–56.

Gregory, R. L. (1970). *The intelligent eye.* London, UK: Weidenfeld & Nicolson.

Gregory, R. L. (1978). *Eye and brain.* New York, NY: McGraw-Hill.

Grenville, B. (2001). *The uncanny: Experiments in cyborg culture.* Vancouver, B.C.: Vancouver Art Gallery; Arsenal Pulp Press.

Grey Walter, W. (1950a). An electro-mechanical animal. *Dialectica, 4*(3), 206–213.

Grey Walter, W. (1950b). An imitation of life. *Scientific American, 182*(5), 42–45.

Grey Walter, W. (1951). A machine that learns. *Scientific American, 184*(8), 60–63.

Grey Walter, W. (1963). *The living brain.* New York, NY: W.W. Norton & Co.

Gridley, M. C., & Hoff, R. (2006). Do mirror neurons explain misattribution of emotions in music? *Perceptual and Motor Skills, 102*(2), 600–602.

Griffith, N. (1995). Connectionist visualization of tonal structure. *Artificial Intelligence Review, 8*(5–6), 393–408.

Griffith, N., & Todd, P. M. (1999). *Musical networks: Parallel distributed perception and performance.* Cambridge, MA: MIT Press.

Griffiths, P. (1994). *Modern music: A concise history* (Rev. ed.). New York, NY: Thames and Hudson.

Griffiths, P. (1995). *Modern music and after.* Oxford, UK; New York, NY: Oxford University Press.

Gritten, A., & King, E. (2011). *New perspectives on music and gesture.* Burlington, VT: Ashgate.

Gross, C. G. (1998). *Brain, vision, memory: Tales in the history of neuroscience.* Cambridge, MA: MIT Press.

Gross, C. G. (2002). Genealogy of the "grandmother cell." *Neuroscientist, 8*(5), 512–518.

Grossberg, S. (1980). How does the brain build a cognitive code? *Psychological Review, 87*, 1–51.

Grossberg, S. (1987). Competitive learning: From interactive activation to adaptive resonance. *Cognitive Science, 11*, 23–63.

Grossberg, S. (1988). *Neural networks and natural intelligence.* Cambridge, MA: MIT Press.

Grossberg, S. (1999). Pitch-based streaming in auditory perception. In N. Griffith & P. M. Todd (Eds.), *Musical networks: Parallel distributed perception and performance* (pp. 117–140). Cambridge, MA: MIT Press.

Grossberg, S., & Rudd, M. E. (1989). A neural architecture for visual motion perception: Group and element apparent motion. *Neural Networks, 2*, 421–450.

Grossberg, S., & Rudd, M. E. (1992). Cortical dynamics of visual motion perception: Short-range and long-range apparent motion. *Psychological Review, 99*, 78–121.

Gurevich, V. (2006). *Electric relays: Principles and applications.* Boca Raton, FL: CRC/Taylor & Francis.

Guthrie, E. R. (1935). *The psychology of learning.* New York, NY: Harper.

Gutin, G., & Punnen, A. P. (2002). *The traveling salesman problem and its variations*. Dordrecht, Netherlands; Boston, MA: Kluwer Academic Publishers.

Guzik, A. L., Eaton, R. C., & Mathis, D. W. (1999). A connectionist model of left–right sound discrimination by the Mauthner system. *Journal of Computational Neuroscience, 6*(2), 121–144.

Haberlandt, K. (1994). *Cognitive psychology*. Boston, MA: Allyn and Bacon.

Hadley, R. F. (1993). Connectionism, explicit rules, and symbolic manipulation. *Minds and Machines: Journal for Artificial Intelligence, Philosophy and Cognitive Science, 3*(2), 183–200.

Hadley, R. F. (1994a). Systematicity in connectionist language learning. *Minds and Machines: Journal for Artificial Intelligence, Philosophy and Cognitive Science, 3*, 183–200.

Hadley, R. F. (1994b). Systematicity revisited: Reply to Christiansen and Chater and Niclasson and van Gelder. *Mind & Language, 9*, 431–444.

Hadley, R. F. (1997). Cognition, systematicity, and nomic necessity. *Mind & Language, 12*, 137–153.

Hadley, R. F., & Hayward, M. B. (1997). Strong semantic systematicity from Hebbian connectionist learning. *Minds and Machines: Journal for Artificial Intelligence, Philosophy and Cognitive Science, 7*, 1–37.

Hamanaka, M., Hirata, K., & Tojo, S. (2006). Implementing "A generative theory of tonal music." *Journal of New Music Research, 35*(4), 249–277.

Hanslick, E. (1957). *The beautiful in music*. New York, NY: Liberal Arts Press. (Original work published 1854)

Hanson, S. J., & Burr, D. J. (1990). What connectionist models learn: Learning and representation in connectionist networks. *Behavioral and Brain Sciences, 13*, 471–518.

Hanson, S. J., & Olson, C. R. (1991). Neural networks and natural intelligence: Notes from Mudville. *Connection Science, 3*, 332–335.

Harkleroad, L. (2006). *The math behind the music*. Cambridge, UK; New York, NY: Cambridge University Press.

Harnad, S. (1987). *Categorical perception*. Cambridge: Cambridge University Press.

Harnad, S. (1990). The symbol grounding problem. *Physica D, 42*(1–3), 335–346.

Harnish, R. M. (2002). *Minds, brains, and computers: An historical introduction to the foundations of cognitive science*. Malden, MA: Blackwell Publishers.

Hartman, E., Keeler, J. D., & Kowalski, J. M. (1989). Layered neural networks with Gaussian hidden units as universal approximation. *Neural Computation, 2*, 210–215.

Haselager, W. F. G. (1997). *Cognitive science and folk psychology: The right frame of mind*. Thousand Oaks, CA: Sage Publications.

Haslinger, B., Erhard, P., Altenmuller, E., Schroeder, U., Boecker, H., & Ceballos-Baumann, A. O. (2005). Transmodal sensorimotor networks during action observation in professional pianists. *Journal of Cognitive Neuroscience, 17*(2), 282–293.

Hastie, R. (2001). Problems for judgment and decision making. *Annual Review of Psychology, 52*, 653–683.

Hatano, G., Miyake, Y., & Binks, M. G. (1977). Performance of expert abacus operators. *Cognition, 5*(1), 47–55.

Haugeland, J. (1985). *Artificial intelligence: The very idea*. Cambridge, MA: MIT Press.

Hauser, M. D., Chomsky, N., & Fitch, W. T. (2002). The faculty of language: What is it, who has it, and how did it evolve? *Science, 298*(5598), 1569–1579.

Haxby, J. V., Hoffman, E. A., & Gobbini, M. I. (2000). The distributed human neural system for face perception. *Trends in Cognitive Sciences, 4*(6), 223–233.

Haxby, J. V., Hoffman, E. A., & Gobbini, M. I. (2002). Human neural systems for face recognition and social communication. *Biological Psychiatry, 51*(1), 59–67.

Hayes, P. J., Ford, K. M., & Agnew, N. (1994). On babies and bathwater: A cautionary tale. *AI Magazine, 15*(4), 15–26.

Hayles, N. K. (1999). *How we became posthuman: Virtual bodies in cybernetics, literature, and informatics.* Chicago, IL: University of Chicago Press.

Hayward, R. (2001). The tortoise and the love-machine: Grey Walter and the politics of electroencephalography. *Science in Context, 14*(4), 615–641.

Heal, J. (1986). Replication and functionalism. In J. Butterfield (Ed.), *Language, mind and logic* (pp. 135–150). Cambridge, UK: Cambridge University Press.

Heal, J. (1996). Simulation and cognitive penetrability. *Mind & Language, 11*(1), 44–67.

Healy, S. (1998). *Spatial representation in animals.* Oxford, UK: Oxford University Press.

Hebb, D. O. (1949). *The organization of behaviour.* New York, NY: John Wiley & Sons.

Hebb, D. O. (1959). A neuropsychological theory. In S. Koch (Ed.), *Psychology: A study of a science: Vol. 1. Sensory, perceptual, and physiological foundations* (pp. 622–643). New York, NY: McGraw-Hill.

Hecht-Nielsen, R. (1987). *Neurocomputing.* Reading, MA: Addison-Wesley.

Hegel, G. W. F. (1931). *The phenomenology of mind* (2nd ed.). (J. B. Bailie, Trans.). London, UK; New York, NY: G. Allen & Unwin, The Macmillan Company. (Original work published 1807)

Heidegger, M. (1962). *Being and time.* (J. Macquarrie & E. Robinson, Trans.). New York, NY: Harper. (Original work published 1927)

Helmholtz, H. (1968). The recent progress of the theory of vision. In R. M. Warren & R. P. Warren (Eds.), *Helmholtz on perception: Its physiology and development* (pp. 61–136). New York, NY: John Wiley & Sons. (Original work published 1868.)

Helmholtz, H., & Ellis, A. J. (1954). *On the sensations of tone as a physiological basis for the theory of music* (A. J. Ellis & H. Margenau, 2nd ed.). New York, NY: Dover Publications. (Original work published 1863)

Helmholtz, H., & Southall, J. P. C. (1962a). *Helmholtz's treatise on physiological optics* (Vol. 1–2, J. P. C. Southall [Ed.]). New York, NY: Dover Publications. (Original work published 1856)

Helmholtz, H., & Southall, J. P. C. (1962b). *Helmholtz's treatise on physiological optics* (Vol. 3, J. P. C. Southall [Ed.]). New York, NY: Dover Publications. (Original work published 1857)

Henriques, G. R. (2004). Psychology defined. *Journal of Clinical Psychology, 60*(12), 1207–1221.

Hermer, L., & Spelke, E. S. (1994). A geometric process for spatial reorientation in young children. *Nature, 370*(6484), 57–59.

Herrnstein, R. J. (1961). Relative and absolute strength of response as a function of frequency of reinforcement. *Journal of the Experimental Analysis of Behavior, 4*(3), 267–272.

Herrnstein, R. J. (1997). *The matching law: Papers in psychology and economics*. New York, NY: Harvard University Press.

Herrnstein, R. J., & Loveland, D. H. (1975). Maximizing and matching on concurrent ratio schedules. *Journal of the Experimental Analysis of Behavior, 24*(1), 107–116.

Hess, R. H., Baker, C. L., & Zihl, J. (1989). The "motion-blind" patient: Low-level spatial and temporal filters. *The Journal of Neuroscience, 9*, 1628–1640.

Hildesheimer, W. (1983). *Mozart*. New York, NY: Vintage Books.

Hildreth, E. C. (1983). *The measurement of visual motion*. Cambridge, MA: MIT Press.

Hillis, W. D. (1985). *The connection machine*. Cambridge, MA: MIT Press.

Hillis, W. D. (1988). Intelligence as emergent behavior, or, the songs of Eden. In S. R. Graubard (Ed.), *The artificial intelligence debate* (pp. 175–189). Cambridge, MA: MIT Press.

Hillis, W. D. (1998). *The pattern on the stone* (1st ed.). New York, NY: Basic Books.

Hinchey, M. G., Sterritt, R., & Rouff, C. (2007). Swarms and swarm intelligence. *Computer, 40*(4), 111–113.

Hinton, G. E. (1986). Learning distributed representations of concepts. Paper presented at The 8th Annual Meeting of the Cognitive Science Society, Ann Arbor, MI.

Hinton, G. E., & Anderson, J. A. (1981). *Parallel models of associative memory*. Hillsdale, NJ: Lawrence Erlbaum Associates.

Hinton, G. E., McClelland, J., & Rumelhart, D. (1986). Distributed representations. In D. Rumelhart & J. McClelland (Eds.), *Parallel distributed processing* (Vol. 1, pp. 77–109). Cambridge, MA: MIT Press.

Hobbes, T. (1967). *Hobbes's* Leviathan. Oxford, UK: Clarendon Press. (Original work published 1651)

Hodges, A. (1983). *Alan Turing: The enigma of intelligence*. London, UK: Unwin Paperbacks.

Hoffman, D. D., & Singh, M. (1997). Salience of visual parts. *Cognition, 63*(1), 29–78.

Hofstadter, D. R. (1979). *Gödel, Escher, Bach: An eternal golden braid*. New York, NY: Basic Books.

Hofstadter, D. R. (1995). *Fluid concepts and creative analogies*. New York, NY: Basic Books.

Holland, J. H. (1992). *Adaptation in natural and artificial systems*. Cambridge, MA: MIT Press.

Holland, O. (2003a). Exploration and high adventure: The legacy of Grey Walter. *Philosophical Transactions of the Royal Society of London Series A: Mathematical Physical and Engineering Sciences, 361*(1811), 2085–2121.

Holland, O. (2003b). The first biologically inspired robots. *Robotica, 21*, 351–363.

Holland, O., & Melhuish, C. (1999). Stigmergy, self-organization, and sorting in collective robotics. *Artificial Life, 5*, 173–202.

Hollerith, H. (1889). An electric tabulating system. *The Quarterly, Columbia University School of Mines, X*(16), 238–255.

Hooker, C. A. (1979). Critical notice: R. M. Yoshida's reduction in the physical sciences. *Dialogue, 18*, 81–99.

Hooker, C. A. (1981). Towards a general theory of reduction. *Dialogue, 20*, 38–59, 201–236, 496–529.

Hoover, A. K., & Stanley, K. O. (2009). Exploiting functional relationships in musical composition. *Connection Science, 21*(2–3), 227–251.

Hopcroft, J. E., & Ullman, J. D. (1979). *Introduction to automata theory, languages, and computation.* Reading, MA: Addison-Wesley.

Hopfield, J. J. (1982). Neural networks and physical systems with emergent collective computational abilities. *Proceedings of the National Academy of Sciences, 79,* 2554–2558.

Hopfield, J. J. (1984). Neurons with graded response have collective computational properties like those of two state neurons. *Proceedings of the National Academy of Sciences, 81,* 3008–3092.

Hopfield, J. J., & Tank, D. W. (1985). "Neural" computation of decisions in optimization problems. *Biological Cybernetics, 52*(3), 141–152.

Horgan, J. (1992). Claude E. Shannon. *IEEE Spectrum, 29*(4), 72–75.

Horgan, J. (1993). The mastermind of artificial intelligence. *Scientific American, 269*(5), 35–38.

Horgan, T., & Tienson, J. (1996). *Connectionism and the philosophy of psychology.* Cambridge, MA: MIT Press.

Horn, B. K. P. (1986). *Robot vision.* Cambridge, MA: MIT Press.

Horn, B. K. P., & Brooks, M. J. (1989). *Shape from shading.* Cambridge, MA: MIT Press.

Horn, B. K. P., & Schunk, B. (1981). Determining optical flow. *Artificial Intelligence, 17,* 185–203.

Hornik, M., Stinchcombe, M., & White, H. (1989). Multilayer feedforward networks are universal approximators. *Neural Networks, 2,* 359–366.

Horwood, H. (1987). *Dancing on the shore.* Toronto, ON: McClelland and Stewart.

Howell, P., Cross, I., & West, R. (1985). *Musical structure and cognition.* London, UK; Orlando, FL: Academic Press.

Hubel, D. H., & Wiesel, T. N. (1959). Receptive fields of single neurones in the cat's striate cortex. *Journal of Physiology, 148,* 574–591.

Humboldt, W. (1999). *On language: On the diversity of human language construction and its influence on the mental development of the human species.* (P. Heath, Trans.). New York, NY: Cambridge University Press. (Original work published 1836)

Hume, D. (1952). *An enquiry concerning human understanding.* La Salle, IL: The Open Court Publishing Company. (Original work published 1748)

Humphreys, G. W., & Bruce, V. (1989). *Visual cognition: Computational, experimental, and neuropsychological perspectives.* Hillsdale, NJ: Lawrence Erlbaum Associates.

Hurley, S. (2001). Perception and action: Alternative views. *Synthese, 129*(1), 3–40.

Huron, D. B. (2006). *Sweet anticipation: Music and the psychology of expectation.* Cambridge, MA: MIT Press.

Husserl, E. (1965). *Phenomenology and the crisis of philosophy.* (Q. Lauer, Trans.). New York, NY: Harper & Row.

Hutchins, E. (1995). *Cognition in the wild.* Cambridge, MA: MIT Press.

Hyvarinen, J., & Poranen, A. (1974). Function of parietal associative area 7 as revealed from cellular discharges in alert monkeys. *Brain, 97,* 673–692.

Iacoboni, M. (2008). *Mirroring people: The new science of how we connect with others* (1st ed.). New York, NY: Farrar, Straus & Giroux.

Ichbiah, D. (2005). *Robots: From science fiction to technological revolution.* New York, NY: Harry N. Abrams.

Inhelder, B., & Piaget, J. (1958). *The growth of logical thinking from childhood to adolescence*. (A. Parsons & S. Milgram, Trans.). New York, NY: Basic Books.

Inhelder, B., & Piaget, J. (1964). *The early growth of logic in the child*. (E.A. Lunzer & D. Papert, Trans.). New York, NY: Harper & Row.

Irvine, M. M. (2001). Early digital computers at Bell Telephone Laboratories. *IEEE Annals of the History of Computing, 23*(3), 22–42.

Isacoff, S. (2001). *Temperament: The idea that solved music's greatest riddle* (1st ed.). New York, NY: Alfred A. Knopf.

Jackendoff, R. (1977). *X-bar syntax: A study of phrase structure*. Cambridge, MA: MIT Press.

Jackendoff, R. (1983). *Semantics and cognition*. Cambridge, MA: MIT Press.

Jackendoff, R. (1987). On beyond zebra: The relation of linguistic and visual information. *Cognition, 26,* 89–114.

Jackendoff, R. (1990). *Semantic structures*. Cambridge, MA: MIT Press.

Jackendoff, R. (2002). *Foundations of language: Brain, meaning, grammar, evolution*. Oxford, UK; New York, NY: Oxford University Press.

Jackendoff, R. (2009). Parallels and nonparallels between language and music. *Music Perception, 26*(3), 195–204.

Jacob, P., & Jeannerod, M. (2003). *Ways of seeing: The scope and limits of visual cognition*. Oxford, UK; New York, NY: Oxford University Press.

Jakobson, L. S., Archibald, Y. M., Carey, D. P., & Goodale, M. A. (1991). A kinematic analysis of reaching and grasping movements in a patient recovering from optic ataxia. *Neuropsychologia, 29*(8), 803–809.

James, W. (1890a). *The principles of psychology* (Vol. 1). New York, NY: Dover Publications.

James, W. (1890b). *The principles of psychology* (Vol. 2). New York, NY: Dover Publications.

Jarvinen, T. (1995). Tonal hierarchies in jazz improvisation. *Music Perception, 12*(4), 415–437.

Jeanne, R. L. (1996). Regulation of nest construction behaviour in *Polybia occidentalis*. *Animal Behaviour, 52,* 473–488.

Jefferies, M. E., & Yeap, W. K. (2008). *Robotics and cognitive approaches to spatial mapping*. Berlin, Germany; New York, NY: Springer.

Jensen, E. M., Reese, E. P., & Reese, T. W. (1950). The subitizing and counting of visually presented fields of dots. *Journal of Psychology, 30*(2), 363–392.

Jevons, W. S. (1870). On the mechanical performance of logical inference. *Philosophical Transactions of the Royal Society of London, 160,* 497–518.

Johnson, M. (2007). *The meaning of the body*. Chicago, IL: University of Chicago Press.

Johnson-Laird, P. N. (1983). *Mental models*. Cambridge, MA: Harvard University Press.

Jonsson, E. (2002). *Inner navigation: Why we get lost and how we find our way*. New York, NY: Scribner.

Jordà, S., Geiger, G., Alonso, M., & Kaltenbrunner, M. (2007). The reacTable: Exploring the synergy between live music performance and tabletop tangible interfaces. Paper presented at the Proceedings of the first international conference on "Tangible and Embedded Interaction" (TEI07), Baton Rouge, Louisiana.

Josephson, M. (1961). *Edison*. New York, NY: McGraw Hill.

Jourdain, R. (1997). *Music, the brain, and ecstasy*. New York, NY: William Morrow & Co.

Jun, S., Rho, S., & Hwang, E. (2010). Music retrieval and recommendation scheme based on varying mood sequences. *International Journal on Semantic Web and Information Systems, 6*(2), 1-16.

Kahneman, D., Treisman, A., & Gibbs, B. J. (1992). The reviewing of object files: Object-specific integration of information. *Cognitive Psychology, 24*(2), 175-219.

Kaltenbrunner, M., Jordà, S., Geiger, G., & Alonso, M. (2007). The reacTable: A collaborative musical instrument. Paper presented at the Proceedings of the Workshop on "Tangible Interaction in Collaborative Environments" (TICE), at the 15th International IEEE Workshops on Enabling Technologies (WETICE 2006), Manchester, UK.

Kamin, L. J. (1968). Attention-like processes in classical conditioning. In M. R. Jones (Ed.), *Miami symposium on the prediction of behavior: Aversive stimulation* (pp. 9-32). Miami, FL: University of Miami Press.

Karsai, I. (1999). Decentralized control of construction behavior in paper wasps: An overview of the stigmergy approach. *Artificial Life, 5*, 117-136.

Karsai, I., & Penzes, Z. (1998). Nest shapes in paper wasps: Can the variability of forms be deduced from the same construction algorithm? *Proceedings of the Royal Society of London Series B-Biological Sciences, 265*(1402), 1261-1268.

Karsai, I., & Wenzel, J. W. (2000). Organization and regulation of nest construction behavior in *Metapolybia* wasps. *Journal of Insect Behavior, 13*(1), 111-140.

Kasabov, N. K. (1996). *Foundations of neural networks, fuzzy systems, and knowledge engineering*. Cambridge, MA: MIT Press.

Katz, B. F. (1995). Harmonic resolution, neural resonance, and positive affect. *Music Perception, 13*(1), 79-108.

Katzko, M. W. (2002). The rhetoric of psychological research and the problem of unification in psychology. *American Psychologist, 57*(4), 262-270.

Kaufman, E. L., Lord, M. W., Reese, T. W., & Volkmann, J. (1949). The discrimination of visual number. *American Journal of Psychology, 62*(4), 498-525.

Kazennikov, O., & Wiesendanger, M. (2009). Bimanual coordination of bowing and fingering in violinists: Effects of position changes and string changes. *Motor Control, 13*(3), 297-309.

Keasar, T., Rashkovich, E., Cohen, D., & Shmida, A. (2002). Bees in two-armed bandit situations: Foraging choices and possible decision mechanisms. *Behavioral Ecology, 13*(6), 757-765.

Keith, G. P., Blohm, G., & Crawford, J. D. (2010). Influence of saccade efference copy on the spatiotemporal properties of remapping: A neural network study. *Journal of Neurophysiology, 103*(1), 117-139.

Kelly, D. M., Spetch, M. L., & Heth, C. D. (1998). Pigeons' (*Columba livia*) encoding of geometric and featural properties of a spatial environment. *Journal of Comparative Psychology, 112*(3), 259-269.

Keysers, C., Wicker, B., Gazzola, V., Anton, J. L., Fogassi, L., & Gallese, V. (2004). A touching sight: SII/PV activation during the observation and experience of touch. *Neuron, 42*(2), 335-346.

Kieras, D. E., & Meyer, D. E. (1997). An overview of the EPIC architecture for cognition and performance with application to human–computer interaction. *Human–Computer Interaction, 12*(4), 391-438.

Kirk, K. L., & Bitterman, M. E. (1965). Probability-learning by the turtle. *Science, 148*(3676), 1484–1485.

Kirkpatrick, S., Gelatt, C. D., & Vecchi, M. P. (1983). Optimization by simulated annealing. *Science, 220*(4598), 671–680.

Kirsh, D. (1992). When is information explicitly represented? In P. P. Hanson (Ed.), *Information, language, and cognition* (pp. 340–365). Oxford, UK: Oxford University Press.

Kitchin, R. M. (1994). Cognitive maps: What are they and why study them? *Journal Of Environmental Psychology, 14*, 1–19.

Kivy, P. (1991). *Sound and semblance: Reflections on musical representation.* Ithaca, NY: Cornell University Press.

Klein, R. (1988). Inhibitory tagging system facilitates visual search. *Nature, 334*, 430–431.

Knobe, J. M., & Nichols, S. (2008). *Experimental philosophy.* Oxford, UK; New York, NY: Oxford University Press.

Knuth, D. E. (1997). *The art of computer programming: Vol. 3. Sorting and Searching* (3rd ed.). Reading, MA: Addison-Wesley.

Ko, B. C., & Byun, H. (2002). Query-by-gesture: An alternative content-based image retrieval query scheme. *Journal of Visual Languages and Computing, 13*(4), 375–390.

Koch, C., & Ullman, S. (1985). Shifts in selective visual attention: Towards the underlying neural circuitry. *Human Neurobiology, 4*, 219–227.

Koch, S. (1959). *Psychology: A study of a science.* New York, NY: McGraw-Hill.

Koch, S. (1969). Psychology cannot be a coherent science. *Psychology Today, 3*(4), 14, 64–68.

Koch, S. (1976). Language communities, search cells, and the psychological studies. In W. J. Arnold (Ed.), *Nebraska symposium on motivation: Conceptual foundations of psychology* (Vol. 23, pp. 477–559). Lincoln, NE: University of Nebraska Press.

Koch, S. (1981). The nature and limits of psychological knowledge: Lessons of a century qua "science." *American Psychologist, 36*(3), 257–269.

Koch, S. (1993). "Psychology" or "the psychological studies." *American Psychologist, 48*(8), 902–904.

Koelsch, S., Kasper, E., Sammler, D., Schulze, K., Gunter, T., & Friederici, A. D. (2004). Music, language and meaning: Brain signatures of semantic processing. *Nature Neuroscience, 7*(3), 302–307.

Koenig, G. M. (1999). PROJECT 1 Revisited: On the analysis and interpretation of PR1 tables. In J. Tabor (Ed.), *Otto Laske: Navigating new horizons* (pp. 53–72). Westport, CT: Greenwood Press.

Koffka, K. (1935). *Principles of gestalt psychology.* New York, NY: Harcourt, Brace & World.

Köhler, W. (1947). *Gestalt psychology: An introduction to new concepts in modern psychology.* New York, NY: Liveright Pub. Corp.

Kohonen, T. (1977). *Associative memory: A system-theoretical approach.* New York, NY: Springer-Verlag.

Kohonen, T. (1984). *Self-organization and associative memory.* New York, NY: Springer-Verlag.

Kohonen, T. (2001). *Self-organizing maps* (3rd ed.). Berlin, Germany; New York, NY: Springer.

Kohonen, T., Laine, P., Tiits, K., & Torkkola, K. (1991). A nonheuristic automatic composing method. In P. M. Todd & D. G. Loy (Eds.), *Music and connectionism* (pp. 229–242). Cambridge, MA: MIT Press.

Kojima, T. (1954). *The Japanese abacus: Its use and theory* (1st ed.). Tokyo, Japan; Rutland, VT: C. E. Tuttle Co.

Kolers, P. (1972). *Aspects of motion perception*. New York, NY: Pergammon Press.

Kolers, P., & Green, M. (1984). Color logic of apparent motion. *Perception, 13*, 149–154.

Kolers, P., & Pomerantz, J. R. (1971). Figural change in apparent motion. *Journal of Experimental Psychology, 87*, 99–108.

Kolers, P., & von Grunau, M. (1976). Shape and colour in apparent motion. *Vision Research, 16*, 329–335.

Konczak, J., van der Velden, H., & Jaeger, L. (2009). Learning to play the violin: Motor control by freezing, not freeing degrees of freedom. *Journal of Motor Behavior, 41*(3), 243–252.

Kosslyn, S. M. (1980). *Image and mind*. Cambridge, MA: Harvard University Press.

Kosslyn, S. M. (1987). Seeing and imagining in the cerebral hemispheres: A computational approach. *Psychological Review, 94*(2), 148–175.

Kosslyn, S. M. (1994). *Image and brain*. Cambridge, MA: MIT Press.

Kosslyn, S. M., Ball, T. M., & Reiser, B. J. (1978). Visual images preserve metric spatial information: Evidence from studies of image scanning. *Journal of Experimental Psychology: Human Perception and Performance, 4*(1), 47–60.

Kosslyn, S. M., Brunn, J., Cave, K. R., & Wallach, R. W. (1984). Individual differences in mental imagery ability: A computational analysis. *Cognition, 18*(1–3), 195–243.

Kosslyn, S. M., Farah, M. J., Holtzman, J. D., & Gazzaniga, M. S. (1985). A computational analysis of mental image generation: Evidence from functional dissociations in split-brain patients. *Journal of Experimental Psychology: General, 114*(3), 311–341.

Kosslyn, S. M., Ganis, G., & Thompson, W. L. (2003). Mental imagery: Against the nihilistic hypothesis. *Trends in Cognitive Sciences, 7*(3), 109–111.

Kosslyn, S. M., & Osherson, D. N. (1995). *An invitation to cognitive science: Vol. 2. Visual cognition* (2nd Ed.). Cambridge, MA: MIT Press.

Kosslyn, S. M., Pascual-Leone, A., Felican, O., Camposano, S., Keenan, J. P., Thompson, W. L. (1999). The role of area 17 in visual imagery: Convergent evidence from PET and rTMS. *Science, 284*, 167–170.

Kosslyn, S. M., & Shwartz, S. P. (1977). A simulation of visual imagery. *Cognitive Science, 1*, 265–295.

Kosslyn, S. M., Thompson, W. L., & Alpert, N. M. (1997). Neural systems shared by visual imagery and visual perception: A positron emission tomography study. *Neuroimage, 6*, 320–334.

Kosslyn, S. M., Thompson, W. L., & Ganis, G. (2006). *The case for mental imagery*. New York, NY: Oxford University Press.

Kosslyn, S. M., Thompson, W. L., Kim, I. J., & Alpert, N. M. (1995). Topographical representations of mental images in area 17. *Nature, 378*, 496–498.

Kremer, S. C. (1995). On the computational powers of Elman-style recurrent networks. *IEEE Transactions on Neural Networks, 6*, 1000–1004.

Krumhansl, C. L. (1984). Independent processing of visual form and motion. *Perception, 13*, 535–546.

Krumhansl, C. L. (1990). *Cognitive foundations of musical pitch*. New York, NY: Oxford University Press.

Krumhansl, C. L. (2005). The geometry of musical structure: A brief introduction and history. *ACM Computers In Entertainments, 3*(4), 1–14.

Krumhansl, C. L., Bharucha, J. J., & Kessler, E. J. (1982). Perceived harmonic structure of chords in three related musical keys. *Journal of Experimental Psychology: Human Perception and Performance, 8*(1), 24–36.

Krumhansl, C. L., & Kessler, E. J. (1982). Tracing the dynamic changes in perceived tonal organization in a spatial representation of musical keys. *Psychological Review, 89*(4), 334–368.

Krumhansl, C. L., & Shepard, R. N. (1979). Quantification of the hierarchy of tonal functions within a diatonic context. *Journal of Experimental Psychology: Human Perception and Performance, 5*(4), 579–594.

Kruskal, J. B., & Wish, M. (1978). *Multidimensional scaling*. Beverly Hills, CA: Sage Publications.

Kube, C. R., & Bonabeau, E. (2000). Cooperative transport by ants and robots. *Robotics and Autonomous Systems, 30*, 85–101.

Kube, C. R., & Zhang, H. (1994). Collective robotics: From social insects to robots. *Adaptive Behavior, 2*, 189–218.

Kuhberger, A., Kogler, C., Hug, A., & Mosl, E. (2006). The role of the position effect in theory and simulation. *Mind & Language, 21*(5), 610–625.

Kuhn, T. S. (1957). *The Copernican revolution*. Cambridge, MA: Harvard University Press.

Kuhn, T. S. (1970). *The structure of scientific revolutions* (2nd ed.). Chicago, IL: University of Chicago Press.

Kurz, M. J., Judkins, T. N., Arellano, C., & Scott-Pandorf, M. (2008). A passive dynamic walking robot that has a deterministic nonlinear gait. *Journal of Biomechanics, 41*(6), 1310–1316.

Kurzweil, R. (1990). *The age of intelligent machines*. Cambridge, MA: MIT Press.

Kurzweil, R. (1999). *The age of spiritual machines*. New York, NY: Penguin.

Kurzweil, R. (2005). *The singularity is near: When humans transcend biology*. New York, NY: Viking.

LaBerge, D., Carter, M., & Brown, V. (1992). A network simulation of thalamic circuit operations in selective attention. *Neural Computation, 4*, 318–331.

Ladd, C. (1883). An algebra of logic. In C. S. Peirce (Ed.), *Studies in logic* (pp. 17–71). Cambridge, MA: Harvard University Press.

Laden, B., & Keefe, B. H. (1989). The representation of pitch in a neural net model of pitch classification. *Computer Music Journal, 13*, 12–26.

Lahav, A., Saltzman, E., & Schlaug, G. (2007). Action representation of sound: Audiomotor recognition network while listening to newly acquired actions. *Journal of Neuroscience, 27*(2), 308–314.

Laporte, G., & Osman, I. H. (1995). Routing problems: A bibliography. *Annals of Operations Research, 61*, 227–262.

Large, E. W., & Kolen, J. F. (1994). Resonance and the perception of musical meter. *Connection Science, 6*, 177–208.

Large, E. W., Palmer, C., & Pollack, J. B. (1995). Reduced memory representations for music. *Cognitive Science, 19*(1), 53–96.

Lavington, S. H. (1980). *Early British computers: The story of vintage computers and the people who built them*. Bedford, MA: Digital Press.

Lawler, E. L. (1985). *The traveling salesman problem: A guided tour of combinatorial optimization*. Chichester, West Sussex, UK; New York, NY: Wiley.

Leahey, T. H. (1987). *A history of psychology* (2nd ed.). Englewood Cliffs, NJ: Prentice Hall.

Leahey, T. H. (1992). The mythical revolutions of American psychology. *American Psychologist, 47*(2), 308–318.

Lee, E. M. (1916). *The story of the symphony.* London, UK: Walter Scott Publishing Co., Ltd.

Lee, J. Y., Shin, S. Y., Park, T. H., & Zhang, B. T. (2004). Solving traveling salesman problems with DNA molecules encoding numerical values. *Biosystems, 78*(1–3), 39–47.

Lee, V. L. (1994). Organisms, things done, and the fragmentation of psychology. *Behavior and Philosophy, 22*(2), 7–48.

Leibniz, G. W. (1902). Leibniz: Discourse on metaphysics, correspondence with Arnauld, and monadology. (G. R. Montgomery, Trans.). La Salle, IL: The Open Court Publishing Co. (Original work published 1714)

Leibovic, K. N. (1969). *Information processing in the nervous system.* New York, NY: Springer-Verlag.

Leman, M. (1991). The ontogenesis of tonal semantics: Results of a computer study. In P. M. Todd & D. G. Loy (Eds.), *Music and connectionism* (pp. 100–127). Cambridge, MA: MIT Press.

Leman, M. (2008). *Embodied music cognition and mediation technology.* Cambridge, MA: MIT Press.

Lepore, E., & Pylyshyn, Z. W. (1999). *What is cognitive science?* Malden, MA: Blackwell Publishers.

Lerdahl, F. (2001). *Tonal pitch space.* New York, NY: Oxford University Press.

Lerdahl, F., & Jackendoff, R. (1983). *A generative theory of tonal music.* Cambridge, MA: MIT Press.

Lerdahl, F., & Krumhansl, C. L. (2007). Modeling tonal tension. *Music Perception, 24*(4), 329–366.

Lettvin, J. Y., Maturana, H. R., McCulloch, W. S., & Pitts, W. H. (1959). What the frog's eye tells the frog's brain. *Proceedings of the IRE, 47*(11), 1940–1951.

Levin, I. (2002). *The Stepford wives.* New York, NY: Perennial. (Original work published 1969)

Levine, M. (1989). *The jazz piano book.* Petaluma, CA: Sher Music Co.

Lévi-Strauss, C. (1966). *The savage mind.* (J.Weightman & D.Weightman, Trans.). Chicago, IL: University of Chicago Press.

Levitan, I. B., & Kaczmarek, L. K. (1991). *The neuron: Cell and molecular biology.* New York, NY: Oxford University Press.

Levitin, D. J. (2006). *This is your brain on music.* New York, NY: Dutton.

Lewandowsky, S. (1993). The rewards and hazards of computer simulations. *Psychological Science, 4,* 236–243.

Lewis, C. I. (1918). *A survey of symbolic logic.* Berkeley, CA: University of California Press.

Lewis, C. I. (1932). Alternative systems of logic. *The Monist, 42*(4), 481–507.

Lewis, C. I., & Langford, C. H. (1959). *Symbolic logic* (2nd ed.). New York, NY: Dover Publications.

Lewis, J. P. (1991). Creation by refinement and the problem of algorithmic music composition. In P. M. Todd & D. G. Loy (Eds.), *Music and connectionism* (pp. 212–228). Cambridge, MA: MIT Press.

Lidov, D. (2005). *Is language a music?: Writings on musical form and signification.* Bloomington, IN: Indiana University Press.

Lieberman, M. D. (2007). Social cognitive neuroscience: A review of core processes. *Annual Review of Psychology, 58,* 259–289.

Lightfoot, D. W. (1989). The child's trigger experience: Degree-o learnability. *Behavioral and Brain Sciences, 12*(2), 321–334.

Lindsay, P. H., & Norman, D. A. (1972). *Human information processing.* New York, NY: Academic Press.

Lippmann, R. P. (1987). An introduction to computing with neural nets. *IEEE ASSP Magazine, April,* 4–22.

Lippmann, R. P. (1989). Pattern classification using neural networks. *IEEE Communications Magazine, November,* 47–64.

Liu, N. H., Hsieh, S. J., & Tsai, C. F. (2010). An intelligent music playlist generator based on the time parameter with artificial neural networks. *Expert Systems with Applications, 37*(4), 2815–2825.

Liversidge, A. (1993). Profile of Claude Shannon. In N. J. A. Sloane & A. D. Wyner (Eds.), *Claude Elwood Shannon: Collected Papers.* New York, NY: IEEE Press.

Livingstone, M., & Hubel, D. (1988). Segregation of form, color, movement and depth: Anatomy, physiology, and perception. *Science, 240,* 740–750.

Lobay, A., & Forsyth, D. A. (2006). Shape from texture without boundaries. *International Journal of Computer Vision, 67*(1), 71–91.

Locke, J. (1977). *An essay concerning human understanding.* London: J. M. Dent & Sons. (Original work published 1706)

Longo, N. (1964). Probability-learning and habit-reversal in the cockroach. *American Journal of Psychology, 77*(1), 29–41.

Longyear, R. M. (1988). *Nineteenth-century Romanticism in music* (3rd ed.). Englewood Cliffs, NJ: Prentice Hall.

Lopez, N., Nunez, M., & Pelayo, F. L. (2007). A formal specification of the memorization process. *International Journal of Cognitive Informatics and Natural Intelligence, 1,* 47–60.

Lorayne, H. (1985). *Harry Lorayne's page-a -minute memory book.* New York, NY: Holt, Rinehart and Winston.

Lorayne, H. (1998). *How to develop a super power memory.* Hollywood, FL: Lifetime Books.

Lorayne, H. (2007). *Ageless memory: Simple secrets for keeping your brain young.* New York, NY: Black Dog & Leventhal Publishers.

Lorayne, H., & Lucas, J. (1974). *The memory book.* New York, NY: Stein and Day.

Lovecraft, H. P. (1933). The dreams in the witch-house. *Weird Tales, 22*(1), 86–111.

Loy, D. G. (1991). Connectionism and musiconomy. In P. M. Todd & D. G. Loy (Eds.), *Music and connectionism* (pp. 20–36). Cambridge, MA: MIT Press.

Luce, R. D. (1986). *Response times: Their role in inferring elementary mental organization.* New York, NY: Oxford University Press

Luce, R. D. (1989). Mathematical psychology and the computer revolution. In J. A. Keats, R. Taft, R. A. Heath & S. H. Lovibond (Eds.), *Mathematical and theoretical systems* (pp. 123–138). Amsterdam, Netherlands: North-Holland.

Luce, R. D. (1999). Where is mathematical modeling in psychology headed? *Theory & Psychology, 9,* 723–737.

Lund, H., & Miglino, O. (1998). Evolving and breeding robots. In P. Husbands & J. A. Meyer (Eds.), *Evolutionary robotics: First European workshop, EvoRobot'98 : Paris, France, April 16–17, 1998, Proceedings* (pp. 192–210). Heidelberg and Berlin, Germany: Springer-Verlag.

Luria, A., Tsvetkova, L., & Futer, J. (1965). Aphasia in a composer. *Journal of Neurological Sciences, 2*, 288–292.

Lynch, J. C., Mountcastle, V. B., Talbot, W. H., & Yin, T. C. T. (1977). Parietal lobe mechanisms for directed visual attention. *Journal of Neurophysiology, 40*, 362–389.

MacCorquodale, K. (1970). On Chomsky's review of Skinner's *Verbal Behavior*. *Journal of the Experimental Analysis of Behavior, 13*(1), 83–99.

MacDorman, K. F., & Ishiguro, H. (2006). The uncanny advantage of using androids in cognitive and social science research. *Interaction Studies, 7*(3), 297–337.

Mach, E. (1959). *The analysis of sensations.* (C. M. Williams, Trans.). New York, NY: Dover. (Original work published 1897)

MacIver, M. A. (2008). Neuroethology: From morphological computation to planning. In P. Robbins & M. Aydede (Eds.), *The Cambridge handbook of situated cognition* (pp. 480–504). New York, NY: Cambridge University Press.

MacKay, D. (1969). *Information, mechanism and meaning.* Cambridge, MA: MIT Press.

Mackenzie, D. (2002). The science of surprise. *Discover, 23*(2), 59–62.

Maestre, E., Blaauw, M., Bonada, J., Guaus, E., & Perez, A. (2010). Statistical modeling of bowing control applied to violin sound synthesis. *IEEE Transactions on Audio Speech and Language Processing, 18*(4), 855–871.

Magnusson, T. (2009). Of epistemic tools: Musical instruments as cognitive extensions. *Organised Sound, 14*(2), 168–176.

Magnusson, T. (2010). Designing constraints: Composing and performing with digital musical systems. *Computer Music Journal, 34*(4), 62–73.

Mammone, R. J. (1993). *Artificial neural networks for speech and vision.* London, UK; New York, NY: Chapman & Hall.

Mandelbrot, B. B. (1983). *The fractal geometry of nature* (Rev. ed.). New York, NY: W.H. Freeman & Co.

Marcus, G. F. (1993). Negative evidence in language acquisition. *Cognition, 46*, 53–85.

Marquand, A. (1885). A new logical machine. *Proceedings of the American Academy of Arts and Sciences, 21*, 303–307.

Marr, D. (1976). Early processing of visual information. *Philosophical Transactions of the Royal Society of London, 275*, 483–524.

Marr, D. (1982). *Vision.* San Francisco, CA: W. H. Freeman.

Marr, D., & Hildreth, E. (1980). Theory of edge detection. *Proceedings of the Royal Society of London, B, 207*, 187–217.

Marr, D., & Nishihara, H. K. (1978). Representation and recognition of the spatial organization of three-dimensional shapes. *Proceedings of the Royal Society of London Series B-Biological Sciences, 200*(1140), 269–294.

Marr, D., Palm, G., & Poggio, T. (1978). Analysis of a cooperative stereo algorithm. *Biological Cybernetics, 28*(4), 223–239.

Marr, D., & Poggio, T. (1979). Computational theory of human stereo vision. *Proceedings of the Royal Society of London Series B-Biological Sciences, 204*(1156), 301–328.

Marr, D., & Ullman, S. (1981). Directional selectivity and its use in early visual processing. *Proceedings of the Royal Society of London, B-Biological Sciences, 211*, 151–180.

Marshall, M. T., Hartshorn, M., Wanderley, M. M., & Levitin, D. J. (2009). Sensor choice for parameter modulation in digital musical instruments: Empirical evidence from pitch modulation. *Journal of New Music Research, 38*(3), 241–253.

Martinez, J. L., & Derrick, B. E. (1996). Long-term potentiation and learning. *Annual Review of Psychology, 47,* 173–203.

Mataric, M. J. (1998). Using communication to reduce locality in distributed multiagent learning. *Journal of Experimental & Theoretical Artificial Intelligence, 10*(3), 357–369.

Maunsell, J. H. R., & Newsome, W. T. (1987). Visual processing in monkey extrastriate cortex. *Annual Review of Neuroscience, 10,* 363–401.

Mays, W. (1953). The first circuit for an electrical logic-machine. *Science, 118*(3062), 281–282.

Mazzoni, P., Andersen, R. A., & Jordan, M. I. (1991). A more biologically plausible learning rule for neural networks. *Proceedings of the National Academy of Sciences of the USA, 88*(10), 4433–4437.

McCawley, J. D. (1981). *Everything that linguists have always wanted to know about logic but were ashamed to ask.* Chicago, IL: University of Chicago Press.

McClelland, J. L. (1986). Resource requirements of standard and programmable nets. In D. Rumelhart & J. McClelland (Eds.), *Parallel distributed processing* (Vol. 1, pp. 460–487). Cambridge, MA: MIT Press.

McClelland, J. L., & Rumelhart, D. E. (1986). *Parallel distributed processing* (Vol. 2). Cambridge, MA: MIT Press.

McClelland, J. L., & Rumelhart, D. E. (1988). *Explorations in parallel distributed processing.* Cambridge, MA: MIT Press.

McClelland, J. L., Rumelhart, D. E., & Hinton, G. E. (1986). The appeal of parallel distributed processing. In D. Rumelhart & J. McClelland (Eds.), *Parallel distributed processing: Vol. 1. Foundations* (pp. 3–44). Cambridge, MA: MIT Press.

McCloskey, M. (1991). Networks and theories: The place of connectionism in cognitive science. *Psychological Science, 2,* 387–395.

McColl, H. (1880). Symbolical reasoning. *Mind, 5*(17), 45–60.

McCorduck, P. (1979). *Machines who think: A personal inquiry into the history and prospects of artificial intelligence.* San Francisco, CA: W.H. Freeman.

McCulloch, W. S. (1988a). *Embodiments of mind.* Cambridge, MA: MIT Press.

McCulloch, W. S. (1988b). What is a number, that a man may know it, and a man, that he may know a number? In W. S. McCulloch (Ed.), *Embodiments of mind* (pp. 1–18). Cambridge, MA: MIT Press.

McCulloch, W. S., & Pitts, W. (1943). A logical calculus of the ideas immanent in nervous activity. *Bulletin of Mathematical Biophysics, 5,* 115–133.

McEliece, R. J., Posner, E. C., Rodemich, E. R., & Venkatesh, S. S. (1987). The capacity of the Hopfield associative memory. *IEEE Transactions on Information Theory, 33*(4), 461–482.

McGeer, T. (1990). Passive dynamic walking. *International Journal of Robotics Research, 9*(2), 62–82.

McLuhan, M. (1994). *Understanding media: The extensions of man.* Cambridge, MA: MIT Press. (Original work published 1957)

McNaughton, B., Barnes, C. A., Gerrard, J. L., Gothard, K., Jung, M. W., Knierim, J. J. (1996). Deciphering the hippocampal polyglot: The hippocampus as a path integration system. *The Journal of Experimental Biology, 199,* 173–185.

McNeill, D. (2005). *Gesture and thought* (e-Pub ed.). Chicago, IL: University of Chicago Press.

Medler, D. A. (1998). A brief history of connectionism. *Neural Computing Surveys, 1,* 18–72.

Medler, D. A., & Dawson, M. R. W. (1994). Training redundant artificial neural networks: Imposing biology on technology. *Psychological Research, 57,* 54–62.

Medler, D. A., Dawson, M. R. W., & Kingstone, A. (2005). Functional localization and double dissociations: The relationship between internal structure and behavior. *Brain and Cognition, 57,* 146–150.

Melhuish, C., Sendova-Franks, A. B., Scholes, S., Horsfield, I., & Welsby, F. (2006). Ant-inspired sorting by robots: the importance of initial clustering. *Journal of the Royal Society Interface, 3*(7), 235–242.

Menary, R. (2008). *Cognitive integration: Mind and cognition unbounded.* New York, NY: Palgrave Macmillan.

Menary, R. (2010). *The extended mind.* Cambridge, MA: MIT Press.

Mendelson, E. (1970). *Schaum's outline of theory and problems of Boolean algebra and switching circuits.* New York, NY: McGraw-Hill.

Menzel, P., D'Aluisio, F., & Mann, C. C. (2000). *Robo sapiens: Evolution of a new species.* Cambridge, MA: MIT Press.

Merleau-Ponty, M. (1962). *Phenomenology of perception.* (C. Smith, Trans.). London, UK; New York, NY: Routledge.

Merriam, E. P., & Colby, C. L. (2005). Active vision in parietal and extrastriate cortex. *Neuroscientist, 11*(5), 484–493.

Merriam, E. P., Genovese, C. R., & Colby, C. L. (2003). Spatial updating in human parietal cortex. *Neuron, 39*(2), 361–373.

Metzinger, T. (2009). *The ego tunnel: The science of the mind and the myth of the self.* New York, NY: Basic Books.

Meyer, D. E., Glass, J. M., Mueller, S. T., Seymour, T. L., & Kieras, D. E. (2001). Executive-process interactive control: A unified computational theory for answering 20 questions (and more) about cognitive ageing. *European Journal of Cognitive Psychology, 13*(1–2), 123–164.

Meyer, D. E., & Kieras, D. E. (1997a). A computational theory of executive cognitive processes and multiple-task performance. 1. Basic mechanisms. *Psychological Review, 104*(1), 3–65.

Meyer, D. E., & Kieras, D. E. (1997b). A computational theory of executive cognitive processes and multiple-task performance. 2. Accounts of psychological refractory-period phenomena. *Psychological Review, 104*(4), 749–791.

Meyer, D. E., & Kieras, D. E. (1999). Précis to a practical unified theory of cognition and action: Some lessons from EPIC computational models of human multiple-task performance. *Attention and Performance Xvii, 17,* 17–88.

Meyer, D. E., Kieras, D. E., Lauber, E., Schumacher, E. H., Glass, J., Zurbriggen, E. (1995). Adaptive Executive Control: Flexible multiple-task performance without pervasive immutable response-selection bottlenecks. *Acta Psychologica, 90*(1–3), 163–190.

Meyer, L. B. (1956). *Emotion and meaning in music.* Chicago, IL: University of Chicago Press.

Milford, M. J. (2008). *Robot navigation from nature.* Berlin, Germany: Springer.

Mill, J. (1829). *Analysis of the phenomena of the human mind.* London, UK: Baldwin and Cradock.

Mill, J., & Mill, J. S. (1869). *Analysis of the phenomena of the human mind: A new edition, with notes illustrative and critical.* (Alexander Bain, Andrew Findlater, and George Grote, Eds.). London, UK: Longmans, Green, Reader, and Dyer.

Mill, J. S. (1848). *A system of logic, ratiocinative and inductive: Being a connected view of the principles of evidence and the methods of scientific investigation.* New York, NY: Harper & Brothers.

Miller, G. A. (1951). *Language and communication* (1st ed.). New York, NY: McGraw-Hill.

Miller, G. A. (2003). The cognitive revolution: a historical perspective. *Trends in Cognitive Sciences, 7*(3), 141–144.

Miller, G. A., Galanter, E., & Pribram, K. H. (1960). *Plans and the structure of behavior.* New York, NY: Henry Holt & Co.

Miller, L. K. (1989). *Musical savants: Exceptional skill in the mentally retarded.* Hillsdale, NJ: Lawrence Erlbaum Associates.

Miller, N. Y. (2009). Modeling the effects of enclosure size on geometry learning. *Behavioural Processes, 80,* 306–313.

Miller, N. Y., & Shettleworth, S. J. (2007). Learning about environmental geometry: An associative model. *Journal of Experimental Psychology: Animal Behavior Processes, 33,* 191–212.

Miller, N. Y., & Shettleworth, S. J. (2008). An associative model of geometry learning: A modified choice rule. *Journal of Experimental Psychology—Animal Behavior Processes, 34*(3), 419–422.

Miller, R. R., Barnet, R. C., & Grahame, N. J. (1995). Assessment of the Rescorla-Wagner model. *Psychological Bulletin, 117*(3), 363–386.

Milligan, G. W., & Cooper, M. C. (1985). An examination of procedures for determining the number of clusters in a data set. *Psychometrika, 50,* 159–179.

Milner, P. M. (1957). The cell assembly: Mark II. *Psychological Review, 64*(4), 242–252.

Minsky, M. L. (1963). Steps toward artificial intelligence. In E. A. Feigenbaum & J. Feldman (Eds.), *Computers And Thought* (pp. 406-450). New York, NY: McGraw-Hill.

Minsky, M. L. (1972). *Computation: Finite and infinite machines.* Englewood Cliffs, NJ: Prentice Hall.

Minsky, M. L. (1981). Music, mind and meaning. *Computer Music Journal, 5*(3), 28–44.

Minsky, M. L. (1985). *The society of mind.* New York, NY: Simon & Schuster.

Minsky, M. L. (2006). *The emotion machine: Commonsense thinking, artificial intelligence, and the future of the human mind.* New York, NY: Simon & Schuster.

Minsky, M. L., & Papert, S. (1969). *Perceptrons: An introduction to computational geometry* (1st ed.). Cambridge, MA: MIT Press.

Minsky, M. L., & Papert, S. (1988). *Perceptrons: An introduction to computational geometry* (3rd ed.). Cambridge, MA: MIT Press.

Mitchell, M. (1996). *An introduction to genetic algorithms.* Cambridge, MA: MIT Press.

Mollenhoff, C. R. (1988). *Atanasoff: Forgotten father of the computer* (1st ed.). Ames, IA: Iowa State University Press.

Monelle, R. (2000). *The sense of music: Semiotic essays*. Princeton, NJ: Princeton University Press.

Monterola, C., Abundo, C., Tugaff, J., & Venturina, L. E. (2009). Prediction of potential hit song and musical genre using artificial neural networks. *International Journal of Modern Physics C, 20*(11), 1697–1718.

Moody, J., & Darken, C. J. (1989). Fast learning in networks of locally-tuned processing units. *Neural Computation, 1*, 281–294.

Moorhead, I. R., Haig, N. D., & Clement, R. A. (1989). An investigation of trained neural networks from a neurophysiological perspective. *Perception, 18*, 793–803.

Moravec, H. (1988). *Mind children*. Cambridge, MA: Harvard University Press.

Moravec, H. (1999). *Robot*. New York, NY: Oxford University Press.

Mori, M. (1970). Bukimi no tani [The uncanny valley]. *Energy, 7*, 33–35.

Morris, E. (Producer), & Morris, E. (Director). (1997). *Fast, cheap & out of control*. [Motion picture]. U.S.A.: Sony Pictures Classics.

Mostafa, M. M., & Billor, N. (2009). Recognition of Western style musical genres using machine learning techniques. *Expert Systems with Applications, 36*(8), 11378–11389.

Motter, B. C., & Mountcastle, V. B. (1981). The functional properties of the light-sensitive neurons of the posterior parietal cortex studied in waking monkeys: Foveal sparing and opponent vector organization. *The Journal of Neuroscience, 1*, 3–26.

Moyer, A. E. (1997). *Joseph Henry: The rise of an American scientist*. Washington, D.C.: Smithsonian Institution Press.

Mozer, M. C. (1991). Connectionist music composition based on melodic, stylistic, and psychophysical constraints. In P. M. Todd & D. G. Loy (Eds.), *Music and connectionism* (pp. 195–211). Cambridge, MA: MIT Press.

Mozer, M. C. (1994). Neural network music composition by prediction: Exploring the benefits of psychoacoustic constraints and multi-scale processing. *Connection Science, 6*, 247–280.

Mozer, M. C., & Smolensky, P. (1989). Using relevance to reduce network size automatically. *Connection Science, 1*, 3–16.

Muller, B., & Reinhardt, J. (1990). *Neural networks*. Berlin, Germany: Springer-Verlag.

Muñoz-Expósito, J. E., García-Galán, S., Ruiz-Reyes, N., & Vera-Candeas, P. (2007). Adaptive network-based fuzzy inference system vs. other classification algorithms for warped LPC-based speech/music discrimination. *Engineering Applications of Artificial Intelligence, 20*(6), 783–793.

Murdock, B. B. (1982). A theory for the storage and retrieval of item and associative information. *Psychological Review, 89*, 609–626.

Nagashima, T., & Kawashima, J. (1997). Experimental study on arranging music by chaotic neural network. *International Journal of Intelligent Systems, 12*(4), 323–339.

Nathan, A., & Barbosa, V. C. (2008). V-like formations in flocks of artificial birds. *Artificial Life, 14*(2), 179–188.

Navon, D. (1976). Irrelevance of figural identity for resolving ambiguities in apparent motion. *Journal of Experimental Psychology: Human Perception and Performance, 2*, 130–138.

Neisser, U. (1967). *Cognitive psychology*. New York, NY: Appleton-Century-Crofts.

Neisser, U. (1976). *Cognition and reality: Principles and implications of cognitive psychology*. San Francisco, CA: W. H. Freeman.

Neukom, H. (2006). The second life of ENIAC. *IEEE Annals of the History of Computing, 28*(2), 4–16.

Newell, A. (1973). Production systems: Models of control structures. In W. G. Chase (Ed.), *Visual information processing* (pp. 463–526). New York, NY: Academic Press.

Newell, A. (1980). Physical symbol systems. *Cognitive Science, 4*, 135–183.

Newell, A. (1982). The knowledge level. *Artificial Intelligence, 18*(1), 87–127.

Newell, A. (1990). *Unified theories of cognition*. Cambridge, MA: Harvard University Press.

Newell, A. (1993). Reflections on the knowledge level. *Artificial Intelligence, 59*(1–2), 31–38.

Newell, A., Shaw, J. C., & Simon, H. A. (1958). Elements of a theory of human problem solving. *Psychological Review, 65*, 151–166.

Newell, A., & Simon, H. A. (1961). Computer simulation of human thinking. *Science, 134*(349), 2011–2017.

Newell, A., & Simon, H. A. (1972). *Human problem solving*. Englewood Cliffs, NJ: Prentice Hall.

Newell, A., & Simon, H. A. (1976). Computer science as empirical inquiry: Symbols and search. *Communications of the ACM, 19*(3), 113–126.

Newport, E. L., Gleitman, H., & Gleitman, L. R. (1977). Mother, I'd rather do it myself: Some effects and noneffects of maternal speech style. In C. Snow & C. Ferguson (Eds.), *Talking to children: Language input and acquisition*. Cambridge, MA: Cambridge University Press.

Nilsson, N. J. (1980). *Principles of artificial intelligence*. Los Altos, CA: Morgan Kaufman.

Nilsson, N. J. (1984). *Shakey the robot*. Menlo Park, CA: Stanford Research Institute.

Niv, Y., Joel, D., Meilijson, I., & Ruppin, E. (2002). Evolution of reinforcement learning in uncertain environments: A simple explanation for complex foraging behaviors. *Adaptive Behavior, 10*(1), 5–24.

Noë, A. (2002). Is the visual world a grand illusion? *Journal of Consciousness Studies, 9*(5–6), 1–12.

Noë, A. (2004). *Action in perception*. Cambridge, MA: MIT Press.

Noë, A. (2009). *Out of our heads* (1st ed.). New York, NY: Hill and Wang.

Nolfi, S. (2002). Power and limits of reactive agents. *Neurocomputing, 42*, 119–145.

Nolfi, S., & Floreano, D. (2000). *Evolutionary robotics*. Cambridge, MA: MIT Press.

Norman, D. A. (1980). Twelve issues for cognitive science. *Cognitive Science, 4*, 1–32.

Norman, D. A. (1993). Cognition in the head and in the world: An introduction to the special issue on situated action. *Cognitive Science, 17*(1), 1–6.

Norman, D. A. (1998). *The invisible computer*. Cambridge, MA: MIT Press.

Norman, D. A. (2002). *The design of everyday things* (1st Basic paperback ed.). New York, NY: Basic Books.

Norman, D. A. (2004). *Emotional design: Why we love (or hate) everyday things*. New York, NY: Basic Books.

Nyman, M. (1999). *Experimental music: Cage and beyond* (2nd ed.). Cambridge, UK; New York, NY: Cambridge University Press.

Oaksford, M., & Chater, N. (1991). Against logicist cognitive science. *Mind & Language, 6*, 1–38.

Oaksford, M., & Chater, N. (1998). *Rationality in an uncertain world: Essays on the cognitive science of human reasoning.* Hove, East Sussex, UK: Psychology Press.

Oaksford, M., & Chater, N. (2001). The probabilistic approach to human reasoning. *Trends in Cognitive Sciences, 5*(8), 349–357.

Oaksford, M., Chater, N., & Stenning, K. (1990). Connectionism, classical cognitive science and experimental psychology. *AI & Society, 4*(1), 73–90.

Ochsner, K. N., & Lieberman, M. D. (2001). The emergence of social cognitive neuroscience. *American Psychologist, 56*(9), 717–734.

Ohkuma, Y. (1986). A comparison of image-induced and perceived Müller-Lyer illusion. *Journal of Mental Imagery, 10*, 31–38.

Ohta, H., Yamakita, M., & Furuta, K. (2001). From passive to active dynamic walking. *International Journal of Robust and Nonlinear Control, 11*(3), 287–303.

Okamoto, A., Tanaka, K., & Saito, I. (2004). DNA logic gates. *Journal of the American Chemical Society, 126*(30), 9458–9463.

O'Keefe, J., & Dostrovsky, J. (1971). The hippocampus as a spatial map: Preliminary evidence from unit activity in the freely moving rat. *Brain Research, 34*, 171–175.

O'Keefe, J., & Nadel, L. (1978). *The hippocampus as a cognitive map.* Oxford, UK: Clarendon Press.

Olazaran, M. (1996). A sociological study of the official history of the perceptrons controversy. *Social Studies of Science, 26*(3), 611–659.

Omlin, C. W., & Giles, C. L. (1996). Extraction of rules from discrete-time recurrent neural networks. *Neural Networks, 9*, 41–52.

O'Modhrain, S. (2011). A framework for the evaluation of digital musical instruments. *Computer Music Journal, 35*(1), 28–42.

O'Regan, J. K., Deubel, H., Clark, J. J., & Rensink, R. A. (2000). Picture changes during blinks: Looking without seeing and seeing without looking. *Visual Cognition, 7*(1–3), 191–211.

O'Reilly, R. C. (1996). Biologically plausible error-driven learning using local activation differences: The generalized recirculation algorithm. *Neural Computation, 8*(5), 895–938.

O'Reilly, R. C., & Munakata, Y. (2000). *Computational explorations in cognitive neuroscience: Understanding the mind by simulating the brain.* Cambridge, MA: MIT Press.

Oreskes, N., Shrader-Frechette, K., & Belitz, K. (1994). Verification, validation, and confirmation of numerical models in the earth sciences. *Science, 263*, 641–646.

Osherson, D. N. (1995). *An invitation to cognitive science* (2nd ed.). Cambridge, MA: MIT Press (3 volume set).

Osherson, D. N., Stob, M., & Weinstein, S. (1986). *Systems that learn.* Cambridge, MA: MIT Press.

Page, M. P. A. (1994). Modeling the perception of musical sequences with self-organizing neural networks. *Connection Science, 6*, 223–246.

Paivio, A. (1969). Mental imagery in associative learning and memory. *Psychological Review, 76*, 241–263.

Paivio, A. (1971). *Imagery and verbal processes.* New York, NY: Holt, Rinehart & Winston.

Paivio, A. (1986). *Mental representations: A dual-coding approach.* New York, NY: Oxford University Press.

Pampalk, E., Dixon, S., & Widmer, G. (2004). Exploring music collections by browsing different views. *Computer Music Journal, 28*(2), 49–62.

Pao, Y.-H. (1989). *Adaptive pattern recognition and neural networks.* Reading, MA: Addison-Wesley.

Papert, S. (1980). *Mindstorms: Children, computers and powerful ideas.* New York, NY: Basic Books.

Papert, S. (1988). One AI or many? *Daedalus, 117*(1), 1–14.

Paradiso, J. A. (1999). The brain opera technology: New instruments and gestural sensors for musical interaction and performance. *Journal of New Music Research, 28*(2), 130–149.

Parker, C. A. C., Zhang, H., & Kube, C. R. (2003). Blind bulldozing: Multiple robot nest construction. Paper presented at the Conference on Intelligent Robots and Systems, Las Vegas, NV.

Parker, L. E. (1998). ALLIANCE: An architecture for fault tolerant multirobot cooperation. *IEEE Transactions on Robotics and Automation, 14*(2), 220–240.

Parker, L. E. (2001). Evaluating success in autonomous multi-robot teams: Experiences from ALLIANCE architecture implementations. *Journal of Experimental & Theoretical Artificial Intelligence, 13*(2), 95–98.

Parkes, A. (2002). Introduction to languages, machines and logic: Computable languages, abstract machines and formal logic. London, UK: Springer.

Parncutt, R., Sloboda, J. A., Clarke, E. F., Raekallio, M., & Desain, P. (1997). An ergonomic model of keyboard fingering for melodic fragments. *Music Perception, 14*(4), 341–382.

Patel, A. D. (2003). Language, music, syntax and the brain. *Nature Neuroscience, 6*(7), 674–681.

Pavlov, I. P. (1927). *Conditioned reflexes.* (G.V. Anrep, Trans.). New York, NY: Oxford University Press.

Peano, G. (1973). *Selected works of Giuseppe Peano.* (H.C. Kennedy, Trans.). Toronto, ON: University of Toronto Press. (Original work published 1889)

Peirce, C. S. (1885). On the algebra of logic: A contribution to the philosophy of notation *American Journal of Mathematics, 7*(2), 180–196.

Pekkilä, E., Neumeyer, D., & Littlefield, R. (2006). *Music, meaning and media.* Imatra and Helsinki, Finland: International Semiotics Institute, Semiotic Society of Finland, University of Helsinki.

Pelaez, E. (1999). The stored-program computer: Two conceptions. *Social Studies of Science, 29*(3), 359–389.

Pelisson, D., Prablanc, C., Goodale, M. A., & Jeannerod, M. (1986). Visual control of reaching movements without vision of the limb, II: Evidence of fast unconscious processes correcting the trajectory of the hand to the final position of a double-step stimulus. *Experimental Brain Research, 62*(2), 303–311.

Peretz, I. (2009). Music, language and modularity framed in action. *Psychologica Belgica, 49*(2–3), 157–175.

Peretz, I., Ayotte, J., Zatorre, R. J., Mehler, J., Ahad, P., Penhune, V. B. (2002). Congenital amusia: A disorder of fine-grained pitch discrimination. *Neuron, 33*(2), 185–191.

Peretz, I., & Coltheart, M. (2003). Modularity of music processing. *Nature Neuroscience, 6*(7), 688–691.

Peretz, I., Cummings, S., & Dube, M. P. (2007). The genetics of congenital amusia (tone deafness): A family-aggregation study. *American Journal of Human Genetics, 81*(3), 582–588.

Peretz, I., & Hyde, K. L. (2003). What is specific to music processing? Insights from congenital amusia. *Trends in Cognitive Sciences, 7*(8), 362–367.

Peretz, I., Kolinsky, R., Tramo, M., Labrecque, R., Hublet, C., Demeurisse, G. (1994). Functional dissociations following bilateral lesions of auditory cortex. *Brain, 117*, 1283–1301.

Peretz, I., & Zatorre, R. J. (2003). *The cognitive neuroscience of music.* Oxford, UK; New York, NY: Oxford University Press.

Peretz, I., & Zatorre, R. J. (2005). Brain organization for music processing. *Annual Review of Psychology, 56*, 1–26.

Perner, J., Gschaider, A., Kuhberger, A., & Schrofner, S. (1999). Predicting others through simulation or by theory? A method to decide. *Mind & Language, 14*(1), 57–79.

Perrett, D. I., Mistlin, A. J., & Chitty, A. J. (1987). Visual neurons responsive to faces. *Trends in Neurosciences, 10*(9), 358–364.

Perrett, D. I., Rolls, E. T., & Caan, W. (1982). Visual neurones responsive to faces in the monkey temporal cortex. *Experimental Brain Research, 47*(3), 329–342.

Peters, S. (1972). The projection problem: How is a grammar to be selected? In S. Peters (Ed.), *Goals of linguistic theory* (pp. 171–188). Englewood Cliffs, NJ: Prentice Hall.

Peterson, C. R., & Beach, L. R. (1967). Man as an intuitive statistician. *Psychological Bulletin, 68*(1), 29–46.

Pfeifer, R., & Scheier, C. (1999). *Understanding intelligence.* Cambridge, MA: MIT Press.

Phattanasri, P., Chiel, H. J., & Beer, R. D. (2007). The dynamics of associative learning in evolved model circuits. *Adaptive Behavior, 15*(4), 377–396.

Piaget, J. (1929). *The child's conception of the world.* (J. Tomlinson & A. Tomlinson, Trans.). London, UK: K. Paul, Trench, Trubner & Co.

Piaget, J. (1970a). *The child's conception of movement and speed.* (G.E.T. Holloway & M.J. MacKenzie, Trans.). London, UK: Routledge & K. Paul.

Piaget, J. (1970b). *Psychology and epistemology.* (A. Rosin, Trans.). Harmondsworth, UK: Penguin Books.

Piaget, J. (1972). *The child and reality.* (A. Rosin, Trans.). Harmondsworth, UK: Penguin Books.

Piaget, J., & Inhelder, B. (1969). *The psychology of the child.* (H. Weaver, Trans.). London, UK: Routledge & Kegan Paul.

Pierce, J. R. (1993). Looking back: Claude Elwood Shannon. *IEEE Potentials, 12*(4), 38–40.

Pike, R. (1984). Comparison of convolution and matrix distributed memory systems for associative recall and recognition. *Psychological Review, 91*, 281–294.

Pinker, S. (1979). Formal models of language learning. *Cognition, 7*, 217–283.

Pinker, S. (1985). *Visual cognition* (1st MIT Press ed.). Cambridge, MA: MIT Press.

Pinker, S. (1994). *The language instinct.* New York, NY: Morrow.

Pinker, S. (1997). *How the mind works.* New York, NY: W.W. Norton.

Pinker, S. (1999). *Words and rules: The ingredients of language* (1st ed.). New York, NY: Basic Books.

Pinker, S. (2002). *The blank slate.* New York, NY: Viking.

Pinker, S., & Prince, A. (1988). On language and connectionism: Analysis of a parallel distributed processing model of language acquisition. *Cognition, 28*, 73–193.

Plantinga, L. (1984). *Romantic music: A history of musical style in nineteenth-century Europe* (1st ed.). New York, NY: W.W. Norton.

Pleasants, H. (1955). *The agony of modern music*. New York: Simon and Schuster.

Poggio, T., & Girosi, F. (1990). Regularization algorithms for learning that are equivalent to multilayer networks. *Science, 247*, 978–982.

Poggio, T., Torre, V., & Koch, C. (1985). Computational vision and regularization theory. *Nature, 317*, 314–319.

Polanyi, M. (1966). *The tacit dimension* (1st ed.). Garden City, NY: Doubleday.

Poldrack, R. A., Clark, J., Pare-Blagoev, E. J., Shohamy, D., Moyano, J. C., Myers, C. (2001). Interactive memory systems in the human brain. *Nature, 414*(6863), 546–550.

Pomerleau, D. A. (1991). Efficient training of artificial neural networks for autonomous navigation. *Neural Computation, 3*, 88–97.

Popper, K. (1978). Natural selection and the emergence of mind. *Dialectica, 32*, 339–355.

Port, R. F., & van Gelder, T. (1995a). It's about time: An overview of the dynamical approach to cognition. In R. F. Port & T. van Gelder (Eds.), *Mind as motion: Explorations in the dynamics of cognition* (pp. 1–43). Cambridge, MA: MIT Press.

Port, R. F., & van Gelder, T. (1995b). *Mind as motion: Explorations in the dynamics of cognition*. Cambridge, MA: MIT Press.

Posner, M. (1978). *Chronometric explorations of mind*. Hillsdale, NJ: Lawrence Erlbaum Associates.

Posner, M. (1991). *Foundations of cognitive science*. Cambridge, MA: MIT Press.

Post, E. L. (1921). Introduction to a general theory of elementary propositions. *American Journal of Mathematics, 43*, 163–185.

Post, E. L. (1936). Finite combinatory processes: Formulation I. *Journal of Symbolic Logic, 1*, 103–105.

Postman, L., & Phillips, L. W. (1965). Short-term temporal changes in free recall. *Quarterly Journal of Experimental Psychology, 17*, 132–138.

Potter, K. (2000). *Four musical minimalists: La Monte Young, Terry Riley, Steve Reich, Philip Glass*. Cambridge, UK; New York, NY: Cambridge University Press.

Pring, L., Woolf, K., & Tadic, V. (2008). Melody and pitch processing in five musical savants with congenital blindness. *Perception, 37*(2), 290–307.

Punnen, A. P. (2002). The traveling salesman problem: Applications, formulations, and variations. In G. Gutin & A. P. Punnen (Eds.), The traveling salesman problem and its variations (pp. 1–28). Dordrecht, Netherlands; Boston, MA: Kluwer Academic Publishers.

Purwins, H., Herrera, P., Grachten, M., Hazan, A., Marxer, R., & Serra, X. (2008). Computational models of music perception and cognition I: The perceptual and cognitive processing chain. *Physics of Life Reviews, 5*(3), 151–168.

Pylyshyn, Z. W. (1973). What the mind's eye tells the mind's brain: A critique of mental imagery. *Psychological Bulletin, 80*, 1–24.

Pylyshyn, Z. W. (1979a). Metaphorical imprecision and the "top-down" research strategy. In A. Ortony (Ed.), *Metaphor and thought* (pp. 420–436). Cambridge, UK: Cambridge University Press.

Pylyshyn, Z. W. (1979b). Rate of mental rotation of images: Test of a holistic analog process. *Memory & Cognition, 7*(1), 19–28.

Pylyshyn, Z. W. (1980). Computation and cognition: Issues in the foundations of cognitive science. *Behavioral and Brain Sciences, 3*(1), 111–132.

Pylyshyn, Z. W. (1981a). The imagery debate: Analogue media versus tacit knowledge. *Psychological Review, 88*(1), 16–45.

Pylyshyn, Z. W. (1981b). Psychological explanations and knowledge-dependent processes. *Cognition, 10*(1–3), 267–274.

Pylyshyn, Z. W. (1984). *Computation and cognition*. Cambridge, MA: MIT Press.

Pylyshyn, Z. W. (1987). *The robot's dilemma: The frame problem in artificial intelligence*. Norwood, NJ: Ablex.

Pylyshyn, Z. W. (1989). The role of location indexes in spatial perception: A sketch of the FINST spatial-index model. *Cognition, 32*, 65–97.

Pylyshyn, Z. W. (1991). The role of cognitive architectures in theories of cognition. In K. VanLehn (Ed.), *Architectures for intelligence* (pp. 189–223). Hillsdale, NJ: Lawrence Erlbaum Associates.

Pylyshyn, Z. W. (1994). Some primitive mechanisms of spatial attention. *Cognition, 50*(1–3), 363–384.

Pylyshyn, Z. W. (1999). Is vision continuous with cognition?: The case for cognitive impenetrability of visual perception. *Behavioral and Brain Sciences, 22*(3), 341–423.

Pylyshyn, Z. W. (2000). Situating vision in the world. *Trends in Cognitive Sciences, 4*(5), 197–207.

Pylyshyn, Z. W. (2001). Visual indexes, preconceptual objects, and situated vision. *Cognition, 80*(1–2), 127–158.

Pylyshyn, Z. W. (2003a). Explaining mental imagery: Now you see it, now you don't: Reply to Kosslyn et al. *Trends in Cognitive Sciences, 7*(3), 111–112.

Pylyshyn, Z. W. (2003b). Return of the mental image: Are there really pictures in the brain? *Trends in Cognitive Sciences, 7*(3), 113–118.

Pylyshyn, Z. W. (2003c). *Seeing and visualizing: It's not what you think*. Cambridge, MA: MIT Press.

Pylyshyn, Z. W. (2006). Some puzzling findings in multiple object tracking (MOT), II. Inhibition of moving nontargets. *Visual Cognition, 14*(2), 175–198.

Pylyshyn, Z. W. (2007). *Things and places: How the mind connects with the world*. Cambridge, MA: MIT Press.

Pylyshyn, Z. W., & Annan, V. (2006). Dynamics of target selection in multiple object tracking (MOT). *Spatial Vision, 19*(6), 485–504.

Pylyshyn, Z. W., & Cohen, J. (1999). Imagined extrapolation of uniform motion is not continuous. *Investigative Ophthalmology & Visual Science, 40*(4), S808.

Pylyshyn, Z. W., Haladjian, H. H., King, C. E., & Reilly, J. E. (2008). Selective nontarget inhibition in multiple object tracking. *Visual Cognition, 16*(8), 1011–1021.

Pylyshyn, Z. W., & Storm, R. (1988). Tracking of multiple independent targets: Evidence for a parallel tracking mechanism. *Spatial Vision, 3*, 1–19.

Quinlan, J. R. (1986). Induction of decision trees. *Machine Learning, 1*, 81–106.

Quinlan, P. (1991). *Connectionism and psychology*. Chicago, IL: University of Chicago Press.

Radford, A. (1981). *Transformational syntax: A student's guide to Chomsky's extended standard theory*. Cambridge, UK; New York, NY: Cambridge University Press.

Raftopoulos, A. (2001). Is perception informationally encapsulated? The issue of the theory-ladenness of perception. *Cognitive Science, 25*(3), 423–451.

Ramachandran, V. S., & Anstis, S. M. (1986). The perception of apparent motion. *Scientific American, 254*, 102–109.

Ramsey, W., Stich, S. P., & Rumelhart, D. E. (1991). *Philosophy and connectionist theory.* Hillsdale, NJ: Lawrence Erlbaum Associates.

Ransom-Hogg, A., & Spillmann, R. (1980). Perceptive field size in fovea and periphery of the light- and dark-adapted retina. *Vision Research, 20*, 221–228.

Rasamimanana, N., & Bevilacqua, F. (2008). Effort-based analysis of bowing movements: Evidence of anticipation effects. *Journal of New Music Research, 37*(4), 339–351.

Ratcliffe, M. (2007). *Rethinking commonsense psychology: A critique of folk psychology, theory of mind and simulation.* Basingstoke, UK; New York, NY: Palgrave Macmillan.

Ratner, L. G. (1992). *Romantic music: Sound and syntax.* New York, NY: Schirmer Books.

Rauschecker, J. P., & Scott, S. K. (2009). Maps and streams in the auditory cortex: nonhuman primates illuminate human speech processing. *Nature Neuroscience, 12*(6), 718–724.

Reddy, M. J. (1979). The conduit metaphor: A case of frame conflict in our language about language. In A. Ortony (Ed.), *Metaphor and thought* (pp. 284–324). Cambridge, UK: Cambridge University Press.

Redish, A. D. (1999). *Beyond the cognitive map.* Cambridge, MA: MIT Press.

Redish, A. D., & Touretzky, D. S. (1999). Separating hippocampal maps. In B. N., K. J. Jeffery & J. O'Keefe (Eds.), *The hippocampal and parietal foundations of spatial cognition* (pp. 203–219). Oxford, UK: Oxford University Press.

Reeve, R. E., & Webb, B. H. (2003). New neural circuits for robot phonotaxis. *Philosophical Transactions of the Royal Society of London Series A–Mathematical Physical and Engineering Sciences, 361*(1811), 2245–2266.

Reich, S. (1974). *Writings about music.* Halifax, NS: Press of the Nova Scotia College of Art and Design.

Reich, S. (2002). *Writings on music, 1965–2000.* Oxford, UK; New York, NY: Oxford University Press.

Reid, T. R. (2001). *The chip: How two Americans invented the microchip and launched a revolution* (Rev. ed.). New York, NY: Random House.

Reitwiesner, G. W. (1997). The first operating system for the EDVAC. *IEEE Annals of the History of Computing, 19*(1), 55–59.

Renals, S. (1989). Radial basis function network for speech pattern classification. *Electronics Letters, 25*, 437–439.

Rescorla, R. A. (1967). Pavlovian conditioning and its proper control procedures. *Psychological Review, 74*(1), 71–80.

Rescorla, R. A. (1968). Probability of shock in presence and absence of CS in fear conditioning. *Journal of Comparative and Physiological Psychology, 66*(1), 1–5.

Rescorla, R. A., & Wagner, A. R. (1972). A theory of Pavlovian conditioning: Variations in the effectiveness of reinforcement and nonreinforcement. In A. H. Black & W. F. Prokasy (Eds.), *Classical conditioning II: Current research and theory* (pp. 64–99). New York, NY: Appleton-Century-Crofts.

Révész, G. E. (1983). *Introduction to formal languages*. New York, NY: McGraw-Hill.

Reynolds, A. G., & Flagg, P. W. (1977). *Cognitive psychology*. Cambridge, MA: Winthrop Publishers.

Reynolds, C. W. (1987). Flocks, herds and schools: A distributed behavioral model. *Computer Graphics, 21*(4), 25–34.

Richards, W. (1988). *Natural computation*. Cambridge, MA: MIT Press.

Richardson, F. C. (2000). Overcoming fragmentation in psychology: A hermeneutic approach. *Journal of Mind and Behavior, 21*(3), 289–304.

Riedel, J. (1969). *Music of the Romantic period*. Dubuque, Iowa: W. C. Brown Co.

Rieser, J. J. (1989). Access to knowledge of spatial structure at novel points of observation. *Journal of Experimental Psychology: Learning Memory and Cognition, 15*(6), 1157–1165.

Ripley, B. D. (1996). *Pattern recognition and neural networks*. Cambridge, UK: Cambridge University Press.

Rips, L., Shoben, E. J., & Smith, E. E. (1973). Semantic distance and verification of semantic relations. *Journal of Verbal Learning and Verbal Behavior, 12*, 1–20.

Rizzolatti, G., & Craighero, L. (2004). The mirror-neuron system. *Annual Review of Neuroscience, 27*, 169–192.

Rizzolatti, G., Fogassi, L., & Gallese, V. (2006). Mirrors in the mind. *Scientific American, 295*(5), 54–61.

Robbins, P., & Aydede, M. (2009). *The Cambridge handbook of situated cognition*. Cambridge, UK; New York, NY: Cambridge University Press.

Robinson, D. L., Goldberg, M. E., & Stanton, G. B. (1978). Parietal association cortex in the primate: Sensory mechanisms and behavioural modulations. *Journal of Neurophysiology, 41*, 910–932.

Robinson, J. (1994). The expression and arousal of emotion in music. In P. Alperson (Ed.), *Musical worlds: New directions in the philosophy of music* (pp. 13–22). University Park, PA: Pennsylvania State University Press.

Robinson, J. (1997). *Music and meaning*. Ithaca, NY: Cornell University Press.

Robinson-Riegler, B., & Robinson-Riegler, G. (2003). *Readings in cognitive psychology: Applications, connections, and individual differences*. Boston, MA: Pearson Allyn & Bacon.

Rochester, N., Holland, J. H., Haibt, L. H., & Duda, W. L. (1956). Tests on a cell assembly theory of the action of the brain, using a large digital computer. *IRE Transactions on Information Theory, IT-2*, 80–93.

Rock, I. (1983). *The logic of perception*. Cambridge, MA: MIT Press.

Rohaly, A. M., & Buchsbaum, B. (1989). Global spatiochromatic mechanisms accounting for luminance variations in contrast sensitivity functions. *Journal of the Optical Society of America A, 6*, 312–317.

Rojas, R. (1996). *Neural networks: A systematic exploration*. Berlin, Germany: Springer.

Romney, A. K., Shepard, R. N., & Nerlove, S. B. (1972). *Multidimensional scaling: Theory and applications in the behavioral sciences, Volume II: Applications*. New York, NY: Seminar Press.

Rosen, C. (1988). *Sonata forms* (Rev. ed.). New York, NY: Norton.

Rosen, C. (1995). *The Romantic generation*. Cambridge, MA: Harvard University Press.

Rosen, C. (2002). *Piano notes: The world of the pianist*. New York, NY: Free Press.

Rosenblatt, F. (1958). The perceptron: A probabilistic model for information storage and organization in the brain. *Psychological Review, 65*(6), 386–408.

Rosenblatt, F. (1962). *Principles of neurodynamics*. Washington, D.C.: Spartan Books.

Rosenfeld, A., Hummel, R. A., & Zucker, S. W. (1976). Scene labeling by relaxation operations. *IEEE Transactions on Systems Man and Cybernetics, 6*(6), 420–433.

Rosner, B. S., & Narmour, E. (1992). Harmonic closure: Music theory and perception. *Music Perception, 9*(4), 383–411.

Ross, A. (2007). *The rest is noise: Listening to the twentieth century* (1st ed.). New York, NY: Farrar, Straus and Giroux.

Rowe, R. (2001). *Machine musicianship*. Cambridge, MA: MIT Press.

Roy, A. (2008). Connectionism, controllers, and a brain theory. *IEEE Transactions on Systems Man and Cybernetics Part A—Systems and Humans, 38*(6), 1434–1441.

Rubel, L. A. (1989). Digital simulation of analog computation and Church's thesis. *Journal of Symbolic Logic, 54*(3), 1011–1017.

Rubinstein, A. (1998). *Modeling bounded rationality*. Cambridge, MA: MIT Press.

Rumelhart, D. E., Hinton, G. E., & Williams, R. J. (1986a). Learning internal representations by error propagation. In D. E. Rumelhart & G. E. Hinton (Eds.), *Parallel distributed processing: Vol. 1. Foundations* (pp. 318–362). Cambridge, MA: MIT Press.

Rumelhart, D. E., Hinton, G. E., & Williams, R. J. (1986b). Learning representations by back-propagating errors. *Nature, 323*, 533–536.

Rumelhart, D. E., & McClelland, J. L. (1985). Levels indeed!: A response to Broadbent. *Journal of Experimental Psychology: General, 114*, 193–197.

Rumelhart, D. E., & McClelland, J. L. (1986a). On learning the past tenses of English verbs. In J. McClelland & D. E. Rumelhart (Eds.), *Parallel distributed processing: Vol. 2. Psychological and Biological Models* (pp. 216–271). Cambridge, MA: MIT Press.

Rumelhart, D. E., & McClelland, J. L. (1986b). PDP models and general issues in cognitive science. In D. E. Rumelhart & J. McClelland (Eds.), *Parallel distributed processing: Vol. 1. Foundations* (pp. 110–146). Cambridge, MA: MIT Press.

Rumelhart, D. E., & McClelland, J. L. (1986c). *Parallel distributed processing: Vol. 1. Foundations*. Cambridge, MA: MIT Press.

Rupert, R. D. (2009). *Cognitive systems and the extended mind*. Oxford, UK; New York, NY: Oxford University Press.

Russell, B. (1993). *Introduction to mathematical philosophy*. New York, NY: Dover Publications. (Original work published 1920)

Ryle, G. (1949). *The concept of mind*. London, UK: Hutchinson & Company.

Safa, A. T., Saadat, M. G., & Naraghi, M. (2007). Passive dynamic of the simplest walking model: Replacing ramps with stairs. *Mechanism and Machine Theory, 42*(10), 1314–1325.

Sahin, E., Cakmak, M., Dogar, M. R., Ugur, E., & Ucoluk, G. (2007). To afford or not to afford: A new formalization of affordances toward affordance-based robot control. *Adaptive Behavior, 15*(4), 447–472.

Sakata, H., Shibutani, H., Kawano, K., & Harrington, T. L. (1985). Neural mechanisms of space vision in the parietal association cortex of the monkey. *Vision Research, 25*, 453–463.

Samuels, R. (1998). Evolutionary psychology and the massive modularity hypothesis. *British Journal for the Philosophy of Science, 49*(4), 575–602.

Sandon, P. A. (1992). Simulating visual attention. *Journal of Cognitive Neuroscience, 2*, 213–231.

Sano, H., & Jenkins, B. K. (1989). A neural network model for pitch perception. *Computer Music Journal, 13*(3), 41–48.

Sapir, S. (2002). Gestural control of digital audio environments. *Journal of New Music Research, 31*(2), 119–129.

Sawyer, R. K. (2002). Emergence in psychology: Lessons from the history of non-reductionist science. *Human Development, 45*, 2–28.

Saxe, J. G. (1868). *The poems of John Godfrey Saxe.* Boston, MA: Ticknor and Fields.

Sayegh, S. I. (1989). Fingering for string instruments with the optimum path paradigm. *Computer Music Journal, 13*(3), 76–84.

Scarborough, D. L., Miller, B. O., & Jones, J. A. (1989). Connectionist models for tonal analysis. *Computer Music Journal, 13*(3), 49–55.

Scassellati, B. (2002). Theory of mind for a humanoid robot. *Autonomous Robots, 12*(1), 13–24.

Schenker, H. (1979). *Free composition* (E. Oster). New York, NY: Longman. (Original work published 1935)

Schlimmer, J. S. (1987). Concept acquisition through representational adjustment. Unpublished doctoral dissertation, University of California Irvine, Irvine, CA.

Schlinger, H. D. (2008). Long good-bye: Why B.F. Skinner's *Verbal Behavior* is alive and well on the 50th anniversary of its publication. *Psychological Record, 58*(3), 329–337.

Schmajuk, N. A. (1997). *Animal learning and cognition: A neural network approach.* Cambridge, UK; New York, NY: Cambridge University Press.

Schneider, W. (1987). Connectionism: Is it a paradigm shift for psychology? *Behavior Research Methods, Instruments, & Computers, 19*, 73–83.

Scholes, S., Wilson, M., Sendova-Franks, A. B., & Melhuish, C. (2004). Comparisons in evolution and engineering: The collective intelligence of sorting. *Adaptive Behavior, 12*(3–4), 147–159.

Scholl, B. J., Pylyshyn, Z. W., & Feldman, J. (2001). What is a visual object?: Evidence from target merging in multiple object tracking. *Cognition, 80*(1–2), 159–177.

Schultz, A. C., & Parker, L. E. (2002). *Multi-robot systems: From swarms to intelligent automata.* Dordrecht, Netherlands; Boston, UK: Kluwer Academic.

Schultz, D. P., & Schultz, S. E. (2008). *A history of modern psychology* (9th ed.). Belmont, CA: Thomson/Wadsworth.

Schuppert, M., Munte, T. F., Wieringa, B. M., & Altenmuller, E. (2000). Receptive amusia: Evidence for cross-hemispheric neural networks underlying music processing strategies. *Brain, 123*, 546–559.

Schwarz, K. R. (1996). *Minimalists.* London, UK: Phaidon.

Scoville, W. B., & Milner, B. (1957). Loss of recent memory after bilateral hippocampal lesions. *Journal of Neurology, Neurosurgery and Psychiatry, 20*, 11–21.

Scribner, S., & Tobach, E. (1997). *Mind and social practice: Selected writings of Sylvia Scribner.* Cambridge, UK; New York, NY: Cambridge University Press.

Searle, J. R. (1980). Minds, brains, and programs. *Behavioral and Brain Sciences, 3*, 417–424.

Searle, J. R. (1984). *Minds, brains and science.* Cambridge, MA: Harvard University Press.

Searle, J. R. (1990). Is the brain's mind a computer program? *Scientific American, 262,* 26–31.

Searle, J. R. (1992). *The rediscovery of the mind.* Cambridge, MA: MIT Press.

Sears, C. R., & Pylyshyn, Z. W. (2000). Multiple object tracking and attentional processing. *Canadian Journal of Experimental Psychology/Revue canadienne de psychologie experimentale, 54*(1), 1–14.

Seashore, C. E. (1967). *Psychology of music.* New York, NY: Dover Publications. (Original work published 1938)

Seidenberg, M. (1993). Connectionist models and cognitive theory. *Psychological Science, 4,* 228–235.

Seidenberg, M., & McClelland, J. (1989). A distributed, developmental model of word recognition and naming. *Psychological Review, 96,* 523–568.

Sejnowski, T. J., & Rosenberg, C. R. (1988). NETtalk: A parallel network that learns to read aloud. In J. A. Anderson & E. Rosenfeld (Eds.), *Neurocomputing: Foundations of research* (pp. 663–672). Cambridge, MA: MIT Press.

Selfridge, O. G. (1956). Pattern recognition and learning. In C. Cherry (Ed.), *Information theory* (pp. 345–353). London, UK: Butterworths Scientific Publications.

Shallice, T. (1988). *From neuropsychology to mental structure.* New York, NY: Cambridge University Press.

Shanks, D. R. (1995). *The psychology of associative learning.* Cambridge, UK: Cambridge University Press.

Shanks, D. R. (2007). Associationism and cognition: Human contingency learning at 25. *Quarterly Journal of Experimental Psychology, 60*(3), 291–309.

Shannon, C. E. (1938). A symbolic analysis of relay and switching circuits. *Transactions of the American Institute of Electrical Engineers, 57,* 713–723.

Shannon, C. E. (1948). A mathematical theory of communication. *The Bell System Technical Journal, 27,* 379–423, 623–656.

Shapiro, L. A. (2011). *Embodied cognition.* New York, NY: Routledge.

Sharkey, A. J. C. (2006). Robots, insects and swarm intelligence. *Artificial Intelligence Review, 26*(4), 255–268.

Sharkey, N. E. (1992). *Connectionist natural language processing.* Dordrecht, Netherlands; Boston, MA: Kluwer Academic Publishers.

Sharkey, N. E. (1997). The new wave in robot learning. *Robotics and Autonomous Systems, 22*(3–4), 179–185.

Sharkey, N. E., & Sharkey, A. (2009). Electro-mechanical robots before the computer. *Proceedings of the Institution of Mechanical Engineers Part C-Journal of Mechanical Engineering Science, 223*(1), 235–241.

Shelley, M. W. (1985). *Frankenstein.* Harmondsworth, Middlesex, England: Penguin Books. (Original work published 1818)

Shepard, R. N. (1984a). Ecological constraints on internal representation: Resonant kinematics of perceiving, imagining, thinking, and dreaming. *Psychological Review, 91*(4), 417–447.

Shepard, R. N. (1984b). Ecological constraints on internal representation: Resonant kinematics of perceiving, imagining, thinking, and dreaming. *Psychological Review, 91*, 417–447.

Shepard, R. N. (1990). *Mind sights: Original visual illusions, ambiguities, and other anomalies.* New York, NY: W. H. Freeman & Co.

Shepard, R. N., & Cooper, L. A. (1982). *Mental images and their transformations.* Cambridge, MA: MIT Press.

Shepard, R. N., & Metzler, J. (1971). Mental rotation of three-dimensional objects. *Science, 171*(3972), 701–703.

Shepard, R. N., Romney, A. K., & Nerlove, S. B. (1972). *Multidimensional scaling: Theory and applications in the behavioral sciences. Volume I: Theory.* New York, NY: Seminar Press.

Shibata, N. (1991). A neural network-based method for chord note scale association with melodies. *Nec Research & Development, 32*(3), 453–459.

Shiffrin, R. M., & Atkinson, R. C. (1969). Storage and retrieval processes in long-term memory. *Psychological Review, 76*(2), 179–193.

Shimansky, Y. P. (2009). Biologically plausible learning in neural networks: A lesson from bacterial chemotaxis. *Biological Cybernetics, 101*(5–6), 379–385.

Shosky, J. (1997). Russell's use of truth tables. *Russell: The Journal of the Bertrand Russell Archives, 17*(1), 11–26.

Siegelmann, H. T. (1999). *Neural networks and analog computation: Beyond the Turing limit.* Boston, MA: Birkhauser.

Siegelmann, H. T., & Sontag, E. D. (1991). Turing computability with neural nets. *Applied Mathematics Letters, 4*, 77–80.

Siegelmann, H. T., & Sontag, E. D. (1995). On the computational power of neural nets. *Journal of Computer and System Sciences, 50*, 132–150.

Simon, H. A. (1969). *The sciences of the artificial.* Cambridge, MA: MIT Press.

Simon, H. A. (1979). Information processing models of cognition. *Annual Review of Psychology, 30*, 363–396.

Simon, H. A. (1980). Cognitive science: The newest science of the artificial. *Cognitive Science, 4*, 33–46.

Simon, H. A. (1982). *Models of bounded rationality.* Cambridge, MA: MIT Press.

Simon, H. A., Egidi, M., & Marris, R. L. (1995). *Economics, bounded rationality and the cognitive revolution.* Aldershot, England; Brookfield, VT: E. Elgar.

Simon, H. A., & Newell, A. (1958). Heuristic problem solving: The next advance in operations research. *Operations Research, 6*, 1–10.

Simons, D. J., & Chabris, C. F. (1999). Gorillas in our midst: sustained inattentional blindness for dynamic events. *Perception, 28*(9), 1059–1074.

Singh, J. (1966). *Great ideas in information theory, language, and cybernetics.* New York, NY: Dover Publications.

Singh, M., & Hoffman, D. D. (1997). Constructing and representing visual objects. *Trends in Cognitive Sciences, 1*(3), 98–102.

Siqueira, P. H., Steiner, M. T. A., & Scheer, S. (2007). A new approach to solve the traveling salesman problem. *Neurocomputing, 70*(4–6), 1013–1021.

Skinner, B. F. (1957). *Verbal behavior*. New York, NY: Appleton-Century-Crofts.

Sloboda, J. A. (1985). *The musical mind: The cognitive psychology of music*. Oxford, UK: Oxford University Press.

Sloboda, J. A., Clarke, E. F., Parncutt, R., & Raekallio, M. (1998). Determinants of finger choice in piano sight-reading. *Journal of Experimental Psychology-Human Perception and Performance, 24*(1), 185–203.

Smiley, J. (2010). *The man who invented the computer: The biography of John Atanasoff, digital pioneer* (1st ed.). New York, NY: Doubleday.

Smith, E. E., & Osherson, D. N. (1995). *An invitation to cognitive science: Vol. 3. Thinking* (2nd ed.). Cambridge, MA: MIT Press.

Smith, J. C., Marsh, J. T., Greenberg, S., & Brown, W. S. (1978). Human auditory frequency-following responses to a missing fundamental. *Science, 201*(4356), 639–641.

Smolensky, P. (1988). On the proper treatment of connectionism. *Behavioral and Brain Sciences, 11*, 1–74.

Smolensky, P., & Legendre, G. (2006). *The harmonic mind: From neural computation to optimality-theoretic grammar*. Cambridge, MA: MIT Press.

Smythe, W. E., & McKenzie, S. A. (2010). A vision of dialogical pluralism in psychology. *New Ideas in Psychology, 28*(2), 227–234.

Snyder, B. (2000). *Music and memory: An introduction*. Cambridge, MA: MIT Press.

Sobel, D. (1999). *Galileo's daughter: A historical memoir of science, faith, and love*. New York, NY: Walker & Co.

Solso, R. L. (1995). *Cognitive psychology* (4th ed.). Boston, MA: Allyn and Bacon.

Sorabji, R. (2006). *Aristotle on memory* (2nd ed.). Chicago: University of Chicago Press.

Sovrano, V. A., Bisazza, A., & Vallortigara, G. (2003). Modularity as a fish (*Xenotoca eiseni*) views it: Conjoining geometric and nongeometric information for spatial reorientation. *Journal of Experimental Psychology-Animal Behavior Processes, 29*(3), 199–210.

Sparshoot, F. (1994). Music and feeling. In P. Alperson (Ed.), *Musical worlds: New directions in the philosophy of music* (pp. 23–36). University Park, PA: Pennsylvania State University Press.

Sperry, R. W. (1993). The impact and promise of the cognitive revolution. *American Psychologist, 48*(8), 878–885.

Squire, L. R. (1987). *Memory and brain*. New York, NY: Oxford University Press.

Squire, L. R. (1992). Declarative and nondeclarative memory: Multiple brain systems supporting learning and memory. *Journal of Cognitive Neuroscience, 4*, 232–243.

Squire, L. R. (2004). Memory systems of the brain: A brief history and current perspective. *Neurobiology of Learning and Memory, 82*(3), 171–177.

Stam, H. J. (2004). Unifying psychology: Epistemological act or disciplinary maneuver? *Journal of Clinical Psychology, 60*(12), 1259–1262.

Standage, T. (2002). *The Turk: The life and times of the famous eighteenth-century chess-playing machine*. New York, NY: Walker & Co.

Stanovich, K. E. (2004). *The robot's rebellion: Finding meaning in the age of Darwin*. Chicago, IL: University of Chicago Press.

Steedman, M. J. (1984). A generative grammar for jazz chord sequences. *Music Perception, 2*(1), 52–77.

Stefik, M. J., & Bobrow, D. G. (1987). T. Winograd, F. Flores, Understanding computers and cognition: A new foundation for design (review). *Artificial Intelligence, 31*(2), 220–226.

Steinbuch, K. (1961). Die lernmatrix. *Kybernetik, 1,* 36–45.

Sternberg, R. J. (1977). Component processes in analogical reasoning. *Psychological Review, 84,* 353–378.

Sternberg, R. J. (1996). *Cognitive psychology.* Fort Worth, TX: Harcourt Brace College Publishers.

Sternberg, R. J. (1999). *The nature of cognition.* Cambridge, MA: MIT Press.

Stevens, C., & Latimer, C. (1992). A comparison of connectionist models of music recognition and human performance. *Minds and Machines: Journal for Artificial Intelligence, Philosophy and Cognitive Science, 2*(4), 379–400.

Stewart, I. (1994). A subway named Turing. *Scientific American, 271,* 104–107.

Stewart, L., von Kriegstein, K., Warren, J. D., & Griffiths, T. D. (2006). Music and the brain: Disorders of musical listening. *Brain, 129,* 2533–2553.

Stibitz, G. R., & Loveday, E. (1967a). The relay computers at Bell Labs: Part I. *Datamation, 13*(4), 35–49.

Stibitz, G. R., & Loveday, E. (1967b). The relay computers at Bell Labs: Part II. *Datamation, 13*(5), 45–50.

Stich, S. P. (1983). *From folk psychology to cognitive science: The case against belief.* Cambridge, MA: MIT Press.

Stich, S. P., & Nichols, S. (1997). Cognitive penetrability, rationality and restricted simulation. *Mind & Language, 12*(3–4), 297–326.

Stillings, N. A. (1995). *Cognitive science: An introduction* (2nd ed.). Cambridge, MA: MIT Press.

Stillings, N. A., Feinstein, M. H., Garfield, J. L., Rissland, E. L., Rosenbaum, D. A., Weisler, S. E. (1987). *Cognitive science: An introduction.* Cambridge, MA: MIT Press.

Stix, G. (1994). Bad apple picker: Can a neural network help find problem cops? *Scientific American, 271,* 44–46.

Stoll, C. (2006). When slide rules ruled. *Scientific American, 295*(5), 80–87.

Stone, G. O. (1986). An analysis of the delta rule and the learning of statistical associations. In D. E. Rumelhart & J. McClelland (Eds.), *Parallel distributed processing: Vol. 1. Foundations* (pp. 444–459). Cambridge, MA: MIT Press.

Strunk, W. O. (1950). *Source readings in music history from classical antiquity through the Romantic Era* (1st ed.). New York, NY: Norton.

Suchman, L. A., Winograd, T., & Flores, F. (1987). Understanding computers and cognition: A new foundation for design (review). *Artificial Intelligence, 31*(2), 227–232.

Suddarth, S. C., & Kergosien, Y. L. (1990). Rule-injection hints as a means of improving network performance and learning time. In L. B. Almeida & C. J. Wellekens (Eds.), *Neural network:, Workshop proceedings (lecture notes in computer science)* (Vol. 412, pp. 120–129). Berlin: Springer Verlag.

Sundberg, J., & Lindblom, B. (1976). Generative theories in language and music descriptions. *Cognition, 4*(1), 99–122.

Susi, T., & Ziemke, T. (2001). Social cognition, artefacts, and stigmergy: A comparative analysis of theoretical frameworks for the understanding of artefact-mediated collaborative activity. *Journal of Cognitive Systems Research, 2,* 273–290.

Sutton, R. S., & Barto, A. G. (1981). Toward a modern theory of adaptive networks: Expectation and prediction. *Psychological Review, 88*(2), 135–170.

Swade, D. D. (1993). Redeeming Charles Babbage's mechanical computer. *Scientific American, 268*, 86–91.

Tarasewich, P., & McMullen, P. R. (2002). Swarm intelligence: Power in numbers. *Communications of the ACM, 45*(8), 62–67.

Tarasti, E. (1995). *Musical signification: Essays in the semiotic theory and analysis of music.* Berlin, Germany; New York, NY: Mouton de Gruyter.

Taylor, W. K. (1956). Electrical simulation of some nervous system functional activities. In C. Cherry (Ed.), *Information theory* (pp. 314–328). London: Butterworths Scientific Publications.

Temperley, D. (2001). *The cognition of basic musical structures.* Cambridge, MA: MIT Press.

Temperley, D. (2007). *Music and probability.* Cambridge, MA: MIT Press.

Tenenbaum, J. M., & Barrow, H. G. (1977). Experiments in interpretation-guided segmentation. *Artificial Intelligence, 8*(3), 241–274.

Teo, T. (2010). Ontology and scientific explanation: Pluralism as an a priori condition of psychology. *New Ideas in Psychology, 28*(2), 235–243.

Terhardt, E., Stoll, G., & Seewann, M. (1982a). Algorithm for extraction of pitch and pitch salience from complex tonal signals. *Journal of the Acoustical Society of America, 71*(3), 679–688.

Terhardt, E., Stoll, G., & Seewann, M. (1982b). Pitch of complex signals according to virtual-pitch theory: Tests, examples, and predictions. *Journal of the Acoustical Society of America, 71*(3), 671–678.

Thagard, P. (1996). *Mind: Introduction to cognitive science.* Cambridge, MA: MIT Press.

Thagard, P. (2005). *Mind: Introduction to cognitive science* (2nd ed.). Cambridge, MA: MIT Press.

Theraulaz, G., & Bonabeau, E. (1995). Coordination in distributed building. *Science, 269*(5224), 686–688.

Theraulaz, G., & Bonabeau, E. (1999). A brief history of stigmergy. *Artificial Life, 5*, 97–116.

Theraulaz, G., Bonabeau, E., & Deneubourg, J. L. (1998). The origin of nest complexity in social insects. *Complexity, 3*(6), 15–25.

Thilly, F. (1900). Locke's relation to Descartes. *The Philosophical Review, 9*(6), 597–612.

Thompson, E. (2007). *Mind in life: Biology, phenomenology, and the sciences of mind.* Cambridge, MA: Belknap Press of Harvard University Press.

Thorpe, C. E. (1990). *Vision and navigation: The Carnegie Mellon Navlab.* Boston, MA: Kluwer Academic Publishers.

Tillmann, B., Jolicoeur, P., Ishihara, M., Gosselin, N., Bertrand, O., Rossetti, Y. (2010). The amusic brain: Lost in music, but not in space. *Plos One, 5*(4).

Tillmann, B., Schulze, K., & Foxton, J. M. (2009). Congenital amusia: A short-term memory deficit for non-verbal, but not verbal sounds. *Brain and Cognition, 71*(3), 259–264.

Todd, P. M. (1989). A connectionist approach to algorithmic composition. *Computer Music Journal, 13*(4), 27–43.

Todd, P. M., & Loy, D. G. (1991). *Music and connectionism.* Cambridge, MA: MIT Press.

Todd, P. M., & Werner, G. M. (1991). Frankensteinian methods for evolutionary music. In P. M. Todd & D. G. Loy (Eds.), *Music and connectionism* (pp. 313–339). Cambridge, MA: MIT Press.

Tolman, E. C. (1932). *Purposive behavior in animals and men*. New York, NY: Century Books.

Tolman, E. C. (1948). Cognitive maps in rats and men. *Psychological Review, 55*, 189–208.

Tourangeau, R., & Sternberg, R. J. (1981). Aptness in metaphor. *Cognitive Psychology, 13*, 27–55.

Tourangeau, R., & Sternberg, R. J. (1982). Understanding and appreciating metaphors. *Cognition, 11*, 203–244.

Touretzky, D. S., & Pomerleau, D. A. (1994). Reconstructing physical symbol systems. *Cognitive Science, 18*, 345–353.

Touretzky, D. S., Wan, H. S., & Redish, A. D. (1994). Neural representation of space in rats and robots. In J. M. Zurada, R. J. Marks & C. J. Robinson (Eds.), *Computational intelligence: Imitating life*. New York, NY: IEEE Press.

Treisman, A. M. (1985). Preattentive processing in vision. *Computer Vision, Graphics, and Image Processing, 31*, 156–177.

Treisman, A. M. (1986). Features and objects in visual processing. *Scientific American, 254*, 114–124.

Treisman, A. M. (1988). Features and objects: The fourteenth Bartlett memorial lecture. *Quarterly Journal of Experimental Psychology, 40A*, 201–237.

Treisman, A. M., & Gelade, G. (1980). A feature integration theory of attention. *Cognitive Psychology, 12*, 97–136.

Treisman, A. M., & Gormican, S. (1988). Feature analysis in early vision: Evidence from search asymmetries. *Psychological Review, 95*, 14–48.

Treisman, A. M., Kahneman, D., & Burkell, J. (1983). Perceptual objects and the cost of filtering. *Perception & Psychophysics, 33*(6), 527–532.

Treisman, A. M., & Schmidt, H. (1982). Illusory conjunctions in the perception of objects. *Cognitive Psychology, 14*(1), 107–141.

Treisman, A. M., Sykes, M., & Gelade, G. (1977). Selective attention and stimulus integration. In S. Dornic (Ed.), *Attention and performance VI*. Hillsdale, NJ: Lawrence Erlbaum Associates.

Trick, L. M., & Pylyshyn, Z. W. (1993). What enumeration studies can show us about spatial attention: Evidence for limited capacity preattentive processing. *Journal of Experimental Psychology: Human Perception and Performance, 19*(2), 331–351.

Trick, L. M., & Pylyshyn, Z. W. (1994). Why are small and large numbers enumerated differently: A limited-capacity preattentive stage in vision. *Psychological Review, 101*(1), 80–102.

Tulving, E. (1983). *Elements of episodic memory*. Oxford, England: Oxford University Press.

Turing, A. M. (1936). On computable numbers, with an application to the *Entscheidungsproblem*. *Proceedings of the London Mathematical Society, Series 2h, 42*, 230–265.

Turing, A. M. (1950). Computing machinery and intelligence. *Mind, 59*, 433–460.

Turino, T. (1999). Signs of imagination, identity, and experience: A Peircian semiotic theory for music. *Ethnomusicology, 43*(2), 221–255.

Turkle, S. (1995). *Life on the screen: Identity in the age of the Internet*. New York, NY: Simon & Schuster.

Turkle, S. (2011). *Alone together: Why we expect more from technology and less from each other* (epub ed.). New York, NY: Basic Books.

Turner-Stokes, L., & Reid, K. (1999). Three-dimensional motion analysis of upper limb movement in the bowing arm of string-playing musicians. *Clinical Biomechanics, 14*(6), 426–433.

Turvey, M. T., Shaw, R. E., Reed, E. S., & Mace, W. M. (1981). Ecological laws of perceiving and acting: In reply to Fodor and Pylyshyn (1981). *Cognition, 9,* 237–304.

Tversky, A. (1977). Features of similarity. *Psychological Review, 84,* 327–352.

Tversky, A., & Gati, I. (1982). Similarity, separability, and the triangle inequality. *Psychological Review, 89,* 123–154.

Tversky, A., & Kahneman, D. (1974). Judgment under uncertainty: Heuristics and biases. *Science, 185*(4157), 1124–1131.

Tye, M. (1991). *The imagery debate.* Cambridge, MA: MIT Press.

Uexküll, J. v. (2001). An introduction to *umwelt. Semiotica, 134*(1–4), 107–110.

Ullman, S. (1978). Two-dimensionality of the correspondence process in apparent motion. *Perception, 7,* 683–693.

Ullman, S. (1979). *The interpretation of visual motion.* Cambridge, MA: MIT Press.

Ullman, S. (1984). Visual routines. *Cognition, 18,* 97–159.

Ullman, S. (2000). *High-level vision: Object recognition and visual cognition.* Cambridge, MA: MIT Press.

Ungerleider, L. G., & Mishkin, M. (1982). Two cortical visual systems. In D. Ingle, M. A. Goodale & R. J. W. Mansfield (Eds.), *Analysis of visual behavior* (pp. 549–586). Cambridge, MA: MIT Press.

Ungvary, T., & Vertegaal, R. (2000). Designing musical cyberinstruments with body and soul in mind. *Journal of New Music Research, 29*(3), 245–255.

Valsiner, J. (2006). Dangerous curves in knowledge construction within psychology: Fragmentation of methodology. *Theory & Psychology, 16*(5), 597–612.

Van den Stock, J., Peretz, I., Grezes, J., & de Gelder, B. (2009). Instrumental music influences recognition of emotional body language. *Brain Topography, 21*(3–4), 216–220.

van der Linden, J., Schoonderwaldt, E., Bird, J., & Johnson, R. (2011). MusicJacket: Combining motion capture and vibrotactile feedback to teach violin bowing. *IEEE Transactions on Instrumentation and Measurement, 60*(1), 104–113.

van Essen, D. C., Anderson, C. H., & Felleman, D. J. (1992). Information processing in the primate visual system: An integrated systems perspective. *Science, 255*(5043), 419–423.

van Gelder, T. (1991). What is the "D" in "PDP"? A survey of the concept of distribution. In W. Ramsey, S. P. Stich & D. E. Rumelhart (Eds.), *Philosophy and connectionist theory* (pp. 33–59). Hillsdale, NJ: Lawrence Erlbaum Associates.

van Hemmen, J. L., & Senn, W. (2002). Hebb in perspective. *Biological Cybernetics, 87,* 317–318.

Varela, F. J., Thompson, E., & Rosch, E. (1991). *The embodied mind: Cognitive science and human experience.* Cambridge, MA: MIT Press.

Vauclair, J., & Perret, P. (2003). The cognitive revolution in Europe: Taking the developmental perspective seriously. *Trends in Cognitive Sciences, 7*(7), 284–285.

Vellino, A. (1987). T. Winograd, F. Flores, Understanding computers and cognition: A new foundation for design (review). *Artificial Intelligence, 31*(2), 213–220.

Vera, A. H., & Simon, H. A. (1993). Situated action: A symbolic interpretation. *Cognitive Science, 17,* 7–48.

Verfaille, V., Depalle, P., & Wanderley, M. M. (2010). Detecting overblown flute fingerings from the residual noise spectrum. *Journal of the Acoustical Society of America, 127*(1), 534–541.

Verfaille, V., Wanderley, M. M., & Depalle, P. (2006). Mapping strategies for gestural and adaptive control of digital audio effects. *Journal of New Music Research, 35*(1), 71–93.

Vico, G. (1988). *On the most ancient wisdom of the Italians*. (L. L. M. Palmer, Trans.). Ithaca: Cornell University Press. (Original work published 1710)

Vico, G. (1990). *On the study methods of our time*. (E. Gianturco, Trans.). Ithaca, NY: Cornell University Press. (Original work published 1708)

Vico, G. (2002). *The first new science*. (L. Pompa, Trans.). Cambridge, UK; New York, NY: Cambridge University Press. (Original work published 1725)

Victor, J. D., & Conte, M. M. (1990). Motion mechanisms have only limited access to form information. *Vision Research, 30,* 289–301.

Vidal, R., & Hartley, R. (2008). Three-view multibody structure from motion. *IEEE Transactions on Pattern Analysis and Machine Intelligence, 30*(2), 214–227.

Vines, B. W., Krumhansl, C. L., Wanderley, M. M., Dalca, I. M., & Levitin, D. J. (2011). Music to my eyes: Cross-modal interactions in the perception of emotions in musical performance. *Cognition, 118*(2), 157–170.

Vines, B. W., Krumhansl, C. L., Wanderley, M. M., & Levitin, D. J. (2006). Cross-modal interactions in the perception of musical performance. *Cognition, 101*(1), 80–113.

Vogt, S., Buccino, G., Wohlschlager, A. M., Canessa, N., Shah, N. J., Zilles, K. (2007). Prefrontal involvement in imitation learning of hand actions: Effects of practice and expertise. *Neuroimage, 37*(4), 1371–1383.

von Bekesy, G. (1928). On the theory of hearing: The oscillation form of the basilar membrane. *Physikalische Zeitschrift, 29,* 793–810.

von Eckardt, B. (1995). *What is cognitive science?* Cambridge, MA: MIT Press.

von Frisch, K. (1974). *Animal architecture* (1st ed.). New York, NY: Harcourt Brace Jovanovich.

von Neumann, J. (1958). *The computer and the brain*. New Haven, CN: Yale University Press.

von Neumann, J. (1993). First draft of a report on the EDVAC (reprinted). *IEEE Annals of the History of Computing, 15*(4), 28–75.

Vulkan, N. (2000). An economist's perspective on probability matching. *Journal of Economic Surveys, 14*(1), 101–118.

Vygotsky, L. S. (1986). *Thought and language* (Rev. ed.). (A. Kozulin, Ed., Trans.). Cambridge, MA: MIT Press.

Walkenbach, J., & Haddad, N. F. (1980). The Rescorla-Wagner theory of conditioning: A review of the literature. *Psychological Record, 30*(4), 497–509.

Walsh-Bowers, R. (2009). Some social-historical issues underlying psychology's fragmentation. *New Ideas in Psychology, 28*(2), 244–252.

Walton, K. (1994). Listening with imagination: Is music representational? In P. Alperson (Ed.), *Musical worlds: New directions in the philosophy of music* (pp. 47–62). University Park, PA: Pennsylvania State University Press.

Waltz, D. (1975). Understanding line drawings of scenes with shadows. In P. H. Winston (Ed.), *The psychology of computer vision* (pp. 19–92). New York, NY: McGraw Hill.

Wanderley, M. M., & Orio, N. (2002). Evaluation of input devices for musical expression: Borrowing tools from HCI. *Computer Music Journal, 26*(3), 62–76.

Wang, Y. (2003). Cognitive informatics: A new transdisciplinary research field. *Brain & Mind, 4*, 115–127.

Wang, Y. (2007). Cognitive informatics: Exploring the theoretical foundations for natural intelligence, neural informatics, autonomic computing, and agent systems. *International Journal of Cognitive Informatics and Natural Intelligence, 1*, i–x.

Wang, Y. (2009). Formal description of the cognitive process of memorization. *Transactions of Computational Science, 5*, 81–98.

Wang, Y., Liu, D., & Wang, Y. (2003). Discovering the capacity of human memory. *Brain & Mind, 4*, 151–167.

Warren, H. C. (1921). *A history of the association psychology*. New York, NY: Charles Scribner's Sons.

Warren, J. (2008). How does the brain process music? *Clinical Medicine, 8*(1), 32–36.

Waskan, J., & Bechtel, W. (1997). Directions in connectionist research: Tractable computations without syntactically structured representations. *Metaphilosophy, 28*(1–2), 31–62.

Wason, P. C. (1966). *Reasoning*. New York, NY: Penguin.

Wason, P. C., & Johnson-Laird, P. N. (1972). *Psychology of reasoning: Structure and content*. London, UK: Batsford.

Wasserman, G. S. (1978). *Color vision: An historical introduction*. New York, NY: John Wiley & Sons.

Watanabe, T. (2010). Metascientific foundations for pluralism in psychology. *New Ideas in Psychology, 28*(2), 253–262.

Watson, J. B. (1913). Psychology as the behaviorist views it. *Psychological Review, 20*, 158–177.

Waugh, N. C., & Norman, D. A. (1965). Primary memory. *Psychological Review, 72*, 89–104.

Webb, B., & Consi, T. R. (2001). *Biorobotics: Methods and applications*. Menlo Park, CA: AAAI Press/ MIT Press.

Wechsler, H. (1992). *Neural networks for perception: Computation, learning, and architectures* (Vol. 2). Boston, MA: Academic Press.

Weizenbaum, J. (1966). Eliza: A computer program for the study of natural language communication between man and machine. *Communications of the ACM, 9*(1), 36–45.

Weizenbaum, J. (1976). *Computer power and human reason*. San Francisco, CA: W.H. Freeman.

Wellman, H. M. (1990). *The child's theory of mind*. Cambridge, MA: MIT Press.

Wells, A. J. (1996). Situated action, symbol systems and universal computation. *Minds and Machines: Journal for Artificial Intelligence, Philosophy and Cognitive Science, 6*(1), 33–46.

Wells, A. J. (2002). Gibson's affordances and Turing's theory of computation. *Ecological Psychology, 14*(3), 141–180.

Werbos, P. J. (1994). *The roots of backpropagation: From ordered derivatives to neural networks and political forecasting*. New York, NY: Wiley.

Wexler, K., & Culicover, P. W. (1980). *Formal principles of language acquisition*. Cambridge, MA: MIT Press.

Wheeler, W. M. (1911). The ant colony as an organism. *Journal of Morphology, 22*(2), 307–325.

Wheeler, W. M. (1926). Emergent evolution and the social. *Science, 64*(1662), 433–440.

Whittall, A. (1987). *Romantic music: A concise history from Schubert to Sibelius*. London, UK: Thames and Hudson.

Wicker, B., Keysers, C., Plailly, J., Royet, J. P., Gallese, V., & Rizzolatti, G. (2003). Both of us disgusted in my insula: The common neural basis of seeing and feeling disgust. *Neuron, 40*(3), 655–664.

Widrow, B. (1962). Generalization and information storage in networks of ADALINE "neurons." In M. C. Yovits, G. T. Jacobi & G. D. Goldsteing (Eds.), *Self-organizing systems 1962* (pp. 435–461). Washington, D.C.: Spartan Books.

Widrow, B., & Hoff, M. E. (1960). Adaptive switching circuits. Institute of Radio Engineers, Wester Electronic Show and Convention, Convention Record, Part 4, 96–104.

Widrow, B., & Lehr, M. A. (1990). 30 years of adaptive neural networks: Perceptron, MADALINE, and backpropagation. *Proceedings Of The IEEE, 78*(9), 1415–1442.

Wiener, N. (1948). *Cybernetics: Or control and communciation in the animal and the machine*. Cambridge, MA: MIT Press.

Wiener, N. (1964). *God & Golem, Inc*. Cambridge, MA: MIT Press.

Wilhelms, J., & Skinner, R. (1990). A notion for interactive behavioral animation control. *IEEE Computer Graphics and Applications, 10*(3), 14–22.

Williams, F. C., & Kilburn, T. (1949). A storate system for use with binary-digital computing machines. *Proceedings of the Institution of Electrical Engineers-London, 96*(40), 81–100.

Williams, J. H. G., Whiten, A., Suddendorf, T., & Perrett, D. I. (2001). Imitation, mirror neurons and autism. *Neuroscience and Biobehavioral Reviews, 25*(4), 287–295.

Williams, M. R. (1993). The origins, uses, and fate of the EDVAC. *IEEE Annals of the History of Computing, 15*(1), 22–38.

Williams, M. R. (1997). *A history of computing technology* (2nd ed.). Los Alamitos, CA: IEEE Computer Society Press.

Wilson, E. O., & Lumsden, C. J. (1991). Holism and reduction in sociobiology: Lessons from the ants and human culture. *Biology & Philosophy, 6*(4), 401–412.

Wilson, M., Melhuish, C., Sendova-Franks, A. B., & Scholes, S. (2004). Algorithms for building annular structures with minimalist robots inspired by brood sorting in ant colonies. *Autonomous Robots, 17*(2–3), 115–136.

Wilson, R. A. (2004). *Boundaries of the mind: The individual in the fragile sciences: Cognition*. Cambridge, UK; New York, NY: Cambridge University Press.

Wilson, R. A. (2005). *Genes and the agents of life: The individual in the fragile sciences: Biology*. New York, NY: Cambridge University Press.

Wilson, R. A., & Keil, F. C. (1999). *The MIT encyclopedia of the cognitive sciences*. Cambridge, MA: MIT Press.

Winograd, T. (1972a). Understanding natural language. *Cognitive Psychology, 3*, 1–191.

Winograd, T. (1972b). *Understanding natural language*. New York, NY: Academic Press.

Winograd, T. (1983). *Language as a cognitive process: Vol. 1. Syntax*. Reading, MA: Addison Wesley Pub. Co.

Winograd, T., & Flores, F. (1987a). On understanding computers and cognition: A new foundation for design: A response to the reviews. *Artificial Intelligence, 31*(2), 250–261.

Winograd, T., & Flores, F. (1987b). *Understanding computers and cognition*. New York, NY: Addison-Wesley.

Wisse, M., Schwab, A. L., & van der Helm, F. C. T. (2004). Passive dynamic walking model with upper body. *Robotica, 22*, 681–688.

Witkin, A. P. (1981). Recovering surface shape and orientation from texture. *Artificial Intelligence, 17*(1–3), 17–45.

Wittgenstein, L. (1922). *Tractatus logico-philosophicus*. (C. K. Ogden, Trans.). New York, NY: Harcourt, Brace & company.

Wood, G. (2002). *Living dolls: A magical history of the quest for artificial life*. London, UK: Faber and Faber.

Wotton, J. M., Haresign, T., & Simmons, J. A. (1995). Spatially dependent acoustic cues generated by the external ear of the big brown bat, *Eptesicus fuscus*. *Journal of the Acoustical Society of America, 98*(3), 1423–1445.

Wotton, J. M., & Simmons, J. A. (2000). Spectral cues and perception of the vertical position of targets by the big brown bat, *Eptesicus fuscus*. *Journal of the Acoustical Society of America, 107*(2), 1034–1041.

Wright, R. D. (1998). *Visual attention*. New York, NY: Oxford University Press.

Wright, R. D., & Dawson, M. R. W. (1994). To what extent do beliefs affect apparent motion? *Philosophical Psychology, 7*, 471– 491.

Wundt, W. M., & Titchener, E. B. (1904). *Principles of physiological psychology* (5th German ed. Trans.). London, UK: Sonnenschein. (Original work published 1873)

Yaremchuk, V., & Dawson, M. R. W. (2005). Chord classifications by artificial neural networks revisited: Internal representations of circles of major thirds and minor thirds. *Artificial Neural Networks: Biological Inspirations - ICANN 2005, Part 1, Proceedings, 3696* (605–610).

Yaremchuk, V., & Dawson, M. R. W. (2008). Artificial neural networks that classify musical chords. *International Journal of Cognitive Informatics and Natural Intelligence, 2*(3), 22–30.

Yates, F. A. (1966). *The art of memory*. Chicago, IL: University of Chicago Press.

Zachary, G. P. (1997). *Endless frontier: Vannevar Bush, engineer of the American century*. New York, NY: Free Press.

Zatorre, R. J. (2005). Neuroscience: Finding the missing fundamental. *Nature, 436*(7054), 1093–1094.

Zatorre, R. J., Chen, J. L., & Penhune, V. B. (2007). When the brain plays music: auditory-motor interactions in music perception and production. *Nature Reviews Neuroscience, 8*(7), 547–558.

Zihl, J., von Cramon, D., & Mai, N. (1983). Selective disturbance of movement vision after bilateral brain damage. *Brain, 106*, 313–340.

Zipser, D., & Andersen, R. A. (1988). A back-propagation programmed network that simulates response properties of a subset of posterior parietal neurons. *Nature, 331*, 679–684.

Zittoun, T., Gillespie, A., & Cornish, F. (2009). Fragmentation or differentiation: Questioning the crisis in psychology. *Integrative Psychological and Behavioral Science, 43*(2), 104–115.

Zuse, K. (1993). *The computer, my life*. Berlin, Germany; New York, NY: Springer-Verlag.

Index

A

acoustic paradigm, 305

action potential, 7, 89, 140

activation function, 129, 140–41, 144–45, 150, 152, 154–56, 159–60, 165, 173, 189–90, 193–94, 348

affordance(s), 12, 205, 209, 219–21, 230, 245, 255–56, 261, 306, 319, 322–23, 392

algorithm, 19, 39, 41, 43–52, 56, 63, 73, 77, 98, 100, 106, 120, 161, 179–80, 182–83

all-or-none law, 140–42, 152, 348

aperture problem, 365–66

arbitrary pattern classifier, 150

Aristotle, 24, 56, 134–35, 226

artifacts, 7, 12, 20, 42–44, 76, 89, 98, 104, 236, 320, 419

artificial intelligence, 3, 6, 10, 93, 96–97, 128, 187, 250, 261, 284, 311, 318, 335, 343, 364

artificial neural networks, 8–9, 125–26, 128–30, 132, 140–41, 148–52, 177–78, 184, 186–90, 201, 203, 207–8, 282, 286–87, 289–91

Ashby, W. Ross, 217–18, 236, 260–61, 401, 421

on Homeostat, 218, 236–37, 260, 421

associationism, 72, 130, 133–36, 139, 148, 150, 187

association, 130, 133, 135–36, 139, 148

automata, 20–21, 60, 65, 72–73, 89, 243–44, 337

B

behavioural objects, 305, 308

behaviourism, 2, 68, 72, 107, 132, 148, 407–10

being-in-the-world, 220, 319, 403

binary logic, 23, 25–26, 28

biologically plausible, 8, 125, 128, 137, 184, 199, 202–3, 282, 286, 333–34, 342, 403, 416

blocking, 191

Boole, G., 21, 23–26

brain imaging, 117–18, 246

Braitenberg, Valentino, 43

Braitenberg vehicle, 216–17, 242

bricolage, 228–29, 236, 422

Brooks, Rodney, 12, 222, 237, 244, 253, 321, 323, 325, 331, 338

C

Cage, John, 269–70, 294–95

Cartesian dualism, 59, 77, 122, 206–7, 255, 260–61, 354, 419

Cartesian philosophy, 20, 58–60, 89, 113, 122, 126, 133, 135, 199, 206, 243, 254, 270, 281–82

central control, 266, 271, 294, 325–26, 331–34, 337, 356

central controller, 8, 200, 329, 331

central executive, 121, 256, 333–34

change blindness, 221, 366

Chinese room argument, 49

Chomsky, Noam, 65, 74, 350–51, 363, 408

Chomsky hierarchy, 73

Clark, Andy, 13, 209, 226, 231, 344–45

classical conditioning, 190–93, 195

classical music, 195, 265–75, 277, 279–81, 283, 285, 287, 289, 291–95, 299, 301, 305, 307–9, 311–13, 315

classical sandwich, 11, 91–92, 147, 201, 207, 215–16, 222–25, 231, 235, 260–62, 285, 329–30, 343

coarse code, 162, 171, 184, 186, 340

coarse coding, 184–86, 340, 356

cognitive dialectic, 399–401, 403, 405, 407, 409, 411, 413, 415, 417, 419, 421, 423

cognitive penetrability, 110–13, 115, 119

cognitive scaffolding, 12, 205, 209, 226–27, 230, 245, 248, 262, 270, 383, 395–96, 411

cognitive vocabulary, 85, 268, 315, 323, 348, 351–56, 359, 398, 411

computer, 4–8, 21–22, 29–30, 33, 36–38, 76–78, 89–90, 93–95, 106, 199–200, 227, 327–31, 334–36, 383, 410–13

analog computer, 30

computable functions, 81–82

computation, 12, 22, 77, 89, 122, 229, 260, 329–30, 335, 359, 371, 375–81, 385, 414–17, 419–20

computer simulation, 4, 37, 55, 93–94, 97, 100, 106, 109, 119, 129, 193, 215–16, 354, 383

digital computer, 1, 4, 6, 8, 21, 26, 29, 76–77, 89–90, 122–23, 199–200, 202, 334–35, 403, 410

conceptualization, 209–10, 226, 261–62, 303

conduit metaphor, 299–303, 308

constitution, 209–10, 262

constraint propagation, 371, 375

context-free grammar, 66–67, 69, 71

contingency, 149, 153–58

contingency theory, 149, 154–55, 157–58

credit assignment problem, 159–61

cybernetics, 4, 12, 40–41, 129, 205–6, 217, 236, 259, 261, 299, 405–7, 410

D

decision tree, 179–83

definite features, 175–77

delta rule, 131, 139, 155–56, 158, 160–61, 165, 291
  generalized delta rule, 131, 158, 160–61, 165, 291

Dennett, Daniel, 5, 47

depictive theory, 108–10, 112, 394

Descartes, R., 55, 57–61, 69, 78, 95, 113, 126, 133, 199, 206, 219–20, 235, 254, 259, 282

designation, 81, 83, 322–23, 352

dialectic, 357, 399–401, 403, 405, 407, 409, 411, 413, 415, 417–19, 421, 423

digital musical instruments, 304–6

direct perception, 12, 322–23, 353, 360

distributed memory, 136, 138, 333, 337

distributed representations, 162, 178, 285, 339, 343, 346–47, 356

dodecaphony, 293

double dissociation, 117, 223–24, 341–42

dynamical systems theory, 259–60, 350, 403

E

EDVAC, 76, 329, 331, 336–37

ELIZA, 95–97

embodiment, 12, 130, 205–6, 208, 210, 216–17, 221–22, 230, 248, 256, 258, 266, 270–71, 282, 383

emergence, 135, 139, 254, 292, 301

empiricism, 126–27, 130, 133, 148, 158–59, 162, 190, 202, 235, 403

enactive perception, 205, 209, 219, 221, 230, 392, 394

equilibrium, 155–58

error evidence, 44–45, 56, 98, 104, 106, 188, 257

evolutionary psychology, 116

extended mind, 12–13, 209, 230–32, 235, 256, 262, 321, 325

extended mind hypothesis, 12, 231–32, 235, 262

F

feature cues, 104–5, 193–94, 239, 241, 243

feature integration theory, 102–3, 380–82, 385, 388

feedback, 4, 12, 74, 131, 161, 205, 211, 216–18, 236, 244–45, 259–61, 291, 393, 405–7, 409

finite state automaton, 70–71, 73, 338

FINST, 385–87, 389–93

Flores, Fernando, 317–20, 323–24, 344

Fodor, Jerry, 47, 71–72, 114–16, 118, 150

folk psychology, 84, 398

formalist's motto, 81–83, 269, 354–55

forward engineering, 188, 206, 210, 218, 235–37, 248, 258, 262–63, 400, 420–22, 424

frame of reference problem, 235, 420

frame problem, 392

functional analysis, 56, 119–22, 235, 262, 420

functional architecture, 48–49, 256–57

function approximation, 149, 151, 348

G

generalized delta rule, 131, 158, 160–61, 165, 291

geometric cues, 104–5, 118, 193–94, 243

Gibson, James J., 12, 220–21, 258, 353, 392, 407, 415

Gold's paradox, 72, 74

H

Heidegger, Martin, 220, 317, 319

hidden unit space, 147, 158, 165–66, 171–72, 182

hierarchical, 46, 64–65, 115, 213, 222, 267, 274–78, 289
  hierarchical organization, 46, 267, 274, 276, 278

Hobbes, Thomas, 20, 59, 134

Hopfield network, 287, 341

Hume, David, 135

I

identification assumption, 410–11
identified in the limit, 73–75
illusory conjunctions, 380–81
incommensurable, 14–15
index projection hypothesis, 395–97
individuation, 385, 388–89
informant learning, 72–74
information processing, 1, 4–10, 12–13, 15–17,
    19, 41, 51–56, 69–73, 122–23, 125, 127–28,
    199–202, 231–32, 259, 330–32
    information processing devices, 51–54, 69–70
    information processing hypothesis, 7, 15–16
    information processing problem, 5, 19, 41, 51,
        56, 71, 77, 113, 257, 259
input-output, 8, 31–32, 34, 39–46, 48, 50, 81–82,
    86, 95, 97, 106, 120, 130–32, 160, 190
    input-output function, 32, 34, 82, 95, 97, 106
    input-output mapping, 32, 40–46, 48, 50, 130–
        32, 151, 160, 162, 190
    input-output relationship, 8, 97, 188
input unit, 129–30, 138–39, 144–47, 151, 158,
    160–61, 164–67, 176, 180, 189–90, 194, 207,
    333, 337, 343
intentionality, 83
intentional stance, 82–85, 251, 268, 323, 352
intermediate state evidence, 44–45, 98–99, 106,
    257, 424
intertheoretic reduction, 159, 178, 183
isotropic processes, 114–16

J

Jackendoff, Ray, 267, 276–79, 351
Jacquard loom, 76, 78–80, 328
James, William, 135–38
jittered density plot, 175–77

K

Kuhn, Thomas, 14–15, 417, 424

L

law of contiguity, 134–36, 190–91

law of habit, 134, 136–38
law of similarity, 134–35, 138
law of uphill analysis and downhill synthesis, 218
learning rule, 8, 131–32, 138, 141, 146, 149, 159–60,
    190–91, 291, 333, 421
Leibniz, Gottfried, 50–51
    and monads, 51
leverage, 206, 248, 262–63
limited order constraint, 365–66
linearly nonseparable problem, 143, 145–46
linearly separable problem, 143
linguistic competence, 351, 357
linguistic performance, 350
locality assumption, 342
local representations, 339, 341–43, 415
Locke, John, 126, 130, 135, 199, 235
logicism, 19, 21–23, 68, 81, 84–85, 133, 199, 203,
    255, 266, 270, 272, 281, 334

M

machine table, 42, 45, 50, 70, 330–32
    machine head, 42–43, 45, 69–71, 80, 85, 152,
        330–31, 347
    machine state, 42–43, 330, 332
mark of the classical, 325–26, 334–35, 337–39, 343,
    345–48, 352, 354–55
mark of the cognitive, 231, 325, 346
Marr, David, 5, 51, 53, 349–52, 355, 391, 415–16
    and the tri-level hypothesis, 349–52, 355
massive modularity hypothesis, 116
matching law, 153
materialism, 59, 84, 113, 122, 255–56, 354
Merleau-Ponty, Maurice, 258, 361
methodological solipsism, 11, 206–8, 255, 261,
    272–73, 317
metrical structure, 276, 278, 288
Mill, John Stuart, 135, 139
Miller, George, 4, 7, 406–8, 411
mind in action, 224, 227, 229–30, 234
Mind reading, 250
minimalism, 83–84, 296–97
    minimalist music, 84, 296–97, 301
modularity, 113–19, 121, 194, 288, 342
    packing problem, 113–14
modules, 56, 114–16, 194, 222–24, 288, 342, 379,
    381–82, 390, 393
motion correspondence, 375–78, 384, 388, 416